Microlepidoptera of Europe

Volume 5

Sjaak (J.C.) Koster & Sergey Yu. Sinev

Momphidae, Batrachedridae, Stathmopodidae, Agonoxenidae, Cosmopterigidae, Chrysopeleiidae

Microlepidoptera of Europe

Volume 5

EDITED BY
P. HUEMER, O. KARSHOLT & L. LYNEBORG

Sjaak (J.C.) Koster & Sergey Yu. Sinev

Momphidae, Batrachedridae, Stathmopodidae, Agonoxenidae, Cosmopterigidae, Chrysopeleiidae

Apollo Books
Stenstrup
2003

Editors: Peter Huemer, Innsbruck; Ole Karsholt, Copenhagen;
and Leif Lyneborg, Copenhagen.

Colour and black and white plates: Sjaak Koster.

Text-figures: Sergey Sinev and Sjaak Koster.

SEM photographs: Twan (A.L.M.) Rutten.

Text formatting: Futura DTP, Frederiksberg.

Printed by: Vinderup Bogtrykkeri A/S.

Published by:
Apollo Books, Kirkeby Sand 19, DK-5771 Stenstrup, Denmark
apollobooks@vip.cybercity.dk – www.apollobooks.com

This publication should be cited as:
Koster, J.C. & Sinev, S.Yu., 2003. Momphidae, Batrachedridae, Stathmopodidae, Agonoxenidae, Cosmopterigidae, Chrysopeleiidae. – In P. Huemer, O. Karsholt and L. Lyneborg (eds): Microlepidoptera of Europe 5: 1-387.

ISBN 87-88757-66-8
ISSN 1395-9506

Authors' addresses:
Sjaak Koster, Van Brederodestraat 53, NL-1759 VG Callantsoog, The Netherlands
sjaak.koster@planet.nl

Sergey Sinev, Systematics of Insects Laboratory, Zoological Institute RAN,
Universitetskaja nab., 1, RU-199034 St. Petersburg, Russia
lepid@zin.ru

Contents

Abstract

The six families, Momphidae, Batrachedridae, Stathmopodidae, Agonoxenidae, Cosmopterigidae, and Chrysopeleiidae, of Europe, North Africa and the Near East are reviewed. A total of 37 genera and 163 species are recognized. Short diagnoses are accompanied by watercolour plates of adults and line drawings of male and female genitalia. For the genera *Cosmopterix*, *Eteobalea*, and *Vulcaniella* keys for identification are also given. Notes on the distribution and biology are added for every species.

One new genus and 10 new species are described: *Gibeauxiella* gen.n. (type species: *Hodgesiella bellaqueifontis* Gibeaux, 1986), *Mompha meridionella* sp.n. (Russia and Greece), *Alloclita deprinsi* sp.n. (Turkey), *Pyroderces tethysella* sp.n. (Spain, Tunisia and Tadjikistan), *Pyroderces wolschrijni* sp.n. (Spain, Morocco), *Vulcaniella gielisi* sp.n. (Spain), *Vulcaniella anatolica* sp.n. (Turkey), *Vulcaniella pontica* sp.n. (Turkey), *Hodgesiella christophi* sp.n. (Turkey), *Sorhagenia fibigeri* sp.n. (Turkey), and *Sorhagenia taurensis* sp.n. (Turkey).

The following new synonymies are established: *Megaceraea* Nye & Fletcher, 1991 syn.n. of *Neomariania* Mariani, 1943; *Coccidiphila rungsella* J. Nel & Brusseaux, 1997 syn.n. of *C. ledereriella* (Zeller, 1850); *Elachista bizonatella* Turati, 1930 syn. n. of *Ascalenia bifasciella* Chrétien, 1915; *Blastobasis sublineatella* Lucas, 1933 syn.n. of *Calycobathra acarpa* Meyrick, 1891.

The following new combinations are introduced: *Gibeauxiella bellaqueifontis* (Gibeaux, 1986) comb.n. and *Gibeauxiella reliqua* (Gibeaux, 1986) comb.n. from *Hodgesiella* Riedl, 1965; *Neomariania oecophorella* (Rebel, 1940) comb.n., *Neomariania scriptella* (Rebel, 1940) comb.n. and *Neomariania incertella* (Rebel, 1940) comb.n. from *Megaceraea* Nye & Fletcher, 1991.

Introduction

General remarks

'The narrow-winged moths' is the name for a very heterogeneous group of gelechioid moths characterized by very slender, lanceolate or linear, fore- and hindwings with long cilia considerably exceeding their width. The first representatives in western Europe became known in the second half of 18th century with the works of Carolus Clerck and Carolus Linnaeus. Clerck published a colour figure of a pretty 'Linnaeus Moth' (*Phalaena linneella* Clerck, 1759), now belonging to the genus *Chrysoclista* Stainton in the family Agonoxenidae, and Linnaeus included the 'Leuwenhoek Moth' (*Phalaena Tinea leuwenhoeckella* Linnaeus, 1761) in the second edition of his 'Fauna Svecica'. This was later transferred to the genus *Pancalia* Stephens in the family Cosmopterigidae. By the end of the 19th century most European genera and species had already been described, but this was not the case for adjacent territories (northern Africa, the Near East, the Caucasus, eastern Europe). Those faunas became better known only more recently. The narrow-winged moths of biogeographical regions other than the Palaearctic remain poorly known.

At the very beginning of the taxonomic study of Microlepidoptera, the species of narrow-winged moths were placed in miscellaneous generic groups such as *Phalaena*, *Tinea*, *Alucita*, *Ornix*, *Anacampsis*, *Oecophora*, *Recurvaria*, and *Elachista*. Separate genera for those moths were described much later. Herrich-Schäffer (1853-1855) was the first to try to clarify the generic assignment of most species known at the time and described many valid genera. Then, a great step forward in the taxonomy of the group was achieved by Rebel, the author of the catalogue of the palaearctic Microlepidoptera (Staudinger & Rebel, 1901), and Spuler (1910), who elaborated the standardized descriptions of the main taxa.

Nevertheless, the systematic position of many genera of narrow-winged moths remained doubtful. For many years it was generally accepted to place most ditrysian Microlepidoptera in the group Tineina, a large taxon without fixed rank. The number of families in it fluctuated considerably, depending on the personal view of various authors. The narrow-winged moths were usually placed in the heterogeneous family Elachistidae uniting practically all 'gelechioid-like' moths with more or less slender hind wings and long cilia (Staudinger & Wocke, 1861). Soon after that, Wocke (1871) was the first to place several genera into his new family Lavernidae, which he left undiagnosed. After that Heinemann (Heinemann & Wocke, [1876]), based mainly on the characters of wing venation, described this family and proposed another two monotypic families, Batrachedrae and Cosmopterygidae. In the third edition of the catalogue of palaearctic Lepidoptera (Staudinger & Rebel, 1901) all these groups were united again and returned to the Elachistidae as subfamily Momphinae (=Laverninae)

Early in the 20th century treatment of a single family of narrow-winged moths, named either Cosmopterygidae or Momphidae in English and German literature, gained more acceptance. The taxonomic structure of the family was first elaborated by Spuler (1898, 1910) who divided it into 3 subfamilies: Momphinae, Psacaphorinae and Cosmopterygisnae. In spite of the very artificial character of this system, it was in use for more than fifty years.

A new period in the systematics of narrow-winged moths began in the second half of the 20th century, when the use of genitalia characters was widely introduced. The first step on the way to a complete overview of the system of the group was made by Hodges (1962). This author not only recognized the family-level rank for both Momphidae and Cosmopterigidae, but also designated the family Walshiidae [described earlier by Mosher (1916) as Chrysopeleiidae] with several palaearctic genera. But even in this more strict sense, the family Momphidae still remained evidently heterogeneous with regard to the genital structures, and Hodges split it into two divisions: the 'group *Mompha'* and 'group *Blastodacna'*. Later on the last group was given family or subfamily rank: Blastodacnidae (Clarke, 1962; Sinev, 1979c; Koçak & Seven, 1993).

In Europe the division of Momphidae sensu Spuler (=sensu lato) was not immediately accepted, and the well-known revision of this group undertaken by the Polish specialist Tadeusz Riedl (1969) was based on the concept of one family. But the taxonomic structure of this family, with three subfamilies (Momphinae, Cosmopteriginae and Walshiinae), already corresponded to Hodges' scheme, although Momphinae still included some genera which were later transferred to other families (Kasy, 1973a, 1973b; Riedl, 1984). This revision, in spite of some annoying mistakes, drew general attention to the narrow-winged moths and laid a good foundation for further studies. Very soon, taking into consideration the rather isolated position of the genera *Blastodacna* Wocke and allied groups and *Batrachedra* Herrich-Schäffer, it was proposed to increase the number of subfamilies of the Momphidae s.l. to five (Bradley et al., 1972). At the same time *Stathmopoda* H.-S. with related genera was excluded from Momphidae (Kasy, 1973b) and placed in the restored family Stathmopodidae, earlier established by Meyrick (1913a). Thus, at the end of the seventies it became rather clear that Momphidae s.l. was very heterogeneous and should be divided into several family groups. But these groups were still regarded as closely related and placed in the system of gelechioid moths side by side. This point of view was most consistently formulated in the latest publications of Riedl (1984, 1994b) who proposed to treat the narrow-winged moths as a complex of six families: Batrachedridae, Blastodacnidae, Chrysopeleiidae, Cosmopterigidae, Momphidae, and Stathmopodidae.

Hodges again made the next very radical step towards perfecting the system of narrow-winged moths. In the introduction to the revision of the Nearctic Cosmopterigidae (Hodges, 1978) he discussed also some general problems of gelechioid phylogeny and classification, taking arguments mainly from the male genital structures. As a result, some groups of the narrow-winged moths appeared to be transferred to other gelechioid families: Stathmopodidae became a tribe Stathmopodini in Oecophoridae, whilst Batrachedridae, Blastodacnidae and Chrysopeleiidae received the rank of subfamilies in the Coleophoridae, Agonoxenidae and Cosmopterigidae, respectively. Hodges recently modified this system again, this time by using a cladistic analysis of various characters (Hodges, 1999). The same viewpoint was supported by Kuznetzov & Stekol'nikov (1984) on the basis of functional morphology of the male genitalia. These authors have gone even farther and removed the different groups of the narrow-winged moths to different superfamilies of their infraorder Coleophoromorpha.

Our own phylogenetic analysis of the gelechioid family-group taxa (Sinev, 1992, 2002b) has shown that Momphidae sensu lato really lacks any reliable synapomorphies and thus cannot be treated as a monophyletic unity. Most of their shared apomorphic characters appear to be either synapomorphies of the gelechioids as a whole, or parallel simplifications due to a similar morphogenetic trend in the process of body size reduction (so-called pumilization). Only Momphidae, Batrachedridae and Stathmopodidae share several synapomorphies, but some of these are shared with other gelechioid families (Coleophoridae, Blastobasidae, Oecophoridae). Agonoxenidae share imaginal and pupal synapomorphies with Elachistidae, Ethmiidae and some other families, whilst Cosmopterigidae and Chrysopeleiidae take rather isolated positions. But each of the six families of narrow-winged moths possesses a number of good autoapomorphies (see morphological diagnoses below), and this is a reason to keep their high taxonomic rank.

In the present volume we are treating these families simply following the old traditions in taxonomy of the narrow-winged moths. Because every family is very peculiar in many respects, each family is described separately regarding morphological and ecological characters. Within the family, for every genus and species a full synonymy is given, including original combination and source of description. The diagnoses of the adults and their genitalia are kept as brief as possible and should be used in combination with the accompanying figures. For some of the most difficult genera keys are also available. The descriptive terminology of genitalic structures generally follows Klots (1970), and more detailed studies of various genera (Riedl, 1969; Sinev, 1985a, 1993b, 1997b, 2002b).

Biological data and information about the immature stages are based both on literature and personal observations. A critical evaluation of host-plant records was undertaken, because they are frequently incorrect. The larval life-style can be generalized as specialized endophagy. The larvae are usually leaf-, rarely stem-miners, sometimes they also feed within twigs, seeds, flower-heads or make stem galls on different groups of green plants. But strict oligophagy or even monophagy is the rule. In some species the larvae are scavengers, inquilines, or even predators on scale insects. Larval and pupal morphological characters are usually not considered. Those of European species are described in widely scattered papers, e.g. larvae by Stainton (1870a, 1870b) and pupae by Patocka (1997). The nomenclature of host-plants mainly follows Tutin et al. (1964-1980).

Information about the distribution usually includes only larger geographical regions. Single countries are listed in the distributional catalogue. Distribution records are mainly based on material examined by us, sometimes on reliable literature sources as well. The use and spelling of geographical names follow as far as possible the 'Times Atlas of the World' (Comprehensive edn. 9, revised, 1994). New records of species for single countries are specified only for extremely rare species.

Specimen labels are, with the exception of those of primary types, cited in a standardized form. Deposition of material is mentioned in brackets.

Collecting methods

There are various methods of collecting narrow-winged moths. Most (but not all!) species are best collected at artificial light in places with rich vegetation, especially at forest edges, meadows, gardens, etc. The individual species are frequently restricted to certain host-plants and their habitats. In such habitats they can often be found at dusk or dawn, flying around trees, shrubs and grasses or dancing on leaves or trunks. It is also possible to collect the moths by sweeping vegetation with a net during daytime, but the results should be checked frequently to avoid damage to these tiny moths. A very few species are active during the daytime, and may be collected on flowers. Searching and breeding of immature stages is the most fruitful collecting method for those species very rarely observed as adults. Besides obtaining important information on bionomics, it also provides fresh specimens, which are easier to identify. Most easy to obtain are the species of which the larvae are leaf-miners, gall-makers or live between spun leaves.

It is strongly recommended that material is brought home alive because adults are easier to set directly after killing. Each specimen has to be kept separately in a small glass tube placed in a cool dark place, preferably a refrigerator. Moths stay alive for a couple of days when put in the vegetable compartment. Avoidance of plastic tubes is recommended because they often cause static electricity and the moths will stick to the walls or lose their scales.

If working in very hot environments e.g. in tropics or deserts when the above mentioned method is impossible, it will be better to kill the moths with ammonia 25%. The glass tubes containing the moths have to be closed with cotton-wool and placed in a firmly closed container. A boll of cotton wool with a few drops of ammonia is put inside the container. The ammonia kills the moths instantly, but they will remain relaxed for at least 24 hours and they can be set afterwards as easy as fresh specimens. This method is less suitable for species with strong red and green colours.

Genitalia preparation

The genitalia preparation technique used for the narrow-winged moths is principally the same as that already described for other gelechioid moths (Bengtsson, 1997; Huemer & Karsholt, 1999). It is mainly based on the standard method of preparation (Robinson, 1976), with further details added by Brown (1997).

We would therefore only like to remind the principal steps of the process:

1) Maceration in 10% potassium hydroxide (KOH) solution with heating for a few minutes, or cold for a couple of hours or overnight.
2) Washing in clean (distilled) water with or without heating.
3) Staining (recommended) with mercurochrome, orange G, eosine or haemaluin in the male genitalia to highlight the sclerotized parts or chlorazol black in the female genitalia to stain the unsclerotized parts.
4) Dissecting and cleaning with micropins and fine brush in clean water or in 70% alcohol.
5) Dehydrating in absolute alcohol.
6) Clearing in Euparal essence (optional).
7) Embedding in Euparal.

However, we urgently recommend to start identification already during dissection, before embedding genitalia in Euparal, or even to store the dissected genitalia temporarily in a vial with a 70% ethanol-glycerine solution (1:1). It is especially important for some representatives of Cosmopterigidae and Chrysopeleiidae with very complicated and more or less asymmetrical genitalia which could hardly be fixed in the standard position needed for confident identification. In many cases identification is only possible with examination of the genitalia from different viewpoints; that is the reason why in many genera the drawings of the male genitalia are given in both lateral and ventral view.

Each of the six families treated here has its own peculiarities in the genital apparatus, which determines some differences in the manner of genitalia spreading, especially for males. The aedeagus is usually removed from the rest of the genitalia capsule to simplify the drawings and to show its inner structures. In Momphidae, for example, the male genitalia are figured in the normal ventral position with aedeagus+anellus lobes separated for better presentation of cornuti. However, this rather delicate job is not absolutely necessary for a proper determination, and we do not recommend it for every specimen examined. In Stathmopodidae, the male genitalia are spread in the normal ventral position except for the tegumen, which is tubular and embedded in the lateral position. But in Cosmopterigidae and Chrysopeleiidae the aedeagus is tightly fused with the bases of the valvae or the anellus and can hardly be severed without considerable damage to the genitalia; in this case it is better left *in situ*.

For the preparation of female genitalia no special actions are usually required. One should, however, be very careful in extracting the genitalia from the abdomen, in order to prevent the cutting of delicate genital ducts and the loss of the bursa copulatrix.

In some cases it is not necessary to make genitalia preparations, as brushing off the scales of the anal tuft will reveal the most important diagnostic features.

Notes on the illustrations

The species dealt with in this book often have a shining and colourful appearance with tufts of raised scales, metallic tubercular scale spots, etc. The silver, golden or coppery reflection of the tubercular scales are especially difficult to photograph. Hence, very well set and undamaged specimens were needed. The raised scale tufts in particular are easily flattened, damaged or rubbed off by careless preparation. For this reason we have chosen to depict the adults in watercolours. The illustrations were made of actual specimens, but in some cases more than one specimen was used to get a proper picture. All species have been depicted at the same scale, which is 8x natural size. The line drawings of male and female genitalia have been reduced to 60-90% of the original drawings.

Acknowledgements

The authors wish to thank the following colleagues for the loan or donation of material, for valuable comments, and for giving access to museum collections:

Dr. David J. L. Agassiz, Gravesend, U.K.; Mr. Ernst Arenberger, Vienna, Austria; Dr. Giorgio Baldizzone, Asti, Italy; Mr. Bengt Å. Bengtsson, Färjestaden, Sweden; Mr. Kees van den Berg, Hoogland, The Netherlands; Dr. Alexey Bidzilya, ZMKU, Kiev, Ukraine; Mr. Willy Biesenbaum, Velbert-Langenberg, Germany; Mr. Ulf Buchsbaum, ZSM, Munich, Germany; Mr. Martin F. V. Corley, Faringdon, U.K.; Dr. Roy Danielsson, MZLU, Lund, Sweden; Dr. Anton L. Cox, Mook, The Netherlands; Dr. Ugo Dall'Asta, MRAC, Tervuren, Belgium; Mr. Willy O. De Prins, Antwerpen, Belgium; Mr. François Desmet, MRAC, Tervuren, Belgium; †Dr. Wolfgang Dierl, ZSM, Munich, Germany; Mr. Günther Ebert, LSNK, Karlsruhe, Germany; Mr. Michael Fibiger, Sorø, Denmark; Dr. Reinhard Gaedike, DEI, Eberswalde, Germany; Mr. Christian Gibeaux, Avon, France; Dr. Cees Gielis, Lexmond, The Netherlands; Dr. Laslo Gozmány, HNHM, Budapest, Hungary; Dr. Axel Hausmann, ZSM, Munich, Germany; Mr. Willem Hogenes, ZMAN, Amsterdam, The Netherlands; Dr. Peter Huemer, TLMF, Innsbruck, Austria; Mr. Hans Huisman, Wezep, The Netherlands; Dr. Lauri Kaila, MZHF, Helsinki, Finland; Mr. Ole Karsholt, ZMUC, Copenhagen, Denmark; †Dr. Friedrich Kasy, NHMW, Vienna, Austria; Mr. Sakari Kerppola, Helsinki, Finland; Prof. Dr. Niels P. Kristensen, ZMUC, Copenhagen, Denmark; Dr. Jean-François Landry, CNC, Ottawa, Canada; Mr. Jan A. W. Lucas, Rotterdam, The Netherlands; Dr. Gérard Luquet, MNHN, Paris, France; Dr. Martin Lödl, NHMW, Vienna, Austria; Dr. S. Löser, LMAD, Düsseldorf, Germany; Dr. Wolfram Mey, ZMB, Berlin, Germany; Dr. Kauri Mikkola, MZHF, Helsinki, Finland; Prof. Dr. Joël Minet, MNHN, Paris, France; Prof. Dr. Clas M. Naumann, ZFMK, Bonn, Germany; Dr. Erik J. van Nieukerken, RMHN, Leiden, The Netherlands; Dr. Gaden Robinson, BMNH, London, U.K.; Dr. Klaus Sattler, BMNH, London, U.K.; Dr. Rob T. A. Schouten, Museon, The Hague, The Netherlands; Mr. Arnold Schreurs, Kerkrade, The Netherlands; Dr. Malcolm Scoble, BMNH, London, U.K.; Mr. Peder Skou, Stenstrup, Denmark; Dr. Wolfgang Speidel, Bonn, Germany; Mr. Reinhard Sutter, Bitterfeld, Germany; Ing. Zdenko Tokár, Michalovce, Slovakia; Mr. Kevin Tuck, BMNH, London, U.K.; Dr. Sandrine A. Ulenberg, ZMAN, Amsterdam, The Netherlands; Mr. Frank Verhoeven, Beerse, Belgium; Mr. M. Vuola, Kaskinen, Finland; Mr. Hugo W. van der Wolf, Nuenen, The Netherlands; Mr. Jacques B. Wolschrijn, Twello, The Netherlands.

The authors are also very grateful to the following foundations for financial support which made it possible to visit the main European museums for studying types and other important material: Alexander Koenig Stiftung, Germany; Carlsberg Foundation, Denmark; Stichting Curt Eisner Fonds, The Netherlands; Deutsche Forschungsgemeinschaft, Germany; Professor Hering Memorial Research Fund, U.K.; Nederlandse Organisatie voor Wetenschappelijk Onderzoek, The Netherlands; Uyttenboogaart-Eliasen Stichting, The Netherlands; Russian Foundation for Basic Research, Russia (grants 98-04-49818, 01-04-49637).

Abbreviations of Museums, Institutions and private Collections

For the abbreviations of museums and institutions the 'Insect and spider collections of the world web site' is mainly followed. This web site is based on Arnett, Samuelson & Nishida (1993).

ABR	coll. Arenberger, Vienna, Austria.
ASZ	coll. Agassiz, Gravesend, U.K.
BDZ	coll. Baldizzone, Asti, Italy.
BMNH	British Museum (Natural History) [=The Natural History Museum], London, U.K.
CNC	Centre for Land and Biological Resources Research, Ottawa, Canada.
COX	coll. Cox, Mook, The Netherlands.
DEI	Deutsches Entomologisches Institut, Eberswalde, Germany.
GBX	coll. Gibeaux, Avon, France.
GLS	coll. Gielis, Lexmond, The Netherlands.
HNHM	Természettudományi Museum Allattára [Hungarian Natural History Museum], Budapest, Hungary.
HSN	coll. Huisman, Wezep, The Netherlands.
ISAK	Institute of Systematics and Evolution of Animals, Krakow, Poland.
KTR	coll. Koster, Callantsoog, The Netherlands.
LCS	coll. Lucas, Rotterdam, The Netherlands.
LMAD	Löbbecke Museum und Aquazoo, Düsseldorf, Germany.
LSNK	Landessamlungen für Naturkunde, Karlsruhe, Germany.
LSUK	Linnean Society, London, U.K.
MMA	Melbourne Museum, Melbourne, Australia.
MNHN	Muséum National d'Histoire Naturelle, Paris, France.
MRAC	Musée Royal de l'Afrique Centrale, Tervuren, Belgium.
MZHF	Finnish Museum of Natural History, Helsinki, Finland.
MZLU	Lund University, Lund, Sweden.
NHMW	Naturhistorisches Museum, Vienna, Austria.
NHRS	Naturhistoriska Riksmuseet, Stockholm, Sweden.
PDW	Plantenziektekundige Dienst, Wageningen, The Netherlands.
RMNH	Nationaal Natuurhistorisches Museum Naturalis, Leiden, The Netherlands.
SRS	coll. Schreurs, Kerkrade, The Netherlands.
STR	coll. Sutter, Bitterfeld, Germany.
TLMF	Tiroler Landesmuseum Ferdinandeum, Innsbruck, Austria.
UMO	University Museum, Oxford, U.K.
VDW	coll. Van der Wolf, Nuenen, The Netherlands.
VHN	coll. Verhoeven, Beerse, Belgium.
WLN	coll. Wolschrijn, Twello, The Netherlands.
ZFMK	Zoologisches Forschungsinstitut und Museum Alexander Koenig, Bonn, Germany.
ZIN	Zoological Institute, Academy of Sciences, St. Petersburg, Russia.
ZMAN	Zoological Museum, Amsterdam, The Netherlands.
ZMHB	Museum für Naturkunde der Humboldt-Universität, Berlin, Germany.
ZMKU	Zoological Museum, Kiev, Ukraine.

ZMUC	Zoologisk Museum, University of Copenhagen, Copenhagen, Denmark.
ZMUH	Zoological Museum, University of Helsinki, Finland.
ZSM	Zoologische Staatssammlung, Munich, Germany.

Check-list of European, North African and Near East Momphidae, Batrachedridae, Stathmopodidae, Agonoxenidae, Cosmopterigidae, and Chrysopeleiidae

MOMPHIDAE Herrich-Schäffer, 1857
Lavernidae Wocke, 1871
Psacaphorinae Spuler, 1898

Mompha Hübner, [1825]
Laverna Curtis, 1839
Lophoptilus Sircom, 1848
Cyphophora Herrich-Schäffer, 1853
Psacaphora Herrich-Schäffer, 1853
Anybia Stainton, 1854
Wilsonia Clemens, 1864
Leucophryne Chambers, 1875

subgenus Mompha Hübner, [1825]
1. conturbatella (Hübner, [1819])
   elegantella (Zetterstedt, 1840)
2. ochraceella (Curtis, 1839)
3. lacteella (Stephens, 1834)
   gibbiferella (Zeller, 1839)
   basipallidella (Stainton, 1849)
4. propinquella (Stainton, 1851)
   paludicolella (Doubleday, 1859) (unavailable)
5. divisella Herrich-Schäffer, [1854]
   decorella auctt., (*nec* Haworth, 1812) (misidentification)
6. jurassicella (Frey, 1881)
   iurassicolella (Reutti, 1898) (emendation)
   subdivisella Bradley, 1951
7. bradleyi Riedl, 1965
8. confusella Koster & Sinev, 1996
9. subbistrigella (Haworth, 1828)
10. sturnipennella (Treitschke, 1833)
    modestella (Eversmann, 1844)
    permutatella Herrich-Schäffer, [1854]
    nodicolella Fuchs, 1902
11. meridionella sp.n.
12. epilobiella ([Denis & Schiffermüller], 1775)
    fulvescens (Haworth, 1828)
    nebulella (Stephens, 1834)

subgenus Anybia Stainton, 1854
13. langiella (Hübner, 1796)
    epilobiella (Römer, 1794) (homonym)
    unipunctella (Duponchel, [1839])
    niveipunctella (Stainton, 1849)
    fulicella Herrich-Schäffer, [1854]

subgenus Cyphophora Herrich-Schäffer, 1853
14. idaei (Zeller, 1839)
idaeella (Duponchel, [1843]) (emendation)
tricristatella (Chambers, 1875)
grandisella (Chambers, 1875)
subiridescens (Walsingham, 1882)
polaris (Sinev, 1986)

subgenus Lophoptilus Sircom, 1848
15. miscella ([Denis & Schiffermüller], 1775)
staintoni (Sircom, 1848)
opacella (Müller-Rutz, 1934)

subgenus Psacaphora Herrich-Schäffer, 1853
16. locupletella ([Denis & Schiffermüller], 1775)
schrankella (Hübner, [1805]) (homonym)
pilipennella (Zetterstedt, 1840)
quadrilobella (Herrich-Schäffer, [1854])
17. maculata (Sinev, 1986)
18. terminella (Humphreys & Westwood, 1845)
patriciella (Stainton, 1849)
chrysargyrella (Herrich-Schäffer, [1854])
engelella Busck, 1906
19. raschkiella (Zeller, 1839)
20. sexstrigella (Braun, 1921)
complexa Svensson, 1982

BATRACHEDRIDAE Heinemann & Wocke, [1876]

Batrachedra Herrich-Schäffer, 1853
Eustaintonia Spuler, 1910
21. praeangusta (Haworth, 1828)
turdipennella (Kollar, 1832)
praeangustella (Doubleday, 1859) (emendation)
clemensella Chambers, 1877
22. pinicolella (Zeller, 1839)
23. parvulipunctella Chrétien, 1915
phragmitidella (Mariani, 1936)
24. amydraula Meyrick, 1916

STATHMOPODIDAE Meyrick, 1913

Stathmopoda Herrich-Schäffer, 1853
Boocara Butler, 1880
Placostola Meyrick, 1887
Erineda Busck, 1909
Agrioscelis Meyrick, 1913
Kakivoria Nagano, 1916
25. pedella (Linnaeus, 1761)
alucitella ([Denis & Schiffermüller], 1775)

cylindrella (Fabricius, 1777)
angustipennella (Hübner, 1796)
cylindricus (Fabricius, 1798) (emendation)
fastuosella (Costa, [1836])

26. auriferella (Walker, 1864)
divisa Walsingham, 1891
ischnotis Meyrick, 1897
crocophanes Meyrick, 1897
theoris (Meyrick, 1906)
tharsalea Meyrick, 1914
adulatrix Meyrick, 1917
cirrhaspis Meyrick, 1922
basiflavella (Matsumura, 1931)

Tortilia Chrétien, 1908
Diadoxastis Meyrick, 1913
Apertodiscus Amsel, 1935

27. flavella Chrétien, 1908
28. trigonella (Zerny, 1935)
29. charadritis (Meyrick, 1924)
stupenda (Turati, 1927)
viatrix Busck, 1934
zernyi (Amsel, 1935)
trissorrhiza (Meyrick, 1939)
30. pallidella Kasy, 1973
31. graeca Kasy, 1981

AGONOXENIDAE Meyrick, 1926
Blastodacnidae Clarke, 1962 (unavailable)
Parametriotidae Capuse, 1971
Blastodacninae Sinev, 1979
Lamprysticinae Lvovsky, 1996

Blastodacna Wocke, [1876]
Sinitinea Yang, 1977

32. hellerella (Duponchel, 1838)
33. rossica Sinev, 1989
34. atra (Haworth, 1828)
putripennella (Zeller, 1839)
35. vinolentella (Herrich-Schäffer, [1854])
36. georgiella Sinev, 1988
37. libanotica Diakonoff, 1939

Spuleria Hofmann, 1898

38. flavicaput (Haworth, 1828)
aurifrontella (Geyer, [1832])
aurocapitella (Bruand, [1851]) (emendation)
flavicapitella (Doubleday, 1859) (emendation)

Heinemannia Wocke, [1876]
Tebenna Hübner, [1825] (homonym)

39. laspeyrella (Hübner, 1796)

40. festivella ([Denis & Schiffermüller], 1775)
kokujevi (Krulikowsky, 1903)
41. albidorsella (Staudinger, 1877)
albidecorella (Hartig, 1951) (emendation)

Dystebenna Spuler, 1910
42. stephensi (Stainton, 1849)
tesselatella (Herrich-Schäffer, [1854])
stephensiella (Doubleday, 1859) (emendation)

Haplochrois Meyrick, 1897
Aetia Chambers, 1880 (homonym)
Rhadinastis Meyrick, 1897
Tetanocentria Rebel, 1902
Platybathra Meyrick, 1911
Parametriotes Kusnezov, 1916
Syntetrernis Meyrick, 1922
43. gelechiella (Rebel, 1902)
44. ochraceella (Rebel, 1903)
45. albanica (Rebel & Zerny, [1932])
46. buvati (Baldizzone, 1985)
47. theae (Kusnezov, 1916)

Chrysoclista Stainton, 1854
Glyphipteryx Curtis, 1827 (rejected ICZN, 1986)
48. linneella (Clerck, 1759)
gemmatella (Costa, [1836])
linnaeella (Zeller, 1839) (emendation)
schaefferella sensu Duponchel, [1839] (misidentification)
obscurilinneella (Bruand, 1859)
49. lathamella (Fletcher, 1936)
bimaculella (Haworth, 1828) (homonym)
razowskii Riedl, 1965
50. splendida Karsholt, 1997
bimaculella auctt., (*nec* Haworth, 1828) (misidentification)
lathamella auctt., (*nec* Fletcher, 1936) (misidentification)
51. zagulajevi Sinev, 1979
52. abchasica (Sinev, 1986)

COSMOPTERIGIDAE Heinemann & Wocke, [1876]
Diplosaridae Meyrick, 1915
Hyposmocomidae Hampson, 1918
Scaeosophides Meyrick, 1922

ANTEQUERINAE Hodges, 1978

Pancalia Stephens, 1829
53. leuwenhoekella (Linnaeus, 1761)
leuvenhoekella: Linnaeus, 1767 (incorrect spelling)
leuwenhöckella: [Denis & Schiffermüller], 1775 (incorrect spelling)
leeuwenhoeckella auctt. (incorrect spelling)

schmidtella (Treitschke, 1833)
mandshuricella Sinev, 1985 (subspecies)
54. schwarzella (Fabricius, 1798)
latreillella Curtis, 1830
55. baldizzonella Riedl, 1994
56. nodosella (Bruand, [1851])
nodosella (Mann, 1854) (homonym)

Euclemensia Grote, 1878
Hamadryas Clemens, 1864 (homonym)
57. woodiella (Curtis, 1830)

Phosphaticola Viette, 1951
58. gemmatella Viette, 1951

Alloclita Staudinger, 1859
Proceleustis Meyrick, 1914
Dhahrania Amsel, 1958
59. recisella Staudinger, 1859
albocinctella (Chrétien, 1915)
orthoclina Meyrick, 1922
inconspicuella (Turati, 1929)
gelechiformis (Turati, 1930)
60. haifensis Rebel, 1911
litorella (Amsel, 1935)
61. delozona Meyrick, 1919
62. brachygrapta Meyrick, 1925
63. deprinsi sp.n.
64. francoeuriae Walsingham, 1905

Gibeauxiella gen. nov.
65. bellaqueifontis (Gibeaux, 1986) comb.n.
66. reliqua (Gibeaux, 1986) comb.n.

Limnaecia Stainton, 1851
Limnoecia Meyrick, 1895 (emendation)
Lymnaecia Dyar, [1903] (emendation)
67. phragmitella Stainton, 1851

SCAEOSOPHINAE Meyrick, 1922

Allotalanta Meyrick, 1913
68. autophaea Meyrick, 1913

COSMOPTERIGINAE Heinemann & Wocke, [1876]

Cosmopterix Hübner, [1825]
Cosmopteryx Zeller, 1839 (emendation)
Lienigia Spuler, 1910
69. zieglerella (Hübner, [1810])
eximia (Haworth, 1828)

70. schmidiella Frey, 1856
mongoliella Sinev, 1980 (subspecies)
71. orichalcea Stainton, 1861
drurella auctt., (*nec* Fabricius, 1775) (misidentification)
druryella Zeller, 1850 (emendation)
singularis Sinev, 1980
72. sibirica Sinev, 1985
73. scribaiella Zeller, 1850
hermsiella E. Hering, 1889
japonica Kuroko, 1960 (subspecies)
74. pulchrimella Chambers, 1875
pulcherrimella Chambers, 1878 (emendation)
parietariae M. Hering, 1931
75. attenuatella (Walker, 1864)
flavofasciata Wollaston, 1879
mimetis Meyrick, 1897
apiculata Meyrick, 1922
antillia Forbes, 1931
76. crassicervicella Chrétien, 1896
flavipes Turati, 1930
superba Gozmány, 1960
dalii Agenjo, 1981
77. turbidella Rebel, 1896
78. pararufella Riedl, 1976
79. coryphaea Walsingham, 1908
donatellae Mariani, 1932
formosa Amsel, 1935
coryphae Agenjo, 1952 (emendation)
80. lienigiella Zeller, 1846
81. salahinella Chrétien, 1907
rufella Turati, 1927
phragmitidis Amsel, 1935

Labdia Walker, 1864
82. semicoccinea (Stainton, 1859)

Pyroderces Herrich-Schäffer, 1853
Lallia Chrétien, 1915
Tenuipenna Amsel, 1959
83. argyrogrammos (Zeller, 1847)
goldeggiella Herrich-Schäffer, 1853
84. caesaris Gozmány, 1957
85. klimeschi Rebel, 1938
86. brosi Riedl, 1969
87. tethysella sp.n.
88. wolschrijni sp.n.
89. apicinotella (Chrétien, 1915)
cinnamomina Turati, 1930
simplicella Amsel, 1959

Anatrachyntis Meyrick, 1915
Lacciferophaga Zagulajev, 1959
Sathrobrota Hodges, 1962

Amneris Riedl, 1993 (homonym)
Euamneris Riedl, 1996
90. rileyi (Walsingham, 1882)
stigmatophora (Walsingham, 1897)
91. badia (Hodges, 1962)
92. coridophaga (Meyrick, 1925)
pygmaella (Turati, 1927)
93. simplex (Walsingham, 1891)
gossypiella (Walsingham, 1906)
hemizopha Meyrick, 1916
repandatella (Legrand, 1966)
flexiloquella (Riedl, 1993)

Coccidiphila Danilevsky, 1950
Batrachedropsis Amsel, 1955
94. gerasimovi Danilevsky, 1950
95. patriciae J. Nel & A. Nel, 2000
96. ledereriella (Zeller, 1850)
unedella (Milliére, 1875)
rungsella J. Nel & Brusseaux, 1997 syn.n.
97. danilevskyi Sinev, 1997
ledereriella auctt., (*nec* Zeller, 1850) (misidentification)
98. kasypinkeri Traugott-Olsen, 1986
99. riedli Traugott-Olsen, 1986
ledereriella auctt., (*nec* Zeller, 1850) (misidentifiaction)

Stagmatophora Herrich-Schäffer, 1853
100. heydeniella (Fischer von Röslerstamm, 1841)

Eteobalea Hodges, 1962
Parastagmatophora Riedl, 1965
101. anonymella (Riedl, 1965)
serratella auctt., (*nec* Treitschke, 1833) (misidentification)
102. serratella (Treitschke, 1833)
sareptensis (Walsingham, 1907)
cinereocapitella (Caradja, 1920)
103. intermediella (Riedl, 1966)
beata auctt., (*nec* Walsingham, 1907) (misidentification)
104. beata (Walsingham, 1907)
105. sumptuosella (Lederer, 1855)
admirandella (Bruand, [1851]) (nomen nudum)
amseli (Osthelder, 1940)
106. teucrii (Walsingham, 1907)
107. klisieckii (Riedl, 1966)
108. tririvella (Staudinger, 1871)
kasyi (Riedl, 1965)
bernhardiella (Kasy, 1973) (subspecies)
109. albiapicella (Duponchel, [1843])
110. alypella (Klimesch, 1946)
111. isabellella (Costa, [1836])
hisabellella (Costa, [1836]) (emendation)
opulentella (Herrich-Schäffer, [1854])
rutilella (Chrétien, 1896)

112. dohrnii (Zeller, 1847)
dohrni auctt. (incorrect spelling)
113. siciliae (Riedl, 1966)
114. thaumatella (Walsingham, 1907)
lactipunctella (Turati, 1924)

Isidiella Riedl, 1965
115. nickerlii (Nickerl, 1864)
nickerli auctt. (incorrect spelling)
116. divitella (Constant, 1885)
tyrrhenica (Amsel, 1939)

Vulcaniella Riedl, 1965
117. pomposella (Zeller, 1839)
118. grandiferella Sinev, 1986
119. cognatella Riedl, 1990
120. extremella (Wocke, 1871)
naviella (Chrétien, 1907)
buhri (M. Hering, 1935)
121. caucasica Sinev, 1986
122. grabowiella (Staudinger, 1859)
123. rosmarinella (Walsingham, 1891)
124. karadaghella Sinev, 1986
125. klimeschi (Riedl, 1966)
126. fiordalisa (Petry, 1904)
gnaphaliella (Chrétien, 1922)
127. glaseri (Riedl, 1966)
128. gielisi sp.n.
129. anatolica sp.n.
130. pontica sp.n.

Ramphis Riedl, 1969
131. ibericus Riedl, 1969
132. libanoticus Riedl, 1969

Hodgesiella Riedl, 1965
133. rebeli (Krone, 1905)
134. rhodorrhisella Kasy, 1970
135. christophi sp.n.

Tolliella Riedl, 1969
136. fulguritella (Ragonot, 1895)
copidobathra (Meyrick, 1936)

CHRYSOPELEIIDAE Mosher, 1916
Walshiidae Hodges, 1962

Sorhagenia Spuler, 1910
Cystioecetes Braun, 1915
Sorhageniella Riedl, 1965
137. rhamniella (Zeller, 1839)
asperipunctella (Bruand, [1851])

138. reconditella Riedl, 1983
139. lophyrella (Douglas, 1846)
rhamniella auctt., (*nec* Zeller, 1839) (misidentification)
tolli Riedl, 1962
140. janiszewskae Riedl, 1962
rhamniella auctt., (*nec* Zeller, 1839) (misidentification)
141. fibigeri sp.n.
142. taurensis sp.n.

Ascalenia Wocke, [1876]
143. vanella (Frey, 1860)
seeboldiella (Ragonot, 1882)
144. vanelloides Gerasimov, 1930
eremella Amsel, 1933 (nomen nudum)
eremella Amsel, 1935
145. antiqua Meyrick, 1925
146. acaciella Chrétien, 1915
tergipunctella (Turati, 1924)
maculatella (Lucas, 1937)
noviciata (Gozmány, 1960)
147. jerichoella Amsel, 1935
148. viviparella Kasy, 1969
149. echidnias (Meyrick, 1891)
signatella Chrétien, 1915
oranella Lucas, 1939
satellita Gozmány, 1960
150. bifasciella Chrétien, 1915
bizonatella (Turati, 1930) syn.n.

Calanesia Sinev, 1990
151. karsholti Sinev, 1990

Calycobathra Meyrick, 1891
152. acarpa Meyrick, 1891
pinguescentella Chrétien, 1915
sublineatella (Lucas, 1933) syn.n.
153. calligoni Sinev, 1979
154. variapennella Sinev, 1984

Bifascia Amsel, 1961
155. nigralbella (Chrétien, 1915)
asteroleuca (Meyrick, 1917)

Bifascioides Kasy, 1968
156. leucomelanellus (Rebel, 1917)
pirastica (Meyrick, 1936)

Gisilia Kasy, 1968
Melnea Clarke, 1986
157. stereodoxa (Meyrick, 1925)
evitans (Meyrick, 1925)
alfieriella (Rebel, 1926)
158. lerautella Gibeaux, 1986

Taxa formerly placed in Momphidae sensu lato

Neomariania Mariani, 1943
Mariania Rebel, 1937 (homonym)
Megaceraea Rebel, 1940 (unavailable)
Megaceraea Nye & Fletcher, 1991 syn.n.
159. partinicensis (Rebel, 1937)
160. rebeli (Walsingham, 1894)
161. scriptella (Rebel, 1940)
162. oecophorella (Rebel, 1940)
163. incertella (Rebel, 1940)

Family MOMPHIDAE

Momphina Herrich-Schäffer, 1857: 58. Type genus: *Mompha* Hübner, [1825].
Lavernidae Wocke, in Staudinger & Wocke, 1871: 318. Type genus: *Laverna* Curtis, 1839.
Psacaphorinae Spuler, 1898: 32. Type genus: *Psacaphora* Herrich-Schäffer, 1853.

General part

Morphology.– Wingspan 7 to 22 mm. Head (Text-fig. 1) covered with densely appressed scales. Antennae shorter than forewings, with scapus bearing bristled pecten. Eyes moderate, ocelli absent. Labial palpi rather long, curved upward, without tufts of scales in front; segment 3 usually shorter than segment 2. Forewings (Text-fig. 2) narrow lanceolate to lanceolate, usually with pronounced tufts of raised scales. In forewings 3 median veins developed; M1 sometimes stalked with R4+5. Hindwings narrow lanceolate or lanceolate linear; M1 stalked with M2 or separate. Hindtibiae with middle spurs at midpoint of tibia or slightly beyond it.

Abdominal tergites II to VII (Text-fig. 3) with specialized lanceolate scales ('spines') deeply embedded and arranged in two large rounded lateral groups; longitudinal sclerotized struts of tergite II short. Male genitalia (Text-fig. 4): uncus long and narrow, articulated with well developed triangular tegumen; gnathos reduced to slightly rough or membranous bulb shifted anteriorly and joined with antero-dorsal processes of valvae (transtilla); vinculum narrow, saccus short, without pronounced medial projection; valvae symmetrical, divided into delicate petal-shaped dorsal lobe (cucullus) and rather heavily sclerotized ventral lobe (sacculus) with stout apical projection; two moderate finger-shaped hairy anellus lobes (valvellae, sometimes treated as juxta lobes) attached to apical portion of aedeagus latero-ventrally; aedeagus considerably shorter than valvae, straight, with desclerotized walls and groups of cornuti in vesica. Female genitalia (Text-fig. 5): ovipositor moderately long or rather short; ductus seminalis merges into medial part of usually broad and more or less sclerotized ductus bursae; bursa copulatrix of moderate or small size, oval, with pair of very characteristic sickle-shaped signa.

Larvae rather short and broad ('fat'), spindle-shaped, slow-moving, without secondary setae, colourless, but becoming deep pink before pupation. Pupae moderately sclerotized, with exposed labial palpi and forefemora; abdominal segments IV to VI movable; apical cremaster unpaired, hooked. Cocoon spindle-shaped, double-layered, with short neck.

Biology.– Mainly associated with woodlands where they are frequently found along fringes of forests, in forest clearings, and especially along roads. Larvae mine leaves, stems, roots or seed-pods, make galls or live between spun leaves and flowers, sometimes changing the life-style from the first to the last instars. At least 90 % of the species with known bionomy associate with onagraceous plants, with only one exception in Europe (Sinev, 1989b). Larvae are usually specialized to particular species of food-plants and even to particular parts of the plants, so that the larval habitat is an important aid to determination. Adults rest on the leaves of their host-plants during the day-time, with the whole body tightly pressed to the substrate; they often fly in dull weather or before sunset, but become especially active early in the morning. In most species adults hibernate under litter, bark of old trees and in other refuges and sometimes they may be collected even in winter.

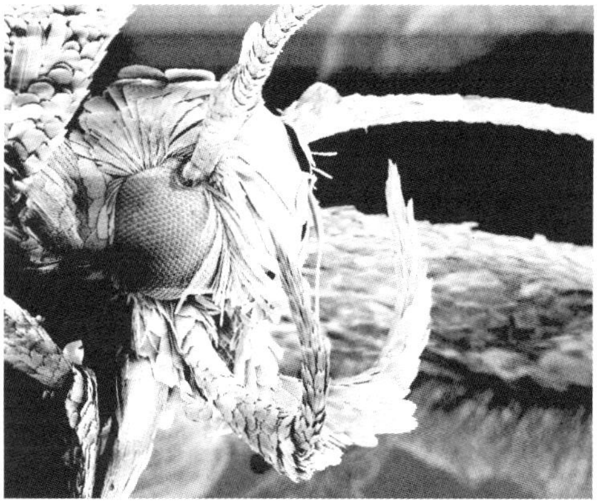

Text-fig. 1. Head of *Mompha sturnipennella* (Treitschke).

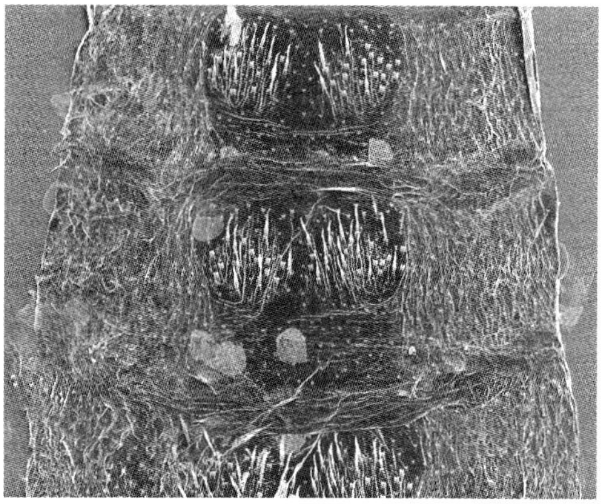

Text-fig. 3. Abdominal tergites of *Mompha sturnipennella* (Treitschke).

DISTRIBUTION.– The family is especially richly represented in North and South America, where several tens of species (many undescribed) occur. More than 30 species are known from the Palaearctic, whereas very few species occur in the Afrotropical and Indo-Australian region. In the region of this study 1 genus, 5 subgenera and 20 species are represented.

SYSTEMATICS.– Morphological characters, both external and genitalic, are rather uniform within the family (at least for the palaearctic representatives), and most species belong to the large genus *Mompha* Hübner. Some characters of wing venation, functional morphology of genitalia, and larval life-style argue in favour of splitting this genus into several groups (up to 5). We are here treating *Mompha* as a single genus with 5 subgenera. The family is most closely related to Blastobasidae and other groups with a weakly sclerotized aedeagus and a more or less well developed sacculus (Stathmopodidae, Oecophoridae, etc.).

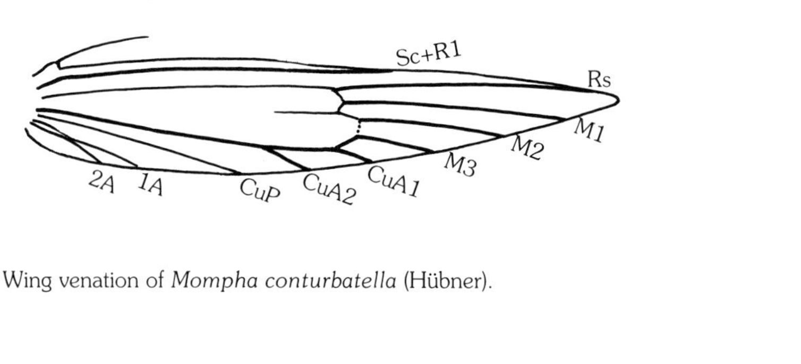

Text-fig. 2. Wing venation of *Mompha conturbatella* (Hübner).

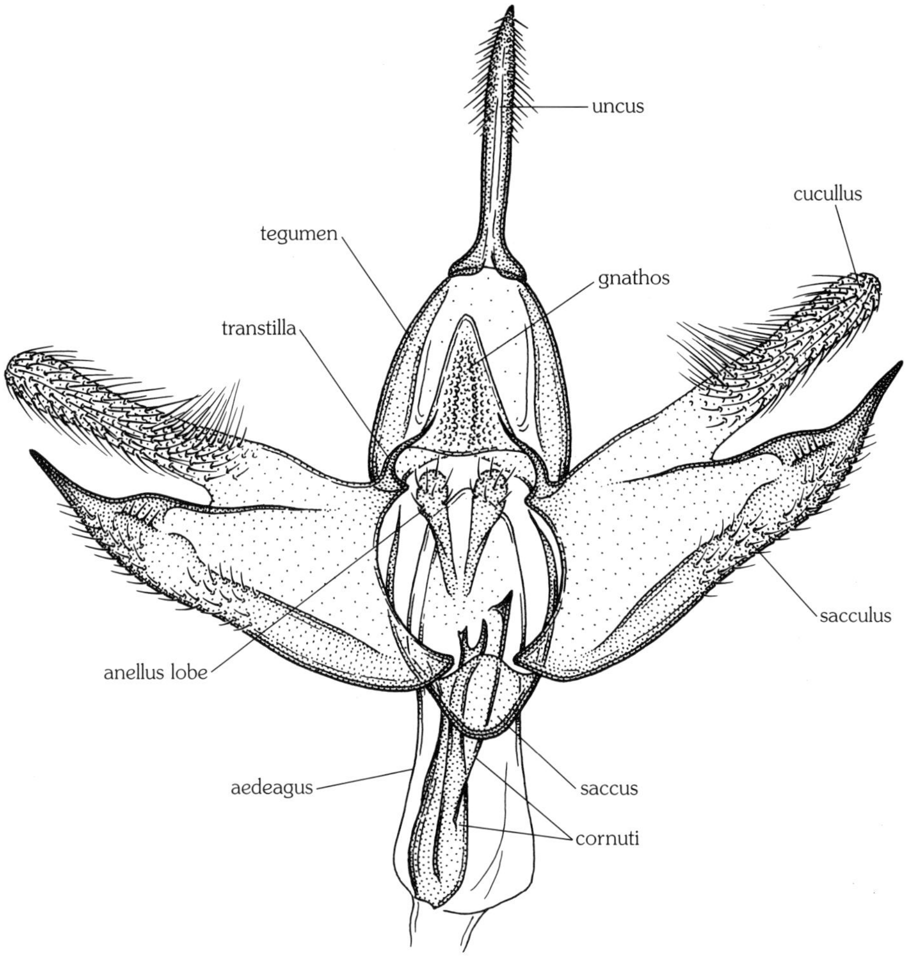

Text-fig. 4. Male genitalia of *Mompha jurassicella* (Frey).

Systematic Treatment of the Genus, Subgenera and Species of Momphidae in Europe, North Africa and the Near East

Mompha Hübner, [1825]

Mompha Hübner, [1825]: 414.
Type species: *Tinea conturbatella* Hübner, [1819], by subsequent designation by Walsingham & Durrant, 1909.
Laverna Curtis, 1839: folio 735.
Type species: *Laverna ochraceella* Curtis, 1839, by original designation.
Lophoptilus Sircom, 1848: 2037.
Type species: *Lophoptilus staintoni* Sircom, 1848 (= *Tinea miscella* [Denis & Schiffermüller], 1775), by monotypy. [Subgenus].

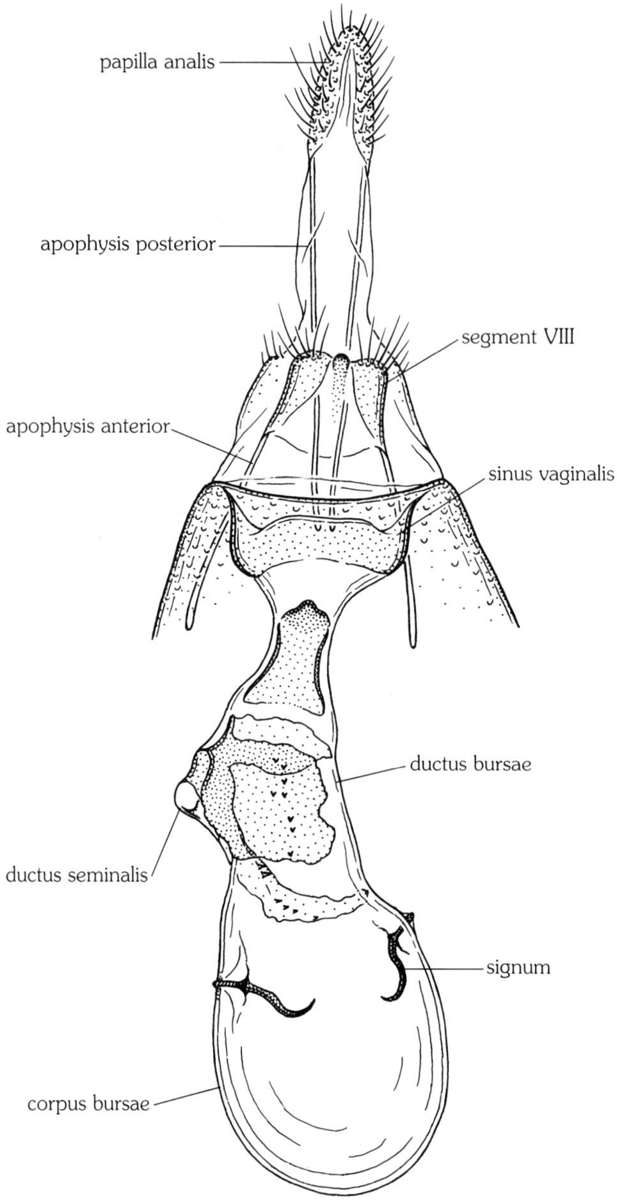

Text-fig. 5. Female genitalia of *Mompha confusella* Koster & Sinev.

Cyphophora Herrich-Schäffer, 1853: 12 (key), 46.
Type species: *Elachista idaei* Zeller, 1839, by monotypy. [Subgenus].
Psacaphora Herrich-Schäffer, 1853: 13 (key), 48.
Type species: *Tinea schrankella* Hübner, 1805 (= *Tinea locupletella* [Denis & Schiffermüller], 1775), by subsequent designation by Walsingham & Durrant, 1909. [Subgenus].
Anybia Stainton, 1854: 244.
Type species: *Tinea langiella* Hübner, 1796 (= *Alucita epilobiella* Römer, 1794), by monotypy. [Subgenus].

Wilsonia Clemens, 1864: 428.
Type species: *Wilsonia brevivittella* Clemens, 1864, by monotypy.
Leucophryne Chambers, 1875c: 210.
Type species: *Leucophryne tricristatella* Chambers, 1875, by monotypy.

Small to medium sized moths with all characters of the family. Several authors have divided the genus *Mompha* into several genera. In Europe the following genera were established: *Mompha* Hübner, [1825], *Anybia* Stainton, 1854, *Psacaphora* Herrich-Schäffer, 1853, *Lophoptilus* Sircom, 1848 and *Cyphophora* Herrich-Schäffer, 1853. They were subsequently downgraded to subgenera by most authors or completely omitted. In this volume we follow Riedl (1969) and Karsholt & Razowski (1996) and treat all these taxa as subgenera of *Mompha* sensu lato. This proposal must be considered as provisional. The taxonomic arrangement of Momphidae might be changed considerably in the near future, following a complete revision of the large number of mostly undescribed and very diverse species in the Nearctic and Neotropical faunas.

In Europe the larvae live on species of Onagraceae (Myrtales) belonging to the genera *Epilobium* L., *Chamaenerion* Adans. and *Circaea* L., sometimes also *Oenothera* L. The only exception is *Mompha miscella*, whose larvae live on *Helianthemum* spp. (Cistaceae).

Subgenus ***Mompha*** Hübner, [1825]

Forewings lanceolate with more or less pronounced tufts of raised scales (from large to almost indistinct). In forewings only R4 and R5 stalked; in hindwings M1 and M2 separate. Male genitalia with prominent and dimply gnathos; aedeagus stout, often with numerous cornuti. Female genitalia with ductus bursae short and wide; antrum enlarged, with sclerotized plates, rims or bands of spicules giving good diagnostic characters.

1. Mompha (Mompha) conturbatella (Hübner, [1819])

Tinea conturbatella Hübner, [1819]: fig. 450.
Lita elegantella Zetterstedt, 1840: 1004.

DIAGNOSIS.– Adult. Wingspan 16-18 mm. Forewings blackish with white costal spot at 4/5, white tornal spot almost beneath it, greyish basal spot and several ochreous and leaden grey spots spread along the wing.

Similar species. The large wingspan and the blackish ground colour with white and ochreous markings are characteristic.

MALE GENITALIA.– Cucullus relatively broad and widely rounded apically, sacculus broad, abruptly narrowed to blunt apex, not reaching the top of cucullus; gnathos small and triangular; anellus lobes short, just reaching the apex of the aedeagus; aedeagus short and thick with small thorn-like cornutus.

FEMALE GENITALIA.– Lamella antevaginalis with two sclerotized lobes; tergite VIII almost quadrangular; sinus vaginalis very wide; ductus bursae short and sclerotized, with sclerotization extending along the wall of corpus bursae.

DISTRIBUTION.– Holarctic. Europe, the Caucasus, Central Asia (local), Siberia, Russian Far East; North America. In southern Europe and Central Asia restricted to mountain areas (Sinev, 1996).

BIOLOGY.– Host plants: *Chamaenerion angustifolium* (L.) Scop. and *C. latifolium* (L.)

Th. Fr. et Lange; also mentioned from *Epilobium montanum* L. (Lhomme, [1948-1949]; Riedl, 1969) but this record should be verified. Univoltine. The larva lives in late spring between tightly spun top leaves of the host plant, boring into the apical part of the growing stem; the leaves are spun together in a vertical position and often somewhat twisted. Pupation takes place in an ochreous brown, spindle-shaped cocoon amongst detritus on the ground. The moths emerge in June and fly to early May of the next spring. Hibernation takes place in the dry stems of the host plant and other shelters. See text-fig. 6.

2. *Mompha (Mompha) ochraceella* (Curtis, 1839)

Laverna ochraceella Curtis, 1839: 735.

Diagnosis.– Adult. Wingspan 14-16 mm. Forewings ochreous yellow, with 4 tufts of raised scales: these are brown to dark brown at 1/5 and 1/2 on dorsum, ochreous in fold near base, and white in fold between two dark brown tufts. Hindwings shining yellowish white.

Similar species. The large wingspan and the ochreous yellow forewings in combination with the yellowish white hindwings are characteristic for the species.

Text-fig. 6. *Mompha conturbatella* Hübner, terminal shoot of *Chamaenerion angustifolium* (L.) Scop. spun up by the larva.

MALE GENITALIA.– Uncus stout, cucullus more than eight times as long as wide; sacculus broad, gradually narrowing after the middle, then abruptly narrowing to the blunt apex, and reaching the top of cucullus; gnathos large, rounded apically; anellus lobes about 1/3 length of aedeagus and widened distally; aedeagus with hook-shaped cornutus.

FEMALE GENITALIA.– Tergite VIII broad, unsclerotized part of its posterior edge distinct, U–shaped; sclerotizations of lamella post- and antevaginalis partly united, broad, distally with U–shaped excavation and protruded laterally; sinus vaginalis large and very flat, ductus bursae funnel-shaped posteriorly with club-shaped sclerotization in the middle, which is more than half as wide as width of ductus, anterior part weakly sclerotized.

DISTRIBUTION.– Western Palaearctic. Europe, to southern Sweden in the north; Morocco, Asia Minor, the Caucasus and Iran.

BIOLOGY.– Host plant: *Epilobium hirsutum* L. Univoltine. The larva lives from summer to autumn under the bark of the stem, making irregular galleries filled with brown frass. In autumn it enters the root and hibernates. In the next spring the larva bores a

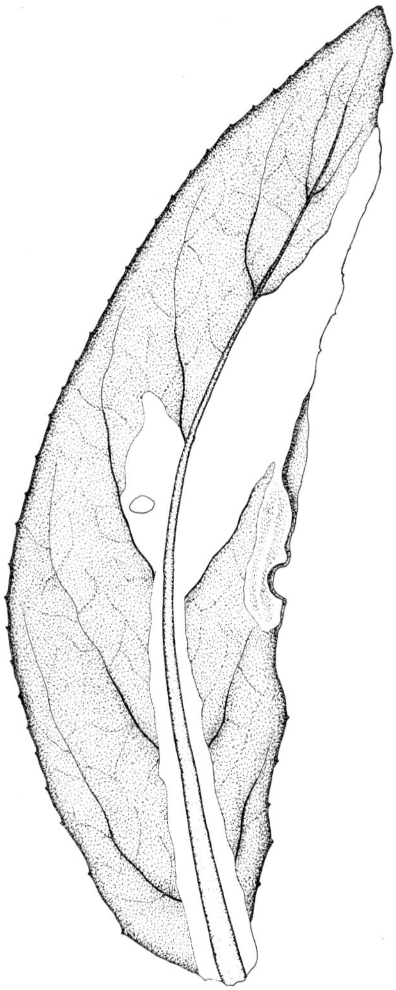

Text-fig. 7. *Mompha ochraceella* (Curtis), larval mine in leaf of *Epilobium hirsutum* L.

gallery in the lower portion of the stem, from where it makes a large blister-like mine on the underside of the lower leaves. Mines are changed, sometimes through the stem. Pupation takes place inside the mine in a whitish silken cocoon. The moths have been recorded from the end of May to the middle of August. In Central Europe rather common in wet places along ditches, streams, lake-shores, etc. See text-fig. 7.

3. ***Mompha (Mompha) lacteella*** (Stephens, 1834)

Anacampsis lacteella Stephens, 1834: 210.
Elachista gibbiferella Zeller, 1839: 212.
Elachista basipallidella Stainton, 1849: 24.

DIAGNOSIS.– Adult. Wingspan 11-13 mm. Head beige, segment 3 of labial palpi with dark brown median and subapical rings. Thorax and tegulae beige. Forewings greyish black with several large irregular orange brown streaks and few white spots; large basal beige spot on dorsum, concolorous with thorax and edged by tufts of raised black scales in fold and near dorsum; large tuft of raised black scales before middle and below fold; white costal spot at 3/4, with opposite white tornal spot bearing raised scales distally; apical area greyish black with several white streaks into cilia. Hindwings shining dark grey.

Similar species. For separation from *propinquella*, q. v.

MALE GENITALIA.– Cucullus about five times as long as wide; sacculus as long as cucullus, rather slender, gradually tapering to blunt apex; gnathos slightly triangular; anellus lobes more than half the length of aedeagus, widened and lobed distally; aedeagus with two cornuti, one short and bifurcate, the other long and very slender.

FEMALE GENITALIA.– Differs from *ochraceella* by the broad and almost square lamella antevaginalis, the reverse V–shaped lamella postvaginalis and the large sclerotization in the anterior part of the ductus bursae.

DISTRIBUTION.– Europe except for the extreme south; Asia Minor, the Caucasus, eastern Transcaucasus, southern Siberia and the Russian Far East (Sinev, 1996).

BIOLOGY.– Host plants: *Epilobium hirsutum* L. and *E. montanum* L. In March and April the larva makes a large blotch mine occupying most of one of the lower leaves; mines are usually changed. The frass in large black grains is spread through the mine or heaped in the middle (Heckford, 1986; Sterling, 1988). Adults fly from mid-June to late July and often come to light.

4. ***Mompha (Mompha) propinquella*** (Stainton, 1851)

Elachista propinquella Stainton, 1851: 8.
Laverna paludicolella Doubleday, 1859: 34; unnecessary substitute name for *Elachista propinquella* Stainton, 1851.

DIAGNOSIS.– Adult. Wingspan 11-13 mm. External appearance very similar to *lacteella*. It can be separated from this species by the white head and thorax, the forewings with dorsal spot white at base, the single dark brown ring on the 3rd segment of labial palpi, and the more brown apical area.

MALE GENITALIA.– Similar to those of *lacteella* but sacculus broader and slightly shorter than cucullus, apically more abruptly tapering to a sharp point; anellus lobes approximately half the length of the aedeagus and narrower distally; aedeagus with shorter cornuti.

FEMALE GENITALIA.– Can be separated from *lacteella* by the more rounded distal edges of the lamella antevaginalis and the shallower V–shaped lamella postvaginalis.

DISTRIBUTION.– Western and Central Europe.

BIOLOGY.– Host plants: *Epilobium hirsutum* L., *E. montanum* L. and *E. palustre* L. (Stainton, 1870a; Emmet, 1979, 1988). The biology most probably does not differ from *lacteella*. The larvae have been recorded from February till April, mining the lower leaves; mines cannot be separated from those of *lacteella* (Hering, 1957). Adults fly from the end of June to mid-September. See text-fig. 8.

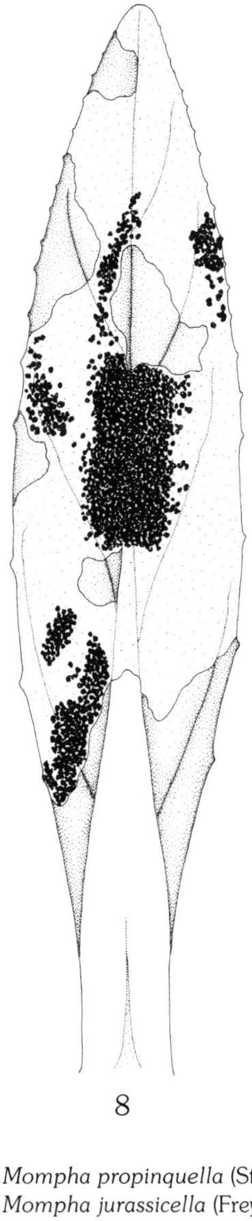

8

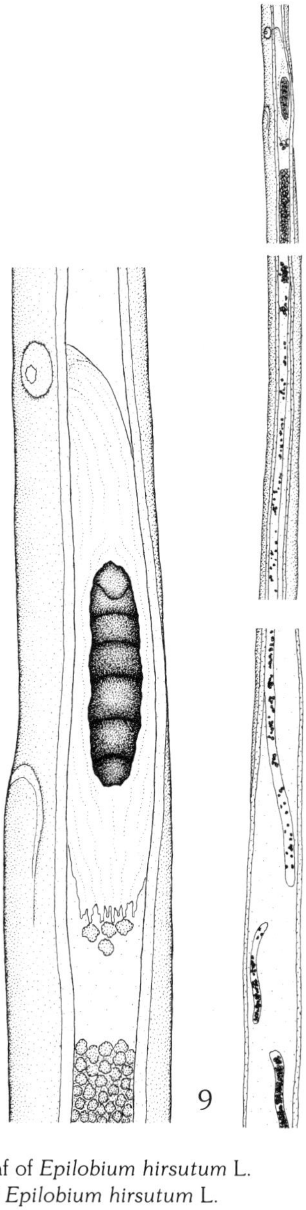

9

Text-fig. 8. *Mompha propinquella* (Stainton), larval mine in leaf of *Epilobium hirsutum* L.
Text-fig. 9. *Mompha jurassicella* (Frey), larval gallery in stem of *Epilobium hirsutum* L.

5. *Mompha (Mompha) divisella* Herrich-Schäffer, [1854]

Mompha divisella Herrich-Schäffer, [1854]: 209.
Mompha decorella auctt. (*nec* Haworth, 1812); misidentification.

DIAGNOSIS.– Adult. Wingspan 10-13 mm. Head ochreous white, strongly irrorated light brown. Thorax light brown. Forewings dark greyish brown, mottled whitish; white inward-oblique fascia at 3/4, often narrowed in the middle; both sides of fascia with light brown streak, inner larger and more irregular; small tufts of slightly raised dark greyish brown-tipped scales at 1/4, 1/2, and 3/4 subdorsally; basal half of dorsal area white, more ochreous towards base. Hindwings shining grey. Female with white fascia broader and abdomen with segment VIII greyish brown ventrally.
Similar species. For separation from *subdivisella*, *bradleyi* and *confusella*, q. v.

MALE GENITALIA.– Cucullus slender, slightly narrowing distally with rounded and slightly broadened tip; sacculus gradually tapering into long apical process, only slightly bent and with blunt apex, reaching at or beyond top of cucullus; anellus lobes small, clavate; aedeagus with three cornuti of almost equal length: left one broad, hooked and covered with microspicules; middle one bifurcate; right one slender, sharply pointed and with small distal hook. For separation from *subdivisella*, *bradleyi* and *confusella*, q. v.

FEMALE GENITALIA.– Tergite VIII twice as broad as long, unsclerotized part of its posterior edge distinct and U–shaped; sinus vaginalis funnel-shaped, as wide as or slightly wider than the largest width of enlarged ductus bursae; inception of ductus seminalis small and oval, placed in posterior part of ductus bursae. For separation from *subdivisella*, *bradleyi* and *confusella*, q. v.

DISTRIBUTION.– Europe, to southern Scandinavia in the north; the Caucasus, Central Asia (Koster & Sinev, 1996).

BIOLOGY.– Host plants: generally restricted to smaller species of the genus *Epilobium*: *E. montanum* L., *E. palustre* (L.) Crantz, *E. lanceolatum* Sebastiani et Mauri and *E. parviflorum* Schreber (Barrett, 1865; Stainton, 1870a; Wakely, 1944; Riedl, 1969; Sinev, 1989b); in Finland also *E. adenocaulon* Haussknecht, a species originating from North America (S. Kerppola, pers. comm.). The records from *E. alpinum* L. and *Chamaenerion angustifolium* (L.) Scop. should be verified. The larva lives in June and July in the main stem, generally near the leaf base, causing a gall and often leading to excessive branching above the gall. Pupation takes place inside the gall in a white cocoon. The moths emerge in August and fly after hibernation till May, but sometimes adults may be collected in December and January during mild days.

6. *Mompha (Mompha) jurassicella* (Frey, 1881)

Laverna jurassicella Frey, 1881: 146.
Laverna iurassicolella Reutti, 1898: 257; unjustified emendation of *Laverna jurassicella* Frey, 1881.
Mompha subdivisella Bradley, 1951: 177.

DIAGNOSIS.– Adult. Wingspan 11-13 mm. Forewings greyish brown with ochreous discal markings, pattern indistinct (especially in male), clearly more brownish grey than in similar species such as *divisella*, *bradleyi* and *confusella*.

MALE GENITALIA.– Sacculus abruptly tapering distally to slightly bent and sharp apex, not reaching the tip of cucullus; aedeagus with three cornuti of different length: left one shortest, very slender with microspicules at distal end, right one longest, slender and apically hooked. The genitalia are similar to those of *divisella* but differ in the shorter and more abruptly tapering sacculus and in the different length of cornuti in the aedeagus.

FEMALE GENITALIA.– Tergite VIII about 1.5 times as broad as long, with indistinctly traced unsclerotized posterior part; sinus vaginalis slightly narrower than the widest part of ductus bursae; posterior part of ductus bursae about as long as anterior part; inception of ductus seminalis small and oval, situated in the posterior part of ductus bursae. The shape and sclerotization of tergite VIII and the length of the posterior part of the ductus bursae are characteristic for the species.

DISTRIBUTION.– Western Europe to Switzerland in the east.

BIOLOGY.– Host plant: *Epilobium hirsutum* L. In early August the larva makes a gallery in the pith, up to 25 cm long, usually unbranched, although occasionally diverticula may lead into larger lateral branches. Up to three such galleries could be found in one plant; they are usually separated vertically. This species hibernates as adult. The moths fly between late August and October, reappearing in March and April (Sterling & Langmaid, 1989; Corley, pers. comm.). See text-fig.9.

REMARKS.– This species, first mentioned from Switzerland, was considered as *incertae sedis* for a long time because of the unknown whereabouts of the type material and insufficient original description, which did not guarantee a certain separation from the other very similar species of the *divisella*-complex (Riedl, 1969). However, the type (designated as lectotype) of *Laverna jurassicella* has been found very recently in the collection of the BMNH (Sinev, 1999) and appears to be conspecific with *Mompha subdivisella*, subsequently described from England.

7. *Mompha (Mompha) bradleyi* Riedl, 1965

Mompha bradleyi Riedl, 1965: 447.

DIAGNOSIS.– Adult. Wingspan 9-11mm. Very similar to *divisella* externally, but forewings somewhat oily shining and markings often rather indistinct, especially in the male.

MALE GENITALIA.– Cucullus basally about twice as wide as distally; sacculus abruptly tapering distally to slightly bent and sharp apex, not reaching tip of cucullus; aedeagus with three cornuti of different length: left one shortest, hooked, distally with microspicules, right one longest, slender and bent. The width of the cucullus base and the shape of cornuti are characteristic for this species.

FEMALE GENITALIA.– Tergite VIII with convex posterior margin and indistinct unsclerotized area; sinus vaginalis distinctly narrower than the widest part of ductus bursae; inception of ductus seminalis large and longitudinally stretched.

DISTRIBUTION.– Western and Central Europe.

BIOLOGY.– Host plant: *Epilobium hirsutum* L. Univoltine. In July the larva causes a conspicuous gall in the thin top or (more often) in a flowering side stalk. When the gall is situated in the top stalk, it leads to a distinct impediment of growth; in the side stalks this is less evident; sometimes the gall has a conspicuous red colour. The moths are on the wing from mid-July and again after hibernation till late May. See text-fig. 10.

Text-fig. 10. *Mompha bradleyi* Riedl, larval gall and cocoon in stem of *Epilobium hirsutum* L.

8. Mompha (Mompha) confusella Koster & Sinev, 1996

Mompha confusella Koster & Sinev, 1996: 146.

DIAGNOSIS.– Adult. Wingspan 11-13 mm. Externally very similar to *divisella*, but forewings more elongate, being 4.9 to 5.2 times as long as wide, and whitish dorsal area near base strongly suffused with brownish grey scales.

Similar species. Can only be separated from *divisella* with certainty by genitalia characters.

MALE GENITALIA.– Cucullus narrow with almost parallel sides, subapically slightly narrowing, tip rounded; sacculus shorter than cucullus, broad at base, abruptly tapering distally to strongly outwardly bent sharp apex; gnathos widely triangular; aedeagus with three cornuti of somewhat different length: left one shortest, bifurcate and covered with microspicules, middle one bifurcate, right one longest, slender and sharply pointed. The strongly outwardly bent apex of the sacculus, the wide triangular gnathos and the shape of the cornuti are characteristic for this species.

FEMALE GENITALIA.– Tergite VIII slightly broader than long, anterior edge rounded, posteriorly with large V–shaped unsclerotized part; sinus vaginalis wide, bowl-shaped, wider than the largest width of ductus bursae; inception of ductus seminalis keyhole-shaped.

DISTRIBUTION.– Central and south-eastern Europe; eastern Transcaucasus (Koster & Sinev, 1996).

BIOLOGY.– Host plants: *Epilobium hirsutum* L. and *Chamaenerion angustifolium* (L.) Scop. According to the labels of two specimens from Austria ('*Epil[obium] hirs[utum]* Anschwell[ung]', and '*Epilob[ium] angustif[olium]* Stengel', the larva is probably a stem gall-maker.

Text-fig. 11. *Mompha subbistrigella* (Haworth), infested seedpod on *Epilobium tetragonum* L.

9. Mompha (Mompha) subbistrigella (Haworth, 1828)

Tinea subbistrigella Haworth, 1828: 581.

DIAGNOSIS.– Adult. Wingspan 7-12 mm. Head grey, mottled light grey, thorax and forewing dark grey.

Similar species. In general, this species has the same wing pattern as *sturnipennella*, but differs by the forewings being without ferruginous streaks in the middle and by the distinct white costal streaks in the apical area.

MALE GENITALIA.– Cucullus broad at base and narrowing distally; sacculus truncate, apex with three or four short and thick teeth; aedeagus with large cornutus ending in an axe-head shape, and another two long and one short and slightly bent cornuti. The shape of the apical part of the sacculus and the cornuti are characteristic for the species.

FEMALE GENITALIA.– Lamella antevaginalis consists of two wedge-shaped sclerites with small sclerotized plate between them; ductus bursae divided into short and narrow posterior (subostial) part and long and wide anterior part, the latter with rim of teeth followed by large sclerotized plate. The shape of the lamella antevaginalis and the ductus bursae are characteristic.

DISTRIBUTION.– North and Central Europe, eastwards to the Caucasus and Central Asia (Sinev, 1989b).

BIOLOGY.– Host plants: smaller species of *Epilobium* like *E. montanum* L., *E. palustre* L., *E. parviflorum* Schreb., and *E. tetragonum* L. The larva lives in a seedpod, eating the unripe seeds from late June to July. An infested seedpod is shorter and curved and has a small hole. Adults can be found throughout the year, but most frequently after hibernation, i.e., in late spring and early summer. See text-fig. 11.

10. Mompha (Mompha) sturnipennella (Treitschke, 1833)

Ornix sturnipennella Treitschke, 1833: 209.
Oecophora modestella Eversmann, 1844: 598.
[Mompha] permutatella Herrich-Schäffer, [1854]: 209, fig. 969; unavailable being originally published as a synonym of *Mompha sturnipennella* (Treitschke, 1833).
Mompha (Laverna) nodicolella Fuchs, 1902: 328.

DIAGNOSIS.– Adult. Wingspan 9-13 mm. Head grey, mottled light grey. Thorax dark grey. Forewings dark grey, basal 2/5 lighter; white, strongly mottled grey and slightly outward oblique fascia before middle, not reaching costa and widening towards dorsum; white inward oblique fascia at 3/4; black streaks in fold, in middle part between both fasciae and in apical area; medial black streak often with white edging on dorsal side, both outer streaks edged by ferruginous streaks or scales; three small black tufts of slightly raised scales near dorsum, first at 1/5, second and third outward white fasciae; apical area with two or three white streaks at costa. Hindwings shining brownish grey.

Similar species. For separation from *subbistrigella*, q. v.

MALE GENITALIA.– Cucullus broad at base, narrowing medially and broadened distally to rounded apex; sacculus gradually tapering to blunt apex with three to five small teeth; uncus slender with broadened subapical part; gnathos rounded; anellus lobes very small with rounded tops; aedeagus with one large curved cornutus, several short and stout cornuti and a bundle of slender cornuti. The shape of the cucullus and the cornuti are characteristic for the species.

FEMALE GENITALIA.– Lamella antevaginalis consists of two wedge-shaped sclerites with large elongate oval sclerotization between them; sinus vaginalis large, bowl-shaped; ductus bursae very short with indistinct sclerotization; corpus bursae with two patches of teeth in posterior part, medially wrinkled and somewhat narrowed. The shape of the lamella antevaginalis and the ductus bursae are characteristic.

DISTRIBUTION.– Holarctic. Europe (except for the south), Siberia and Russian Far East. Also reported from Canada (Sinev, 1996).

BIOLOGY.– Host plant: *Chamaenerion angustifolium* (L.) Scop. Univoltine (in the north) or bivoltine. In May and June the larva makes a spindle-shaped gall, usually in the

Text-fig. 12. *Mompha sturnipennella* (Treitschke), larval gall in stem of *Chamaenerion angustifolium* (L.) Scop.

flowering part of the stem; several galls can be found in one stem. The larvae of the second generation can be found in July and August and they sometimes produce galls too, but more often they live in the seedpods. Only a small hole indicates the presence of a larva in a seedpod. Pupation takes place outside the mine or seedpod (Emmet, 1982). Adults fly from July till May of the next year and are most frequently observed after hibernation. See text-fig. 12.

11. *Mompha (Mompha) meridionella* sp.n.

Diagnosis.– Adult. Wingspan 11-12 mm. Head with shining white frons and ochreous grey vertex, mottled light grey; collar ochreous grey; antennae dark grey with indistinct lighter annulations, especially in basal half, scape ochreous grey dorsally, whitish ventrally; labial palpi white, segment 1 very short, segment 2 as long as segment 3, light grey ventrally and laterally on outside, segment 3 with narrow dark grey basal ring, wide irregular ochreous grey medial ring and blackish tip. Thorax and tegulae ochreous grey, with some light grey scales posteriorly. Legs dark grey, banded white at end of segments, tibiae with white medial and apical bands, tarsi white at end of segments, spurs white. Forewings ochreous grey, with darker grey outer part; large light grey spot at base, not reaching dorsum and strongly mottled ochreous grey; white, slightly outward oblique fascia before middle, not reaching costa and widening towards dorsum, strongly mottled with ochreous grey, especially in dorsal half; white inward oblique fascia at 3/4, somewhat narrowed in middle; blackish streaks in fold of basal area, in middle between both fasciae, and in apical area beyond outer fascia, all streaks edged by ferruginous streaks or scales; three small blackish tufts of slightly raised scales near dorsum, first at 1/5, second and third on outer sides of white fasciae, the middle outwardly edged ferruginous; apical area with two or three white streaks and white dorsal spot, mottled with ochreous grey; cilia grey, lighter at apex, with two dark transverse lines. Hindwings shining brownish grey; cilia grey. Abdomen dorsally grey, with segments ochreous spotted anteriorly and banded light grey posteriorly, ventrally with segments having broad, shining white bands posteriorly; anal tuft white, mixed brownish.

Similar species. The species is closely related to *sturnipennella* and differs by the more ochreous grey colour and by the stronger ferruginous lining of the blackish streaks on the forewings.

Male genitalia.– Uncus slender, widened distally and with small apical hook; cucullus broader at base, slightly narrowing in medial part and broadened again to rounded apex; sacculus abruptly tapering to blunt apex with two small apical teeth; gnathos broad, distally flattened; anellus lobes small and rounded, distally setose; aedeagus with one large cornutus and a bundle of five shorter, but stout cornuti. Differs from *sturnipennella* by the more parallel-sided cucullus, the broad and abruptly tapered apex of the sacculus, and by the absence of a bundle of slender cornuti.

Female genitalia.– Lamella antevaginalis consists of two wedge-shaped sclerites, with very large, pear-shaped sclerotization between them; sinus vaginalis large, bowl-shaped; ductus bursae short, almost completely occupied by large sclerotized plate; corpus bursae narrowed anteriorly, with deep indentation in middle of posterior half, edged by several patches of teeth, and with two sickle-shaped signa. The large pear-shaped sclerotization of the lamella antevaginalis, the strongly sclerotized plate in the ductus bursae, and the deep indentation in the middle of the corpus bursae are characteristic.

Distribution.– Southern Europe: Northern Caucasus and Greece.

Biology.– Host plant unknown. The larva is possibly a gall-maker on *Epilobium* L. Adults have been collected in late May (Caucasus) and in late July (Greece).

Material.– Holotype. ♂, '[Russia, Northern Caucasus] Kabardino-Balkaria, zapov[ednik] Verkh[nij] Balk[ar], Zagulajev, 28.V.1988', on the underside: 'ushchelie, dnem, lesnaya polyana' (ZIN).

Paratypes. ♀, '[Russia, Northern Caucasus] Kabardino-Balkarsk[ij] zapovednik, les, Zagulajev, 23.V.1990', on the underside: 'kordon Chegem-2, sosna, bereza, raznotravie' (ZIN); 3 ♀, 'Greece, Smolikas, 10 km W. Ag. Paraskevi by Konitsa, 600 m, 23.vii.1990, leg.M. Fibiger' (ZMUC).

Etymology.– The name describes the southern distribution of this species in Europe.

12. *Mompha (Mompha) epilobiella* ([Denis & Schiffermüller], 1775)

Tinea epilobiella [Denis & Schiffermüller], 1775: 141.
Recurvaria fulvescens Haworth, 1828: 554.
Cleodora nebulella Stephens, 1834: 221.

Diagnosis.– Adult. Wingspan 10-13 mm. Head yellowish white, mottled with ochreous. Thorax ochreous. Forewings ochreous, greyish on costa; indistinct yellowish spot on costa at 3/4 and similar spot on tornus; two small tufts of slightly raised dark brown scales on dorsum at 1/5 and 1/2. Hindwings shining grey.

Male genitalia.– Sacculus truncate, with group of small teeth apically; aedeagus with one hooked and several straight cornuti. These characters separate the species from *subbistrigella*.

Female genitalia.– Tergite VIII with rounded posterior edge; ductus bursae with narrow and short posterior and abruptly widened anterior parts, both with peculiar large sclerotized plate bearing strongly sclerotized rims in lower section.

Distribution.– Europe to southern Fennoscandia in the north; Asia Minor, the Caucasus, submontane regions of Central Asia. A very common species. Recently found in the eastern part of North America, probably introduced (Sinev, 1996).

Biology.– Host plants: *Epilobium hirsutum* L., rarely also *E. montanum* L., *E. palustre* L., *Chamaenerion angustifolium* (L.) Scop., and *Oenothera* sp. Host plants belonging to other families have been mentioned repeatedly: *Pulicaria dysenterica* Gaertn. (Asteraceae) (Fischer von Röslerstamm, 1841), *Lythrum salicaria* L. (Lythraceae), and *Eupatorium cannabinum* L. (Asteraceae) (Ivinskis, 1982), but these records need confirmation. Two, sometimes three generations per year. The larvae live from mid-May to June and again in July and early August between the spun top leaves, often gregariously (up to 50 larvae in one shoot). In first instars they usually feed as leaf-miners or borers of the flower buds and apical portion of shoots (Fischer von Röslerstamm, 1841; Stainton, 1870a; Hering, 1957). Pupation takes place in a white silky cocoon between the leaves. Adults hibernate and can be found throughout the year, but are most abundant in July and August.

Subgenus ***Anybia*** Stainton, 1854

Forewings without tufts of raised scales; vein M1 connate with stalk of R4 and R5. In hindwings M1 and M2 separate. Male genitalia with gnathos weakly sclerotized and without dimples. Female genitalia with antrum sclerotized and broad.

13. Mompha (Anybia) langiella (Hübner, 1796)

Tinea langiella Hübner, 1796: 69; pl. 27, fig. 187.
Alucita epilobiella Römer, 1794: 162; junior secondary homonym of *Tinea epilobiella* [Denis & Schiffermüller], 1775.
Adela unipunctella Duponchel, in Godart, [1839]: 374.
Elachista niveipunctella Stainton, 1849: 25.
Mompha fulicella Herrich-Schäffer, [1854]: 209.

DIAGNOSIS.– Adult. Wingspan 9-11 mm. Head dark brown with some purplish reflection, frons white; antennae with basal third dark brown, medial third grey annulated dark brown, apical third grey. Thorax and forewings dark brownish black, with purplish reflection. Forewings with irregular white spot beyond middle, sometimes also with few white scales at 1/3 in fold; costa in apical half and around apex rarely mottled with white scales. Hindwings greyish brown. Abdomen brownish black with anal tuft grey in male and white in female.

Similar species. The dark and smooth forewings with more or less pronounced white spots are very characteristic for this species.

MALE GENITALIA.– Uncus slender with rounded top; cucullus slender, widest subapically, apex bluntly pointed; sacculus longer than cucullus, narrow, with sharp and slightly bent apex; gnathos broad and weakly sclerotized; anellus lobes small and rounded; aedeagus rather slender, with large cornutus, wide at base, narrow and bent at apex.

FEMALE GENITALIA.– Ostium very wide, bowl-shaped; posterior third of ductus bursae wide, weakly sclerotized and with two short, heavily sclerotized rims, anterior part narrowed and with large sclerite; inception of ductus seminalis at 1/3 of ductus bursae; corpus bursae oval, slightly narrowing anteriorly and with two large sickle-shaped signa.

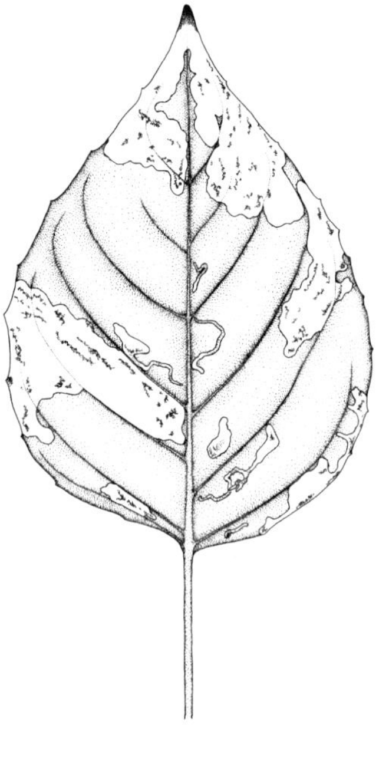

Text-fig. 13. *Mompha langiella* (Hübner), larval mines in leaf of *Circaea lutetiana* L.

DISTRIBUTION.– Europe, but not in the Mediterranean area; the Caucasus.

BIOLOGY.– Host plants: *Circaea lutetiana* L. and *C. alpina* L., *Epilobium hirsutum* L., sometimes *E. montanum* L., *E. parviflorum* Schreb., and *Chamaenerion angustifolium* (L.) Scop. Univoltine. In June and July the larva mines the leaves, rarely also the seedpods (Stainton, 1870a). The greenish white mine starts as a winding gallery with the black frass in an irregular or interrupted line in the middle. Later on the larva leaves this mine and starts another one in the vicinity or on a new leaf. The second mine very soon changes into a blotch or several blotches connected to each other. Frass initially in a broad irregular line, then dispersed. Larvae sometimes rather numerous on the host plant, and mines often occupying the whole leaf. Pupa in a whitish cocoon attached to a leaf or to leaf litter on the ground, occasionally inside the mine. Adults fly from August to April, hibernating in thatch (Emmet, 1979), but also in houses, hollows, and other shelters. See text-fig. 13.

Subgenus *Cyphophora* Herrich-Schäffer, 1853

Contains the largest species of Momphidae. Male genitalia with wide setose gnathos. Forewings with large tufts of raised scales. In forewings only R4 and R5 stalked. In hindwings M1 stalked with M2. Female genitalia with poorly developed antrum, long and narrow ductus bursae and signa located on slightly sclerotized plates. Larvae are root-borers.

14. *Mompha (Cyphophora) idaei* (Zeller, 1839)

Elachista idaei Zeller, 1839: 211.
Butalis idaeella Duponchel, in Godart, [1843]: 462; unjustified emendation of *Elachista idaei* Zeller, 1839.
Laverna grandisella Chambers, 1875b: 296.
Leucophryne tricristatella Chambers, 1875c: 211.
Laverna subiridescens Walsingham, 1882: 194.
Cyphophora polaris Sinev, 1986: 28.

DIAGNOSIS.– Adult. Wingspan 17-20 mm. Head dark brown with yellowish white frons; antennae shining dark brown. Thorax dark brown. Forewings yellowish ochreous to ochreous brown, basal half of dorsum lighter; greyish brown spot at base extending into streak along costa to white costal spot at 3/4; grey spots at 1/4 and 1/2, connected with costal streak; two subdorsal streaks with tuft of dark brown and yellow raised scales before 1/2 and 2/3; grey line around apex; several patches of greyish brown and white scales spread over the wing. Hindwings dark greyish brown. Abdomen with anal tuft brownish grey in male and whitish yellow in female.

MALE GENITALIA.– Uncus large, bottle-shaped; cucullus short and wide, apex rounded; sacculus about as long as cucullus, slender, gradually tapering distally, apex pointed; gnathos large and heavily spiculate; anellus lobes short, spatulate and setose; aedeagus short and stout with hook-shaped cornutus and spiculate area.

FEMALE GENITALIA.– Tergite VIII with convex posterior margin; lamella antevaginalis trapezoid, lamella postvaginalis with two large and slightly bent setose lobes; ostium wide, shallowly bowl-shaped; ductus bursae twice as long as corpus bursae, narrow, with square sclerite bearing lateral rims just below ostium; inception of ductus seminalis near posterior end of longitudinal sclerite at 1/3 of ductus bursae; corpus bursae oval, with sickle-shaped signa embedded in middle of irregular sclerite.

DISTRIBUTION.– Holarctic. Widely distributed and very common in northern Europe, southwards mainly restricted to mountain areas; the Caucasus, mountains of Central Asia (up to 2900 m), Siberia, Russian Far East. Also known from Canada and U.S.A. (Sinev, 1996).

BIOLOGY.– Host plants: *Chamaenerion angustifolium* (L.) Scop. and *C. latifolium* (L.) Th. Fr. et Lange; *Epilobium rosmarinifolium* Haencke has also been mentioned (Lhomme, [1948-1949]; Riedl, 1969). Univoltine. The larva lives from late summer to April under the bark of the rootstock; rather often several larvae can be found together. The moths fly in May and June, especially in sandy places along roads, railways, etc; in the daytime they sit on leaves of the host plant and can easily be swept.

Subgenus **Lophoptilus** Sircom, 1848

Forewings with tufts of raised scales. In forewings vein CuP stalked with 1A + 2A. In hindwings vein M1 and M2 stalked. Male genitalia without pronounced gnathos. Female genitalia with ductus bursae long and narrow, with a loop just before corpus bursae. Only a single species in Europe; its larvae feed on Cistaceae.

15. Mompha (Lophoptilus) miscella ([Denis & Schiffermüller], 1775)

Tinea miscella [Denis & Schiffermüller], 1775: 142.
Lophoptilus staintoni Sircom, 1848: 2037.
Tebenna opacella Müller-Rutz, 1934: 121.

DIAGNOSIS.– Adult. Wingspan 7-9 mm. Head grey, mottled with dark grey, with frons creamy white; antennae grey, annulated light grey in basal half (in female with 5 apical white segments each divided by 3 grey segments). Thorax grey, mottled with dark grey. Forewings ochreous brown with grey blotch at base; grey outward oblique fascia at 1/3, widening towards dorsum; grey geniculate fascia at 2/3 and similar fascia at apex, not reaching costa; two patches of raised blackish scales at outer edge of inner and middle fasciae subdorsally; small white spot in fold at 1/3, similar spot above fold at 1/2 and white costal spot at 3/4 (white spots on forewings larger in female). Hindwings light brownish grey.

MALE GENITALIA.– Cucullus parallel-sided, apex rounded; sacculus longer than cucullus, basal half rather broad, distal half long, narrow and slightly bent, apex pointed; uncus slender and pointed; anellus lobes short and rounded; aedeagus slender with one long bent cornutus and two short cornuti.

FEMALE GENITALIA.– Tergite VIII broad, weakly sclerotized; lamella antevaginalis semicircular, lamella postvaginalis with two setose lobes; ostium bowl-shaped, anterior edge concave, with square sclerite; ductus bursae narrow, more than twice as long as corpus bursae, with loop anteriorly; inception of ductus seminalis just below ostium; corpus bursae egg-shaped, with sickle-shaped signa embedded transversely in middle of semicircular sclerite.

DISTRIBUTION.– Europe, north to southern Fennoscandia; Asia Minor, North Africa.

BIOLOGY.– Host plants: *Helianthemum nummularium* (L.) Mill., *H. polifolium* D.C., *H. canum* (L.) Baumg., and *H. ovatum* (Viv.) Dun. (Riedl, 1969), in southern Europe also *Cistus* sp. (Cistaceae) (Hering, 1957). Bivoltine, often with overlapping generations. The larvae feed from October to April and in June and July, mainly as leaf miners. Mine starts as a gallery almost filled with frass, later changing into a blotch often occupying the whole leaf with frass dispersed and heaped. According to Lhomme ([1948-1949]) the larvae of the second generation also live inside the flowers and seeds. Pupation in a cocoon spun to leaf litter. The moths are found from the end of April to the beginning of October, but are most abundant from May to the first half of June and again from the second half of July to the end of August. See text-fig. 14.

Subgenus ***Psacaphora*** Herrich-Schäffer, 1853

Forewings with peculiar tufts of raised, more or less shining scales. In forewings vein CuP stalked with 1A + 2A; in hindwings veins M1 and M2 stalked. Male genitalia with gnathos very weakly developed, membranous. Female genitalia with rather short ovipositor and slightly modified posterior margin of sternite VII.

16. Mompha (Psacaphora) locupletella ([Denis & Schiffermüller], 1775)

Tinea locupletella [Denis & Schiffermüller], 1775: 141.
Tinea schrankella Hübner, [1805]: pl. 38, fig. 264; junior primary homonym of *Tinea schrankella* Goeze, 1783.
Adela pilipennella Zetterstedt, 1840: 1008.
Psacaphora quadrilobella Herrich-Schäffer, [1854]: 216.

Diagnosis.– Adult. Wingspan 9-12 mm. Head with frons silvery grey, vertex and neck tuft blackish brown with leaden grey gloss; antennae completely dark brown in male, with apical quarter white in female. Thorax blackish brown with leaden grey gloss. Forewings bright orange; leaden grey blotch at base from dorsum almost to costa, edged by large blackish spot; costa with dark brown streak and white spot at 3/4; two leaden grey subcostal spots at 1/4 and 1/2, outer spot edged underneath by small white spot; two similar grey spots on dorsum, first before 1/2 and often connected to first costal, distally edged by tuft of raised blackish scales, and second at tornus and prolonged towards costa, distally edged by white spot; apex beyond white costal spot blackish brown. Hindwings greyish brown.

Similar species. Differs from *terminella* by the larger wingspan and grey coloration of the second subcostal and both dorsal spots of the forewings.

Male genitalia.– Cucullus parallel-sided, apex rounded; sacculus tapering, apically abruptly narrowed to a curved apex, ventral margin with a triangular hump at one third; uncus slender and pointed; gnathos membranous; anellus lobes long, pointed and outwardly bent; aedeagus stout with one large rod-like spiculate cornutus.

Female genitalia.– Sternite VII weakly sclerotized, posterior margin with large V–shaped excavation; sternite VIII with sclerotized and setose lobes, sterigma large and bulbous, wedge-shaped anteriorly and with two pointed projections postero-laterally; ostium wide, with peg-like lateral projections; ductus bursae narrowed towards corpus bursae and with some folds in anterior part; inception of ductus seminalis at 1/3 of ductus bursae; corpus bursae egg-shaped with two sickle-shaped signa.

Distribution.– Northern Europe and mountain regions of Central and southern Europe to north-western Spain. Eastwards known from Transbaikalia and Kuril Islands (Sinev, 1996).

Biology.– Host plants: *Epilobium alsinifolium* Vill., *E. palustre* L., *E. montanum* L.,

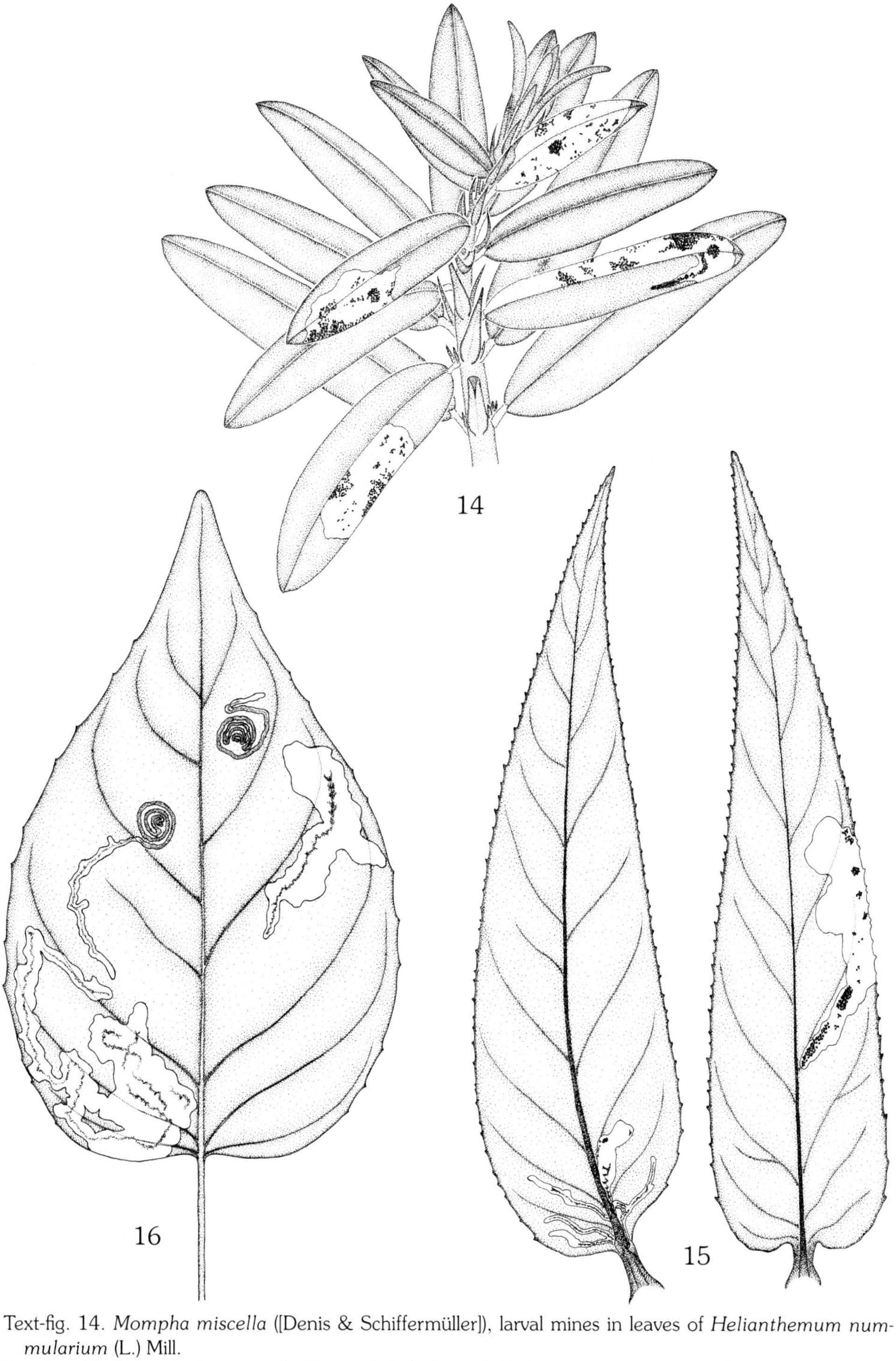

Text-fig. 14. *Mompha miscella* ([Denis & Schiffermüller]), larval mines in leaves of *Helianthemum nummularium* (L.) Mill.

Text-fig. 15. *Mompha locupletella* ([Denis & Schiffermüller]), larval mines in leaves of *Epilobium montanum* L.

Text-fig. 16. *Mompha terminella* ([Humphreys & Westwood]), larval mines in leaf of *Circaea lutetiana* L.

and *E. lanceolatum* Seb. et Maur. Bivoltine, but univoltine in the north. From the end of April to May and from July to early August the larvae live as leaf miners, in spring in the leaf-rosette. Pupation in a whitish cocoon spun to surrounding vegetation or leaf litter on the ground. The moths of the first generation fly from the second half of May till early July, second generation from August to early September. The species prefers localities where the food plants grow in shady places in woods or ditches. See text-fig. 15.

17. Mompha (Psacaphora) maculata (Sinev, 1986)

Psacaphora maculata Sinev, 1986: 26.

DIAGNOSIS.– Adult. Wingspan 12-13 mm. Head with frons shining pale grey, vertex and neck tuft shining greyish brown; antennae greyish brown, apical part of about ten segments greyish white. Thorax and tegulae greyish brown with strong bronzy gloss. Forewings shining dark brown with slightly outward oblique orange yellow fascia near base, similar coloured spot in fold at 1/4; four white, yellow-edged spots, first on dorsum at 2/5, second near costa at 1/2, third on tornus and fourth on costa at 4/5; cilia greyish brown. Hindwings shining brownish grey.

MALE GENITALIA.– Resembles those of *locupletella*, but differs by the longer uncus and the aedeagus having a larger cornutus.

FEMALE GENITALIA.– Can be distinguished from *locupletella* by the equal length of the apophyses, the larger sclerotized and setose lobes of sternite VIII, the bent posterior margins of the excavation in sternite VII, and by the wider ductus seminalis.

DISTRIBUTION.– Georgia, Azerbaijan (Nakhichevan), Armenia.

BIOLOGY.– Host plant unknown. Moths have been collected from the beginning of June to mid-August in mountain regions at elevations from 400 to 2300 m above sea level.

18. Mompha (Psacaphora) terminella (Humphreys & Westwood, 1845)

Glyphipteryx terminella Humphreys & Westwood, 1845: 216.
Elachista patriciella Stainton, 1849: 25.
Psacaphora chrysargyrella Herrich-Schäffer, [1854]: 216.
Mompha engelella Busck, 1906: 123.

DIAGNOSIS.– Adult. Wingspan 7-9 mm. Head dark brown with purplish gloss, frons ochreous white; antennae dark brown, apical area of approximately eight segments white. Thorax blackish brown with purplish gloss. Forewings dark orange; leaden grey blotch from base to 1/3, reaching dorsum near base and connected to costa, with blackish spot inside; narrow brown costal streak from blackish spot to white costal spot at 3/4; leaden grey subdorsal spot with golden gloss below fold, distally edged by tuft of raised blackish scales, similar spot near tornus; small white subcostal spot opposite tornal spot; apex blackish brown with shining leaden grey streak on termen. Hindwings dark greyish brown.

Similar species. The smaller wingspan, the blackish spots inside the grey basal spot and two subdorsal spots distinguish this species from *locupletella*.

MALE GENITALIA.– Cucullus short and broad, basally one and a half times as wide as apically, apex rounded; sacculus shorter than cucullus, narrow, apex rather blunt;

uncus long with hooked apex; gnathos in form of spiculate band; anellus lobes short, truncate and setose; aedeagus short and stout with long and pointed cornutus and spiculate area.

FEMALE GENITALIA.– Tergite VIII rectangular, weakly sclerotized; lobes of sternite VIII large, heavily sclerotized; sterigma bulbous, tapering anteriorly; ostium wide, funnel-shaped; ductus bursae S–shaped, about one and a half times as long as corpus bursae, with sclerotized widest part connected to ostium and narrowing anteriorly; inception of ductus seminalis at 1/5 of ductus bursae; corpus bursae oval with two large sickle-shaped signa.

DISTRIBUTION.– North-western and Central Europe, Russian Far East. Has also been recorded from the eastern coast of North America (Sinev, 1996).

BIOLOGY.– Host plants: *Circaea lutetiana* L. and *C. alpina* L. Univoltine. The larva mines leaves from the second half of August to the first half of September. The mine starts as a narrow whitish gallery and forms several circles or semicircles around the egg site; later it turns into a greenish white and very conspicuous gallery with frass in a narrow black line; sometimes both galleries join and form a large blotch mine, but the lines of frass show the original gallery (Stainton, 1870b; Koster, 2002a). The larva often leaves the original spiral mine and starts another in the vicinity or in a new leaf. Pupation takes place in a whitish cocoon spun to a leaf or leaf litter on the ground, occasionally inside the mine. The moths fly in July and August. See text-fig. 16.

19. Mompha (Psacaphora) raschkiella (Zeller, 1839)

Elachista raschkiella Zeller, 1839: 211.

DIAGNOSIS.– Adult. Wingspan 7-11mm. Head blackish brown with frons creamy white; antennae dark grey. Thorax blackish brown. Forewings orange brown with blackish brown basal area and brown costal streak; four leaden grey spots, two subcostally and two on dorsum, latter outwardly edged by tuft of raised blackish scales; white costal spot at 3/4; blackish brown spot at apex edged by shining leaden grey streak. Hindwings dark brownish grey.

Similar species. For separation from *sexstrigella*, q. v.

MALE GENITALIA.– Cucullus widest at base, slightly curved, apex rounded; sacculus slightly shorter than cucullus, tapering distally, apex curved, ventral margin with large sclerotized hump; uncus slender and pointed; gnathos absent; anellus lobes broad basally, narrow and pointed apically; aedeagus short with three cornuti of different shapes. Differs from *sexstrigella* by the presence of large hump on the ventral margin of sacculus and by the lack of bifurcate sclerotized dorsal margin of sacculus.

FEMALE GENITALIA.– Differs from *sexstrigella* by the almost square and heavily setose sterigma, the bulbous shape of the posterior part of ductus bursae and the narrow inception of ductus seminalis at 1/3.

DISTRIBUTION.– Europe, except from the extreme southern and northern parts; also the Caucasus, mountains of Central Asia, southern Siberia eastwards to the Sakhalin I. (Sinev, 1996).

BIOLOGY.– Host plant: *Chamaenerion angustifolium* (L.) Scop., probably also *Epilobium hirsutum* L. (Riedl, 1969). Usually bivoltine, but univoltine in the north. The larvae mine leaves from the end of May to the end of July and from the end of August to September. The early mine starts as a narrow gallery, often follows the midrib, and is almost filled with frass; this mine is usually surrounded with reddish leaf tissues. Later on the mine gradually, or sometimes abruptly, widens to a blotch with the frass dispersed. Pupation takes place in a greyish white cocoon spun amongst detritus on the ground; pupa hibernates. The moths fly from early May to late October. See text-fig. 17.

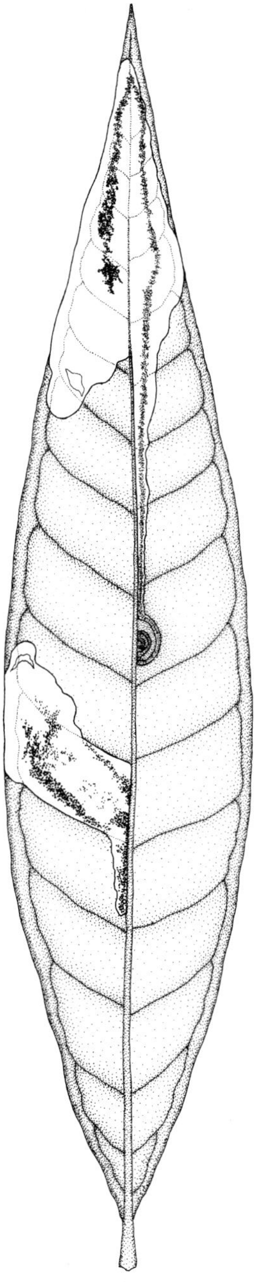

Text-fig. 17. *Mompha raschkiella* (Zeller), larval mines in leaf of *Chamaenerion angustifolium* (L.) Scop.

20. *Mompha (Psacaphora) sexstrigella* (Braun, 1921)

Psacaphora sexstrigella Braun, 1921: 5.
Mompha complexa Svensson, 1982: 297.

DIAGNOSIS.– Adult. Wingspan 10-11 mm. Very similar to *raschkiella* externally, but all four leaden grey spots on the forewing confluent. Forewings with brownish orange coloration reduced to small spots at 1/5, before 1/2 and at 3/4, and two streaks in apical area; apex outwardly edged by dark brown streak with four to six white spots. Hindwing shining light grey.

Similar species. The absence of large orange spots on the forewings in combination with the lighter hindwings distinguishes this species from *raschkiella*.

MALE GENITALIA.– Cucullus widest subbasally, gradually tapering distally, apex rounded; sacculus as long as cucullus, rather abruptly tapering to a blunt and curved apex, dorsal margin heavily sclerotized from base to 2/3, apically bifurcate; uncus slender and pointed; gnathos membranous; anellus lobes long and pointed; aedeagus slender with three cornuti, the middle one large and hook-shaped. For separation from *raschkiella*, q. v.

FEMALE GENITALIA.– Sternite VII rather heavily sclerotized, rugose laterally, posterior margin pronouncedly concave on either side of ostium, and lateral margins appear to be produced posteriorly; sterigma with two tapering lobes united posteriorly; ostium slightly bowl-shaped; posterior half of ductus bursae bottle-shaped and sclerotized, with unsclerotized band in widest part, anterior half unsclerotized, narrowed towards corpus bursae and with wide inception of ductus seminalis at 1/3; corpus bursae with two sickle-shaped signa. For separation from *raschkiella*, q. v.

DISTRIBUTION.– Holarctic. Northern Fennoscandia, Russia (Kola Peninsula, Altai, southern Siberia, Sakhalin I.); Canada and western part of the U.S.A (Sinev, 1996; Koster & Harrison, 1997).

BIOLOGY.– Host plant: *Chamaenerion angustifolium* (L.) Scop. Univoltine. The larvae mine leaves in July-August. The mine starts as a gallery and ends as a blotch. Pupation takes place inside the mine (Itämies & Kyrki, 1979; Svensson, 1982; Sinev, 1986). The moths fly from the second half of June to the first half of August.

Family BATRACHEDRIDAE

Batrachedrae Heinemann & Wocke, [1876]: 524. Type genus: *Batrachedra* Herrich-Schäffer, 1853.

General part

MORPHOLOGY.– Wingspan 10 to 15 mm. Head (Text-fig. 18) covered with densely appressed scales. Antennae about as long as forewings; their basal segment short, with a single bristle. Eyes moderate, ocelli absent. Labial palpi rather long, curved upward, without tufts of scales in front; segment 3 slightly shorter than segment 2. Forewings (Text-fig. 19) linear, with two peculiar dark dots in fold. In forewings R5 reduced; M1 and M2 very inconspicuous at bases. Hindwings very narrow, linear; M1 and M2 stalked, outer margin of wing with 4 veins only. Hindtibiae with middle pair of spurs noticeably before midpoint of tibia.

Abdominal tergites II to VII (Text-fig. 20) with paired longitudinally arranged rows of deeply embedded specialized lanceolate scales ('spines'); sclerotized struts (venulae) of tergite II moderate. Male genitalia (Text-fig. 21): uncus long and narrow, articulated with well developed triangular tegumen; gnathos unpaired, well sclerotized, as long as uncus; vinculum narrow, saccus short, without pronounced medial projection; valvae rather wide, hairy, unilobate, with slightly thickened but not separated sacculus; anellus lobes small, digitate; aedeagus much longer than valvae, very narrow, without cornuti. Female genitalia (Text-fig. 22): ovipositor moderately long; ductus seminalis merges into medial part of ductus bursae; ductus bursae long and narrow, with numerous denticles and slightly broadened antrum; bursa copulatrix rounded, with large lamellar signum covered by small spines.

Larvae long and slender, cylindrical, without secondary setae, yellowish, very mobile. Pupae with exposed labial palpi and forefemora; abdominal segments IV to VI movable; apical cremaster not developed, only small crochets present at the apex of abdomen. Cocoon spindle-shaped, double-layered, and covered with frass and other detritus.

BIOLOGY.– These moths are mainly associated with woodlands, but also sometimes very common in parks and avenues. The larvae live in silky galleries between or inside the reproductive organs of various plants, usually dicotyledonous, but often also monocotyledonous and even conifers. In the tropics and subtropics some species are known as pests of cultivated palms; exceptionally, myrmecophily occurs. Adults are often found on trunks of trees in a very peculiar resting position with raised anterior part of the body and forelegs directed backwards ('frog-moth'). When disturbed, they run quickly over the bark or jump away. Rather easily attracted to light.

DISTRIBUTION.– The family is not very species-rich, and its main diversity is found in America and the Afrotropical region. Only about 10 species are known from the Palaearctic, with 1 genus and 4 species in Europe and northern Africa.

SYSTEMATICS.– A very specialized group of gelechioid moths, with the following two autapomorphies: wings extremely narrow with reduced venation, and resting position very peculiar. Most of the genital characters (except the sometimes enormously long aedeagus) and morphology of the head are rather plesiomorphic for Gelechioidea. Unspecialized larval life style, trophic relations with generative organs of mostly trees, as well as scanty of species and absence of evident geographic centres of diversity speak in favour of the viewpoint that the family represents a small advanced group of

Text-fig. 18. Head of *Batrachedra praeangusta* (Haworth).

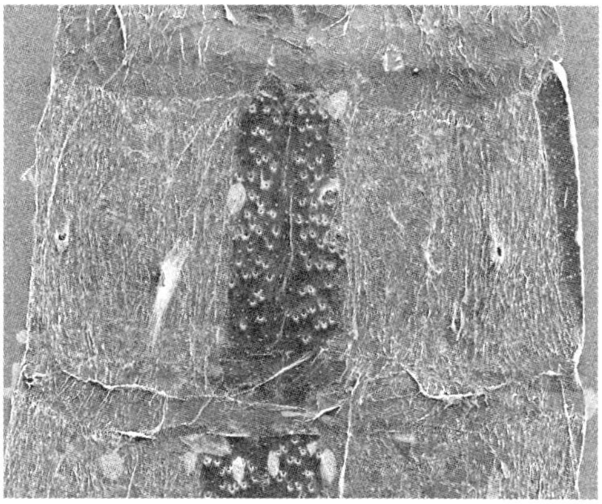

Text-fig. 20. Abdominal tergites of *Batrachedra praeangusta* (Haworth).

gelechioid moths with some very peculiar characters. Thus, we treat Batrachedridae as a monophyletic and well separated family, closely related to the Coleophoridae.

Systematic Treatment of the Genus and Species of Batrachedridae in Europe, North Africa and the Near East

Batrachedra Herrich-Schäffer, 1853

Batrachedra Herrich-Schäffer, 1853: 14 (key), 54.
Type species: *Ornix turdipennella* Kollar, 1832 (= *Gracillaria praeangusta* Haworth, 1828), by monotypy.
Batrachedra Stainton, 1854: 225 (key), 230; homonym and junior synonym of *Batrachedra* Herrich-Schäffer, 1853.
Type species: *Gracillaria praeangusta* Haworth, 1828, by subsequent designation by Walsingham, in Sharp, 1907c.
Eustaintonia Spuler, 1910: 388.
Type species: *Cosmopteryx pinicolella* Zeller, 1839, by monotypy.

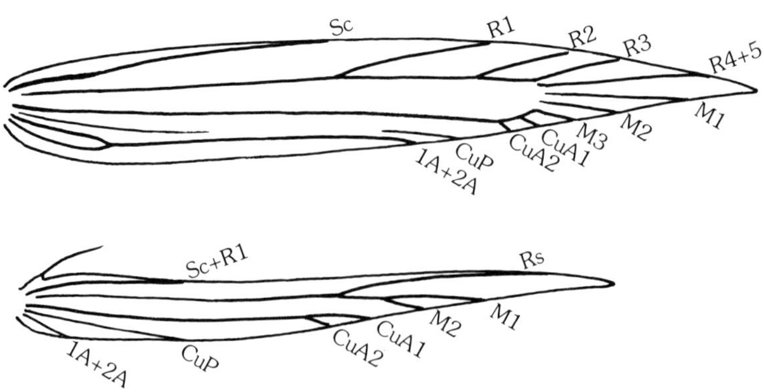

Text-fig. 19. Wing venation of *Batrachedra praeangusta* (Haworth).

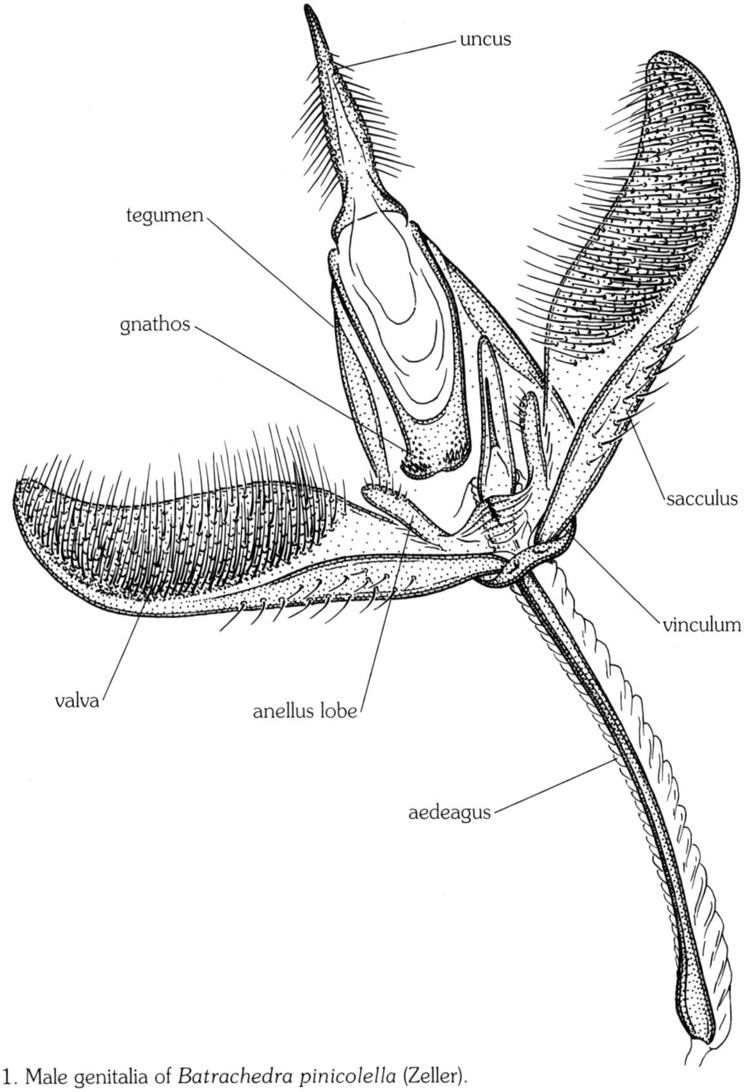

Text-fig. 21. Male genitalia of *Batrachedra pinicolella* (Zeller).

DIAGNOSIS.– Head smoothly scaled; antennae 4/5 length of forewing with basal segment short; labial palpi porrect, segment 2 sometimes more or less thickened with raised scales on lower side, segment 3 slightly shorter than 2 and angled upwards. Forewings narrow, linear or linear lanceolate, with only 4 radial veins. Hindwings very narrow, linear. The members of this genus are often taken to be *Coleophora* species.

MALE GENITALIA.– Uncus long and pointed, sometimes hooked. Tegumen large, narrowing distally. Gnathos large, as long as uncus, spatulate. Valvae long and simple, rounded or pointed distally, sometimes with a small clasper, sacculus well defined. Anellus lobes prominent, bulbous to spatulate. Vinculum narrow. Aedeagus from rather short and compact to very long and slender, with or without cornuti.

FEMALE GENITALIA.– Apophyses of almost equal length, apophyses anteriores often forked basally. Antrum funnel-shaped. Ductus bursae narrow, partially spiculate inside. Corpus bursae ovoid or oval, often with large spiculate plate-like signum.

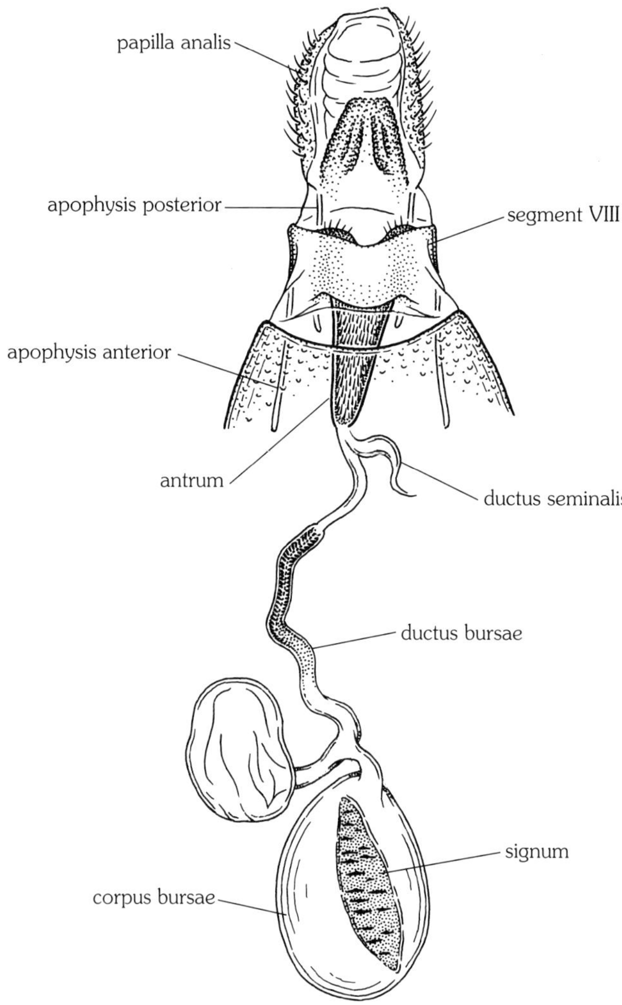

Text-fig. 22. Female genitalia of *Batrachedra pinicolella* (Zeller).

DISTRIBUTION.– A large genus represented in all parts of the world, but most abundant in the Nearctic, Neotropical and Oriental regions.

BIOLOGY.– The larvae are miners or live in seeds, catkins or buds. *Batrachedra parvulipunctella*, however, is carnivorous and lives on coccids (Hemiptera).

21. *Batrachedra praeangusta* (Haworth, 1828)

Gracillaria praeangusta Haworth, 1828: 530.
Ornix turdipennella Kollar, 1832: 99.
Batrachedra praeangustella Doubleday, 1859: 34; unjustified emendation of *Gracillaria praeangusta* Haworth, 1828.
Batrachedra clemensella Chambers, 1877: 134.

DIAGNOSIS.– Adult. Wingspan 14-15 mm. Head, thorax and tegulae white, mixed grey; antennae greyish, annulated dark grey, with two white rings in apical area. Forewings ground colour white, heavily mottled with dark grey scales; broad blackish subbasal streak in fold, interrupted in middle and distally by yellowish white spots; indistinct blackish spot on dorsum near base nearly connected to streak in fold; narrow blackish streak medially from 2/5 to 3/4, interrupted by yellowish white spot in middle; broad yellowish white streak almost from base to 2/5 and yellowish white spots on tornus and distal to narrow blackish streak; apical area with three costal and two dorsal yellowish white streaks into termen; cilia grey, mixed white around apex. Hindwings shining grey, lighter towards base.

Similar species. The species can be distinguished from *pinicolella* and *parvulipunctella* by white and grey forewings.

MALE GENITALIA.– Uncus very slender; gnathos spatulate, apex pointed; anellar lobes large and oval; valvae broad with rounded apex and small clasper, sacculus rather narrow, reaching beyond middle; aedeagus very long and narrow, about three times as long as valva, slightly curved.

FEMALE GENITALIA.– Antrum short, unsclerotized, gradually turning into ductus bursae; ductus bursae with posterior part tapering and sclerotized, middle part unsclerotized, bent and connected with large, balloon-shaped ductus seminalis, anterior part slightly widening towards corpus bursae and spiculate; corpus bursae egg-shaped, signum small with coarse spines.

DISTRIBUTION.– Europe except for the extreme north and south-east; the Caucasus, Central Asia, southern Siberia, Russian Far East (Sinev, 1982). North America (Hodges, 1966).

BIOLOGY.– Host plants: *Populus alba* L., *P. tremula* L., *Salix alba* L., and *S. caprea* L. (Salicaceae). Univoltine. The larva lives from April to June in the female catkins of willows and poplars; in first instars it mines flowers and seeds, later in silky galleries between the seeds. Sometimes larvae also use buds, boring into them at the base and hollowing them completely; frass is partly ejected and visible as a small pile in the middle of the bud. The last instars can also feed between young leaves. The full-fed larva leaves its web for pupation, which usually takes place in cracks of the bark of the host plant. The very slender bright brown pupa lies in a narrow felt-like cocoon covered with particles of lichens (Schefer-Immel, 1958). Flight period from the second half of June to early September; the moths can be found running on the trunks of trees. They are easily attracted to light.

22. *Batrachedra pinicolella* (Zeller, 1839)

Cosmopteryx pinicolella Zeller, 1839: 210.

DIAGNOSIS.– Adult. Wingspan 9-13 mm. Head yellow, medially mixed grey; antennae yellowish brown, ringed white in apical area, more distinctly in female. Thorax shining yellow, strongly mixed with grey brown scales, tegulae yellow with some brown scales anteriorly. Forewings yellow with scattered dark brown scales, especially on costa and dorsum; dark brown spot in fold at 1/3, and a similar spot before apex; cilia grey, with greyish brown streak on costal margin at 3/4. Hindwings shining grey.

Similar species. Differs from *parvulipunctella* by the somewhat smaller size and the yellow forewings with a greyish brown streak in the costal cilia towards apex.

MALE GENITALIA.– Uncus broad at base, strongly narrowing towards apex; gnathos large with rounded tip; anellar lobes short and club-shaped; valvae broad, apex rounded; aedeagus about 1/4 longer than valva and slightly bent.

Female genitalia.– Antrum funnel-shaped and covered with microspicules; ductus bursae with anterior and posterior thirds unsclerotized, middle third with rows of spicules; corpus bursae egg-shaped; signum about 1/3 of bursa length, shaped as a sclerotized longitudinal plate covered with stout ridges and short teeth.

Distribution.– Europe, in the south mainly in mountainous regions; the Caucasus, southern Siberia, Russian Far East.

Biology.– Host plants: *Abies* spp., *Picea* spp., and *Pinus* spp. (Pinaceae). Univoltine. The larva can be found from September to May. After hatching in autumn, it mines a needle from the base, later on it makes a silken gallery on the surface of a twig; from this gallery it mines halfway into the needle (Wood, 1887; Hering, 1957; Patocka, 1960). Prior to pupation the larva constructs a rough spindle-shaped cocoon in the vicinity of the feeding place, attached to a twig and some needles. The cocoon is covered with small pieces of bark and excrement, which provide a very good camouflage. Pupa slender, with long forewing sheaths and a group of hooked spines at the end of the abdomen (Baer, 1910). The moths fly from mid-June to August and can easily be beaten from spruce. The species has been reported as a pest on various ornamental coniferous trees in northern Caucasus (Bolov & Sinev, 1989). See text-fig. 23.

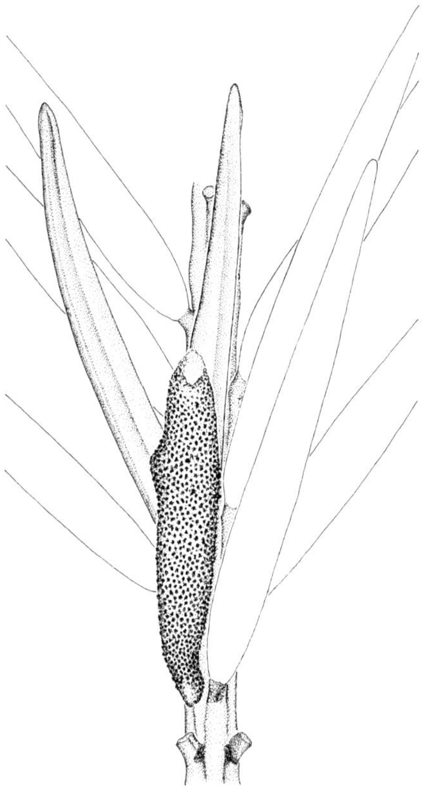

Text-fig. 23. *Batrachedra pinicolella* (Zeller), cocoon on *Pinus* sp.

23. *Batrachedra parvulipunctella* Chrétien, 1915

Batrachedra parvulipunctella Chrétien, 1915: 350.
Eustaintonia phragmitidella Mariani, 1936: 97.

DIAGNOSIS.– Adult. Wingspan 11-15 mm. Head ochreous yellow, sometimes mixed grey; antennae yellowish brown, apex with five indistinct brown streaks of different size. Thorax and tegulae yellowish ochreous. Forewings yellowish ochreous mottled with brown scales, more densely on costa and dorsum; indistinct brown streak just above fold from base to tornus, and similar streak between fold and costa, ending on dorsum just before apex; small brown spot in lower streak beyond 1/3, and similar spot in the upper streak in apical area; cilia yellowish ochreous, lighter around tornus towards base. Hindwings shining yellowish white.

Similar species. Resembles *pinicolella*, but somewhat larger and can be distinguished by the yellowish ochreous forewings with two indistinct longitudinal brown streaks.

MALE GENITALIA.– Differs from *pinicolella* and *praeangusta* by the tapering valvae with sharp and slightly upwardly bent apex and by elongated anellar lobes.

FEMALE GENITALIA.– Antrum funnel-shaped, next part of ductus bursae weakly sclerotized, followed by short unsclerotized section, remaining part with rows of spicules; corpus bursae oval; signum of almost same length as bursa, in form of large, weakly sclerotized plate covered with sclerotized ridges. Differs from *pinicolella* by the shape of the sclerotization in the ductus bursae and the very large signum.

DISTRIBUTION.– Southern Europe, northern Africa (Tunisia). Recently found in England (Heckford, 1994), where a specimen had probably been transported by strong southerly winds.

BIOLOGY.– Commensal in the colonies of *Aclerda berlesei* Buffa (Homoptera: Aclerdidae) or other species of coccids, which occur on *Phragmites australis* (Cav.) Trin. ex Steud. and *Arundo donax* L. (Poaceae). Probably bivoltine. The larvae live in June-July and from autumn to spring in white silky cases with openings on both sides, feeding on the waxy secretions of coccids. The case covers a group of females or larvae of the coccid, thus the larva has no need to leave it completely while feeding and can retract at the slightest sign of disturbance. The slender pupa can be found in a white, slightly transparent cocoon in the leaf-sheaths of the plant (Mariani, 1936, 1937). Adults have been found in May and June and again in July and August.

24. *Batrachedra amydraula* Meyrick, 1916

Batrachedra amydraula Meyrick, 1916b: 30.

DIAGNOSIS.– Adult. Wingspan 11-12 mm. Head ochreous grey, with lateral white streak from base of antenna, widening towards neck tuft; antennae ochreous grey, indistinctly annulated brown, basal four segments ochreous grey. Thorax ochreous grey, laterally yellowish white; tegulae yellowish white, mottled with ochreous grey. Forewings pale yellow to yellowish white, dorsal half to fold and costal half above fold densely covered with brown scales, leaving medial pale yellow streak at 1/3 with only few scattered brown scales; indistinct brown streak in fold just before middle, and brown spots at 3/5 and 4/5 in middle of pale yellowish streak; cilia ochreous grey. Hindwings shining pale grey.

Similar species. The smaller size and the forewings with the contrasting longitudinal pale yellow streak distinguish the species from *parvulipunctella*.

MALE GENITALIA.– Differs from *praeangusta*, *pinicolella* and *parvulipunctella* by the strongly tapered apex of the gnathos and the short aedeagus, which is just slightly longer than the valvae.

FEMALE GENITALIA.– Resemble those of *pinicolella*, but differ from this species by the large signum with coarse spines.

DISTRIBUTION.– North Africa, Near and Middle East to India (Zhang, 1994).

BIOLOGY.– Host plant: *Phoenix dactylifera* L. (Arecaceae). The larva chews a hole adjacent to the stem end, through which it penetrates the young fruit and damages the flesh and the soft seed. Injury inflicted to the trees by this pest results in considerable fruit drop and serious losses in yield (Gharib, 1968; Blumberg, 1975; Sinev, 1994c). The moths fly from March to August in three generations. One of the most important pests of the date palm, known as the lesser date moth.

Family STATHMOPODIDAE

Stathmopodidae Meyrick, 1913a: 310. Type genus: *Stathmopoda* Herrich-Schäffer, 1853.

General part

The family was first established by Meyrick (1913a) without description, but with the name clearly derived from the valid generic name *Stathmopoda* H.-S. This is enough reason (ICZN (1999: article 12.2.4) to keep the authorship of Meyrick for Stathmopodidae, not Janse (1917), as usually accepted in the recent literature.

MORPHOLOGY.– Wingspan 7 to 15 mm (up to 30 mm in tropical species). Head (Text-fig. 24) smooth, often with more or less bulging frons. Antennae slightly shorter than forewings, usually long ciliate in male; scapus without bristled pecten. Eyes rather large, ocelli absent. Labial palpi long, recurved, covered with densely appressed scales; segment 3 acuminate, of equal length with segment 2. Forewings (Text-fig. 25) lanceolate linear or linear with very acute apex, usually yellowish with oblique brown fasciae. In forewings the bases of radial veins are approximated to the end of cell; M1 stalked with R4+5; M2 sometimes absent. Hindwings very narrow, linear, with very long ciliae; usually all veins represented, but M3, Cu1 and Cu2 are very short. Tibiae and tarsi of hindlegs usually brightly coloured ('bright-legged moths') and densely covered, especially on upper side, with piliform scales, with tufts of long bristles at base of spurs and at the end of 1st tarsal segment.

Abdominal tergites II to VII (Text-fig. 26) with specialized lanceolate scales ('spines') deeply embedded and arranged in a moderately wide or narrow transverse band along the posterior margin of each tergite; longitudinal sclerotized struts (venulae) of tergite II short. Male genitalia (Text-fig. 27): uncus rather long, widely articulated with well developed triangular tegumen; gnathos of equal length with uncus, beak-shaped, widely attached to tegumen; saccus short, without pronounced medial projection; valvae symmetrical, with rather large anterodorsal projections, subdivided into floppy and hairy dorsal part (cucullus) and slightly sclerotized ventral part (sacculus) separated from each other by more or less pronounced fold; juxta hardly visible and attached to the apical part of aedeagus beneath; anellus lobes in a form of very small hairy tubercles or absent; aedeagus rather large, often pear-like, with narrow apical projection and heavily sclerotized thorns or ridges in vesica. Female genitalia (Text-fig. 28): ovipositor moderately long; ductus seminalis very long, usually with intricate folds, dilations and irregular granulations, fuses with bursa copulatrix in region of its transition into ductus bursae; ductus bursae rather short and straight, with more or less wide slightly folded antrum; bursa copulatrix large, oval, with a pair of lamellar, often slightly asymmetrical signa and a group of spines at the border with ductus seminalis.

Larvae rather short and broad ('fat'), spindle-shaped, moderately mobile, without secondary setae, colourless, but become darker before pupation. Pupae weakly sclerotized, with exposed forefemora but concealed labial palpi; abdominal segments IV to VI usually movable; apical cremaster tuberculate, with a group of stout curved crochets. Cocoon spindle-shaped, double-layered, with rather long exit tube.

BIOLOGY.– The representatives of the family prefer woodlands but in the south can also be found in the open landscape with isolated groups of trees. Larvae usually consume reproductive organs of plants (including sporangia of ferns) making silky galleries inside or between them. In the tropics several species are also known as scavengers or

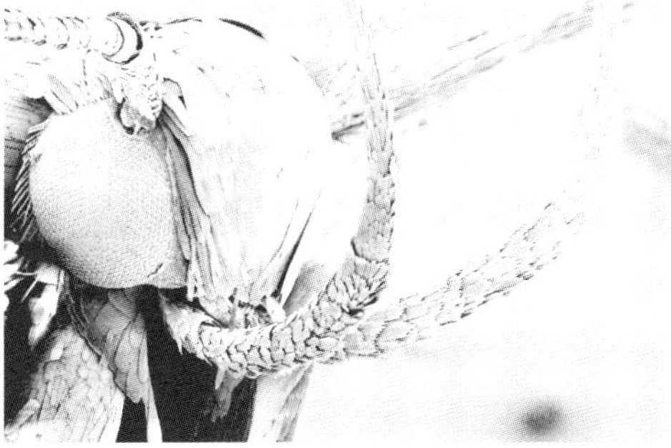

Text-fig. 24. Head of *Stathmopoda pedella* (Linnaeus).

Text-fig. 26. Abdominal tergites of *Stathmopoda pedella* (Linnaeus).

predators in colonies of scale insects. Some species are important pests of cultivated plants (fig, persimmon, carob, etc.). Adults rest on the leaves of their host-plants in a very peculiar position, with hindlegs stretched along the sides of the body or even raised above it. Very easily attracted to light.

DISTRIBUTION.– The family is well represented in the Indo-Australian and Afrotropical regions, with very few species recorded for South America and the temperate zones of the Holarctic. More than 20 species are known from the Palaearctic (mainly from the extreme south and south-east), with 2 genera and 7 species in the region of this study.

SYSTEMATICS.– While keeping rather archaic larval life-style, bright-legged moths have rather specialized morphology and behaviour. Most of them have extremely narrow and long wings with very long fringes, ciliate antennae in males, peculiarly 'armed' hindlegs, and also a characteristic genital duct morphology in female. Adults are not only very remarkable in repose, but also have a curious hobbling gait, being functionally 'tetrapod' moths (Sinev, 1988b). The taxonomy of the family should be revised. Most of the genera fall in two groups of subfamily rank; those with rounded head, ciliate antennae in males and trophic relations mainly with various angiosperms (*Stathmopoda* H.-S. and allies), and those with more or less anteriorly protruded or dorsoventrally compressed head, non-ciliate antennae in males and trophic relations with ferns (*Cuprina* Sin. and tropical allies).

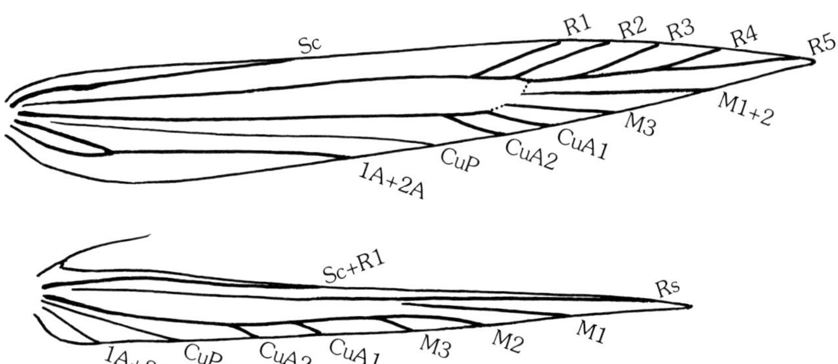

Text-fig. 25. Wing venation of *Stathmopoda pedella* (Linnaeus).

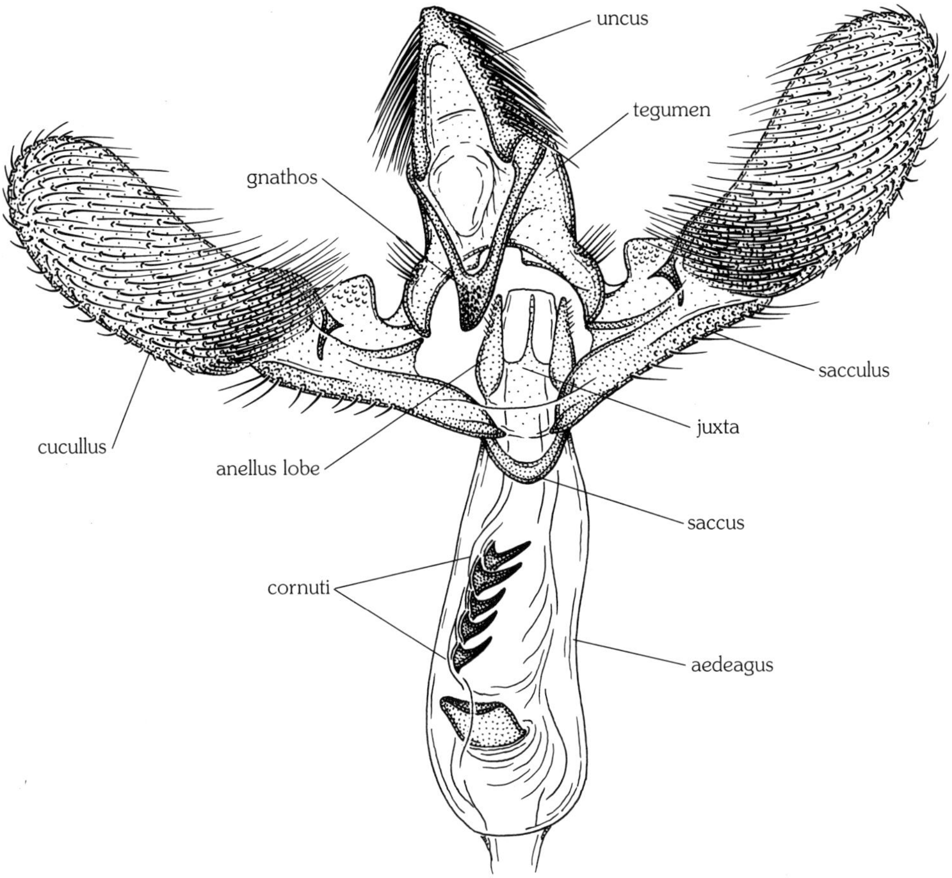

Text-fig. 27. Male genitalia of *Stathmopoda pedella* (Linnaeus).

Systematic Treatment of the Genera and Species of Stathmopodidae in Europe, North Africa and the Near East

Stathmopoda Herrich-Schäffer, 1853

Stathmopoda Herrich-Schäffer, 1853: 14 (key), 54.
Type species: *Phalaena pedella* Linnaeus, 1761, by subsequent designation by Meyrick, 1914b.
Boocara Butler, 1880: 562.
Type species: *Boocara skelloni* Butler, 1880, by original designation.
Placostola Meyrick, 1887: 280.
Type species: *Placostola diplaspis* Meyrick, 1887, by monotypy.
Erineda Busck, 1909: 94.
Type species: *Erineda elyella* Busck, 1909, by original designation.
Agrioscelis Meyrick, 1913b: 96.
Type species: *Agrioscelis tacita* Meyrick, 1913, by original designation.
Kakivoria Nagano, 1916: 138.
Type species: *Kakivoria flavofasciata* Nagano, 1916, by monotypy.

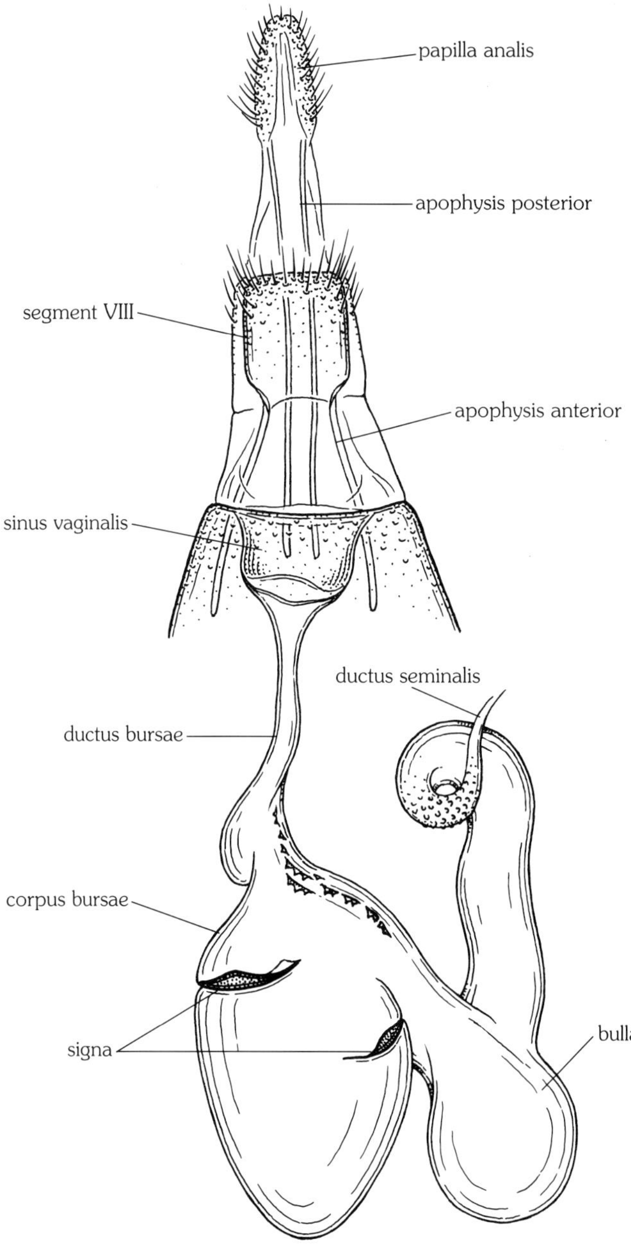

Text-fig. 28. Female genitalia of *Stathmopoda pedella* (Linnaeus).

DIAGNOSIS.– Head smoothly scaled; antennae about 3/4 length of forewings, ciliate in male; labial palpi cylindrical, porrect, segment 3 strongly angled upwards. Forewings narrowly lanceolate, tapering towards apex and often with a yellowish or brownish pattern. Hindwings very narrow, linear.

MALE GENITALIA.– Uncus and gnathos of equal size. Tegumen bent. Valvae with swollen costa, cucullus short and rounded, sacculus well developed or weak. Vinculum rounded; saccus absent. Aedeagus cylindrical or tapering distally, with apical hook or spatulate projection; cornuti present or absent.

FEMALE GENITALIA.– Apophyses anteriores from 1/2 to 2/3 length of apophyses posteriores. Antrum wide, funnel- or bowl-shaped. Ductus bursae rather wide, often wrinkled. Corpus bursae with one or two signa; ductus seminalis usually long and slender, often with dilations of different form; the region of transition between ductus seminalis and corpus bursa (bulla) usually with variable teeth or spicules.

DISTRIBUTION.– The genus has a worldwide distribution mainly in tropical and subtropical areas, the Oriental region being the centre of distribution.

BIOLOGY.– Only known for a very few species. The larvae live in flowers, fruits and shoots or on dried plant material.

25. *Stathmopoda pedella* (Linnaeus, 1761)

Phalaena Tinea pedella Linnaeus, 1761: 367.
Tinea alucitella [Denis & Schiffermüller], 1775: 144.
Tinea cylindrella Fabricius, 1777: 295.
Tinea angustipennella Hübner, 1796: 69, pl. 29, fig. 197.
Ypsolophus cylindricus Fabricius, 1798: 69; unjustified emendation of *Tinea cylindrella* Fabricius, 1777.
Oecophora fastuosella Costa, [1836]: [279].

DIAGNOSIS.– Adult. Wingspan 14-17 mm. Head light yellow, patagia blackish; antennae dorsally grey, apical third yellowish white, strongly ciliate in male. Thorax yellow, anterior third and posterior part greyish brown; tegulae yellow, edged greyish brown. Forewings yellowish ochreous to golden yellow, suffused with light brown scales, dark brown at base, often with leaden gloss; two large dark brown transverse spots (or incomplete bands) at 1/3 and 2/3, inner spot often connected to basal area by costal and dorsal streaks and to outer spot by narrow brown stria in middle of cell; cilia yellow on costa, greyish brown on dorsum. Hindwings shining brownish grey.

MALE GENITALIA.– Tegumen rather short and wide, narrowing basally; valvae with short triangular anterodorsal projection; sacculus slightly raised, in length more than one third of valva; cucullus broad, oval; aedeagus tapering abruptly into acuminate apex; vesica with short and thick cornuti and additional more or less distinct field of minute spinules.

FEMALE GENITALIA.– Antrum large, bowl-shaped; corpus bursae with two penellipsoidal signa of different size with minutely dentate margin; bulla in form of short broad tube with vesicular dilation in middle, forming loop with minutely sculptured wall at transition with narrow part of seminal duct, and carrying series of irregular spinose lamellae that often fuse in region of junction with bursa.

DISTRIBUTION.– Europe except the extreme north and south-east, in the southern part in mountain areas; the Caucasus, Russian Far East. Recently introduced to North America (Hodges et al., 1983).

BIOLOGY.– Host plants: *Alnus glutinosa* (L.) Gaerth. and *A. incana* Willd. (Betulaceae). Univoltine. The larva lives from August to the end of September in the unripe fruits of alder, mainly eating the seeds; affected fruits have brownish spots and the extrusion of larval frass. Pupation takes place in October in an elongate silky cocoon spun among detritus on the ground (Zeller, 1868; Stainton, 1868, 1870b). The moths fly from the end of June to July and they are easily attracted to light; in repose they sit on the upper or underside of the leaves of old alder trees and can be beaten from the lower branches.

26. *Stathmopoda auriferella* (Walker, 1864)

Gelechia? auriferella Walker, 1864: 1022.
Stathmopoda divisa Walsingham, 1891a: 121.
Stathmopoda ischnotis Meyrick, 1897: 324.
Stathmopoda crocophanes Meyrick, 1897: 324.
Aeoloscelis theoris Meyrick, 1906: 410.
Stathmopoda tharsalea Meyrick, 1914a: 199.
Stathmopoda adulatrix Meyrick, 1917: 61.
Stathmopoda cirrhaspis Meyrick, 1922b: 585.
Chrysoclista basiflavella Matsumura, 1931: 1094.

DIAGNOSIS.– Adult. Wingspan 9-13 mm. Head with frons shining yellowish white and vertex ochreous; antennae yellowish white, sometimes with indistinct ochreous annulations. Thorax and tegulae ochreous, thorax with broad longitudinal brown streak dorsally. Forewings ochreous, with shining yellowish white streak at base of dorsum; small dark brown costal spot near base followed by thin brown streak on costa and fold; broad dark brown, slightly inward oblique and medially narrowed fascia before middle of wing; subapical area ochreous brown, darker towards dorsum; cilia ochreous brown on costa, ochreous on dorsum. Hindwings shining yellowish white, yellowish grey towards dorsum and apex.

Variation. The dark brown fascia can vary in width and shape.

MALE GENITALIA.– Tegumen slender, slightly bent; valvae oval, ventral edge somewhat angular, costa large, rounded and spiculate, sacculus stout, reaching beyond middle of valva, spiculate at base; anellus lobes rounded; aedeagus rather long, with apical projection large and pointed, cornuti in bundle of rods and spicules.

FEMALE GENITALIA.– Antrum slightly longer than wide, funnel-shaped; corpus bursae with two small and narrow signa of different size and rows of spicules; bulla with scattered spicules; ductus seminalis very long, narrow and coiled.

DISTRIBUTION.– Widespread in the Oriental and Afrotropical regions. In the Palaearctic region only in the parts with a subtropical climate: Middle East, Far East of Russia.

BIOLOGY.– The larvae live in silken galleries among various types of decaying vegetable matter; sometimes recorded from living flowers, seeds or shoots, injured by other invertebrates (Fletcher, 1920; Kasy, 1973a; Sinev, 1994a). Known as a pest of minor importance on agricultural crops. The moths fly from June to December, probably in two or three generations.

Tortilia Chrétien, 1908

Tortilia Chrétien, 1908: 201.
Type species: *Tortilia flavella* Chrétien, 1908, by monotypy.
Diadoxastis Meyrick, 1913b: 78.
Type species: *Diadoxastis parathicta* Meyrick, 1913, by monotypy.
Apertodiscus Amsel, 1935a: 303.
Type species: *Apertodiscus zernyi* Amsel, 1935 (= *Isorrhoa charadritis* Meyrick, 1924), by monotypy.

DIAGNOSIS.– Small yellowish moths often with a forewing pattern of brownish fascia or spots. Very similar to species of the genus *Stathmopoda*, differing by the shorter forewings and the very short ciliation of the male antennae.

MALE GENITALIA.– Differ from those of *Stathmopoda* by more or less quadrangular valvae lacking pronounced anterodorsal dilation.

FEMALE GENITALIA.– Corpus bursae with only one signum in the anterior part and with rows or patches of long spines in the posterior part.

DISTRIBUTION.– The genus includes about ten species, restricted to arid and semiarid regions of the Old World.

BIOLOGY.– Only known for two species. The larvae mainly live in dried flowers, fruits and leaves of various plants.

27. *Tortilia flavella* Chrétien, 1908

Tortilia flavella Chrétien, 1908: 202.

DIAGNOSIS.– Adult. Wingspan 9-11 mm. Head shining white on frons and yellowish white on vertex; neck tuft shining brown; antennae with scapus brown apically, flagellum pale yellow in basal half and shining ochreous brown in apical half. Thorax and tegulae shining pale yellow. Forewings pale yellow, with brown triangular or trapezoidal fascia from costa to dorsum before the middle; apical 1/3 brown, sometimes with yellow streak, and often connected just before dorsum with outer edge of fascia; cilia pale yellowish ochreous. Hindwings shining light grey.

MALE GENITALIA.– Resemble *graeca*, but differ by the more slender valvae and the bottle-shaped aedeagus with a strong apical hook.

FEMALE GENITALIA.– Resemble those of *charadritis*, but differ by the very long row of short spines in the ductus bursae and the smaller patches of spines in the corpus bursae.

DISTRIBUTION.– Spain (Granada), northern Africa and Near East (Syria and Israel).

BIOLOGY.– The larvae have been recorded from the flowers and fruits of the gum-tree (*Acacia* sp.) (Mimosaceae) (Chrétien, 1908) and from dried pomegranates (*Punica granatum* L.) (Punicaceae), hanging in the trees from the preceding season (Chrétien, 1917). The moths fly from May to September, probably in two generations.

REMARKS.– *Tortilia flavella* belongs to a complex of several externally similar species distributed from North Africa through the Middle East to India.

28. *Tortilia trigonella* (Zerny, 1935)

Stathmopoda trigonella Zerny, 1935: 149.

DIAGNOSIS.– Adult. Wingspan 9 mm. Differs from *flavella* by presence of a brown costal spot near the base of the forewing and the pale yellow basal segment of the antennae. The underside of the forewing is pale yellow and shows the upperside brown pattern, although more weakly.

MALE GENITALIA.– Not available.

FEMALE GENITALIA.– Resemble those of *graeca*, but differ by the much smaller signum and one long and one short row of spines in the corpus bursae.

DISTRIBUTION.– Morocco.

BIOLOGY.– Unknown. The adults have been collected in June (Zerny, 1935).

29. Tortilia charadritis (Meyrick, 1924)

Isorrhoa charadritis Meyrick, 1924: 65.
Stagmatophora stupenda Turati, 1927: 341.
Tortilia viatrix Busck, 1934: 69.
Apertodiscus zernyi Amsel, 1935: 303.
Stathmopoda trissorrhiza Meyrick, 1939: 58.

DIAGNOSIS.– Adult. Wingspan 7-9 mm. Head pale yellow; antennae shining yellowish brown, slightly serrate in distal half. Thorax and tegulae brown anteriorly, pale yellow posteriorly. Forewings pale yellow; brown basal spot from costa to fold; irregular, broad trapezoidal brown fascia from costa to dorsum before 1/2; brown, very outwardly oblique streak at apex, often scattered brown scales on dorsum near base and in apical half; cilia brownish on costa, yellowish grey at apex. Hindwings shining greyish white.

Similar species. The size of the brown markings on the forewing can vary considerably. Differs from *graeca* by the almost entirely brown thorax and the less sharply defined of the forewings.

MALE GENITALIA.– Very similar to *graeca*, but differs by the absence of a tuberculate part of the sacculus and ampulla, and by the straight and cylindrical aedeagus.

FEMALE GENITALIA.– Antrum wide; corpus bursae with a small signum; bulla with four patches of long spines: one small and three large.

DISTRIBUTION.– North Africa and Middle East to western Pakistan. Also imported to the eastern U.S.A. (Kasy, 1973a).

BIOLOGY.– The larvae have been found causing damage to dried leaves of *Cassia* sp. (Mimosaceae) which were imported to the U.S.A. from Sudan (Busck, 1934). Also recorded feeding on dried fruit of *Zizyphus* sp. (Rhamnaceae) (Meyrick, 1939). Adults have been collected from May to July.

30. Tortilia pallidella Kasy, 1973

Tortilia pallidella Kasy, 1973a: 242.

DIAGNOSIS.– Adult. Wingspan 6-9 mm. Head shining white, neck-tuft yellowish orange; antennae ochreous, basal 1/5 white. Thorax shining pale yellowish grey, with yellowish orange scale tufts latero-posteriorly; tegulae shining pale yellowish grey, with pale yellow spot anteriorly. Forewings shining pale yellowish grey with lilac gloss; small dark grey costal spot near base, edged by yellowish orange spot; narrow white subcostal streak at 1/3; inwardly bent, dorsally thickened yellowish orange fascia at 1/4; cilia yellowish white. Hindwings shining white with yellowish tinge. Abdomen shining grey; anal tuft shining white.

Similar species. Easily distinguished from other species of *Tortilia* by pale yellowish grey forewing with lilac gloss.

MALE GENITALIA.– Tegumen, uncus and gnathos distinctly shorter than valvae; valvae with cucullus distally rounded and slightly bent upwards, costa small and flat, sacculus short and narrow; aedeagus straight, tapering distally, with strong apical hook.

FEMALE GENITALIA.– Antrum wide and spiculate; corpus bursae with signum shorter than half corpus length; bulla with four patches of spines of different size: two elongate and two short.

DISTRIBUTION.– Near and Middle East (Israel, southern Iran, Pakistan).

BIOLOGY.– Unknown. The only specimen known from Israel was collected in August; in Iran many specimens were collected in the first half of April under old *Acacia*-trees (Mimosaceae), and the only specimen from Pakistan was collected at the end of February or early March (Kasy, 1973a).

31. Tortilia graeca Kasy, 1981

Tortilia graeca Kasy, 1981: 30.

DIAGNOSIS.– Adult. Wingspan 8-10 mm. Head pale yellow; antennae shining yellowish brown, slightly serrate in distal half. Thorax dark brown, laterally pale yellow; tegulae pale yellow. Forewings pale yellow; dark brown basal spot from costa to fold and greyish brown basal streak on dorsum; dark brown trapezoidal fascia from costa to dorsum before 1/2 and dark brown, outwardly oblique fascia at apex; cilia yellowish brown. Hindwings shining light grey in basal half, grey and coarser scaled in apical half. Abdomen greyish brown, anal tuft pale yellow.

Similar species. Differs from *charadritis* by the laterally yellow thorax and the more distinct pattern of the forewings.

MALE GENITALIA.– Tegumen, uncus and gnathos large, as long or slightly longer than valva; valvae more or less rectangular, costa appears as little hump, sacculus with tubercular part in middle, ampulla small, V–shaped; anellus lobes spiculate; aedeagus slightly tapering and bent. The tubercular part of the sacculus and the bent aedeagus are charasteristic.

FEMALE GENITALIA.– Antrum small and shallow; corpus bursae with signum that is more than half as long as corpus; bulla with two patches of spines of different size. The large signum and only two patches of spines distinguish this species from *charadritis*.

DISTRIBUTION.– Greece (including Crete), Turkey.

BIOLOGY.– Unknown. The moths fly in June and July and come to light.

Family AGONOXENIDAE

Agonoxenidae Meyrick, 1926: 245. Type genus: *Agonoxena* Meyrick, 1921.
Blastodacnidae Clarke, 1962: 125; unavailable as proposed without description.
Parametriotidae Capuse, 1971: 55. Type genus: *Parametriotes* Kusnezov, 1916 (= *Haplochrois* Meyrick, 1897).
Blastodacninae Sinev, 1979c: 594. Type genus: *Blastodacna* Wocke, [1876].
Lamprysticinae Lvovsky, 1996: 432. Type genus: *Lamprystica* Meyrick, 1914.

General part

MORPHOLOGY.– Adult. Wingspan 8 to 20 mm. Head (Text-fig. 29) covered with more or less densely appressed scales. Eyes moderate in size, ocelli absent. Antennae slightly shorter or longer than forewing, scapus usually with bristled pecten. Labial palpi long or of moderate length, curved upward, often with tuft of rough scales in front of segment 2; segment 3 usually more or less shorter than segment 2. Forewings (Text-fig. 30) narrowly lanceolate or lanceolate, usually with 2 large and sometimes several small tufts of raised scales or with tubercular silvery spots. Forewings with 2 or 3 median veins; M1 sometimes stalked with R4+5; bases of all Cu veins approximate at hind angle of cell. Hindwings narrowly lanceolate or lanceolate linear; M1 usually long stalked with M2.

Abdominal tergites (Text-fig. 31) without any specialized scales; longitudinal sclerotized struts of tergite II (venulae) long. Male genitalia (Text-fig. 32): uncus represented by two hairy lobes ('tegminal lobes') or completely reduced; tegumen well developed, sometimes very long; gnathos paired, with two peculiar spiny knobs of different size or with sclerotized lateral arms; saccus moderately long to very long, stick-shaped; valvae symmetrical, often with reduced ventral area and long dorsocaudal projection, occasionally membranous petal-shaped (*Haplochrois*); juxta more or less well developed, lying between two large sclerotized anellus lobes which sometimes replace valvae both in a morphological and functional sense; aedeagus long, well sclerotized, usually curved, in a few cases with group of spines in vesica. Female genitalia (Text-fig. 33): ovipositor moderately long or rather short, with apices of apophyses slightly broadened; ductus seminalis attached not far from ostium; ductus bursae long, slender, sometimes spirally twisted; bursa copulatrix of moderate to small size, egg-shaped, with or without a more or less invaginated spiny signum.

Larva cylindrical, moderately mobile, with sparse secondary setae. Pupa with thin cuticula, exposed labial palpi and forefemora, and diverged apices of antennae; abdominal segments III to VI movable; two leg-like appendages subapically attached to ventral surface of abdomen. Cocoon usually absent, and pupa is able to move inside larval shelter.

BIOLOGY.– Mainly distributed in woodlands where they can be found along edges of forests, among shrubs, and in gardens. Some species are especially common in old city parks. Larvae are usually twig- or bark-borers, sometimes gall-makers of various woody, rarely herbaceous plants of various families, in first instars sometimes mining leaves or stems. Adults rest on trunks and branches of their host-plants during daytime, body tightly appressed to the substrate; some of them are easily attracted to light. Hibernation takes place in the larval stage.

DISTRIBUTION.– The family is especially well represented in tropical regions where many species were erroneously described in other groups of Microlepidoptera. More than

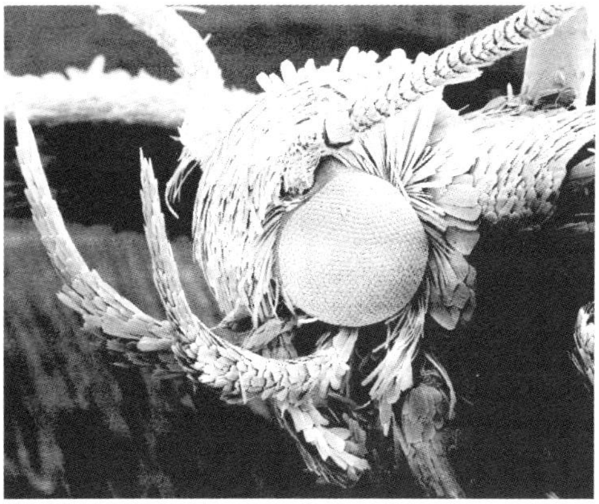

Text-fig. 29. Head of *Blastodacna atra* (Haworth).

Text-fig. 31. Abdominal tergites of *Blastodacna atra* (Haworth).

50 species are known from the Palaearctic, with 6 genera and 21 species in the region of study, mainly in Central and southern Europe.

SYSTEMATICS.– A number of characters of both adults and immatures show plesiomorphic states, demonstrating that the Agonoxenidae is an archaic family of gelechioid moths. Very few apomorphies can be mentioned, such as the considerably enlarged valvellae (anellus lobes), the more or less weakened valvae, and the peculiar leg-shaped appendages of the pupa. This last character represents a remarkable synapomorphy for Agonoxenidae together with families like Ethmiidae, Hypertrophidae, etc. The taxonomy of the family at suprageneric level should be revised. Sometimes several subfamilies (or tribes) are recognized, i.e. Parametriotinae Capuse, 1971 (prior proposal as Parametriotidae), Blastodacninae (Kuznetzov & Stekol'nikov, 1984; Koçak & Seven, 1993) and Lamprysticinae (Lvovsky, 1996).

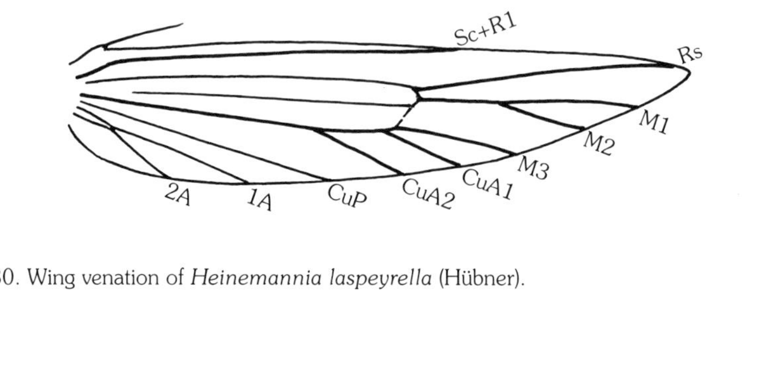

Text-fig. 30. Wing venation of *Heinemannia laspeyrella* (Hübner).

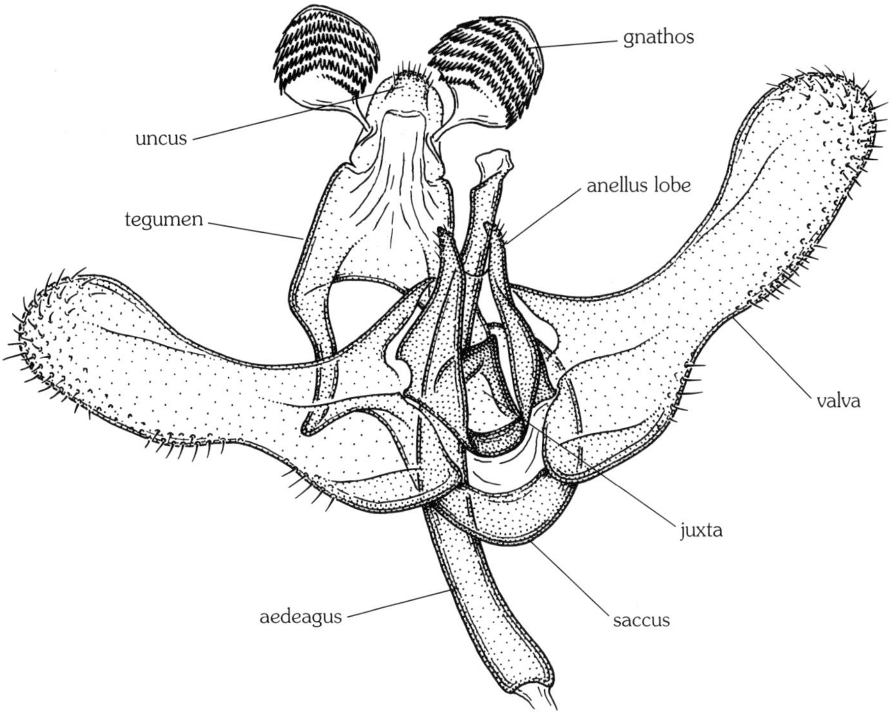

Text-fig. 32. Male genitalia of *Heinemannia laspeyrella* (Hübner).

Systematic Treatment of the Genera and Species of Agonoxenidae in Europe, North Africa and the Near East

Blastodacna Wocke, [1876]

Blastodacna Wocke, in Heinemann, [1876]: 428.
Type species: *Alucita hellerella* Duponchel, in Godart, 1838, by subsequent designation by Fletcher, 1928.
Sinitinea Yang, 1977: 262.
Type species: *Sinitinea pyrigalla* Yang, 1977, by original designation.

Diagnosis.– Head smoothly scaled; antennae 4/5 to just slightly shorter than length of forewing; labial palpi porrect, segment 3 angled upwards, segment 2 distinctly thickened by coarse and spread scales on lower side. Forewings lanceolate with two large tufts of raised scales. Hindwings about half width of forewings, narrow lanceolate, acutely pointed.

Male genitalia.– Uncus very small, tuberculate. Gnathos with paired arms, very short to almost half the length of tegumen; each arm ending in knob-like appendage covered with rows of spines. Tegumen large, long and narrowing distally. Valvae long, narrow, wider in distal half, reaching beyond tegumen. Anellus lobes large and sclerotized, slightly shorter than tegumen. Vinculum rather broad, with short triangular saccus. Aedeagus long, stout, slightly bent in middle, with rows of spines in vesica.

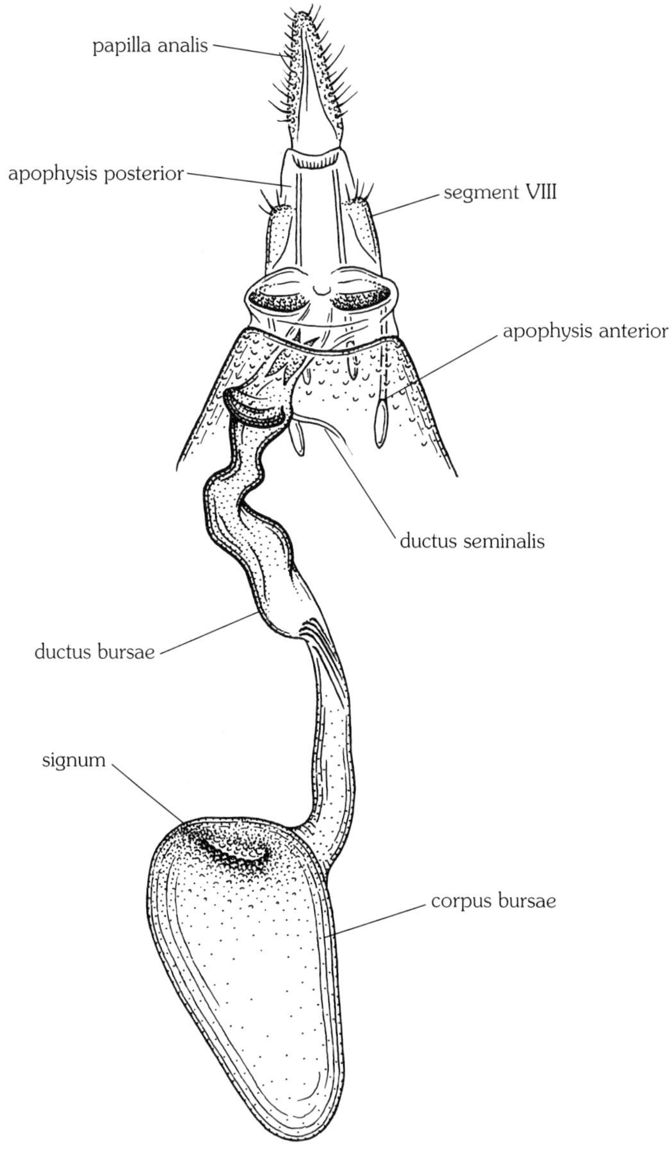

Text-fig. 33. Female genitalia of *Blastodacna rossica* Sinev.

FEMALE GENITALIA.– Apophyses of nearly equal length, slightly dilated apically. Ostium surrounded by sclerotized genital plate with lateral triangular, conical or ribbon-like protrusions. Antrum moderately wide and long, more or less sclerotized, followed by bulbous and often wrinkled section with ductus seminalis opening. Corpus bursae oval or egg-shaped, with single signum in form of rough invagination, or signum absent.

DISTRIBUTION.– Holarctic, most species occur in the Palaearctic region.

BIOLOGY.– The larvae are twig-borers, gall-makers, or live inside the berries of ligneous Rosaceae. Some species can be pests on fruit trees.

REMARKS.– The genus can be split into two homogeneous groups of species. The first group includes *atra*, *hellerella*, and *rossica*; it is characterized by the more contrasting pattern on the rather long and narrow forewings, long and backwardly curved gnathos arms with rather small apical knobs in the male genitalia, and large and distinct genital plate around ostium in the female genitalia. Another group includes *vinolentella* and *georgiella*; it is characterized by shorter, broader and more uniformly coloured forewings, very short gnathos arms with large apical knobs in the male genitalia, and genital plate reduced to a small sclerotized ridge in the female genitalia. These two species-groups probably have the validity of subgenera.

32. *Blastodacna hellerella* (Duponchel, 1838)

Alucita hellerella Duponchel, in Godart, 1838: 218.

DIAGNOSIS.– Adult. Wingspan 10-13 mm. Forewings fuscous black and white, without yellowish orange streaks, with only few ochreous brown scales forming a moderately distinct subbasal stripe; dorsal margin more or less purely white, with very scarce dark irroration.

Similar species. For differences from *atra* and *rossica*, q. v.

MALE GENITALIA.– Gnathos with two long and thin arms each of which terminates in small spiny knob. Anellus lobes moderately narrow, with rounded apex and lack of pronounced ventral projection. Aedeagus with rather high caecal ridge and 3-4 groups of spines in vesica; one group is 1.5-2 times larger than others. Very similar to *atra* and *rossica* and can be separated mainly by the anellus lobes which are distinctly narrowed in apical half and have a rounded apex.

FEMALE GENITALIA.– Genital plate around ostium triangular, distinctly tapered laterally, slightly spiculate. Corpus bursae egg-shaped, signum horn-like and minutely granular. Can be separated from *atra* and *rossica* by the laterally tapered genital plate and the slightly sclerotized and smaller bulbous section of the ductus bursae.

DISTRIBUTION.– Europe, except the north; Caucasus.

BIOLOGY.– Host plants: *Crataegus laevigata* (Poir.) DC and *C. monogyna* Jacq. Also *Prunus spinosa* L. (Lhomme, [1948-1949]), *P. domestica* L. and *Malus* Mill. (Rosaceae) (Grabe, 1955) were mentioned, but these records need confirmation; most of them evidently concern *B. atra* (Sinev, 1994b). The data about great damage to apple trees in Japan (Matsumura, 1898) appear to be erroneous and should be attributed to *Argyresthia conjugella* Zeller, 1839 (Yponomeutidae). Univoltine. The larva lives from the end of summer to autumn in the berries; the infested berries have a small hole in the side. Pupation takes place in late autumn in a loose cocoon in rotten wood or in the soil (Diakonoff, 1939a). Adults fly from May to August.

33. *Blastodacna rossica* Sinev, 1989

Blastodacna rossica Sinev, 1989a: 4.

DIAGNOSIS.– Adult. Wingspan 9-13 mm. Forewings fuscous black and white, with strong blackish brown suffusion and indistinct orange brown streak in fold; dorsal margin dirty white, mixed up with rather dense dark.

Similar species. Very similar to *atra* and *hellerella*, but differs by the blackish brown thorax and the darkened dorsal margin of the forewings.

MALE GENITALIA.– Gnathos with two long and thin arms terminating in small spiny knobs. Anellus lobes moderately narrow, with pointed apex and small but distinct teeth on the ventral margin. Aedeagus without pronounced caecal ridge; vesica with 3-4 groups of spines, one of which is 3-4 times larger than others. Very similar to *atra* and *hellerella* but can be separated mainly by the downwards-bent anellus lobes with a strongly tapered apex.

FEMALE GENITALIA.– Differs from *atra* and *hellerella* by the large, rounded genital plate with four transverse ridges at each side of ostium, the granular sclerotization of the major part of the surface of corpus bursae and the adjoining part of ductus bursae, and the lanceolate signum.

DISTRIBUTION.– The Crimea; western Transcaucasus, Central Asia (Sinev, 1989b).

BIOLOGY.– Host plant: *Malus* spp., probably also *Pyrus communis* L. (Rosaceae). Univoltine. The larvae bore in the young twigs from the second half of summer to autumn, making a tunnel up to a length of 4 cm; its presence is revealed by brownish frass on the surface of a twig near the entrance hole. Hibernation usually takes place as a full-fed larva; pupation in the following spring. The moths are flying from late May to August in Europe and from April to May in Central Asia (Plotnikov, 1926; Sinev, 1994b).

34. Blastodacna atra (Haworth, 1828)

Recurvaria atra Haworth, 1828: 553.
Elachista putripennella Zeller, 1839: 211.

DIAGNOSIS.– Adult. Wingspan 10-13 mm. Thorax creamy white with variable dark brown suffusion. Forewings dark brown, more greyish brown on costa; creamy white dorsal streak from base to tornus, suffused with dark brown and yellow ochreous scales, narrowed in the middle by a dark brown subdorsal scale-tuft; second scale-tuft at tornus; ochreous streaks in fold and in disc, the latter with several inwardly and outwardly oblique creamy white lines to costa; small creamy white line before apex; cilia dark grey around apex, grey towards base, with two transverse dark lines around apex.

Similar species. Very similar to *hellerella* and *rossica*, but can be distinguished by the light yellowish orange markings of the forewings.

MALE GENITALIA.– Gnathos arms long and narrowing distally, curved backwards to half of tegumen, apical knobs with two rows of spines. Anellus lobes more than half as long as valva, widest in basal half, with narrow apex and triangular pointed eminence on ventral margin at 3/4. Aedeagus long and curved, slightly narrowing distally; vesica with 4 similar groups of spines.

FEMALE GENITALIA.– Genital plate rectangular, spiculate laterally and anteriorly. Anterior part of ductus bursae with a rectangular sclerotized plate, followed by a wide bulbous unsclerotized section. Corpus bursae oval, with granulated cap-shaped signum.

DISTRIBUTION.– Europe, except the north.

BIOLOGY.– Host plant: *Malus* Mill. (Rosaceae). Univoltine. The oval egg is laid singly close to the axil of a leaf. The larva mines the pith of a twig from August to May or early June, and feeds continuously during the winter period. The presence of a larva is indicated by a small heap of frass ejected from the entrance hole on a twig. The boring activity causes the death of the shoots and flowers and cracks in the bark. Pupation takes place in May-June inside a twig. Adults fly from the end of May to

early September. The species can be a serious pest on apple trees (Miles, 1930; Balachowsky, 1966; Sinev, 1994b).

REMARKS.– The dark brown suffusion of the vertex, thorax and dorsal streak can vary considerably. The original description of *atra* was based on a melanistic form mainly occurring in England. The light form was described by Zeller (1839) as *putripennella* and subsequently considered a variation.

35. Blastodacna vinolentella (Herrich-Schäffer, [1854])

Tebenna vinolentella Herrich-Schäffer, [1854]: 214.

DIAGNOSIS.– Adult. Wingspan 9-12 mm. Head shining dark brown; antennae shining dark brown, more greyish and distinctly annulated in apical half. Thorax and tegulae shining dark brown. Forewings shining dark brown, with indistinct greyish white, mixed light brown costal spot at 3/4; two large scale tufts above dorsum, first at 1/3, second and largest, inwardly edged with shining greyish white scales, at 2/3; indistinct narrow and sinuate greyish white incomplete fascia between scale tufts; similar subapical, often interrupted fascia from costal spot towards dorsum; cilia dark brown around apex, greyish brown towards base, with one indistinct transverse line. Hindwings shining brownish grey.

Similar species. The relatively broad blackish brown forewings with very few whitish scales separate this species from the other species of the genus except *georgiella*; for separation from the latter, q. v.

MALE GENITALIA.– Gnathos arms very short, distal knobs large and spherical with five or six rows of spines. Tegumen rather short, strongly tapering distally. Valvae about twice as long as tegumen, slender, very broad at base, apex rounded and slightly widened. Anellus lobes fused in basal half, distal half tapering to a blunt apex, inwardly curved and with a strong fold dorsally at inner side. Aedeagus long with bent and slender distal half, with a patch of spines in vesica.

FEMALE GENITALIA.– Genital plate reduced to a small, shallowly V–shaped sclerotization. Antrum funnel-shaped, narrowing to ostium. Ductus bursae with rounded sclerotized plate before antrum. Corpus bursae egg-shaped, signum slightly granulated.

DISTRIBUTION.– Central and southern Europe, from Holland to Greece (very local).

BIOLOGY.– Unkown. Although *Vitis vinifera* L. (Vitaceae) was mentioned as host plant (Diakonoff & Doets, 1938; Diakonoff, 1939a), there is no evidence that the species has been reared from this plant. Diakonoff simply followed Herrich-Schäffer ([1854]) who, when describing the species, noted: '...im Juli an Weinstöcken findet.' The records of larvae from the buds of *Malus* Mill. and young shoots of *Crataegus monogyna* Jacq. (Rosaceae) (Heinemann & Wocke, [1876]; Barrett, 1898, etc.) may have been caused by confusion with the melanistic form of *B. atra* (Sinev, 1994b). Adults fly from early June to early August.

36. Blastodacna georgiella Sinev, 1988

Blastodacna georgiella Sinev, 1988a: 15.

DIAGNOSIS.– Adult. Wingspan 11-12 mm. Very similar to *vinolentella*. Differs from this species by the greyish frons and the presence of a small white spot, which borders the inner scale tuft inwardly.

MALE GENITALIA.– Unknown.

FEMALE GENITALIA.– Differs from *vinolentella* by the wider and shallower antrum and by the granulation of posterior part of corpus bursae and adjacent part of ductus bursae.

DISTRIBUTION.– Western Transcaucasia: Georgia (Sinev, 1988a).

BIOLOGY.– Host plant unknown. Adults have been found from early June to the second half of July.

37. *Blastodacna libanotica* Diakonoff, 1939

Blastodacna libanotica Diakonoff, 1939b: 78.

DIAGNOSIS.– Adult. Wingspan 14-15 mm. Head shining white; antennae white, annulated grey. Thorax grey with two greyish white spots anteriorly and greyish white posterior band; tegulae grey with white posterior edges. Forewings dark greyish brown, paler on dorsum at base; white costal streak suffused with ochreous and brown scales; reddish brown spot in discal area, bordered by triangular figure of white lines with extension towards dorsum; two dark brown tufts of raised scales, first below fold before middle, second and largest on dorsum beyond the white triangle; indistinct interrupted white line in apical area; cilia ochreous grey, dark brown with two blackish transverse lines at apex. Hindwings shining light grey.

Similar species. The large wingspan, the whitish costa and the white triangular figure of the forewings are characteristic.

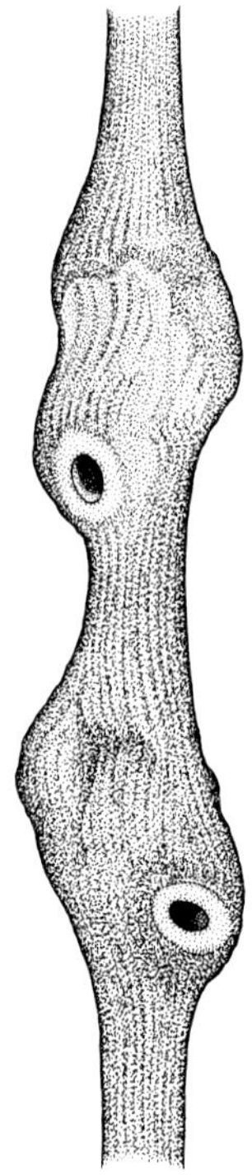

Text-fig. 34. *Blastodacna libanotica* Diakonoff, galls in a twig of *Pyrus* L.

MALE GENITALIA.– Uncus very short, rounded distally. Gnathos arms long and narrowing distally, curved backwards to middle of tegumen; apical knob mushroom-shaped with four rows of spines. Tegumen long and narrowing distally. Valvae long, narrowest part at base, gradually widening towards apex, tip rounded. Anellus lobes broad, bifurcate, costal arm longer, pointed, widest part before apex, dorsal arm parallel-sided, curved inwardly at apex. Aedeagus stout, tapering distally, with two parallel rows of four large thorns in vesica.

FEMALE GENITALIA.– Apophyses distinctly widening apically. Genital plate more or less trapezoidal, strongly spiculate except for the central part. Antrum very wide, concave and spiculate. Ductus bursae with bulbous section directly connected to antrum. Corpus bursae without signum.

DISTRIBUTION.– Asia Minor and the Near East (Diakonoff, 1939b).

BIOLOGY.– Host plant: *Pyrus* L. (Rosaceae). The larva makes a gallery in a thin twig in summer and early autumn. The gallery starts at an angle of about 30°, then directed upward. This activity causes the development of an elongate gall of 11.5-13.5 mm in length and 7.3-8.6 mm in width. The gall is more or less constricted in the middle and with a rounded opening at the lower end. There can be more than one gall in a twig. Prior to pupation the larva enlarges the exit-hole and polishes a circle around it. Pupation takes place inside the gall at the end of autumn. Adults appear from February to May (Talhouk, 1948; Zangheri, 1966). See text-fig. 34.

Spuleria Hofmann, 1898

Spuleria Hofmann, 1898: 230.
Type species: *Tinea aurifrontella* Geyer, in Hübner, [1832], by monotypy.

DIAGNOSIS.– Head smoothly scaled; antennae 4/5 length of forewing; labial palpi porrect. Forewings lanceolate, more or less unicolorous dark, with two indistinct tufts of raised scales. Hindwings 2/3 width of forewings, lanceolate and pointed.

MALE GENITALIA.– Uncus very short. Gnathos arms short, apical knobs large and spherical and covered with rows of spines. Tegumen large, narrowing distally. Valvae long and narrow with rounded apex. Anellus lobes large and broad, acutely pointed at apex. Aedeagus long and straight, tubular, tapering distally, with rows of spines in vesica.

FEMALE GENITALIA.– Apophyses posteriores about one and a half times as long as apophyses anteriores. Lamella vaginalis weakly sclerotized. Antrum funnel-shaped or tubular, long and often sclerotized. Ductus bursae short, as long as or shorter than corpus bursae. Corpus bursae oval; signum granular or absent.

DISTRIBUTION.– Palaearctic, with one species in Europe and two in the Russian Far East.

BIOLOGY.– Only known for the type species, which is a twig borer on Rosaceae.

38. *Spuleria flavicaput* (Haworth, 1828)

Porrectaria flavicaput Haworth, 1828: 536.
Tinea aurifrontella Geyer, [1832]: pl. 70, fig. 469.
Rösler[s]tammia aurocapitella Bruand, [1851]: 78; unjustified emendation of *Tinea aurifrontella* Geyer, [1832].

Chrysoclista flavicapitella Doubleday, 1859: 34; unjustified emendation of *Porrectaria flavicaput* Haworth, 1828.

DIAGNOSIS.– Adult. Wingspan 12-14mm. Head bright yellow, with a narrow streak of dark brown scales laterally; antennae dark brown, slightly serrate in apical half. Thorax and tegulae dark brown with purplish gloss. Forewings shining dark brown with some bluish suffusion; two small and indistinct subdorsal tufts of raised scales at 1/3 and 2/3; cilia dark brown. Hindwings shining grey-brown, cilia grey-brown.

MALE GENITALIA.– See genus description.

FEMALE GENITALIA.– Ostium with two sclerotized and curved lateral rims. Antrum sclerotized and narrowed in middle. Ductus bursae and posterior part of corpus bursae finely granulate; signum elongate.

DISTRIBUTION.– Europe; Asia Minor.

BIOLOGY.– Host plants: *Crataegus monogyna* Jacq. and *C. laevigata* (Poir.) DC. (Rosaceae). The larva lives in summer and early autumn in a soft side-twig. It starts boring from the fork into the pith in the direction of the tip. The gallery is about 3 cm long, at the end it turns towards the surface of the twig, usually on the underside. The exit-hole is oval and partly covered with a thin layer of bark. Before pupation the larva closes the gallery by a web not far from the exit-hole (Sorhagen, 1885). Pupation takes place in the gallery in late autumn. Adults fly in May and June in the morning sunshine around the branches of hawthorn. It can also be disturbed by beating hawthorn during day, or found sitting in flowers of hawthorn (Karsholt, pers. comm.).

Heinemannia Wocke, [1876]

Heinemannia Wocke, in Heinemann, [1876]: 428.
Type species: *Tinea laspeyrella* Hübner, 1796, by subsequent designation by Fletcher, 1929.
Tebenna Hübner, [1825]: 414, *nec Tebenna* Billberg, 1820; homonym.
Type species: *Tinea festivella* [Denis & Schiffermüller], 1775, by subsequent designation by Fletcher, 1929.

DIAGNOSIS.– Head smoothly scaled; antennae only slightly shorter than forewings; labial palpi long, cylindrical and strongly bent upwards, segment 2 broadened by rough scales ventrally. Forewings broad lanceolate with two large tufts of raised scales. Hindwings lanceolate, about 1/4 narrower than the forewings.

MALE GENITALIA.– Uncus absent. Gnathos arms very short, ending in a large to very large knob covered with rows of spines. Tegumen long, narrowing distally. Valvae long, broad at base and narrowed in middle part, with widely rounded apex. Anellus lobes large, about half as long as valvae, tapering distally and with several blunt spines at apex. Aedeagus long, narrow and tubular, slightly bent, with or without cornuti.

FEMALE GENITALIA.– Papillae anales long and narrow. Apophyses posteriores about 1/4 longer than apophyses anteriores. Ostium wide, bowl or funnel-shaped. Antrum with a pair of short sclerotized ridges. Ductus bursae long and narrow, slightly widening towards corpus bursae, with widest part more or less granulated. Corpus bursae oval or egg-shaped, without signum.

DISTRIBUTION.– Palaearctic, to south-western Siberia in the east.

BIOLOGY.– Only known for one species. Larvae feed in the seed pods of Fabaceae.

REMARKS.– The male and female genitalia show only very small interspecific differences. It is easier to identify the species on external characters.

39. *Heinemannia laspeyrella* (Hübner, 1796)

Tinea laspeyrella Hübner, 1796: 60, pl. 13, fig. 90.

DIAGNOSIS.– Adult. Wingspan 17-21 mm. Head light yellow. Thorax yellow; tegulae dark greyish brown, inwardly lined yellow. Forewings deep yellow, with dark brown costal streak, paler towards apex; dark brown oval spot before middle from costal streak almost to fold; tuft of raised dark brown scales opposite and below fold; another dark brown spot at tornus with a tuft of raised dark brown scales inside; costal streak at base and both dark brown spots and scale tufts very narrowly edged shining white; cilia greyish brown. Hindwings shining dark greyish brown.

Similar species. The yellow head and thorax in combination with the deeply yellow forewings distinguishes *laspeyrella* from *festivella* and *albidorsella*.

MALE GENITALIA.– See genus description. Can be separated from *festivella* and *albidorsella* by the gnathos arms, which have nearly the same width as the distal part of the valva. Aedeagus with rows of spicules in vesica.

FEMALE GENITALIA.– See genus description. Tergite VIII rectangular, about one and a half time as long as wide. Ostium bowl-shaped. The area of junction between ductus bursae and corpus bursae with very minute granulation. Corpus bursae egg-shaped.

DISTRIBUTION.– North, Central and eastern Europe; Ural, southern Siberia.

BIOLOGY.– Host plants: *Lathyrus pisiformis* L., *Orobus vernus* L., *Trifolium* sp. (Fabaceae) (Sinev, 1986, 1994b); also swept as adult from *Orobus niger* L. (Liška et al., 1992). Univoltine. The larva lives in July and first half of August inside a pod where it feeds on the seeds; 1 to 6 larvae can be found in one pod. The full-fed larva hibernates in the bark of neighbouring trees. Adults have been collected from mid-May to the beginning of July.

40. *Heinemannia festivella* ([Denis & Schiffermüller], 1775)

Tinea festivella [Denis & Schiffermüller], 1775: 319.
Borkhausenia kokujevi Krulikowsky, 1903: 31.

DIAGNOSIS.– Adult. Wingspan 14-18mm. Head shining white; antennae pale grey with indistinct darker annulations. Thorax shining white; tegulae white, with a dark greyish brown spot anteriorly. Forewings pale yellow, distally darker, with dark brown costal streak, paler towards apex, interrupted at 3/4 by whitish spot; dark brown oval spot before middle, from costal streak almost to fold; large tuft of raised dark brown scales opposite and below fold; second dark brown spot at tornus with tuft of raised dark brown scales inside; costal streak at base and both dark brown spots and scale tufts edged shining white; shining white line in apical area at dorsal side and shining white spot at apex; cilia greyish yellow with indistinct light brown line around apex. Hindwing shining pale grey.

Similar species. The species can be distinguished from *laspeyrella* and *albidorsella* by the pale yellow ground colour of the forewings in combination with the white head and thorax.

MALE GENITALIA.– Can be separated from *laspeyrella* and *albidorsella* by the width of the gnathos arms, which are distinctly wider than the distal part of the valva.

FEMALE GENITALIA.– Tergite VIII almost square, basal edge rounded. Ostium funnel-shaped. The area of junction between ductus bursae and corpus bursae with distinct granulation. Corpus bursae oval.

DISTRIBUTION.– Central and southern Europe, to Sweden in the north; Asia Minor, Middle East, Central Asia.

BIOLOGY.– Host plant unknown. Lhomme ([1948-1949]) cited an unpublished record of Chrétien, who collected the adults from *Solidago virgaurea* L. (Asteraceae). Later several authors erroneously attributed this record to the larva (Hruby, 1964; Riedl, 1969). In Central Asia the moths have often been collected among shrubs of *Lonicera* L. (Caprifoliaceae). Adults fly from May to first half of August.

41. *Heinemannia albidorsella* (Staudinger, 1877)

Laverna albidorsella Staudinger, 1877: 206.
Mompha albidecorella Hartig, 1951: 146; unjustified emendation of *Laverna albidorsella* Staudinger, 1877.

DIAGNOSIS.– Adult. Wingspan 13-16 mm. Head shining white; antennae unicolorous brownish grey. Thorax shining white; tegulae white, with a dark greyish brown spot anteriorly. Forewings white, yellowish white in fold and greyish brown in apical area; broad greyish brown costal streak, broadly interrupted at 3/4 by a large white streak; subcostal greyish brown oval spot before middle, reaching fold and narrowly edged white on costal side, with tuft of raised blackish brown scales underneath, on dorsal side of fold; similar spot on tornus, with tuft of raised scales in distal half and white edging connected to white costal streak; brown streaks between both oval spots and beyond outer one; outer streak the largest, more reddish brown and distally edged white; blackish brown subcostal streak from inner oval spot to apex, interrupted by white costal streak and white edging of outer brown streak; cilia greyish yellow with broad greyish brown line around apex. Hindwings shining pale grey.

Similar species. Differs from *laspeyrella* and *festivella* by the forewings with white ground colour, the larger white costal streak at 3/4 and the two brownish spots in the distal half.

MALE GENITALIA.– Differs from *laspeyrella* and *festivella* by the width of the gnathos arms, which are narrower than the distal part of the valva.

FEMALE GENITALIA.– Tergite VIII almost square, slightly narrowing distally. Ostium bowl-shaped. The area of junction between ductus bursae and corpus bursae with distinct granulation. Corpus bursae egg-shaped.

DISTRIBUTION.– Southern France, Corsica, Sardinia.

BIOLOGY.– Host plant unknown. The moths have been collected from May to early July.

Dystebenna Spuler, 1910

Dystebenna Spuler, 1910: 386.
Type species: *Elachista stephensi* Stainton, 1849, by monotypy.

DIAGNOSIS.– Head smoothly scaled; antennae 4/5 of forewing length; labial palpi porrect, with segment 3 angled upwards. Forewings rather broadly lanceolate with two distinct tufts of raised scales. Hindwings slightly broader than half width of forewings, lanceolate, acutely pointed.

MALE GENITALIA.– Uncus absent. Gnathos arms short, with large apical knobs. Tegumen rectangular. Valvae bilobed, with broad ventral and long and pointed dorsal parts. Anellus lobes broad, with lateral hooked protrusion. Vinculum V–shaped, with long and pointed saccus. Aedeagus long and straight, tubular, apex hooked, with a row of spicules in vesica.

FEMALE GENITALIA.– Apophyses posteriores almost twice as long as apophyses anteriores. Tergite VIII narrowing distally. Lamella postvaginalis with two sclerotized plates. Antrum bowl-shaped. Ductus bursae short, widening towards corpus bursae. Corpus bursae egg-shaped; signum a long and granulated band.

DISTRIBUTION.– A monotypic genus restricted to Europe and the Caucasus.

BIOLOGY.– The larvae feed under the bark of Fagaceae trees.

42. *Dystebenna stephensi* (Stainton, 1849)

Elachista stephensi Stainton, 1849: 24.
Tebenna tesselatella Herrich-Schäffer, [1854]: 215.
Laverna stephensiella Doubleday, 1859: 34; unjustified emendation of *Elachista stephensi* Stainton, 1849.

DIAGNOSIS.– Adult. Wingspan 8-9 mm. Head shining white, neck tuft suffused with brown scales; antennae white, annulated light brown, about 5 apical segments white. Thorax and tegulae white, heavily mottled with brown anteriorly. Forewings grey and white, with brown spot on costa at base, prolonged as a suffused broad and oblique band to middle of wing; brown costal streaks at 1/2 and at apex; two dark brown spots with raised scales on dorsum at 1/3 and 2/3; narrow orange brown streak from middle of base to inner dorsal spot; orange brown spot below costal streak, connected with an orange brown streak at apex; cilia light grey, with two brown transverse lines. Hindwings shining grey.

MALE GENITALIA.– As described for the genus.

FEMALE GENITALIA.– As described for the genus.

DISTRIBUTION.– Central Europe, England, southern Sweden, the Crimea; western Transcaucasia.

BIOLOGY.– Host plants: *Quercus* L. and *Castanea* Mill. (Fagaceae). Univoltine. The larva feeds under the living bark of large and old trees (Emmet & Johnson, 1979; Sinev, 1986). Pupation also takes place in the bark. Adults fly from the end of June to early September and can often be found in the crevices on the trunks of the food plants, where they are easily caught in spider webs.

Haplochrois Meyrick, 1897

Haplochrois Meyrick, 1897: 299 (key), 310.
Type species: *Haplochrois chlorometalla* Meyrick, 1897, by original designation.

Aetia Chambers, 1880: 186, *nec Aetia* Agassiz, 1847: 10; homonym.
Type species: *Aetia bipunctella* Chambers, 1880, by monotypy.
Rhadinastis Meyrick, 1897: 299 (key), 311.
Type species: *Rhadinastis microlychna* Meyrick, 1897, by original designation.
Tetanocentria Rebel, 1902: 107.
Type species: *Tetanocentria gelechiella* Rebel, 1902, by monotypy.
Platybathra Meyrick, 1911: 78.
Type species: *Platybathra ganota* Meyrick, 1911, by monotypy.
Parametriotes Kusnezov, 1916: 628.
Type species: *Parametriotes theae* Kusnezov, 1916, by original designation.
Syntetrernis Meyrick, 1922b: 574.
Type species: *Syntetrernis xiphodes* Meyrick, 1922, by original designation.

DIAGNOSIS.– Head smoothly scaled; antennae slightly shorter than forewings; labial palpi slender and angled upwards. Forewings very narrowly lanceolate, smooth, often without any pattern. Hindwings less than half width of forewings, linear and acutely pointed. The members of this genus are often taken to be *Coleophora* species.

MALE GENITALIA.– Uncus composed of two rounded and partly united lobes. Tegumen very long and rather narrow, about two to three times length of uncus, with rounded excavation at anterior margin. Gnathos arms short and bent, ending in a spherical knob covered with rows of spines. Valvae broad and very short, rounded or nearly square, membranous. Anellus with complicated sclerotization; anellus lobes of various size, shield-shaped, rounded or elongate and supplied with thorns of various length and arrangement. Vinculum broad with a long and narrow stick-like saccus. Aedeagus tubular, straight or bent, with an oval manica; cornuti absent.

FEMALE GENITALIA.– Apophyses very short; apophyses posteriores three to six times as long as apophyses anteriores, the latter often curved. Genital plate with sclerotized curved rim of different shape, sometimes separated in middle. Antrum weakly pronounced. Ductus bursae very long, narrow, gradually widening towards corpus bursae and coiled. Corpus bursae small, rounded or oval, with one more or less funnel-shaped signum. The shape of the genital plate and signum is of diagnostic importance.

DISTRIBUTION.– Mainly Pantropic, with only few species reaching the south of the temperate zone.

BIOLOGY.– Only known for very few species. The larva is a leaf miner in first instars, in later instars a shoot borer.

43. Haplochrois gelechiella (Rebel, 1902)

Tetanocentria gelechiella Rebel, 1902: 107.

DIAGNOSIS.– Adult. Wingspan 12-14 mm. Head creamy white, pale beige laterally and posteriorly; antennae pale ochreous, annulated brown. Thorax and tegulae pale beige grey. Forewings shining pale beige grey, strongly mottled with brownish grey scales; cilia pale grey around apex, more ochreous grey towards dorsum. Hindwings shining grey.

Similar species. The species resembles *albanica*, but differs by the more intensely beige grey colour of the forewings, lighter hindwings, and by the presence of a dark brown apical streak on the underside of the forewing.

MALE GENITALIA.– Similar to those of *albanica* but distal knobs of gnathos with coarser spines; valvae more square and anellus lobes broader and rounded with a semicircular excavation anteriorly.

FEMALE GENITALIA.– Differs from *albanica* by the more slender apophyses anteriores, by the narrower and U–shaped rim of the genital plate and by the shape of signum, which is surrounded with a stronger sclerotized ridge.

DISTRIBUTION.– Greece.

BIOLOGY.– Unknown. The few known specimens have been collected in July.

44. *Haplochrois ochraceella* (Rebel, 1903)

Tetanocentria ochraceella Rebel, 1903: 99.

DIAGNOSIS.– Adult. Wingspan 10-15 mm. Head pale ochreous yellow; antennae yellowish white, annulated brown. Thorax and tegulae ochreous yellow. Forewings ochreous yellow, with scattered ochreous brown scales at apex; cilia ochreous yellow. Hindwings grey, coarsely scaled in male, smoother in female.

MALE GENITALIA.– Gnathos with distal knobs less than twice length of tegumen. Valvae small and rounded. Anellus lobes elongate, distally broadened and triangular, with three rows of short and straight spikes at inner side. Saccus narrowing distally, about 2/3 of tegumen length. Aedeagus straight, with curved, finger-shaped apical protrusion.

FEMALE GENITALIA.– Apophyses very short and almost straight. Genital plate broadly semicircular with strong lateral ridges. Ductus bursae with two loops before connection with corpus bursae. Corpus bursae almost circular; signum irregularly triangular, with central spiculate ridge.

DISTRIBUTION.– Southern Europe; in recent decades it has been found locally in Central Europe.

BIOLOGY.– Unknown. Adults have been collected from early June to early August.

45. *Haplochrois albanica* (Rebel & Zerny, [1932])

Tetanocentria albanica Rebel & Zerny, [1932]: 152.

DIAGNOSIS.– Adult. Wingspan 11-12 mm. Head ochreous; antennae whitish, annulated brown. Thorax and tegulae ochreous, mixed brownish anteriorly. Forewings shining ochreous brown, strongly mottled with grey-tipped brown scales, denser towards apex; cilia greyish brown. Hindwings shining greyish brown.

Similar species. Resembles *gelechiella* but differs by the ochreous brown forewings and the darker hindwings.

MALE GENITALIA.– Gnathos with distal knobs rather large, almost half the length of tegumen. Valvae rectangular with rounded edges. Anellus lobes united ventrally, shield-shaped, with two small patches of short and blunt spikes. Saccus slender, narrowing in middle and as long as tegumen. Aedeagus straight, with long and slightly bent finger-shaped protrusion. For differences from *gelechiella*, q. v.

FEMALE GENITALIA.– Apophyses very short, stout and curved. Genital plate with narrow almost semicircular rim. Ductus bursae with two loops before connection with corpus bursae. Corpus bursae slightly elongate; signum irregularly pointed on upper and lower side and with several protrusions, all covered with spicules.

DISTRIBUTION.– Southern Europe, in recent decades also found very locally in Central Europe.

BIOLOGY.– Unknown. Adults have been collected from June to August.

46. *Haplochrois buvati* (Baldizzone, 1985)

Aetia buvati Baldizzone, 1985: 135.

DIAGNOSIS.– Adult. Wingspan 13-18 mm. Head pale ochreous with some greyish scales intermixed; antennae pale ochreous, annulated grey, slightly serrate apically, ciliate in male. Thorax and tegulae pale ochreous, strongly mottled with brownish grey, lighter posteriorly. Forewings narrowing distally, shining light ochreous grey, strongly mottled with brownish grey, especially at base and along costa; irregular and indistinct dark fascia at 1/2 and 3/4; cilia ochreous grey on costa, grey towards base. Hindwings shining light grey.

Similar species. Differs from *albanica* and *gelechiella* by the larger wingspan and the two indistinct fasciae of the forewings.

MALE GENITALIA.– As described in the genus description. Gnathos with rather small distal knobs, about 1/4 length of tegumen. Valvae rounded triangular. Anellus lobes large, about the length of tegumen, strongly widened distally and here covered with numerous long and slightly curved spikes. Saccus short and stout, its length about 2/3 of tegumen. Aedeagus strongly curved.

FEMALE GENITALIA.– Apophyses anteriores relatively long, narrow and only slightly bent. Genital plate with shallow V–shaped rim and two large egg-shaped wrinkled sclerotizations. Ductus bursae sinuate towards corpus bursae. Corpus bursae very small; signum elongate with series of sharp projections.

DISTRIBUTION.– South-western Europe.

BIOLOGY.– Unknown. Adults have been collected from mid-August to mid-October.

47. *Haplochrois theae* (Kusnezov, 1916)

Parametriotes theae Kusnezov, 1916: 628.

DIAGNOSIS.– Adult. Wingspan 9-12 mm. Head ochreous grey, grey brown posteriorly; antennae ochreous grey with darker annulations, slightly serrate. Thorax and tegulae greyish brown. Forewings greyish brown, somewhat darker at base; two small dark grey streaks, first in the fold before the middle, second at 3/4 and almost in middle of wing; several dark grey spots around apex; cilia grey with ochreous tinge. Hindwings shining greyish brown.

Similar species. Can be distinguished from other species of the genus by the smaller wingspan and the two dark markings on the forewings.

MALE GENITALIA.– Gnathos with distal knobs large, half the length of tegumen. Valvae almost circular. Anellus lobes united ventrally, small, shield-shaped, with two small patches of rather large spikes. Saccus slender, narrowing distally, about 3/4 length of tegumen. Aedeagus straight, basal half somewhat globular.

FEMALE GENITALIA.– Apophyses very short, hardly curved. Genital plate divided into two elongate triangular lateral parts. Ductus bursae strongly sinuate towards corpus bursae. Corpus bursae oval and strongly wrinkled together with anterior part of ductus bursae; signum straight, horn-shaped and spiculate.

DISTRIBUTION.– Western Transcaucasia and Asia Minor; China. Very probably originated from South East Asia from were it was introduced together with the tea plant at the end of 19th century (Stepanov, 1966; Sinev, 1979c; Liu & Shen, 1981).

BIOLOGY.– Host plants: *Thea sinensis* Sims. and *Camellia japonica* L. (Theaceae). A serious pest of the tea plant (Demokidov, 1916; Dzhashi, 1958, 1969; Bozan & Aslitürk, 1975; Sinev, 1994b). Univoltine. At the end of summer and in autumn the young larva makes a small blotch mine in a leaf. Sometimes up to 40 mines can be found in one leaf. During the second half of March it leaves the mine and bores into the twig of a fresh shoot, making a gallery in the pith. During the second half of April the infested shoots can easily be recognized by the withered terminal leaves. Before pupation the larvae gnaws an exit-hole and lines it with a silken web. Pupation takes place in the gallery from the end of May to late July. Adults fly from June till August.

Chrysoclista Stainton, 1854

Chrysoclista Stainton, 1854: 225 (key), 240.
Type species: *Phalaena linneella* Clerck, 1759, by subsequent designation by Fletcher, 1928.
Glyphipteryx Curtis, 1827: 152; unjustified emendation of *Glyphipterix* Hübner, [1825].

DIAGNOSIS.– Head smoothly scaled; antennae 4/5 to just slightly shorter than length of forewing, sometimes slightly serrate distally; labial palpi cylindrical and porrect, with segment 3 slightly angled upwards. Forewings lanceolate and with orange ground colour, often with some distinct tubercules of shining metallic scales and patches of raised scales. Hindwings about 2/3 width of forewing, lanceolate, often acutely pointed.

MALE GENITALIA.– Uncus absent. Gnathos arms sclerotized, often dentate. Tegumen short and broad, with small socii. Valvae broad and simple, with rounded or truncated apex. Anellus lobes large, about half length of aedeagus. Aedeagus long, tubular, bent in middle, without cornuti.

FEMALE GENITALIA.– Apophyses posteriores about one and half to two times length of apophyses anteriores. Tergite VIII with sclerotization of diagnostic importance. Antrum rather small, with or without sclerotization. Ductus bursae long and narrow. Corpus bursae elongated, with or without signum.

DISTRIBUTION.– Holarctic, with a few rather locally distributed species.

BIOLOGY.– The larva lives under the bark of *Tilia* L. (Tiliaceae) and *Salix* L. (Salicaceae).

REMARKS.– The International Commission on Zoological Nomenclature designated *Glyphipteryx* Curtis, 1827 as an unjustified emendation of *Glyphipterix* Hübner, [1825] and placed *Chrysoclista* Stainton, 1854 on the Official list of generic names in zoology (ICZN, 1986).

48. *Chrysoclista linneella* (Clerck, 1759)

Phalaena linneella Clerck, 1759: pl. 12, fig. 8.
Elachista gemmatella Costa, [1836]: [278].
Elachista linnaeella Zeller, 1839: 212; unjustified emendation of *Phalaena linneella* Clerck, 1759.
Oecophora schaefferella sensu Duponchel, in Godart, [1839]: 482, pl. 306, fig. 4, (*nec* Linnaeus, 1758); misidentification.

Lita linneella var. *obscuri-linneella* Bruand, 1859: 662.

DIAGNOSIS.– Adult. Wingspan 10-13mm. Head dark brown; antennae shining dark greyish brown, about 10 apical segments white. Thorax and tegulae shining greyish brown. Forewings bright orange, broadly edged shining blackish brown, narrowest on costa; wing pattern of three rounded tubercular silvery spots edged blackish brown, first on dorsum at 1/3 with tuft of raised blackish brown scales, second on costa before 1/2, third on dorsum at 2/3; two small silvery streaks, one in fold near base, another on costa opposite outer dorsal spot; cilia dark greyish brown. Hindwings brownish grey with shining bronze reflection.

Variation. In some specimens, the orange ground colour of the forewings is partly or even completely replaced by dark fuscous. In female, the silvery spots and streaks on forewings are larger.

MALE GENITALIA.– Gnathos arms very short, rounded and strongly dentate. Tegumen longer than wide. Valvae truncate and slightly widened distally. Anellus lobes strongly widening distally with a large semicircular subapical excavation at inner side, apex dentate. Aedeagus strongly curved, apex furcate. The truncated valvae and the shape of the anellus lobes are characteristic.

FEMALE GENITALIA.– Apophyses posteriores about two times as long as the length of apophyses anteriores. Tergite VIII with sclerotization laterally concave. Genital plate with triangular sclerotization. Antrum rather wide with rectangular, weakly sclerotized dilation. Ductus bursae narrow, gradually widening towards corpus bursae, less than one and a half times the length of corpus bursae. Corpus bursae oval, widening distally.

DISTRIBUTION.– Europe to southern Fennoscandia in the north; Asia Minor (?). Introduced to the eastern part of the U.S.A. and Canada, probably in the first half of the 20th century (Klots, 1942; Koster, 2002c).

BIOLOGY.– Host plants: *Tilia cordata* Mill. and *T. europaea* L. (Tiliaceae). The old records of *Fagus* L. (Fagaceae) and *Malus* Mill. (Rosaceae) (Sorhagen, 1885, 1886) need confirmation. Univoltine. The larva lives from August to May under the bark of a trunk where it makes galleries and rounded chambers. The presence of a larva can easily be observed by bright brown frass well visible in the crevices of the bark, especially in May. It probably prefers old trees and those weakened by annual cutting. Pupation takes place inside the larval gallery (Stainton, 1859a). Lhomme ([1948-1949]), however, mentioned that adults were also obtained from dried leaves, indicating that the larva should leave its gallery prior to pupation. Adults fly from the end of May till August and can be found resting or running on the trunks of the food plant. They are extremely rare in the wild, but rather often very abundant in cities with lime-tree avenues.

49. *Chrysoclista lathamella* (Fletcher, 1936)

Glyphipteryx lathamella Fletcher, 1936: 504; replacement name for *Tinea bimaculella* Haworth, 1828.

Tinea bimaculella Haworth, 1828: 575; junior primary homonym of *Tinea bimaculella* Thunberg, 1794.

Chrysoclista razowskii Riedl, 1965: 456.

DIAGNOSIS.– Adult. Wingspan 11-13 mm. Differs from *linneella*, *splendida* and *abchasica* by the dark costal and dorsal edgings of the forewings; these edgings are broadly connected before the middle, thus dividing the orange ground colour into two spots; basal spot smaller and completely situated below fold.

Similar species. For separation from *zagulajevi*, q. v.

MALE GENITALIA.– Resembles *splendida*, but gnathos arms more straight and slightly widening distally, with many more thorns. Valvae more strongly narrowing apically. Anellus lobes not as slender and about half the length of valvae. Aedeagus only slightly bent.

FEMALE GENITALIA.– Similar to *splendida*, but sclerotization of tergite VIII reduced to two curved narrow lateral bands, antrum narrowly funnel-shaped and ductus bursae almost twice as long as corpus bursae.

DISTRIBUTION.– North and Central Europe (very locally).

BIOLOGY.– Host plants: *Salix alba* L., *S. caprea* L., probably also *S. fragilis* L. and *S. cinerea* L. (Salicaceae). Univoltine. The larval mode of life is quite similar to that of *linneella*. Adults have been collected from June to August. Very occasionally they can be plentiful on the trunks of old willow trees growing close to the waterside.

REMARKS.– For many years two different species were united under the name *lathamella* (=*bimaculella*). A careful investigation of the type specimen of *C. lathamella* in BMNH led to the conclusion that it belongs to the smaller and more widely distributed species, subsequently described as *C. razowskii* (Riedl, 1965), and not to the larger species that occurs in Central Europe only. The latter was finally described as *C. splendida* (Karsholt, 1997).

50. *Chrysoclista splendida* Karsholt, 1997

Chrysoclista splendida Karsholt, 1997: 30.
Chrysoclista bimaculella auctt., (*nec* Haworth, 1828); misidentification.
Chrysoclista lathamella auctt., (*nec* Fletcher, 1936); misidentification.

DIAGNOSIS.– Adult. Wingspan 14-17 mm. The species resembles *linneella*, but differs by the larger wingspan and by the forewings with the blackish brown costal streak distinctly widened from base to silvery costal spot, the small silvery spot near base, and also by the silvery spots on dorsum, both having a tuft of raised blackish brown scales.

MALE GENITALIA.– Gnathos arms long, bent, wide at base and strongly tapering distally, apex with four large thorns. Tegumen short, its width larger than its length. Valvae broadest at base, gradually narrowing to a rounded apex. Anellus lobes long and slender, about 2/3 the length of valva. Aedeagus bent in an angle of almost 90°. Differs from *lathamella* by the pointed gnathos arms, more slender anellus lobes, and by the strongly bent aedeagus.

FEMALE GENITALIA.– Apophyses posteriores about one and a half times as long as apophyses anteriores. Sclerotization of tergite VIII laterally convex. Antrum simple, funnel-shaped. Ductus bursae narrowed and bent before middle. Corpus bursae elongated and slightly narrowed medially.

DISTRIBUTION.– Central Europe.

BIOLOGY.– Host plants: *Salix* L. (Salicaceae). Several specimens were accidentally reared by Kasy (1954) from branches and trunks of *Salix alba* L. and *S. purpurea* L. affected by the gall midge *Rhabdophaga saliciperda* (Duf.). Rebel (1925) found plenty of specimens on the trunks of old *Salix alba* L. and *S. fragilis* L. Adults have been found from late May to June on the trunks and stumps of old willow trees usually growing in the vicinity of water. However, several specimens in the collection of the BMNH in London, according to their labels, were collected in Vienna in September and October (Karsholt, pers. comm.). This, one of the most beautiful European moths, was not seen for more than fifty years even in the known collecting sites, like parks and the neighbourhood of Budapest.

51. Chrysoclista zagulajevi Sinev, 1979

Chrysoclista zagulajevi Sinev, 1979a: 110.

DIAGNOSIS.– Adult. Wingspan 10 mm. Resembles *lathamella*, but differs by more reduced orange ground colour of the forewings, split into two spots of nearly equal size, and the presence of a narrow orange subcostal streak near base.

MALE GENITALIA.– Very similar to *abchasica*, but differs by the flatter apex of the valvae and the more slender anellus lobes.

FEMALE GENITALIA.– Unknown.

DISTRIBUTION.– Western Transcaucasia (Georgia: Adzharia).

BIOLOGY.– Unknown. Only known from the holotype, which was collected in mid-July.

52. Chrysoclista abchasica (Sinev, 1986)

Glyphipteryx abchasica Sinev, 1986: 31.

DIAGNOSIS.– Adult. Wingspan 9 mm. Very similar to *linneella*. It differs by the more irregular dark edging of the forewings and the narrow silvery costal streak between base and silver costal spot.

MALE GENITALIA.– Gnathos arms short and broad, strongly dentate. Valvae widening distally with rounded apex. Anellus lobes half the length of valvae, with pointed apex and subapical teeth. Aedeagus strongly bent, with row of cornuti. Can be distinguished from *zagulajevi* by the different shape of the valvae and the anellus lobes.

FEMALE GENITALIA.– Unknown.

DISTRIBUTION.– Western Transcaucasia (Georgia: Abchasia).

BIOLOGY.– Unknown. The holotype was collected at the beginning of July.

Family COSMOPTERIGIDAE

Cosmopterygidae Heinemann & Wocke, [1876]: 520. Type genus: *Cosmopterix* Hübner, [1825].
Diplosaridae Meyrick, 1915: 339. Type genus: *Diplosara* Meyrick, 1883.
Hyposmocomidae Hampson, 1918: 386. Type genus: *Hyposmocoma* Butler, 1881.
Scaeosophides Meyrick, 1922a: 48. Type genus: *Scaeosopha* Meyrick, 1914. [Subfamily].
Antequerinae Hodges, 1978: 16. Type genus: *Antequera* Clarke, 1941. [Subfamily].

General part

MORPHOLOGY.– Wingspan 7 to 21 mm. Head (Text-fig. 35) often strongly protruded at frons, covered with very densely appressed scales. Labial palpi long, thin, sharply curved upward; segment 2 smooth; segment 3 distinctly pointed and usually longer than segment 2, rarely of same length. Antennae shorter than forewings; scapus long, with or without bristled pecten. Eyes large, often with bright red pigment; ocelli absent. Forewings (Text-fig. 36) lanceolate or narrowly lanceolate to linear, with long-pointed apex, usually with shining metallic markings on surface; 3 median veins developed; M1 and sometimes also M2 stalked with R4+5. Hindwings lanceolate or narrowly lanceolate to linear, often with more or less reduced venation; length of cilia several times width of wing. Hindtibiae with middle pair of spurs usually beyond midpoint.

Abdominal tergites (Text-fig. 37) without specialized scales, but intersegmental membrane sometimes bears peculiar transversal rows of small spinules; longitudinal sclerotized struts (venulae) of tergite II very short or absent. Male genitalia (Text-fig. 38): uncus absent and substituted by pair of sclerotized and remarkably, sometimes sharply, asymmetrical arms of gnathos (brachia); tegumen well developed or slightly reduced; vinculum very narrow, true saccus absent, but sometimes with pseudosaccus on basis of sternite VIII; valvae symmetrical or slightly asymmetrical, with more or less pronounced pedicel and broadened hairy apical part (cucullus); anellus sclerotized and fused with walls of aedeagus; anellus lobes (valvellae) of very different size, usually asymmetrical and right one may be reduced; aedeagus rather massive, stout, rarely curved, fused with anellus and connected to valvae bases by caulis; cornuti absent. One of the most peculiar characters of family, the presence of unique ventro-lateral (pleural) lobes of segment VIII, sometimes of a very complicated shape, which cover the valvae bases or the whole valvae from the outside. Female genitalia (Text-fig. 39): ovipositor moderately long or rather short, rarely very long or even piercing; ostium always surrounded by sclerotized plate, sometimes strongly outwardly protruding (sterigma); ductus seminalis often merges directly into bursa copulatrix; ductus bursa usually very narrow and membranous; bursa copulatrix oval or rounded, often rather large, with paired, rarely unpaired signa of different form and size.

Larvae of various shape, mostly moderately long and slightly dorso-ventrally compressed, slow-moving, but other larvae are sometimes rather long, cylindrical and more mobile; without secondary setae, colourless or with longitudinal lines; they usually become deep pink before pupation. Pupae usually barrel-shaped, rather sclerotized, with labial palpi and forefemora concealed, and immobile due to the wing cases which almost reach to the end of abdomen; rarely the end of abdomen slightly movable, with curved crochets at extreme apex. Cocoon mostly comparatively short and broad, densely covered with detritus; exceptionally, pupation takes place inside the mine without pronounced cocoon.

Text-fig. 35. Head of *Cosmopterix zieglerella* (Hübner).

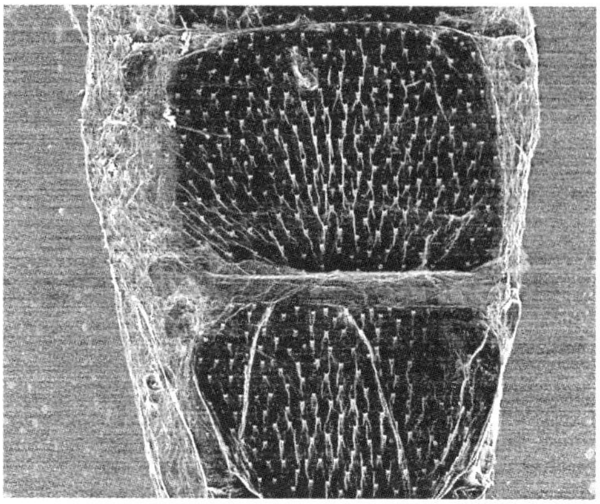

Text-fig. 37. Abdominal tergites of *Pancalia nodosella* (Bruand).

BIOLOGY.– The representatives of the family prefer more or less open landscapes (meadows, steppes), sparse forests, along river banks, clearings along roads, etc. Larvae usually mine leaves or sometimes cause galls on mainly herbaceous plants and shrubs, rarely feed on flowers, seeds or fruits. The host plants belong to many groups of both monocotyledonous and dicotyledonous plants, but especially to Rosaceae, Asteraceae, Lamiaceae and Poaceae. Larvae of many species in the tropics use decaying plant substances; some of them are inquilines in galls made by wasps or predators in colonies of scale insects. Adults rest in a peculiar position, with hind legs appressed to slightly or considerably (up to 70°) raised to posterior part of the body. Most species are active before sunset, when they often dance on the leaves (preferably of their host-plants) making very peculiar circular passes. As a rule, the moths are easily attracted to light, but some species with diurnal activity can be collected only on their host plant or on flowering grasses.

DISTRIBUTION.– The most abundant family of the narrow-winged moths, and especially richly represented in the tropics of both Old and New World, where thousands of species (mostly undescribed) are known (Sinev, 2002a). On some Pacific archipelagos (Hawaii, Marquesas, Rapa, Fiji) Cosmopterigidae are a dominant group of Microlepidoptera, with some genera counting hundreds of species (Clarke, 1972, 1986;

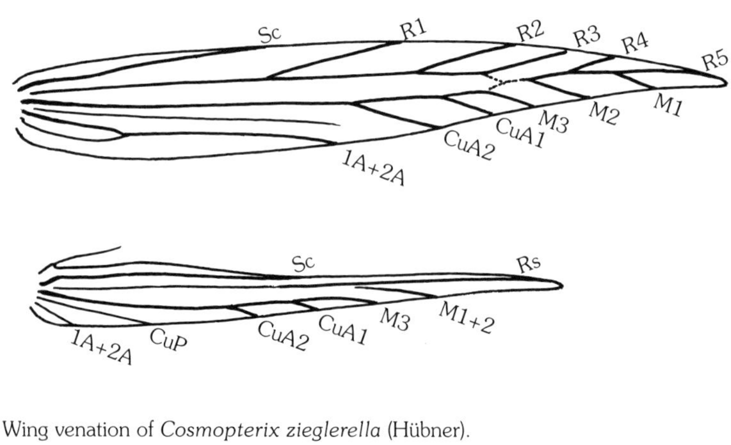

Text-fig. 36. Wing venation of *Cosmopterix zieglerella* (Hübner).

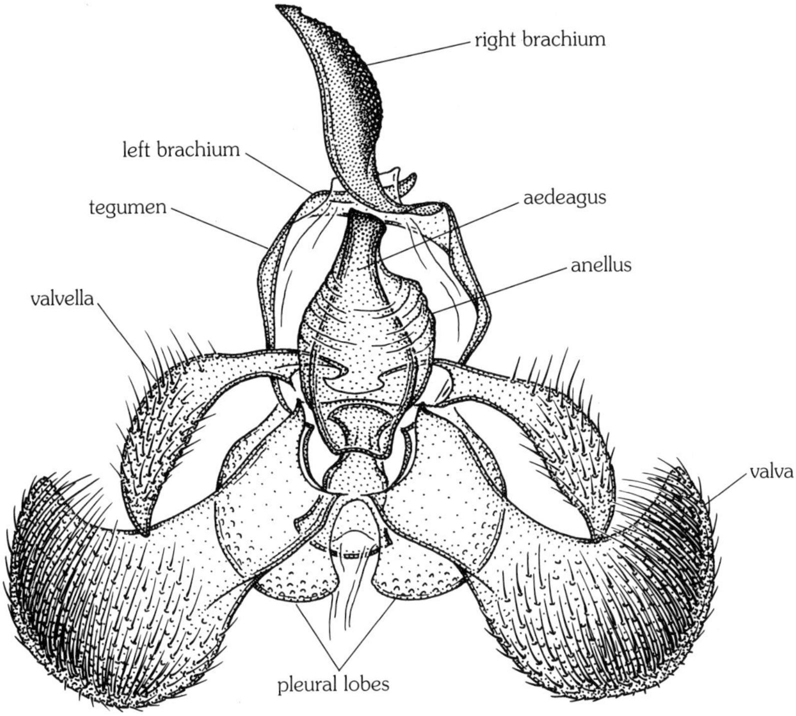

Text-fig. 38. Male genitalia of *Cosmopterix schmidiella* Frey.

Zimmerman, 1978). More than 200 species are recorded from the Palaearctic, with 19 genera and 85 species in the region of study.

SYSTEMATICS.– The Cosmopterigidae is a very specialized family of gelechioid moths. Its evolutionary success is evidently caused by the trophic relations with young and flourishing groups of angiosperm plants, especially herbaceous ones. The family is characterized by a large number of apomorphic traits, although many of them have appeared only at later phylogenetic stages, being autapomorphies of subfamilies or genus-groups. If the asymmetry of male genitalia, the uncus being replaced by paired sclerotized gnathos arms, the sclerotization of anellus and its fusing with aedeagus, the presence of ventrolateral lobes, etc., are synapomorphic for all representatives, the degree and even polarity of asymmetry, as well as most features of wing venation, are characteristic for some groups. The subfamilies Antequerinae and Scaeosophinae are rather archaic according to the relatively broad wings with complete venation and slight asymmetry of male genitalia. The latter subfamily (mainly represented in the tropics of the Old World) has a very peculiar autapomorphy, an aberrant hindwing venation with considerably shortened cell bearing a peculiar, more or less pronounced, transparent 'window'. The Cosmopteriginae is morphologically the most advanced subfamily, richly represented in nearly all regions of the world except the polar and subpolar areas. More than 95% of the known species belong to this subfamily.

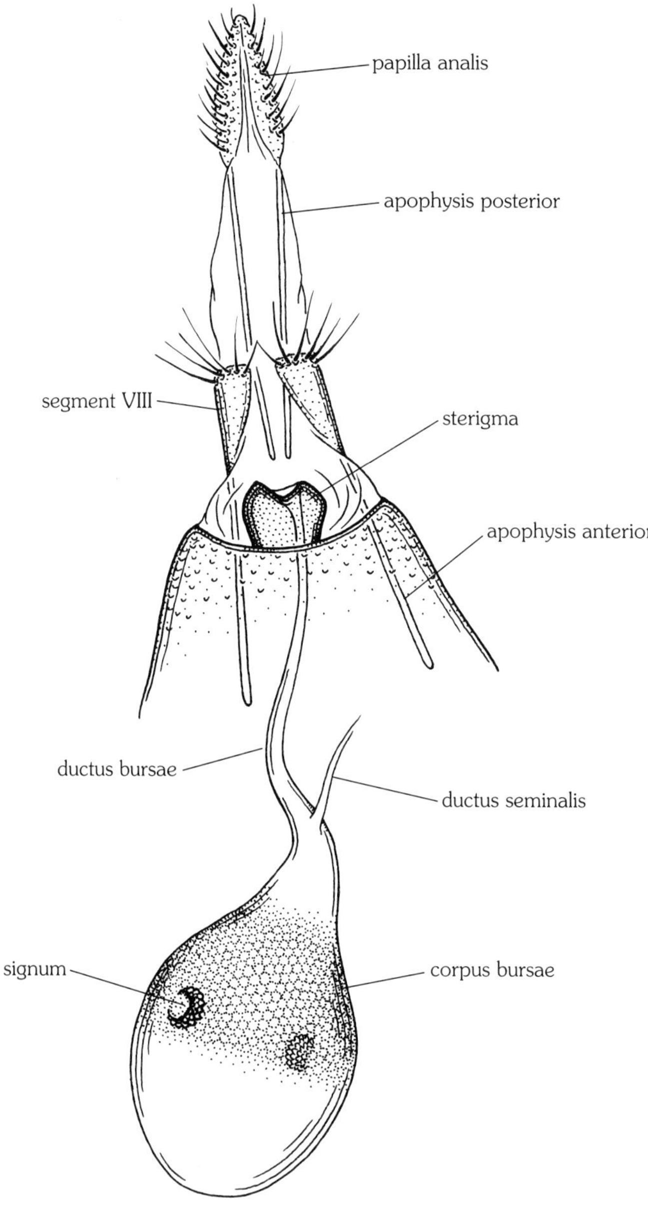

Text-fig. 39. Female genitalia of *Pancalia schwarzella* (Fabricius).

Systematic Treatment of the Genera and Species of Cosmopterigidae in Europe, North Africa and the Near East

Subfamily Antequerinae Hodges, 1978

Pancalia Stephens, 1829

Pancalia Stephens, 1829: 49.
Type species: *Phalaena leuwenhoekella* Linnaeus, 1761, by subsequent designation by Curtis, 1830.

DIAGNOSIS.– Head smoothly scaled, rounded. Antennae 3/4 to 4/5 length of forewings, flattened; in male uniformly coloured, in female with white apical portion and often thickened with blackish rough scales in middle. Labial palpi flattened laterally, porrect, segment 3 angled upwards and reaching above head. Forewings lanceolate, ground colour orange-brown with silvery or pale golden metallic markings, inner ones usually forming outward oblique fascia. Hindwings slightly narrower than forewings.

MALE GENITALIA.– Slightly asymmetrical. Brachia long and stout, of equal length and shape. Tegumen broad, well sclerotized. Valvae rather broad, often slightly narrowed at basal third; left valva often slightly larger than right one. Valvellae asymmetrical; left valvella always larger than right one. Aedeagus bent and pointed, without cornuti. The shape of the valvellae, valvae and the aedeagus are of diagnostic value.

FEMALE GENITALIA.– Apophyses posteriores one and a half time as long as apophyses anteriores. Sterigma more or less plane, of various shape. Ductus bursae long and narrow, widening towards corpus bursae. Corpus bursae oval or egg-shaped, with paired signa of two granular dimples.

DISTRIBUTION.– Mainly Palaearctic, especially richly represented in eastern Asia (Sinev, 1985a); a single species is known from the Oriental region.

BIOLOGY.– The larvae live in silky galleries on species of Violaceae. They feed on the outside, in first instars also as miners inside stem and rootstock.

REMARKS.– The authorship of Stephens was rejected by Gaedike (1967), who erroneously treated *Pancalia* Stephens, 1829 as nomen nudum and accepted the authorship of Curtis (1830).

53. ***Pancalia leuwenhoekella*** (Linnaeus, 1761)

Phalaena Tinea leuwenhoekella Linnaeus, 1761: 361.
Phalaena Tinea leuvenhoekella: Linnaeus, 1767: 897; unjustified emendation of *Phalaena Tinea leuwenhoekella* Linnaeus, 1761.
Phalaena Tinea leuwenhöckella [Denis & Schiffermüller], 1775: 141; unjustified emendation of *Phalaena Tinea leuwenhoekella* Linnaeus, 1761.
Oecophora schmidtella Treitschke, 1833: 167.
Pancalia leeuwenhoeckella auctt.; incorrect subsequent spelling of *Phalaena Tinea leuwenhoekella* Linnaeus, 1761.
Pancalia leuwenhoekella mandshuricella Sinev, 1985a: 817. Subspecies.

DIAGNOSIS.– Adult. Wingspan 10-12 mm. Head shining bronzy brown with purplish reflection; antennae shining bronzy brown, unthickened in female, with subapical area of about 8-10 segments dorsally white and with 5 apical segments bronzy brown in both sexes. Thorax and tegulae shining bronzy brown with purplish tint. Forewings shining orange-brown, irregularly edged blackish brown with purplish reflection, with three costal and four dorsal tubercular silvery or pale golden metallic spots: first costal spot at 1/5, inward oblique and reaching fold; second costal spot before 1/2; third costal spot at 3/4, very outward oblique, changing to white into the cilia; first dorsal spot before wing margin, opposite the first costal; second dorsal spot small, inwardly of the first costal; third dorsal spot on tornus, perpendicular to dorsum; fourth dorsal spot partly on termen, between third costal and apex. Hindwings shining brown with bronzy tint.

Similar species. Differs from other European species by the smaller wingspan, the white subapical area of antennae in male, unthickened antennae in female, and by the third dorsal spot situated perpendicular to dorsum.

MALE GENITALIA.– Valvae slender, bent upwards in middle, left valva slightly longer and stouter than right one. Valvellae slender, left lobe about one and a half times as long as right lobe. Aedeagus with right-angled bend, distal part tubular.

FEMALE GENITALIA.– Sterigma wide, shallowly bowl-shaped, with posterior edge heavily sclerotized. Ductus bursae long and gradually widening towards corpus bursae. Corpus bursae oval, slightly pectinate; signa large.

DISTRIBUTION.– Europe, except for northern Fennoscandia; Asia Minor, the Caucasus, south-western Siberia (Sinev, 1984b). In the Russian Far East represented by the allopatric subspecies *mandshuricella* (Sinev, 1985a).

BIOLOGY.– Host plants: *Viola canina* L., *V. hirta* L., *V. tricolor* (L.) Wittr. The larva lives from June to August. In first instars it mines the leaf petioles, ejecting frass through a small hole or depositing it in the centre of the channel. Later the larva makes an opaque gallery of silk and detritus between the roots of the food plant and eats a large patch of bark from its underground stem (Fletcher, 1893). Pupation in August in a firm cocoon of silk and detritus on the ground. The moths fly in the sunshine around the food plants from late April till June and sometimes also in a second generation in July-August. Prefers more or less open and dry landscapes, especially dry meadows and bushy slopes.

54. *Pancalia schwarzella* (Fabricius, 1798)

Tinea schwarzella Fabricius, 1798: 493.
Pancalia latreillella Curtis, 1830: 304.

DIAGNOSIS.– Adult. Wingspan 11-16 mm. Antennae completely shining bronzy brown, without white segments in male; in female antennae gradually thickening to middle, where they are twice as broad as at base, with white subapical section of about 5 segments followed by bronzy brown apical part of about 15 segments. Forewings ochreous-brown to orange brown, with first costal spot connected to first dorsal spot, forming an inward oblique, pale golden metallic fascia, not reaching dorsum; third dorsal spot outward oblique.

Similar species. For separation from the very closely related *nodosella* and *baldizzonella*, q. v.

MALE GENITALIA.– Left valva one and a half times as wide as right valva and slightly longer, with straight ventral margin. Left valvella wide, more than half as long and wide as left valva, right valvella shorter and narrower. Aedeagus curved, distally gradually narrowing to sharp apex. The stout left valva in combination with the broad left valvella is characteristic.

FEMALE GENITALIA.– Sterigma large and square, with rounded corners. Ductus bursae narrow in posterior part, more than twice as wide close to corpus bursae. Corpus bursae elongate with two well-defined signa. The large sterigma in combination with the abruptly widened ductus bursae is characteristic.

DISTRIBUTION.– Europe, except for the south-east, in the southern part mainly in the mountainous regions up to 2600 m above sea level; Central Asia (at high elevations in mountains only), Siberia, Transbaikalia, Kamchatka (Sinev, 1984a).

BIOLOGY.– Host plants: *Viola canina* L. and *V. hirta* L. The immature stages are insufficiently known, but should be similar to *leuwenhoekella*. *P. schwarzella* (as *latreillella*) was considered to be a form of *leuwenhoekella* by several authors in the past, therefore the descriptions in the old literature are not reliable and could refer to one or the other. Flight period from the end of April to mid-June, at higher altitudes to the end of July. Moths are active in sunshine both in the morning and afternoon and visit yellow flowers such as *Taraxacum* spp., *Hieracium* spp. and the yellow centres of *Bellis perennis* L. (Asteraceae), *Lotus corniculatus* L. (Fabaceae), *Fragaria vesca* L. (Rosaceae), etc. (Fletcher, 1935).

55. *Pancalia baldizzonella* Riedl, 1994

Pancalia baldizzonella Riedl, 1994a: 87.

DIAGNOSIS.– Adult. Wingspan 15 mm. Forewings with brownish yellow ground colour, the only character by which this species can be distinguished from *schwarzella* externally.

MALE GENITALIA.– No reliable differences from *schwarzella*.

FEMALE GENITALIA.– Very similar to *schwarzella*, differing by the smaller and more regularly square sterigma.

DISTRIBUTION.– Only known from the type locality: Mt. Pollino, Lucania, Italy.

BIOLOGY.– Unknown. The adults have been collected in July.

REMARKS.– The species is most likely only a form of *schwarzella*. The moths of *Pancalia* tend to fade from the bright orange ground colour to a more yellowish colour during their lifetime, which may be the explanation of the paler colour of both type specimens. Also the late collecting date at rather high altitudes points in this direction. No clear differences between *baldizzonella* and *schwarzella* can be observed in genitalia structures. On the genital slide of the male holotype, the left valvella appears long and narrow, but it is actually folded lengthwise by embedding. We prefer to keep *baldizzonella* as a separate species until additional material from the type locality becomes available.

56. *Pancalia nodosella* (Bruand, [1851])

Oecophora nodosella Bruand, [1851]: 79.

Oecophora nodosella Mann, 1854: 586; junior primary homonym and synonym of *Oecophora nodosella* Bruand, [1851].

DIAGNOSIS.– Adult. Wingspan 12-14 mm. Differs from *schwarzella* and *baldizzonella* by the following characters: forewings with orange-brown area more pronounced; inward oblique fascia, formed by first costal and first dorsal spot, reaches dorsum or nearly so; second subdorsal spot reduced to few pale golden metallic scales or absent. Female with antennae more abruptly thickened, being up to three times the basal thickness, and scales in this part more erect.

MALE GENITALIA.– Valvae of almost equal length, left one slightly broader and with deeply concave ventral margin, whereas ventral margin of right valva is almost straight. Valvellae short, left one about 1/3 of the length of left valva and right one about 1/6 of the length of right valva. Aedeagus with sharply pointed apex bearing series of spicules. The short valvellae, the wide ventral excavation of the left valva and the spiculate apex of the aedeagus are characteristic.

FEMALE GENITALIA.– Sterigma weekly sclerotized, ostium posteriorly edged by slightly bent, heavily sclerotized rim. Ductus bursae gradually widening towards corpus bursae with full turn in part between ductus seminalis and corpus bursae. Corpus bursae egg-shaped with two very small and indistinct signa. The shape of sterigma and coiled ductus bursae in combination with the small signa are characteristic.

DISTRIBUTION.– Southern and Central Europe, north to the sandy coastal dunes of Holland and Latvia; the Caucasus, Central Asia to Kirghizia in the east.

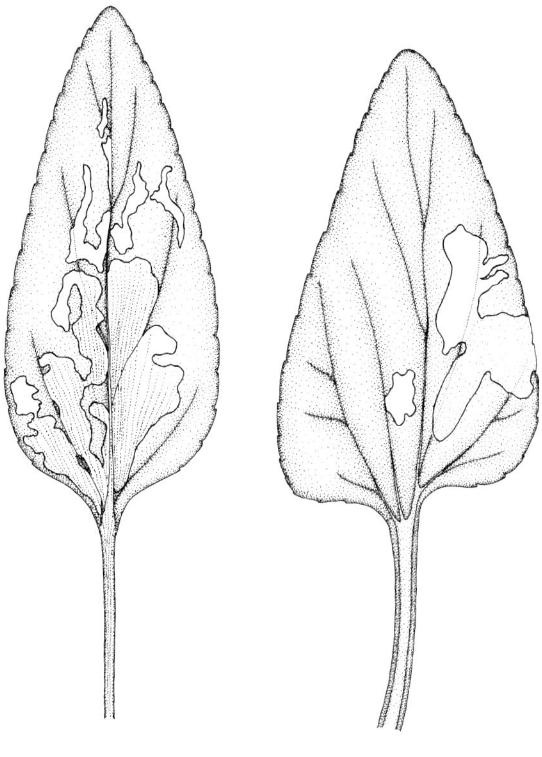

Text-fig. 40. *Pancalia nodosella* (Bruand), larval mines in leaves of *Viola curtisii* (Forst.).

BIOLOGY.– Host plant: *Viola curtisii* (Forst.). The larva feeds in June and July, first mining the petiole of a leaf and often making a blotch mine inside the leaf; later it skeletonizes the upper side of the leaf. Sometimes the larva also mines in the flower stalk and causes a long, whitish and rather conspicuous gallery. In last instars the larva bores into the stem, ejecting the frass through the entrance hole; this ochreous or light brown frass in a leaf-axil indicates the presence of a larva. At this stage the epidermis of the stem is eaten, causing the plant to shrivel. The feeding area is surrounded by a light web, in which the larva moves quickly forwards and backwards in its silken galleries. Later these galleries reach into the ground, where the bark of the roots is eaten. Pupation takes place at the end of summer between the roots of the food plant in a tough silken cocoon covered with sand. Adults have been found from the second half of April to early June flying in sunshine around the food plants (Koster, pers. obs.). See text-fig. 40.

Euclemensia Grote, 1878

Euclemensia Grote, 1878: 69.

Type species: *Hamadryas bassettella* Clemens, 1864, by monotypy (of *Hamadryas* Clemens, 1864).

Hamadryas Clemens, 1864: 422; homonym of *Hamadryas* Hübner, [1806].

DIAGNOSIS.– Head smoothly scaled; antennae 3/5 to 3/4 length of forewings; labial palpi recurved, extending to or beyond vertex, smoothly scaled. Forewings lanceolate, with tubercular patches of scales. Hindwings as broad as forewings, narrowing towards apex from one-half. Abdomen tapering posteriorly.

MALE GENITALIA.– Very slightly asymmetrical. Brachia long and broad, tapering distally. Valvae long and narrow, slightly bent. Valvellae long and narrow, tapering distally. Vinculum broad, without saccus. Aedeagus long, straight or bent and with one or two large cornuti.

FEMALE GENITALIA.– Apophyses posteriores longer than apophyses anteriores. Ostium wide, funnel-shaped. Ductus bursae short, with heavily sclerotized antrum. Corpus bursae divided into two parts of unequal size, without signa.

DISTRIBUTION.– Holarctic and Neotropic.

BIOLOGY.– The larvae of the American species are parasitoids of armoured scales and live inside the females of *Kermes* Latr., Kermesidae, Homoptera (Hodges, 1978).

57. *Euclemensia woodiella* (Curtis, 1830)

Pancalia woodiella Curtis, 1830: 304.

DIAGNOSIS.– Adult. Wingspan 17 mm. Head shining bronze-brown with greenish tinge; antennae shining dark brown with purplish gloss, subapical area of about 9 segments white posteriorly. Thorax and tegulae shining bronze-brown with greenish tinge. Forewings bright orange, at base dark brown on costa and dorsum, at costal side with club-shaped spot; dark brown costal streak from 1/2 to 4/5, widest towards base and enclosing small orange subcostal spot; dorsal side of wing from 1/3 towards and around apex dark brown, divided by orange fascia beyond the middle; yellow-orange costal streak at 4/5; four tubercular spots with strong bluish and purplish reflection, first as distal end of the club-shaped spot, second and third on either side of the fold at basal edge of dark brown dorsal section, fourth and smallest below orange subcostal spot; narrow dark streak with purplish and greenish reflection along fold, distally of both tubercular spots, extending around termen; cilia greyish brown. Hindwings shining greyish brown with orange gloss.

Male genitalia.– As in genus description. Aedeagus bent, with two cornuti.

Female genitalia.– Unknown.

Distribution.– England: Kersall Moor near Manchester (type locality). Not found since 1829 and evidently extinct.

Biology.– Unknown. The larva should be associated with *Kermes*, as in both closely related species from North America. Adults have been collected in mid-June.

Remarks.– If the larva of *woodiella* should really feed on *Kermes*, it is very unlikely that this species originates from Europe. The females of the few European species of *Kermes*, in contradiction to the large American ones, do not exceed 5 mm in diameter and cannot supply enough food for a larva. The species was probably imported with young trees from North America in the 19th century.

Phosphaticola Viette, 1951

Phosphaticola Viette, 1951: 58.
Type species: *Phosphaticola gemmatella* Viette, 1951, by original designation.

Diagnosis.– Head large and smoothly scaled; antennae 3/4 length of forewings; labial palpi long and recurved. Forewings rather broad, slightly tapering towards apex, with large tubercular patches of metallic scales. Hindwings as broad as forewings, with blunt apex.

Male genitalia.– Brachia long and asymmetrical, longer than tegumen; left brachium longer than right one and with apical protrusion. Valvae rather short, narrowly spatulate. Valvellae mace-shaped and of equal size. Aedeagus large, bent in middle. Juxta sparsely covered with small spines.

Female genitalia.– Apophyses posteriores about one and a half times as long as apophyses anteriores. Anterior margin of tergite VIII with V–shaped excavation. Sterigma triangular and strongly sclerotized. Ductus bursae narrow, half the length of the corpus bursae. Corpus bursae very large, pear-shaped with two elongate signa.

Distribution.– Northern Africa.

Biology.– Unknown.

58. *Phosphaticola gemmatella* Viette, 1951

Phosphaticola gemmatella Viette, 1951: 59.

Diagnosis.– Adult. Wingspan 11-12 mm. Head ochreous grey, laterally white; antennae greyish brown, basal half annulated white. Thorax and tegulae ochreous grey, paler posteriorly. Forewings yellowish grey, more or less intermixed with greyish scales, especially at dorsal part; basal 1/4 ochreous grey, followed by indistinct white fascia; large greyish brown spot on costa at 3/4, reaching tornus and narrowing towards it, followed by broad white fascia; apical part yellowish mixed with ochreous, on costa and dorsum bordered with greyish brown; four shining bronze-brown tubercular spots with golden and blue reflections and blackish mounting, first below fold at 1/3, second subcostal, slightly more towards apex, third in middle at 2/3, fourth and smallest above tornus; cilia whitish, with broad greyish brown line just before termen. Hindwings shining orange-grey, with costal margin shining white basally; cilia ochreous brown. Abdomen dorsally pale yellow, segment I and basal half of segment II greyish.

MALE GENITALIA.– See genus description.

FEMALE GENITALIA.– See genus description.

DISTRIBUTION.– Tunisia.

BIOLOGY.– Unknown. The moths have been collected at the beginning of February.

Alloclita Staudinger, 1859

Alloclita Staudinger, 1859: 247.
Type species: *Alloclita recisella* Staudinger, 1859, by monotypy.
Proceleustis Meyrick, 1914c: 267.
Type species: *Proceleustis paraphracta* Meyrick, 1914, by monotypy.
Dhahrania Amsel, 1958: 78.
Type species: *Dhahrania fasciella* Amsel, 1958, by original designation.

DIAGNOSIS.– Medium sized moths. Head smoothly scaled; antennae 3/4 length of forewings; labial palpi long and recurved. Forewings rather broad, lanceolate, with well-defined darkened basal area and one or more, often very indistinct fasciae. Hindwings slightly narrower than forewings, lanceolate and acutely pointed.

MALE GENITALIA.– Brachia more or less asymmetrical, as long as or longer than tegumen; left brachium longer than right one. Tegumen broad, square or slightly narrowed distally. Valvae laciniate, with broadened base and rounded apex. Valvellae rather long and narrow, digitate. Aedeagus long and bent, distally tapering. Anellus forms a long tube around aedeagus.

FEMALE GENITALIA.– Apophyses posteriores slightly longer than apophyses anteriores. Ostium edged with sclerotized band and partly hidden by membranous fold between sternites VI and VII. Ductus bursae membranous, widening towards corpus bursae. Corpus bursae oval with or without signa.

DISTRIBUTION.– From Spain eastwards throughout the Mediterranean area to India, Mongolia and south-east Asia; Africa (Sinev, 1993b).

BIOLOGY.– Probably similar to *Pancalia*. Only known from one species, the larvae of which live in sandy galleries attached to the crown of the root of *Pulicaria* Gaertn. (Asteraceae), feeding under the bark.

59. *Alloclita recisella* Staudinger, 1859

Alloclita recisella Staudinger, 1859: 247.
Borkhausenia albocinctella Chrétien, 1915: 344.
Alloclita orthoclina Meyrick, 1922b: 547.
Tinea inconspicuella Turati, 1929: 127.
Mompha gelechiformis Turati, 1930: 82.

DIAGNOSIS.– Adult. Wingspan 9-15 mm. Head shining ochreous white; antennae dark brown, annulated greyish white. Thorax and tegulae ochreous white, strongly mottled with grey, tegulae ochreous white posteriorly. Forewings ochreous white suffused greyish brown, basal 1/6 greyish brown, dark grey towards inward oblique outer edge; in middle of costa wide greyish brown fascia, often narrowing towards dorsum or interrupted in middle, with more or less distinct dark grey band from costa at inner edge, bending in middle towards apex; apical area greyish brown, separated from middle fascia by narrow ochreous white band; cilia ochreous grey. Hindwings shining grey, coarsely scaled. Abdomen dark brown, segments greyish brown banded posteriorly, anal tuft greyish brown.

Variation. The pattern of forewings can vary considerably from very light with distinct fasciae to almost completely greyish brown with fascia just indicated by dark brown edges.

MALE GENITALIA.– Left brachium slightly longer than tegumen, widened and apically bent; right brachium with dorsal protrusion before 1/2, apex flattened. Tegumen square. Valvae spatulate. Valvellae slender and bent. Anellar tube broad at base, gradually tapering to pointed apex.

FEMALE GENITALIA.– Apophyses of almost equal length. Sterigma a weakly sclerotized oval plate. Ostium with semicircular band, strongly sclerotized laterally. Ductus bursae about half as long as corpus bursae. Corpus bursae with two crescent-shaped and thorny signa.

DISTRIBUTION.– Mediterranean, from Spain and Portugal to Cyprus and Israel.

BIOLOGY.– The immature stages are unknown. Adults have been collected in May and again from July till September.

60. Alloclita haifensis Rebel, 1911

Alloclita haifensis Rebel, 1911: 153.
Mompha litorella Amsel, 1935a: 303.

DIAGNOSIS.– Adult. Wingspan 13-15 mm. Head shining yellowish white, more ochreous white posteriorly, laterally dark brown; antennae yellowish white with very indistinct pale ochreous annulations. Thorax yellowish white, more ochreous anteriorly, postero-laterally with two blackish brown spots; tegulae yellowish white, anteriorly irrorated dark brown. Forewings yellowish white intermixed with ochreous and pale brown scales; basal quarter dark brown, paler towards base, with outer edge somewhat prolonged towards apex below fold; broad irregular dark greyish brown median fascia, narrowed in middle on both sides; apical third dark greyish brown, inwardly prolonged to costa and dorsum and connected with middle fascia; cilia pale grey, strongly intermixed with scattered dark brown scales. Hindwings pale brownish grey, coarsely scaled; cilia pale grey, basally pale ochreous. Abdomen brownish (according to the original description).

MALE GENITALIA.– Resembles *recisella*, but differs by the longer left brachium, the more strongly bent valvellae, and the right-angled bend of the aedeagus.

FEMALE GENITALIA.– Unknown.

DISTRIBUTION.– Near and Middle East (Israel, Palestine).

BIOLOGY.– Unknown. Adults have been collected in dune areas in May.

61. Alloclita delozona Meyrick, 1919

Alloclita delozona Meyrick, 1919: 238.

DIAGNOSIS.– Adult. Wingspan 11-13 mm. Head shining pale brown; antennae ochreous white with dark brown annulations. Thorax and tegulae pale brown with dark brown-tipped scales. Forewings yellowish white, basal one-fifth dark brown, mottled with reddish brown; broad shining dark grey median fascia, narrowed in middle, with dark brown and reddish brown scales on margins; apical part shining dark grey sometimes with white spot at extreme apex; cilia pale grey, more ochreous at base and intermixed with scattered dark brown scales, with distinct dark line. Hindwings shining white, ochreous towards apex, end of veins marked with ochreous grey scales; cilia ochreous grey. Females are usually darker than males.

MALE GENITALIA.– Left brachium somewhat longer than tegumen, slightly bent apically. Valvae rather narrow. Valvellae long, slender and club-shaped.

FEMALE GENITALIA.– Apophyses posteriores one quarter longer than apophyses anteriores. Ostium with large X–shaped sclerotization. Ductus bursae very long, about eight times the length of corpus bursae and with couple of turns. Corpus bursae covered with indistinct pectinations and with two crescent-shaped and thorny signa.

DISTRIBUTION.– Near and Middle East from Egypt to India.

BIOLOGY.– Unknown. In Saudi Arabia specimens have been collected in February and March.

62. Alloclita brachygrapta Meyrick, 1925

Alloclita brachygrapta Meyrick, 1925: 213.

DIAGNOSIS.– Adult. Wingspan 11-12 mm. Head shining yellowish white; antennae dark brown, annulated greyish white. Thorax and tegulae pale ochreous; tegulae yellowish white apically. Forewings ochreous white, with ochreous brown spot on costa at base; ochreous brown arcuate fascia from costa at 1/5 to fold; basal part of forewings pale ochreous brown, darker towards fascia; ochreous brown spot on costa at 2/5 connected to ochreous brown longitudinal streak in middle; apical part suffused with ochreous and ochreous brown scales into cilia, dorsal half from longitudinal streak towards apex with greyish shade; cilia yellowish white. Hindwings shining yellowish white.

MALE GENITALIA.– Can be distinguished from *recisella*, *haifensis* and *deprinsi* by the almost straight left brachium and by the short and broadly based valvae.

FEMALE GENITALIA.– Not available.

DISTRIBUTION.– Egypt: Sinai, Romani (type series only).

BIOLOGY.– Unknown. Adults have been collected in September.

REMARKS.– The description is based on two specimens in the BMNH. Both specimens are in a rather bad condition. Meyrick (1925), in his original description, mentioned the existence of four specimens including at least one female. The deposition of these two additional specimens is unknown to us.

63. Alloclita deprinsi sp. n.

DIAGNOSIS.– Adult. Wingspan 11-13 mm. Head with frons white, vertex and neck-tuft shining ochreous white, dark brown laterally, collar ochreous white dorsally, dark brown laterally. Antennae greyish brown with darker annulations, scape dark brown with white apical ring. Labial palpi with segment 1 roughly scaled, dark brown on outside, inner side white, segment 2 white with dark brown basal and subapical rings, widest ventrally, both rings sometimes interrupted on inner side, segment 3 ochreous white with blackish brown basal and subapical rings and some ochreous irroration. Thorax ochreous white, strongly irrorated with reddish brown and dark grey scales; tegulae blackish brown anteriorly, ochreous white posteriorly. Forelegs and midlegs blackish brown, hindlegs with femur white, tibia and tarsi grey; tibiae of all legs with white subbasal, medial and apical rings, tarsi of all legs with white basal and apical rings on segment 1, other segments with white apical rings, spurs ochreous white. Forewings shining grey with inward oblique white fascia at 1/5; white costal spot at 2/5 and larger similar spot at 3/4; white dorsal spot opposite first costal, sometimes united with white fascia on dorsum inwardly; second white dorsal spot at tornus and white apical spot into cilia; blackish brown spot at base above fold; dorsum at base and inner side of fascia edged with reddish brown and blackish brown scales; fascia outwardly edged with ochreous, reddish brown and blackish scales; outer sides of first costal and dorsal spot and inner sides of second costal and dorsal spot edged likewise; two medial blackish brown spots before 1/2 and at 2/3, both edged ochreous white and reddish brown; cilia grey with ochreous tinge, strongly intermixed with dark grey scales around apex, dorsally with grey terminal cilia line. Hindwings shining grey; cilia ochreous grey. Abdomen yellow ochreous, ventrally lighter, with anal tuft white.

MALE GENITALIA.– Left brachium distinctly longer than tegumen, slender, bent at right angle distally, with apex pointed; right brachium narrow at base, gradually widening to the middle, with distal part gradually narrowing to the rounded apex, like a banana. Tegumen square, widening distally. Valvellae slender and straight. Valvae spatulate, with wide base and rounded apex. Aedeagus slender and slightly tapering distally. Anellar tube long and slender, distal half tapering, apex pointed, dorsally with short setae.

FEMALE GENITALIA.– Unknown.

DISTRIBUTION.– Asia Minor.

BIOLOGY.– Unknown. All specimens have been collected in May and August.

MATERIAL.– Holotype. ♂, 'Turkey, St. 2388, Icel 10 m, 8 km SSE Kazanli, 10.viii.1997, leg. W. de Prins' 'Genitalia on slide JCK 4269' (ZMAN).

Paratypes. 14 ♂, same data as holotype (ZMAN, RMNH, ZIN); 4 ♂, '20.v.[19]69, Asia min., Dünen 25 km SW v. Mersin' 'leg. Arenberger' (LSNK).

ETYMOLOGY.– Dedicated to Mr. Willy de Prins, Antwerp, Belgium, one of the collectors.

64. Alloclita francoeuriae Walsingham, 1905

Alloclita francoeuriae Walsingham, 1905: 126.

DIAGNOSIS.– Adult. Wingspan 15 mm. Head shining white on frons, shining yellowish white to pale ochreous posteriorly; antennae shining white with pale ochreous brown annulations. Thorax and tegulae ochreous brown, yellowish white posteriorly. Forewings yellowish white irrorated with ochreous and some pale brown scales; basal 1/5 dark brown, ochreous towards base, outer edge outwardly bent in middle; dark brown, outwardly bent fascia at 1/3, broadly interrupted with yellow ochreous subcostally; indistinct outward oblique pale brown fascia at 2/3, interrupted with yellow ochreous in middle, with dark brown spot in middle on inner edge; apical area yellow ochreous dorsally, irrorated dark brown, with apex dark brown on costal side and prolonged into termen by scattered dark brown scales; cilia ochreous grey, mixed white, especially at tornus. Hindwings shining yellowish white; cilia pale ochreous. Abdomen shining pale yellow, first segment shining white, anterior half of second segment pale yellowish grey, anal tuft shining yellowish white.

Variation. In dark forms, the yellowish white parts of forewings are almost completely covered with brown scales, and only yellow ochreous spots indicate the fasciae.

MALE GENITALIA.– Brachia distinctly longer than tegumen, slender and with somewhat hooked apices; left brachium just slightly longer than right one. Tegumen short, narrowing distally. Valvae short, with very wide base and almost square cucullus bearing long setae. Valvellae short, asymmetrical: left valvella rectangular, right one spatulate. Aedeagus slender, apex widened and rounded. Anellar tube rather short, stout, with very short setae on its tapered distal part.

FEMALE GENITALIA.– Ostium with strongly sclerotized ventral band. Ductus bursae narrow and smooth in posterior half, anterior part widening towards corpus bursae, strongly wrinkled and with several loops. Corpus bursae oval; signa as spiculate rims with irregular sheet-shaped sclerotization.

DISTRIBUTION.– North Africa (Algeria, Tunisia, Morocco).

BIOLOGY.– Host plant: *Pulicaria (=Francoeuria) crispa* Forssk. In April the larva burrows under the woolly bark of a stem; after eating out the inside of the leaves it makes sand galleries attached to the crown of the root (Walsingham, 1905). Pupation in a dense white cocoon, covered with sand and frass particles. Adults have been collected at light from the end of February to early April; one female has been bred in July.

REMARKS.– The species has forewings longer and narrower than in any other species of the genus; the male genitalia are somewhat similar to those of the genus *Pancalia*. It should probably be regarded as representing a separate genus or subgenus. Leraut (1980) erroneously mentioned the species for France as *Aristotelia francoeuriae* in Gelechiidae; in the second edition of the check-list (Leraut, 1997) it has been excluded.

Gibeauxiella gen. nov.

Type species: *Hodgesiella bellaqueifontis* Gibeaux, 1986, by present designation.

DIAGNOSIS.– Head smooth-scaled, neck tuft appressed to head. Antennae about 3/4 length of forewings, flagellum simple. Labial palpi slender and porrect, with segment 3 angled upwards and segment 2 4/5 of its length. Forewings lanceolate, with pointed apex and patches of slightly raised scales; venation complete, veins R4 and R5 stalked for less than half of their length. Hindwings slightly narrower than forewings, with acutely pointed apex and length of dorsal cilia about one and a half times width of wing; veins M2 and M3 fused, Cu1 and Cu2 shortly stalked. Abdominal tergite VIII sclerotized, with short lateral protrusions and with oval lobe posteriorly; plate of abdominal sternite VIII curved laterally.

MALE GENITALIA.– Brachia of unequal length, left brachium shorter than right one. Tegumen rather short and broad. Valvae simple, scarcely setose. Valvellae short and rounded, right valvella longer than left one; another pair of slightly asymmetrical short lobes between valvae and aedeagus. Aedeagus short, bent and tapering distally; cornuti absent.

FEMALE GENITALIA.– Unknown.

DISTRIBUTION.– France.

BIOLOGY.– Unknown. The pupa of the type species was found in a tree-fungus.

REMARKS.– The genus belongs to the subfamily Antequerinae according to the wing venation and male genitalia. Both known species are represented by a single specimen only, and they were collected on the same day in forêt de Fontainebleau, Seine-et Marne, France (Gibeaux, 1986). The origin of this new genus is mysterious. It is hard to believe that such striking species have remained unobserved for so long time in this very well explored part of Europe. A possible explanation is that they have been accidently imported in this environment.

65. *Gibeauxiella bellaqueifontis* (Gibeaux, 1986) comb.n.

Hodgesiella bellaqueifontis Gibeaux, 1986: 174.

DIAGNOSIS.– Adult. Wingspan 9 mm. Head with frons shining yellowish white changing into ochreous white with light brown-tipped scales on vertex. Antennae brown, annulated creamy white. Thorax shining brown with an ochreous spot posteriorly; tegulae shining brown, laterally and posteriorly ochreous. Forewings ground colour brown mixed dark brown; small creamy white spot in middle at extreme base; outward bent creamy white fascia at 1/5, narrowing dorsally and not reaching dorsum; two creamy white costal spots, first at 1/2, inward oblique and almost in middle of wing, second at 4/5, smaller and more rounded; two creamy-white spots on dorsum, first irregular and in alignment with first costal spot, second small and opposite second costal spot; very small ochreous white spot at apex; three small tufts of raised scales, first in basal area below fold, bordered by fascia, two opposite tufts on either side of fold between fascia and first costal spot; small ochreous patch outside connected to first dorsal spot; some ochreous scales at edges of creamy white markings; cilia dark brownish grey around apex, paler towards dorsum, with distinct dark brown transverse line. Hindwings shining brownish grey. Abdomen dorsally greyish brown, anal tuft lighter, ventrally whitish mixed brownish.

Similar species. For differences with *reliqua* see under that species.

MALE GENITALIA.– Right brachium almost straight, gradually tapering distally; left brachium narrowing apically and pointed. Valvae pointed, with convex dorsal margin and wide excavation on ventral margin beyond middle. Valvellae distinctly asymmetrical, both with four or five strong setae; right valvella digitate with rounded tip; left valvella slightly club-shaped, about half length of right one. Additional lobes appear (on slide) asymmetrical: right lobe triangular, left one narrow and shorter. Aedeagus short, arched, slightly narrowing distally.

FEMALE GENITALIA.– Unknown.

DISTRIBUTION.– France.

BIOLOGY.– The only specimen was found at the beginning of May, emerging from the tree-fungus *Ganoderma applanatum* (Pers. ex Wallr.) Pat. (Gibeaux, 1986).

66. Gibeauxiella reliqua (Gibeaux, 1986) comb.n.

Hodgesiella reliqua Gibeaux, 1986: 176.

DIAGNOSIS.– Adult. Wingspan 11 mm. Head with frons shining yellowish white changing into creamy white with ochreous-tipped scales on vertex; antennae brown, annulated creamy white. Thorax shining brown; tegulae shining brown, ochreous laterally and posteriorly. Forewings shining brown, roughly scaled; small creamy white spot at extreme base; outward bent creamy white fascia at 1/5, not reaching dorsum; two creamy white costal spots, first at 1/2, second and largest at 4/5; large round creamy white spot on tornus; cilia dark greyish brown, covered with dark brown scales around apex, paler towards dorsum. Hindwings greyish brown with golden gloss; cilia greyish brown. Abdomen greyish brown, shining, anal tuft clear grey.

Similar species. Differs from *bellaqueifontis* by the forewings lacking the second creamy white spot on the dorsum, the apical spot, the ochreous markings, and the raised scales. Hindwings slightly darker.

MALE GENITALIA.– Right brachium bent before apex; left brachium abruptly narrowing distally with sharp apex. Valvae lobate, without pronounced lobes or excavations, very slightly asymmetrical; right valva with small tooth at apex. Valvellae spatulate and slightly outwardly bent; right valvella larger than left one. Additional lobes of equal length; right lobe elongate triangular, left one parallel-sided. Aedeagus short, arched, gradually tapering distally.

FEMALE GENITALIA.– Unknown.

DISTRIBUTION.– France.

BIOLOGY.– The only specimen was collected at the beginning of May on the fallen trunk of an old beech tree, *Fagus sylvatica* L. (Fagaceae) (Gibeaux, 1986).

Limnaecia Stainton, 1851

Limnaecia Stainton, 1851: 4.
Type species: *Limnaecia phragmitella* Stainton, 1851, by monotypy.
Limnoecia Meyrick, 1895: 674; unjustified emendation.
Lymnaecia Dyar, [1903]: 537; unjustified emendation.

DIAGNOSIS.– Rather large species. Head covered with tightly appressed scales; labial palpi very long and slender, recurved. Forewings lanceolate, apex pointed; hindwings more than half as wide as forewings.

MALE GENITALIA.– Brachia markedly asymmetrical; right brachium long, pointed and heavily sclerotized, two and a half times as long as left one. Valvae concave with short and rounded cucullus; right valva wider at base. Valvellae strongly asymmetrical; left valvella spatulate, longer than valva, apically with patch of short spines, right one short and bent, heavily sclerotized, fused with right valva. Aedeagus short, tapering apically, proximal part bulbous.

FEMALE GENITALIA.– Papillae anales fused and slightly sclerotized. Apophyses very long and stout, heavily sclerotized. Sterigma tubular, slightly bent and sclerotized. Ductus bursae slightly narrowing towards corpus bursae, distal half sclerotized. Corpus bursae egg-shaped, with two signa, each bearing pair of short spines.

DISTRIBUTION.– Cosmopolitan. Several tens of species have been described from the tropical regions, but most of them are not congeneric with *Limnaecia phragmitella*, the type species of the genus.

BIOLOGY.– The larva of the type species feeds in the pith and flower-heads of *Typha* L. (Typhaceae).

67. Limnaecia phragmitella Stainton, 1851

Limnaecia phragmitella Stainton, 1851: 4.

DIAGNOSIS.– Adult. Wingspan 16-22 mm. Head ochreous; antennae yellowish white, annulated brown, apical part almost white. Thorax and tegulae ochreous. Forewings shining ochreous with pale yellow costal margin; indistinct greyish brown line following fold with white elongate spot near its distal end; two medial greyish brown, white-edged spots, inner before 1/2, outer at 3/4, both spots connected by greyish brown streak which starts as narrow line just beyond base; veins in apical area indistinctly indicated by greyish brown, sometimes white-edged lines; series of whitish dots around apex; cilia pale ochreous, more yellow ochreous towards dorsum. Hindwings shining grey. Abdomen shining brownish grey, segments with yellowish white bands posteriorly.

MALE GENITALIA.– See genus description.

FEMALE GENITALIA.– See genus description.

DISTRIBUTION.– Practically worldwide; can be found wherever its host plant grows.

BIOLOGY.– Host plants: *Typha angustifolia* L. and *T. latifolia* L. The young larva mines in the pith of a leaf-sheath; later it goes into the flower-spike or enters the thin stems (Stainton, 1870a; Hering, 1957). The hibernation takes place in the larval stage inside the flower-spike (often gregariously) or inside the leaf-sheaths or stems (Grabe, 1955). In spring an infested flower-spike can be recognized by the large portions of down hanging out from it, but the spun webs of the larvae prevent the seeds from being dispersed by the wind. Pupation takes place in June, in an elongate whitish cocoon usually inside the flower-spike. Adults fly from the end of June to the end of August.

Subfamily Scaeosophinae Meyrick, 1922

Allotalanta Meyrick, 1913

Allotalanta Meyrick, 1913b: 114.
Type species: *Allotalanta autophaea* Meyrick, 1913, by monotypy.

DIAGNOSIS.– Medium sized moths. Head smoothly scaled, with slightly raised scales laterally; antennae 3/4 length of forewings; labial palpi long, slightly flattened and recurved. Forewings broad and lanceolate, unicolorous or with spots and streaks. Hindwings as broad as forewings, apex rounded.

MALE GENITALIA.– Very slightly asymmetrical. Brachia symmetrical or asymmetrical, of different length and shape. Tegumen large, broad at base, narrowing distally. Valvae simple, short and broad or long and tapering distally. Valvellae long, more or less symmetrical. Aedeagus very long, circularly bent. Abdominal tergite VIII with sclerotized armament; abdominal sternite VIII with long bristles.

FEMALE GENITALIA.– Apophyses posteriores about twice as long as apophyses anteriores. Sterigma formed by heavily sclerotized plate with lateral protrusions. Ductus bursae long and coiled, membranous. Corpus bursae oval, with two small signa surrounded by pectinations.

DISTRIBUTION.– The genus is mainly distributed in the Afrotropical and Oriental regions, with only one species in the Palaearctic.

BIOLOGY.– Unknown.

68. Allotalanta autophaea Meyrick, 1913

Allotalanta autophaea Meyrick, 1913b: 114.

DIAGNOSIS.– Adult. Wingspan 19-20 mm. Head ochreous-white, pale ochreous posteriorly; antennae greyish ochreous; labial palpi ochreous, long, recurved and cylindrical. Thorax and tegulae greyish ochreous. Forewings shining greyish ochreous, finely mottled with greyish brown; cilia greyish ochreous. Hindwings shining pale ochreous, cilia pale ochreous. Abdomen greyish ochreous.

MALE GENITALIA.– Brachia furcate, with inner arms more slender and twice as long as outer ones. Valvae short and broad, cucullus rounded, ventral margin slightly concave. Valvellae about 3/4 the length of valva, tapering distally, apex hooked. Aedeagus very long, in loop-like curve; distal part sclerotized with truncated apex, membranous proximal part with sclerotized rim on inner side; cornuti spicular, in long tapering band. Tergite VIII sclerotized and distally furcate, both extremities with bundle of setae; sterniteVIII with bundles of setae, smaller in medial part of posterior margin and twice as large on hump in middle of segment.

FEMALE GENITALIA.– Unknown.

DISTRIBUTION.– Turkey.

BIOLOGY.– Unknown. Adults have been collected in June.

Subfamily Cosmopteriginae Heinemann & Wocke, [1876]

Cosmopterix Hübner, [1825]

Cosmopterix Hübner, [1825]: 424.
Type species: *Tinea zieglerella* Hübner, [1810], by subsequent designation by the International Commission on Zoological Nomenclature, 1969: 150 (Opinion 866).
Cosmopteryx Zeller, 1839: 210; unjustified emendation.
Lienigia Spuler, 1910: 389.
Type species: *Cosmopteryx lienigiella* Zeller, in Lienig & Zeller, 1846, by monotypy.

DIAGNOSIS.– Head smoothly scaled; antennae 3/4 to 4/5 length of forewings, often slightly serrate distally, with species-specific alternation of dark and white sections in apical part; labial palpi cylindrical, porrect, apical segment strongly angled upwards and often reaching well above head. Forewings narrowly lanceolate with long and very narrowly protruding apex and very characteristic yellow or orange fascia bordered by tubercular metallic fasciae or spots beyond the middle. Hindwings almost linear.

MALE GENITALIA.– Brachia strongly asymmetrical; left brachium very short and slightly sclerotized, often inconspicuous; right brachium long, heavily sclerotized and of diagnostic value. Tegumen short with sclerotized rim at anterior margin. Valvae rounded, with rather narrow proximal part and widened and densely haired distal part. Valvellae usually symmetrical, long and bent ventrally, sometimes with dorsal projection. Aedeagus short and often bulbous, tapering apically; its basal part with typical sclerotized flange. Segment VIII with two oval pleural lobes covering bases of valvae.

FEMALE GENITALIA.– Apophyses posteriores one and a half to two times as long as apophyses anteriores. Sterigma heavily sclerotized, tubular, often strongly projected, with ostium at posterior end. Ductus bursae long and narrow, sometimes with sclerotization before entrance of corpus bursae. Corpus bursae oval with two triangular, sometimes very minute signa. Sternite VII with median projection of posterior margin, usually surrounded with specialized black scales laterally. The shape of the sterigma is of diagnostic value.

DISTRIBUTION.– Worldwide, except for polar regions. More than 200 species are already described, most of them from the tropical regions.

BIOLOGY.– The larvae are leaf-miners on various grasses, herbs or shrubs from the families Cannabaceae, Poaceae, Fabaceae, Urticaceae, etc. Most of them are oligophagous or even monophagous, therefore the knowledge of host plant combined with the shape of the mine can be of diagnostic importance. The larva often constructs a hiding place of silk inside the mine into which it withdraws when not feeding. Hibernation takes place at the larval stage in a cocoon spun inside or outside the mine.

Key to species of *Cosmopterix*

1.	Forewings ground colour yellowish brown	2
–	Forewings ground colour darker	3
2.	Forewings with medial line in basal area narrow and not reaching yellow fascia	***lienigiella***
–	Forewings narrower and with medial line in basal area broad and reaching yellow fascia	***salahinella***
3.	Forewings pale brown or olivaceous brown	4
–	Forewings dark brown or blackish brown	6
4.	Forewings olivaceous brown	5
–	Forewings pale brown, both silver metallic fascia with strong pinkish gloss. Small species from the Canary Islands	***turbidella***
5.	Male genitalia: right brachium club-shaped with long and sharp apical process, valvae angular. Female genitalia: sterigma with long and narrow base and oval apical extension around ostium	***coryphaea***
–	Male genitalia: right brachium apically furcate, valvae rounded. Female genitalia: sterigma spindle-shaped, with rounded apex.	***pararufella***
6.	Forewings with broad golden fascia or costal streak in basal area	7
–	Forewings with white, silvery or golden lines in basal area	9

7.	Forewings with large oblique golden spot from base of costa towards dorsum	***orichalcea***
–	Forewings with broad transverse golden fascia in basal area	8
8.	Forewings with white apical line broadly interrupted in middle	***zieglerella***
–	Forewings with white apical line not interrupted in middle	***schmidiella***
9.	Forewings with outer metallic fascia interrupted and orange fascia often medially prolonged towards apex	10
–	Forewings with outer metallic fascia not interrupted	***pulchrimella***
10.	Forewings with narrow white lines in basal and apical area	11
–	Forewings with broad golden lines in basal and apical area	***sibirica***
11.	Antennae with interrupted white line from base. Abdomen dorsally not completely dark brown	12
–	Antennae with uninterrupted white line from base. Abdomen dorsally dark brown	***scribaiella***
12.	Forewings narrow, about 10 times as long as wide. Antennae with interrupted white line from base to 2/3. Abdomen dorsally brown, spotted orange-brown	***attenuatella***
–	Forewings wider, about 8 times as long as wide. Antennae with an interrupted white line from base to 1/3. Abdomen dorsally with segments II-V ochreous-yellow to orange-brown	***crassicervicella***

69. *Cosmopterix zieglerella* (Hübner, [1810])

Tinea zieglerella Hübner, [1810]: pl. 44, fig. 306.
Gracillaria eximia Haworth, 1828: 532.

Diagnosis.– Adult. Wingspan 8-11 mm. Head shining golden metallic, shining dark brown posteriorly, laterally with thin white line above eye. Antennae shining dark brown with subapical section of about 3 white segments and apical section of about five white segments; labial palpi white, segment 2 with shining dark brown dorsal line, segment 3 with shining dark brown ventral line and dark brown tip. Thorax shining dark brown with golden metallic medial line; tegulae golden metallic. Forewings shining dark brown; outwardly oblique golden metallic fascia in basal area; bright orange fascia beyond middle, narrowing towards dorsum, bordered by irregular narrow line of dark brown scales, followed by inner and outer pale golden metallic fasciae, often with purplish gloss; outer fascia with small white costal spot; apical line reduced to silvery spot with bluish gloss in middle and shining white spot at apex extending into cilia; cilia dark brown. Hindwings shining dark greyish-brown.

Similar species. Very similar to *schmidiella*; for differences see the key.

Male genitalia.– Right brachium spatulate, gradually tapering distally and ending in blunt apex, three times longer than left brachium. Valvae rounded, upper and lower margin concave, caudal margin convex. Valvellae slightly bent, narrowed near base, widest part beyond middle, distally tapering to blunt apex. Aedeagus rather flat with central part bulbous, apically tubular and slightly bent downwards.

Female genitalia.– Posterior edge of sternite VII rather straight. Sternite VIII about twice as broad as long. Ostium bursae round with semilunar sclerotization and short sclerotized ridge underneath. Sterigma inverted pear-shaped, elongate. Ductus bursae slightly shorter than corpus bursae. Corpus bursae with two small crescent-shaped signa of different size.

DISTRIBUTION.– Europe; Asia Minor, the Caucasus, central and East Asia to Japan (Riedl, 1994b).

BIOLOGY.– Host plants: *Humulus lupulus* L., *H. japonicus* Sieb. et Zucc. (Cannabaceae) (Sorhagen, 1885), *Boehmeria nipononivea* Koidz. (Urticaceae) (Kuroko, 1960; Sinev, 1994d). The egg is laid on the leaf close to the midrib or another large vein. The larva can be found in July–September. The young larva makes an irregular linear mine with a silk-lined gallery, longitudinal to the leaf-vein, which serves as a shelter between feedings. From this place the larva later produces clear branches in all directions causing a very irregular yellowish white blotch-mine with dispersed frass. The larva can change mines, and in large leaves more than one mine can be found. In autumn it leaves the mine and makes a white elongate cocoon spun to detritus on the ground. Here it hibernates prior to pupation. Flight period from the end of May till July. The adults can be beaten from the food plant and sometimes come to light. See text-fig. 41.

70. *Cosmopterix schmidiella* Frey, 1856

Cosmopteryx schmidiella Frey, 1856: 257.
Cosmopteryx schmidiella mongoliella Sinev, 1980: 380. Subspecies.

DIAGNOSIS.– Adult. Wingspan 9-10 mm. Head golden metallic. Antennae shining dark brown with subapical and apical section of about 6 white segments each, separated by dark brown section of about 10 segments. Thorax shining dark brown with golden metallic medial line, tegulae golden metallic. Forewings pattern similar to *zieglerella*,

Text-fig. 41. *Cosmopterix zieglerella* (Hübner), larval mine in leaf of *Humulus lupulus* L.

but with uninterrupted silvery metallic apical line with purplish gloss, becoming shining white on cilia at apex. Hindwings shining greyish brown.

Similar species. Differs from *orichalcea* by the single white subapical section of the antennae and the metallic golden outwardly oblique basal fascia of the forewings.

MALE GENITALIA.– Right brachium spatulate and twisted, distally ending in sharp apex with subapical teeth, about three times as long as left one. Valvae narrow at base, rounded, upper margin strongly concave distally, lower margin slightly concave, caudal margin strongly convex. Valvellae narrowed near base, bent in middle with shallow hump on upper margin, apically tapering, apex rounded. Aedeagus bottle-shaped, posterior part flange-shaped ventrally.

FEMALE GENITALIA.– Posterior edge of sternite VII with rounded protrusion in middle. Segment VIII about twice as broad as long. Ostium square with rounded edges and semicircular sclerotized rim. Sterigma elongated oval with broad medial ridge. Ductus

Text-fig. 42. *Cosmopterix schmidiella* Frey, larval mines in leaves of *Vicia sepium* L.

bursae slightly shorter than corpus bursae. Corpus bursae with circular wrinkles and two crescent-shaped signa of equal size.

Distribution.– Central and south-west Europe, England (now extinct); Middle East (northern Iran). In Mongolia, Korea and Japan represented by the subspecies *mongoliella* (Sinev, 1980).

Biology.– Host plants: *Vicia sepium* L., *V. pisiformis* L., *Orobus tuberosus* L. (=*Lathyrus montanus* Bernh.), *O. niger* L., *O. vernus* L. (Fabaceae) (Frey, 1856; Stainton, 1870b; Hering, 1957; Sinev, 1994d). The larva makes a large blotch mine on the underside of a leaf in August and September.The mine often occupies the whole leaf, and the parenchyma becomes entirely devoured. The resting larva lies under the midrib in a spun hiding place and is difficult to observe. Many larvae can colonize a single plant. Pupation takes place in May, after hibernation in an elongate whitish semi-transparent cocoon. Adults are flying from the end of May till mid-July. See text-fig. 42.

71. Cosmopterix orichalcea Stainton, 1861

Cosmopteryx orichalcea Stainton, 1861: 90.
Cosmopteryx drurella auctt., (*nec* Fabricius, 1775); misidentification.
Cosmopteryx druryella Zeller, 1850: 196; unjustified emendation and misidentification of *Tinea drurella* Fabricius, 1775.
Cosmopterix singularis Sinev, 1980: 382.

Diagnosis.– Adult. Wingspan 8-10mm. Head shining golden metallic anteriorly and shining dark brown posteriorly, with thin white line above eyes. Antennae shining dark brown, with two white subapical sections, inner of 3 segments and outer of 2 segments, separated by 2 brown segments; apex of about 5 segments white. Thorax shining dark brown with golden metallic medial line; tegulae shining golden metallic. Forewings shining dark brown; large triangular golden metallic streak at costa from base, pointing towards dorsum, but not reaching it; yellow orange fascia strongly narrowing towards dorsum; silvery metallic line with purplish gloss in apical area, sometimes narrowed in middle, shining white in cilia at apex; cilia dark brown. Hindwings shining greyish brown. Abdomen shining light brown dorsally, ventrally dark grey with pale golden gloss, segments with white bands posteriorly; anal tuft grey, mixed whitish.

Similar species. The two white subapical sections of the antennae and the large golden metallic basal streak of the forewings distinguish this species from *zieglerella* and *schmidiella*.

Male genitalia.– Right brachium spatulate, gradually tapering distally, apex truncate and slightly curved downwards. Valvae sickle-shaped with strongly concave upper margin and strongly convex lower margin. Valvellae slightly bent, apical half twice as wide as basal half, distally tapering to blunt apex. Aedeagus bottle-shaped, posterior end widening distally and with small ventral flange.

Female genitalia.– Posterior edge of sternite VII arched. Segment VIII about one and a half times as broad as long. Ostium rounded, with semicircular sclerotized rim. Sterigma elongated oval with heavily sclerotized medial ridge. Ductus bursae about as long as corpus bursae and with small sclerotization before entrance into corpus bursae. Corpus bursae with circular wrinkles and two minute crescent-shaped signa of different size.

Distribution.– Palaearctic, except for the northernmost part.

BIOLOGY.– Host plants: *Phalaris arundinacea* L., *Anthoxanthum odoratum* L., *Festuca arundinacea* Schreb., *Milium effusum* L., *Hierochloe odorata* Whib., *H. australis* R. et Sch. (Poaceae) (Stainton, 1870b; Hering, 1889; Emmet, 1968; Sinev, 1994d). Probably bivoltine. Larvae in June-July and in August-September, mining the leaves. The mine starts as a straight gallery with a pile of frass at the beginning; later on the gallery widens and often become more irregular. Sometimes the whole tip of the leaf is occupied and afterwards the gallery points downwards, and the frass is partly ejected or spread throughout the mine in several heaps. The mines can be changed. Pupation takes place in a tough white elongate cocoon on a leaf or among debris (after hibernation in the second generation). Adults are flying in May-June, also in July-August in the south. Can be swept from its food plants during daytime and is easily attracted to light. See text-fig. 43.

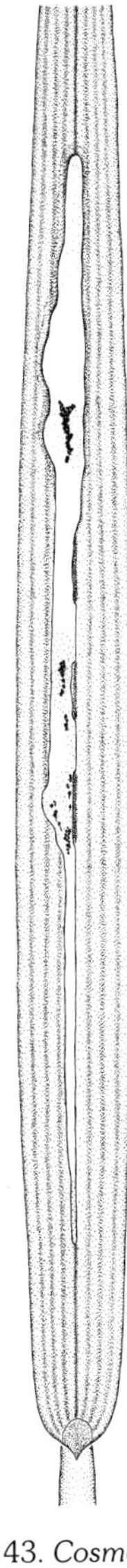
43

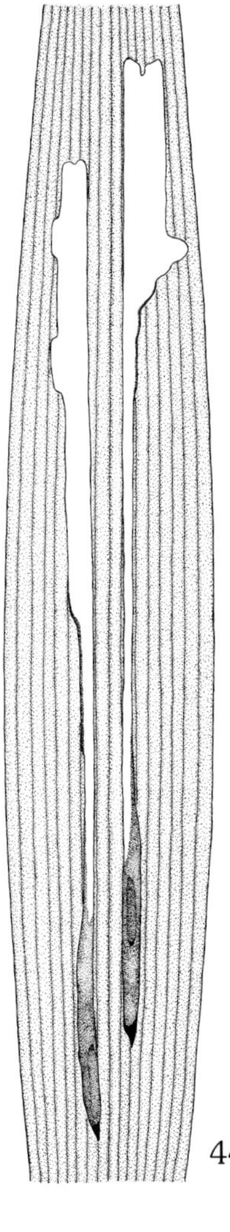
44

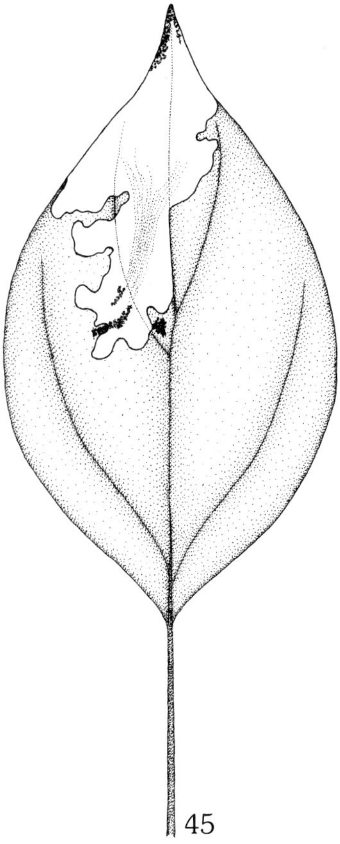
45

Text-fig. 43. *Cosmopterix orichalcea* Stainton, larval mine in leaf of *Phalaris arundinacea* L.
Text-fig. 44. *Cosmopterix scribaiella* Zeller, larval mines in leaf of *Phragmites australis* (Cav.) Trin. ex Steud.
Text-fig. 45. *Cosmopterix pulchrimella* Chambers, larval mine in leaf of *Parietaria officinalis* L.

72. *Cosmopterix sibirica* Sinev, 1985

Cosmopterix sibirica Sinev, 1985b: 93.

DIAGNOSIS.– Adult. Wingspan 11-13 mm. Head shining dark brown, medially and laterally lined white. Antennae shining dark brown, with short white line from base to one-tenth, two white subapical sections, inner of 4 and outer of 2 segments, and white apex of about 8 segments. Thorax shining dark brown with white medial line; tegulae shining dark brown, edged pale golden metallic inwardly. Forewings shining dark brown, with three broad spindle-shaped pale golden metallic lines in basal area, costal from base to 2/3 of basal area, median above fold from 1/5 to 4/5 (of basal area), dorsal just below fold from 2/5 and ending at or just beyond median; broad yellow orange fascia beyond 1/2 with short extension towards apex in middle, strongly narrowing towards dorsum, inwardly edged by slightly irregular tubercular pale golden metallic fascia, outwardly edged by tubercular pale golden metallic costal and dorsal spots; dorsal spot about twice as large as costal one and situated closer to base; costal spot outwardly edged with white streak into cilia; both fascia and spots inwardly irregularly edged blackish brown; broad pale golden metallic apex line on dorsum; cilia dark brown, paler on dorsum towards base, white at apex. Hindwings shining brownish grey, cilia greyish brown.

Similar species. The three broad golden basal fasciae and the golden metallic apical streak of the forewings are characteristic.

MALE GENITALIA.– Right brachium long and stout, apically narrowing and with slightly bent tip. Valvae slender and sickle-shaped. Valvellae slightly bent and broad in distal half. Aedeagus large, bottle-shaped. Resembles *orichalcea*, but differs by the shape of the right brachium and the more bulky aedeagus.

FEMALE GENITALIA.– Unknown.

DISTRIBUTION.– Eastern Europe (southern Ural); southern Siberia and Mongolia.

BIOLOGY.– Unknown. Adults have been collected in June.

73. *Cosmopterix scribaiella* Zeller, 1850

Cosmopteryx scribaiella Zeller, 1850: 197.
Cosmopteryx hermsiella E. Hering, 1889: 308.
Cosmopterix scribaiella japonica Kuroko, 1960: 9. Subspecies.

DIAGNOSIS.– Adult. Wingspan 10-11 mm. Head shining brown with medial and two lateral white lines. Antennae shining brown with white line on anterior edge from base to middle and two white subapical sections, inner of 3 segments and outer of 2 segments, separated by 2 two brown segments, apex of about 5 segments white. Thorax shining brown with white medial line, tegulae shining brown, inner sides lined white. Forewings shining dark brown; basal area with four narrow white lines, subcostal from base to just beyond half and bending inwardly, short medial above fold, subdorsal about as long as or slightly shorter than medial but situated further from base, and short dorsal line on dorsum at 1/5; bright orange fascia beyond middle, narrowing towards dorsum and medially prolonged towards apex, bordered at inner edge by slightly inwardly oblique, tubercular pale golden metallic fascia, not reaching costa and bordered at outer edge by two tubercular pale golden metallic spots, dorsal about twice as large as costal, small subcostal patch or streak of black scales on outer side of pale golden metallic fascia, pale golden metallic spots irregularly lined blackish on inner side, white costal spot apical of costal tubercular spot; shining white apical line in alignment with orange extension of fascia; cilia brown. Hindwings greyish brown.

Similar species. Differs from *attenuatella* and *crassicervicella* by the uninterrupted line on the antennae, the short dorsal line on the forewings not reaching base and the dark brown abdomen. For differences with *pulchrimella*, q. v.

MALE GENITALIA.– Right brachium stout, triangular with sharp, protruded apex and very heavily sclerotized, more than twice as long as left arm. Valvae rounded, narrow at base, upper and lower margins slightly concave, caudal margin convex. Valvellae long, bent in middle, almost parallel-sided, apically tapering to blunt point. Aedeagus cylindrical, apically narrowing and laterally bent, posterior part widening distally and with small ventral flange.

FEMALE GENITALIA.– Posterior edge of sternite VII trapezoidal. Segment VIII longer than wide. Ostium rounded with crescent-shaped sclerotization. Sterigma oval, with long sclerotized extension distally. Ductus bursae slightly shorter than corpus bursae, distally with two sclerotized ridges. Corpus bursae oval with two small crescent-shaped signa of equal size.

DISTRIBUTION.– North and Central Europe; northern Africa, Asia Minor, Central Asia. In the Russian Far East and in Japan represented by the subspecies *japonica* (Kuroko, 1960).

BIOLOGY.– Host plant: *Phragmites australis* (Cav.) Trin. ex Steud. (Poaceae) (Herms, 1888; Amsel, 1930; Hering, 1957). The larva mines a leaf from August till autumn and hibernates inside the mine till April. The mine starts as a straight gallery, later widening irregularly; the frass is heaped in the beginning of the mine and also partly ejected. The larva spins a tube in the basal part of the mine and uses this as a hiding place. Usually a leaf contains only one mine, but sometimes up to ten mines can be found, giving the plant a withered appearance visible from some distance. Pupation in May in a cocoon in a narrow part of the mine (Koster, 2002b). Flight period from early June to the end of October. Adults can be swept from common reed and also come to light. See text-fig. 44.

74. *Cosmopterix pulchrimella* Chambers, 1875

Cosmopteryx pulchrimella Chambers, 1875c: 231.
Cosmopteryx pulcherrimella Chambers, 1878: 137; unjustified emendation of *Cosmopteryx pulchrimella* Chambers, 1875.
Cosmopteryx parietariae M. Hering, 1931: 68.

DIAGNOSIS.– Adult. Wingspan 6-8 mm. The species differs from *scribaiella*, *attenuatella* and *crassicervicella* by the uninterrupted outer tubercular pale golden metallic fascia and the three short silvery metallic basal streaks of the forewings.

MALE GENITALIA.– Right brachium spatulate, about four times as long as left one, distally gradually tapering, apex blunt. Valvae rounded, upper and lower margins concave, caudal margin convex. Valvellae slightly bent, narrowed near base, widest part in middle, sharply pointed distally. Aedeagus slender, anterior part straight and slightly bent downwards, posterior part short and straight.

FEMALE GENITALIA.– Posterior edge of sternite VII slightly concave. Segment VIII about two and a half times as broad as long. Ostium rounded with cup-shaped sclerotization and weakly sclerotized ridge underneath. Sterigma oval. Ductus bursae slightly longer than corpus bursae and gradually widening towards corpus bursae. Corpus bursae oblong with two tiny crescent-shaped signa.

DISTRIBUTION.– Mediterranean area, from Portugal to western Transcaucasus, north to Switzerland and Hungary; Azores, Canary Islands, and Madeira. Recently found in England (Parsons, 2002). Widely spread in the U.S.A. (Hodges, 1978).

BIOLOGY.– Host plants: *Parietaria officinalis* L., *P. pennsylvanica* Muhl., *Pilea pumila* (L.) Gray. (Urticaceae) (Busck, 1906; Hering, 1931). Several overlapping generations. The leaf mine starts as an irregular gallery usually at the main rib, soon leading to an irregular blotch. Inside the mine the larva spins a web and this often causes contraction of the leaf. It also constructs a silk-lined gallery, which serves as a shelter when the larva is not feeding. Frass partly inside the mine, but most of it is removed through a hole at the beginning of the mine. The black pellets of frass on the underside of a leaf indicate the presence of a larva. The mines are changed very often. Pupation takes place inside the mine. Adults are flying from late April to the beginning of autumn. See text-fig. 45.

75. *Cosmopterix attenuatella* (Walker, 1864)

Gelechia attenuatella Walker, 1864: 1019.
Cosmopteryx flavofasciata Wollaston, 1879: 438.
Cosmopteryx mimetis Meyrick, 1897: 339.
Cosmopteryx apiculata Meyrick, 1922b: 568.
Cosmopteryx antillia Forbes, 1931: 356.

DIAGNOSIS.– Adult. Wingspan 9-10 mm. Similar to *crassicervicella* but differs by the interrupted white line on antennae running from base to 2/3. Forewings more elongate, with length 10 times width; medial and subdorsal white lines in basal area almost opposite each other. Abdomen dorsally brown, segments spotted orange brown.

MALE GENITALIA.– Right brachium hook-shaped, apex flattened and sharp, about three times as long as left one. Valvae with upper margin strongly concave towards apex, lower margin slightly concave, caudal margin convex. Valvellae slender, curved and pointed. Aedeagus arrow-shaped, very narrow, anterior part slightly bent upwards, posterior part short and widening distally.

FEMALE GENITALIA.– Posterior edge of sternite VII straight. Segment VIII about twice as broad as long. Ostium bursae elongate. Sterigma jar-shaped. Ductus bursae very narrow, more than twice as long as corpus bursa, entering corpus bursae on anterior side. Corpus bursae elongate, signa absent.

DISTRIBUTION.– Canary Islands, Madeira. All records from continental Europe belong to *crassicervicella* (Sinev, 1997b). Widely distributed in the tropics and subtropics of both the Old and New World (Hodges, 1978).

BIOLOGY.– Host plants: *Scirpus* L., *Cyperus rotundus* L. (Cyperaceae). Probably bivoltine in the south of the temperate zone. The larva mines a leaf starting near the tip, then moving downwards. The mine occupies the whole leaf with black frass inside; its older part becomes brown and shrivelled, whilst the freshly eaten portion is pale yellow. Pupation occurs in the mine in a whitish elongated cocoon (Fletcher, 1933). In the tropics adults can be found throughout the year (Hodges, 1978).

76. *Cosmopterix crassicervicella* Chrétien, 1896

Cosmopteryx crassicervicella Chrétien, 1896: 105.

Cosmopteryx flavipes Turati, 1930: 84.
Cosmopteryx superba Gozmány, 1960: 419.
Cosmopteryx dalii Agenjo, 1981: 50.

DIAGNOSIS.– Adult. Wingspan 8-9 mm. Antennae with interrupted white line from base to 1/3, two white sections of 2 segments each at 3/4 separated by 2 brown segments, and subapical white ring of 3 segments. Forewings shining brown; five narrow white lines in basal area, costal from 1/3 (of basal area) to inner metallic fascia, subcostal from base and bending inwardly, short medial above fold, subdorsal, slightly longer than medial and further from base, and dorsal from base to beyond 1/2; shining white apex line, narrowed or interrupted in middle; cilia brown. Hindwings shining greyish brown. Abdomen dorsally brown, segments II-V dorsally ochreous yellow to orange brown.

Similar species. Differs from *attenuatella* by the shorter interrupted white line on the antennae, the broader forewings with length 8 times width and the short costal line.

MALE GENITALIA.– Right brachium hook-shaped, apex rounded, about eight times as long as left one. Valvae nearly triangular with upper margin concave, caudal margin slightly convex. Valvellae slender, narrowed near base, curved, sharply pointed apically. Aedeagus very slender bottle-shaped, anterior part slightly bent and with sharp apex, posterior part very short and slightly widening distally.

FEMALE GENITALIA.– Posterior edge of sternite VII very shallowly arched. Segment VIII about three times as broad as long. Ostium rounded with crescent-shaped sclerotization. Sterigma oval. Ductus bursae one and a half times as long as corpus bursae and with weakly sclerotized band beyond middle. Corpus bursae egg-shaped, signa very small or absent.

DISTRIBUTION.– Southern Europe (Spain, Portugal, Greece); Asia Minor, Transcaucasus, northern Africa and Near East.

BIOLOGY.– Host plant: *Cyperus* L. (Cyperaceae). Bivoltine. The larva mines the entire leaf to the apex. The mine is folded at the place where the larva is resting in a silky gallery. Pupation takes place inside the mine in a spun chamber causing the leaf to fold. Adults fly from mid-April to mid-June and again from the end of August to the end of October (Chrétien, 1917; Hering, 1957).

77. Cosmopterix turbidella Rebel, 1896

Cosmopteryx turbidella Rebel, 1896: 135.

DIAGNOSIS.– Adult. Wingspan 7-8 mm. Head pale brown, with two longitudinal white lines laterally. Antennae shining brown, with short white line from base changing into interrupted line to 1/2; three white sections of 2 segments each at 2/3 and one white subapical section of 2 segments. Thorax shining pale brown with white medial line; tegulae shining light brown, lined white inwardly. Forewings shining pale brown; five narrow white lines in basal area, costal very short and just inward of yellow fascia, subcostal from base to 1/4 (of basal area), distally bending sharply inwards, medial and subdorsal above and below fold and very short, latter slightly further from base, dorsal as narrow white edging near base; bright yellow, dorsally mixed brownish, fascia beyond middle, narrowing towards dorsum, bordered at inner and outer edges by tubercular silvery metallic fascia with strong pinkish gloss; inner fascia not reaching costa, with blackish spot on outer side; outer fascia inwardly edged brown at costa and dorsum, with small white costal spot; short shining white apex line; cilia brown around apex, greyish brown towards dorsum. Hindwings shining greyish brown. Abdomen dorsally shining brown.

Similar species. Resembles *pulchrimella*, but differs by the lighter brown coloration of the forewings, the five unequal white streaks in the basal area, the pinkish reflection of the silvery metallic fasciae, and the short uninterrupted apical line.

MALE GENITALIA.– Right brachium spatulate with rounded apex, more than twice as long as left one. Valvae short and broad, upper margin concave, lower and caudal margins almost straight. Valvellae almost as long as valvae, parallel-sided, rounded distally, apex covered with short and stout setae. Aedeagus bottle-shaped, strongly narrowed apically, posterior part very long and widening distally.

FEMALE GENITALIA.– Posterior edge of sternite VII trapezoidal with posterior margin slightly concave. Segment VIII slightly broader than long. Ostium oval with crescent-shaped sclerotization. Sterigma lozenge-shaped, basally protruded, distal walls stronger sclerotized and prolonged. Ductus bursae slightly longer than corpus bursae, curved nearly at a right angle before entering corpus bursae. Corpus bursae strongly wrinkled and with two small crescent-shaped signa of equal size.

DISTRIBUTION.– Canary Islands. Occurs on walls and in ditches, especially in shaded places to the upper edge of the dry zone.

BIOLOGY.– Host plants: *Parietaria officinalis* L., *P. debilis* Forst., *P. arborea* Gaud., and *Forsskaolea angustifolia* Retz. (Urticaceae). The larva firstly causes a short linear mine which usually starts at the midrib and runs more or less obliquely to the edge of the leaf. The mine then changes into an irregular blotch, with frass ejected from the underside at the beginning of the mine. Pupation takes place among detritus on the ground (Hering, 1927; Klimesch, 1983). Adults fly from March till July, but most abundant in April and May.

78. Cosmopterix pararufella Riedl, 1976

Cosmopterix pararufella Riedl, 1976: 190.

DIAGNOSIS.– Adult. Wingspan 11 mm. Very similar to *coryphaea* externally, but differs slightly by the forewings with the dorsal white line starting from the base, and by the yellow fascia which has a somewhat larger extension towards apex. Safe identification only possible by examination of the genitalia.

MALE GENITALIA.– Right brachium very narrow at base, about twice as long as left one, widening into furcate and strongly sclerotized process. Valvae rounded, upper margin concave, lower margin almost straight, caudal margin strongly convex. Valvellae short, bent as right angle in middle, almost parallel-sided, apically tapering to blunt point. Aedeagus pitcher-shaped, posterior part slightly widening distally and with ventral flange. Differs from *coryphaea* by the furcate right brachium, the rounded valvae, and the more slender valvellae.

FEMALE GENITALIA.– Posterior edge of sternite VII arched. Segment VIII slightly wider than long. Ostium narrow with broad semicircular sclerotization. Sterigma wide, slightly tapering towards ostium, distally narrowed and with long weakly sclerotized extension. Ductus bursae slightly shorter than corpus bursae, with sclerotized ring distally. Corpus bursae oval, with two rounded signa bearing crescent-shaped sclerotizations.

DISTRIBUTION.– Spain (Almeria), Corsica, Cyprus; northern Africa.

Biology.– Host plant: *Saccharum officinarum* L. (Poaceae), evidently also on some local grasses. Probably bivoltine. According to El-Serwy (pers. comm.), in Egypt the larva mines the leaves of sugarcane in April-May and in September-October. The mine starts as a line, then turning into a blotch. Pupation takes place inside the mine. Adults have been found in May and June and again in September.

79. *Cosmopterix coryphaea* Walsingham, 1908

Cosmopteryx coryphaea Walsingham, 1908: 964.
Cosmopteryx donatellae Mariani, 1932: 38.
Cosmopteryx formosa Amsel, 1935a: 304.
Cosmopteryx coryphae Agenjo, 1952: 81; unjustified emendation of *Cosmopteryx coryphaea* Walsingham, 1908.

Diagnosis.– Adult. Wingspan 8-10 mm. Head olivaceous brown with medial and two lateral white lines. Antennae shining light brown with white line from base to beyond 1/2 and two subapical sections of 2 white segments each separated by 2 light brown segments, apex white. Thorax olivaceous brown with white medial line; tegulae light brown lined white inwardly. Forewings olivaceous brown; basal area with five narrow white lines, costal from 1/4 (of basal area) to yellow fascia, subcostal from base to 1/3, bends halfway slightly from costa, medial straight from base almost to yellow fascia, subdorsal from 1/4 to yellow fascia, bending from dorsum, dorsal starting near base following dorsum to 1/4; broad yellow fascia in middle, distally extending towards apex, proximally with small extension in middle; four silvery metallic, tubercular spots, two on inner edge of yellow fascia and edged outwardly with blackish dot, costal slightly larger than dorsal, two in middle of yellow fascia, dorsal about twice as large as costal, both brownish edged inwardly; white costal spot outwards tubercular silvery spot; shining white apical line; cilia ochreous grey. Hindwings shining brownish grey. Abdomen dorsally ochreous.

Similar species. Resembles *pararufella* superficially but genitalia are characteristic.

Male genitalia.– Right brachium club-shaped and with sharply pointed process from middle, about four times as long as left one. Valvae almost square, upper margin concave, lower margin slightly convex, caudal margin straight. Valvellae short and stout, bent as right angle in middle, apically tapering to blunt point. Aedeagus pitcher-shaped, posterior part slightly widening distally and with ventral flange. Differs from *pararufella* by the club-shaped and acutely pointed right brachium, the square-built valvae, and the stout valvellae.

Female genitalia.– Posterior edge of sternite VII arched. Segment VIII slightly wider than long. Ostium rounded with semicircular sclerotization. Sterigma large and oval around ostium, distally with long and narrowing extension ending in bulbous apex. Ductus bursae slightly shorter than corpus bursae, distally with some sclerotized patches. Corpus bursae oval, two circular signa with anchor-like sclerotizations.

Distribution.– Southern Europe; northern Africa, Near East, Canary Islands.

Biology.– Host plant: *Phragmites australis* (Cav.) Trin. ex Steud. (Poaceae) (Amsel, 1935a). The larva mines the leaf in the same way as *lienigiella* (Hering, 1957). Pupation takes place inside the mine. Adults have been collected from the end of February till June.

80. Cosmopterix lienigiella Zeller, 1846

Cosmopteryx lienigiella Zeller, in Lienig & Zeller, 1846: 298.

DIAGNOSIS.– Adult. Wingspan 10-13 mm. Head yellowish brown with one medial and two lateral white lines. Antennae shining greyish brown with short basal white line, apical part sometimes more or less annulated white. Thorax shining yellowish brown with white medial line; tegulae shining yellowish brown edged white inwardly. Forewings shining yellowish brown; basal area with five narrow white lines, costal from 1/4 (of basal area) almost to inner silvery metallic fascia, subcostal from base to 1/3, bending halfway slightly from costa, medial straight, from base to almost 1/2, short subdorsal from 1/3 and ending beyond medial, dorsal from base, following dorsum to about 1/2; yellow fascia beyond middle, narrowing towards dorsum, bordered by inner and outer tubercular silvery metallic fasciae; inner fascia not reaching costa, sometimes interrupted in middle and with subcostal patch of black scales on outer side; outer fascia interrupted in middle, with subcostal and subdorsal patches of black scales on inner side; white costal spot beyond outer silvery metallic fascia, lengthened towards apex; white, slightly bent, apical line; cilia from yellowish white on costa to yellowish brown on dorsum. Hindwings yellowish grey. Abdomen dorsally yellowish brown.

Similar species. Similar to *salahinella* but differs by the broader forewings with the narrow medial line in basal area not reaching the yellow fascia.

MALE GENITALIA.– Right brachium hook-shaped, apex strongly prolonged and pointed, about eight times as long as left one. Valvae semicircular, upper margin slightly concave, lower margin almost straight. Valvellae long, asymmetrical, bent as right angle in middle, almost parallel-sided, sharply pointed apically; right lobe with large hook-shaped dorsal hump, left one with small irregular dorsal hump. Aedeagus pitcher-shaped, posterior part slightly widening distally and with ventral flange.

FEMALE GENITALIA.– Posterior edge of sternite VII strongly arched. Segment VIII one and half times wider than long. Ostium egg-shaped with semilunar sclerotization. Sterigma lozenge-shaped, distal walls stronger sclerotized. Ductus bursae distinctly shorter than corpus bursae, distally with some sclerotizations. Corpus bursae oval with two small crescent-shaped signa.

DISTRIBUTION.– Europe; northern Africa, Russian Far East, Korea and Japan.

BIOLOGY.– Host plant: *Phragmites australis* (Cav.) Trin. ex Steud. (Poaceae). Univoltine. The larva mines a leaf and can be found from September to spring, hibernating inside the mine. The mine starts as a gallery but soon afterwards it widens into an elongate blotch. Frass is heaped at the beginning, later irregularly dispersed. The larva usually mines in an upward direction, sometimes downwards. It is distinctly visible inside the mine, for it does not make a shelter of silk as many other species of the genus *Cosmopterix*. Leaves can contain several mines. Pupation inside the mine in May. Adults fly from the end of May till mid-October and can be swept from the food plant but also come to light. See text-fig. 46.

81. Cosmopterix salahinella Chrétien, 1907

Cosmopteryx salahinella Chrétien, 1907: 306.
Cosmopteryx rufella Turati, 1927: 341.
Cosmopteryx phragmitidis Amsel, 1935a: 304.

DIAGNOSIS.– Adult. Wingspan 10-12 mm. Resembles *lienigiella* but differs by the narrower forewings with the broad medial line in basal area from base to yellow fascia and by the genitalia.

Variation. Some specimens from Tunisia have a less pronounced medial white line.

MALE GENITALIA.– Resembles very much those of *lienigiella*, but differs by the short and rounded hump on the right brachium and the less pronounced hump on the right valvella.

FEMALE GENITALIA.– Can be separated from *lienigiella* by the lesser arched posterior edge of sternite VII and the smooth and pointed posterior part of the ostium.

DISTRIBUTION.– Tunisia, Libya, Israel, eastwards to Saudi Arabia and Iran (Riedl, 1994b).

BIOLOGY.– Host plant: *Phragmites australis* (Cav.) Trin. ex Steud. (Poaceae). The larva mines a leaf and can be found from summer till February. Mines cannot be distinguished from those of *scribaiella*. Frass is piled in the lower, narrower part and partly ejected. Pupation inside the mine (Hering, 1957). Adults emerge from late January to late May (Amsel, 1935a, 1958).

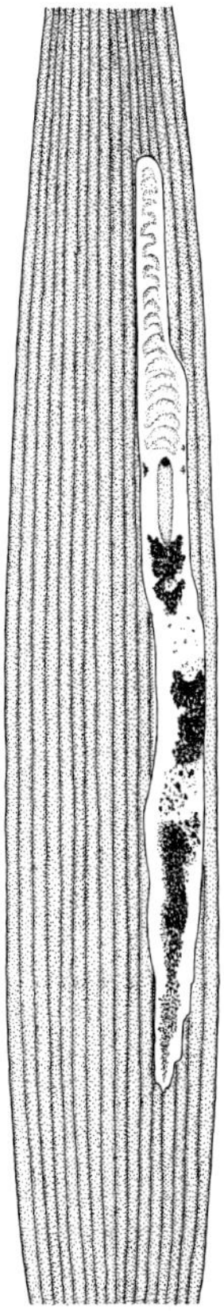

Text-fig. 46. *Cosmopterix lienigiella* Zeller, larval mine in leaf of *Phragmites australis* (Cav.) Trin. ex Steud

Labdia Walker, 1864

Labdia Walker, 1864: 823.
Type species: *Labdia deliciosella* Walker, 1864, by monotypy.

DIAGNOSIS.– Head smoothly scaled; antennae 3/4 to 4/5 length of forewings; labial palpi cylindrical, curved in shape of an arch, segment 3 strongly angled upwards and often reaching well above head. Forewings narrowly lanceolate with variable pattern of lines and dots. Hindwings linear or linear lanceolate.

MALE GENITALIA.– Brachia of unequal length, right one often considerably longer. Valvae long and narrow. Valvellae sharply asymmetrical, left one long digitate. Aedeagus rather short, concealed within very long tubular anellus. Segment VIII with elongated rounded lobes.

FEMALE GENITALIA.– Apophyses posteriores up to one and half times as long as apophyses anteriores. Sterigma oval or elongate, tubular. Ductus bursae long and narrow. Corpus bursae oval with or without signa.

DISTRIBUTION.– Mainly Oriental and Afrotropical regions. Very few species in the Palaearctic, some of them very recently imported.

BIOLOGY.– Larvae have been reported from various plants, but probably use mostly decaying tissues or detritus.

82. Labdia semicoccinea (Stainton, 1859)

Cosmopteryx semicoccinea Stainton, 1859: 123.

DIAGNOSIS.– Adult. Wingspan 11-14 mm. Head beige brown; antennae dark brown, dorsally with white line, apex with 3 white sections. Thorax orange brown, dorsally with two lateral grey lines; tegulae orange brown, lined grey inwardly. Forewings orange brown to beyond 1/2, ochreous yellow in apical part, narrowly lined light yellow on costa and dorsum; series of grey lines of different width, first bordering ochreous yellow costal line, second from costa near base, splitting halfway into two lines, lower one following fold, third from base just above fold to middle of orange brown area, fourth more brownish grey and follows light yellow dorsal line, halfway with branch towards grey line on fold; light yellow spot at one-half of inner side of fourth grey line; narrow brown costal line from 1/3 to 3/4, abruptly broadened in apical part and divided by shining greyish white line; end of first grey line with blackish subcostal spot, basally and distally bordered greyish white; inwardly oblique shining grey streak at dorsal side, opposite blackish spot; similar but more curved streak at tornus, inwardly connected to brownish grey outwardly oblique line; cilia ochreous yellow at costa changing into greyish brown towards apex and around dorsum, with broad ochreous yellow streak at tornus. Hindwings shining grey, brownish tinged towards apex; cilia ochreous grey. Abdomen shining light grey, segments I-III dorsally ochreous. Male with distinct bundle of light yellow androconial setae attached to metathorax just beneath hindwings, curved inwardly over the abdomen and partly covering first three abdominal segments.

MALE GENITALIA.– Right brachium very long, tapering distally with flattened and strongly sclerotized apex; left brachium broad and bifurcate, half length of right one. Valvae lanceolate, tapering distally, strongly narrowed before middle. Left valvella very long and narrow, slightly curved. Aedeagus short, concealed within very slender, upwardly bent and apically hooked anellar tube.

FEMALE GENITALIA.– Posterior edge of sternite VII with deep excavation. Ostium narrow; sterigma protuding into long tube, strongly sclerotized. Ductus bursae as long as corpus bursae, very narrow. Corpus bursae long, pear-shaped; signa absent.

DISTRIBUTION.– Georgia: Abkhasia (probably introduced); Japan, South Korea. Oriental and Australian Regions. In The Netherlands several times accidentally bred from imported plants and trees from Sri Lanka, Taiwan and Japan.

BIOLOGY.– Larvae have been found in the stems of *Cajanus indicus* Spreng. (Fabaceae) (Fletcher, 1920) and *Polyscias* sp. (Araliaceae) (Ulenberg & Van Frankenhuyzen, 1987). In The Netherlands also reared from imported *Zamia* sp. and *Prunus donarium* (Sieb.) (Rosaceae). In Georgia the adults were collected at light in July; the imported specimens hatched throughout the year.

Pyroderces Herrich-Schäffer, 1853

Pyroderces Herrich-Schäffer, 1853: 12 (key), 47.
Type species: *Pyroderces goldeggiella* Herrich-Schäffer, 1853 (= *Cosmopteryx argyrogrammos* Zeller, 1847), by monotypy.
Lallia Chrétien, 1915: 351.
Type species: *Lallia apicinotella* Chrétien, 1915, by monotypy.
Tenuipenna Amsel, 1959: 68.
Type species: *Tenuipenna simplicella* Amsel, 1959, by original designation.

DIAGNOSIS.– Head slightly protruded frontally; eyes with red pigmentation; labial palpi very long and recurved, smoothly scaled. Forewings narrowly lanceolate, apex lengthened and pointed; ground colour varying from yellowish to reddish brown with more or less distinct fasciae or spots. Hindwings very narrow, linear.

MALE GENITALIA.– Tegumen wide, somewhat bulging with strengthening ridge mid-dorsally, right half larger than left half. Brachia well sclerotized and slightly bent; right brachium about one and a half time as long as left one. Valvae simple, spatulate to long and narrow. Valvellae strongly asymmetrical; left valvella long and slender, right one rudimentary or absent. Aedeagus very short, tapering apically, posteriorly with sclerotized ring, often flange-shaped; anellar tube (manica) very long, tapering distally, often with bent apex. Lateral lobes of segment VIII rounded or elongate.

FEMALE GENITALIA.– Apophyses posteriores slightly longer than apophyses anteriores. Ostium very small and surrounded by sclerotized rim; sterigma slightly asymmetrical. Ductus bursae long and narrow. Corpus bursae oval or elongate; two signa of different shape and often with rows of small sclerotized plates.

DISTRIBUTION.– Palaearctic, but only reaching Central Europe to the north; Africa.

BIOLOGY.– Larvae live in the seed heads of Asteraceae, where they eat seeds or predate on other insect inhabitants.

83. Pyroderces argyrogrammos (Zeller, 1847)

Cosmopteryx argyrogrammos Zeller, 1847a: 37.
Pyroderces goldeggiella Herrich-Schäffer, 1853: 47, pl. 13, figs 29, 30.

DIAGNOSIS.– Adult. Wingspan 11-17 mm. Head white, laterally ochreous brown. Antennae ochreous grey with white line from base to beyond 1/4, apical two thirds serrate.

Thorax ochreous brown with white medial line; tegulae ochreous brown, lined white inwardly. Forewings ochreous brown with silky white markings and longitudinal basal line before dorsum; inwardly oblique fascia at 1/3, interrupted in middle; irregular subcostal streak before 1/2, bifurcated towards apex; two spots in middle of dorsum, one spot in disc, some subcostal spots in apical half and an interrupted line from tornus to apex, most streaks and spots irregularly edged dark brown; cilia ochreous brown around apex, yellow ochreous on tornus, greyish towards dorsum. Hindwings shining grey, cilia ochreous grey. Abdomen shining ochreous grey, segments white banded posteriorly.

Variation. The silky white markings and the dark brown edging of the forewings are variable.

Similar species. For differences with *caesaris*, q. v.

MALE GENITALIA.– Tegumen rounded. Right brachium bent, left one almost straight. Valvae spatulate, upper margin slightly concave, lower margin convex. Aedeagus truncate with flange-shaped basal ring bearing two lateral projections, manica tapering distally, tip bent downwards. Left valvella long and slender, slightly narrowed in middle and bent, right one very short and blunt.

FEMALE GENITALIA.– Ostium with sclerotized ring and opening on left side of asymmetrical sterigma. Ductus bursae membranous, slightly shorter than corpus bursae. Corpus bursae oval; signa with two rows of irregular spines of different shape and size.

DISTRIBUTION.– South and Central Europe, to central Germany in the north (Biesenbaum, 2000); the whole Mediterranean region, from Canary Islands to Iran and Central Asia (Sinev, 1984c). Very abundant in steppe and forest-steppe areas.

BIOLOGY.– Host plants: *Carlina* spp., *Centaurea* spp., and *Carduus* spp., also on *Cnicus benedictus* L., *Scolymus hispanicus* L., *Carthamus lanatus* L. and *C. tinctorius* L., *Pycnomon acarna* (L.) Cass., etc. (Constant, 1883; Walsingham, 1907a; Rungs, 1979; Batra et al., 1981; Sinev, 1993a, 1994d). Bivoltine, but in the south and during especially warm seasons a third generation is possible. The larva bores into the floral disk, eating mainly the seeds. Different generations use different host plants. Hibernation and pupation takes place in a light cocoon between the down of the seeds. The damaged seed heads are very conspicuous, especially in winter, as they remain unopened. The moths can be found during nearly the whole warm season. In the Mediterranean, specimens occur irregularly from the end of April to the end of September.

REMARKS.– *Pyroderces goldeggiella* was a Fischer von Röslerstamm manuscript name made nomenclaturally available by Herrich-Schäffer (1853) when describing the new genus.

84. *Pyroderces caesaris* Gozmány, 1957

Pyroderces caesaris Gozmány, 1957: 132.

DIAGNOSIS.– Adult. Wingspan 13-15 mm. Forewings ochreous brown with silky white markings and two narrow longitudinal basal lines in fold and along dorsum; inwardly oblique fascia at 1/3, similar, but less oblique, fascia at 1/2, not reaching costa; very outward oblique streak connected to outer edge of medial fascia; some longitudinal costal streaks towards apex and line from tornus to apex, branched in middle towards dorsal cilia, most streaks and spots irregularly edged dark brown; cilia ochreous yellow around apex, ochreous grey to grey towards dorsum.

Similar species. Differs from *argyrogrammos* by the forewings with a more distinct and continuous fascia, the yellow ochreous cilia at apex, and the tibiae of the hindlegs of which the apical white ring is inwardly edged brown.

MALE GENITALIA.– Tegumen trapezoidal. Right brachium bent downward, left brachium bent inward. Valvae elongated, upper margin slightly concave with a shallow hump in middle, lower margin almost straight. Aedeagus truncate with sclerotized basal ring, manica slightly tapering distally and straight. Left valvella long and slender, slightly bent inward, right one very short. Resembles *argyrogrammos* but differs by the shape of the tegumen, valvae and manica.

FEMALE GENITALIA.– Sterigma almost circular. Ostium in left upper side of sterigma, with sclerotized lobate protrusion. Ductus bursae slightly longer than corpus bursae. Corpus bursae oval; signa formed by two rounded weakly sclerotized plates with pecten of tongue-shaped processes of different size.

DISTRIBUTION.– Mediterranean, from Spain to Ukraine (Crimea) and Turkey.

BIOLOGY.– Host plants: *Echinops ruthenicus* M.B., *Centaurea orientalis* L., *C. salonitana* Vis., *Cirsium sublaniflorum* Sojak. (Sinev, 1993a). Univoltine. The larva bores into the floral disk, eating mainly the young seeds. After hibernation, the larva uses the last years's seeds, but rather often also eats the pupae of other phytophagous insects (Coleoptera: Curculionidae; Diptera: Tephritidae). Pupation takes place inside the seed head between the down of the seeds. Adults can be found from the end of May till the beginning of autumn; such a long flight period is explained by a long and asynchronous development of larvae.

85. Pyroderces klimeschi Rebel, 1938

Pyroderces klimeschi Rebel, 1938: 5.

DIAGNOSIS.– Adult. Wingspan 13-17 mm. Head yellowish white. Antennae with basal two thirds white, annulated brown, apical third divided into 3 brown and 3 white sections. Thorax yellowish white, dorsally with wide pale brown streak, somewhat narrowed in middle and with whitish posterior spot; tegulae brown. Forewings with basal area brown, bordered by very inward oblique ochreous white streak from costa to beyond fold, remaining part of wing ochreous brown, darker in apical area; dorsal white streak, often mixed yellowish, from base to apex, narrowed at tornus; dark brown subapical spot on ochreous white streak at 1/5, similar spot above and below fold as edging of brown basal area; costa from 1/4 to 2/3 with greyish irroration; indistinct greyish streak in middle at 1/2, extending towards apex; indistinct white spot above tornus; ochreous white streak on costa in apical area; cilia ochreous brown at apex to ochreous yellow at dorsum. Hindwings shining ochreous grey, darker towards apex Abdomen ochreous grey, segments with lighter posterior bands and last two segments yellowish white.

Variation. In males the forewings are more uniformly coloured and without the dark brown spots.

Similar species. Differs from the other treated species by the yellowish white head and the pale brown medial line on the white thorax.

MALE GENITALIA.– Tegumen long, tapering distally. Both brachia strongly bent; right brachium about twice as long as left one, tip hooked. Valvae very long, slender and slightly bent. Aedeagus short, truncated, with sclerotized basal ring; manica strongly tapering distally, apical third bent downwards. Left valvella narrowly club-shaped, right valvella in form of rounded hump.

FEMALE GENITALIA.– Ostium in middle of rounded sterigma, with sclerotized ring; sterigma with lateral protrusion at right side. Ductus bursae shorter than corpus bursae. Corpus bursae elongate, narrowed anteriorly; signa formed by two irregular oval scaly plates.

DISTRIBUTION.– Poland, Slovakia, Italy, Austria, Hungary, Romania and Spain (Mallorca).

BIOLOGY.– Unknown. Adults have been collected in May and June.

86. *Pyroderces brosi* Riedl, 1969

Pyroderces brosi Riedl, 1969: 740.

DIAGNOSIS.– Adult. Wingspan 13 mm. Head white, laterally yellowish brown. Antennae ochreous grey with white line from base to 1/5. Thorax yellowish brown; tegulae yellowish brown, white-lined laterally and posteriorly. Forewings yellowish white with pale yellowish brown suffusion; short white basal streak in fold; basal area yellowish brown to 1/4, paler on dorsum, edged by very indistinct white outward oblique fascia; yellowish brown streak above fold from 1/3 to 2/3, apically wider and darker; narrow yellowish brown streak on dorsum at apex; cilia ochreous at apex, yellowish white along dorsum. Hindwings shining grey.

MALE GENITALIA.– Resembles *argyrogrammos* and *caesaris* but differs by the rather stout and parallel-sided right brachium and the slightly bent manica; from *argyrogrammos* also by the bent left brachium and from *caesaris* by the shorter and spatulate valvae.

FEMALE GENITALIA.– Unknown.

DISTRIBUTION.– Italy.

BIOLOGY.– Unknown. The holotype was collected in the first third of June.

REMARKS.– Description is based on a single known specimen, which is rather worn.

87. *Pyroderces tethysella* sp. n.

DIAGNOSIS.– Adult. Wingspan 9-12 mm. Head shining white on frons, vertex and neck tuft yellowish white, suffused with brown scales; collar shining yellowish white dorsally, reddish brown laterally. Antennae with basal two-thirds ochreous white, annulated dark brown, apical third divided into three brown sections of 4 or 5 segments; scapus white with dark brown subapical ring.Labial palpi white, segment 1 very short, segment 2 three-quarters of length of segment 3, strongly mottled greyish brown on outside, segment 3 with indistinct dark brown subapical ring. Thorax yellowish white with reddish brown posterior spot; tegulae reddish brown. Forelegs dark grey, tibiae with subbasal, medial and subapical white rings, tarsal segment 1 with white subbasal and apical rings, segment 2 with white basal ring, segment 5 entirely white; midlegs with femur white, otherwise as forelegs, but tarsal segments 4 and 5 white; hindlegs as midlegs, spurs white, mottled with dark grey on outside in basal half. Forewings reddish brown; strongly mixed dark brown with whitish streak from 1/5 of costa to fold and again to costa at 3/4 and into termen; whitish streak on dorsum from base to 1/3, bending towards fold and ending in coalition with costal streak, below fold on inside edged by dark brown spot; short whitish streak at 1/2 from dorsum along fold to tornus, inwardly edged dark brown; small whitish, dark brown suffused, spot in discal area, indistinctly prolonged as narrow whitish streak on dorsum in apical area; short dark brown streak in costal cilia just before apex; cilia around apex from dark brown, to reddish brown and whitish at dorsal side, with two short but distinct dark brown cilia lines, ochreous grey towards dorsum. Hindwings shining grey, cilia ochreous yellow. Abdomen dorsally shining dark grey, first two segments with reddish brown tinge, ventrally shining white; anal tuft white.

Variation. One female specimen with a darker purplish brown ground colour.

MALE GENITALIA.– Tegumen wide, tapering distally, right part longer than left. Right brachium bent, apex broader and flattened, about twice as long as more strongly bent left one. Valvae long, narrow, gradually bent upwards and with setose basal lobe. Aedeagus very short, almost square, apex shortly humped; manica in basal half cylindrical, apical half tapering, apex very slightly bent. Left valvella long, slender, narrowly club-shaped, and slightly shorter than aedeagus; right valvella very short, rounded.

Similar species. Resembles *klimeschi*, but differs by the longer and narrower left valvella.

FEMALE GENITALIA.– Apophyses anteriores about two-thirds as long as apophyses posteriores. Sterigma large, bulbous, tapering anteriorly and with bent sclerotized rim in middle. Ductus bursae narrow, slightly longer than corpus bursae and with sclerotized band in posterior half. Corpus bursae elongate; signa with two irregular rows of spikes. The large sterigma is characteristic.

DISTRIBUTION.– Spain, Tunisia and Central Asia (Tadzhikistan).

BIOLOGY.– Unknown. The adults have been collected from mid-August to early October.

MATERIAL.– Holotype. ♂, 'South. Tadzhikistan, Tigrovaya Balka, env. Dzhilikul 19.08. [19]90, R. Noreika' 'Genitalia on slide JCK 4839' (ZIN).

Paratypes. 12 ♂ and 6 ♀, labelled as holotype (ZIN); 2 ♂, 'Espana, Prov. Alicante, La Marina, 16.ix.1998, Wolschrijn' (coll. Wolschrijn), 1 ♂, ditto, but genitalia on slide JCK 4591 (coll. Koster), 4 ♂, 'Espana-Alicante, La Marina-Elche, Platje el Pinet, 22/25.ix.2000, J. Wolschrijn' (coll. Wolschrijn); 1 ♀, 'Espana, Prov. Alicante, La Marina, 16.ix.1998, Wolschrijn' 'Genitalia on slide JCK 4595' (coll. Koster); 6 ♂, 'Mauretania, Tunesia-Sud, Oase Metouia (Gabes), 28.9.-10.10.1980, leg. M. u. W. Glaser (LSNK).

ETYMOLOGY.– The name is derived from Tethys, an ancient mediterranean ocean spreaded from the Atlantic to Central Asia in the east.

88. ***Pyroderces wolschrijni*** sp. n.

DIAGNOSIS.– Adult. Wingspan 7-11 mm. Head shining ochreous white on frons, vertex; neck tuft and collar shining reddish brown. Antennae dark grey with white anterior line to 1/3, slightly serrated in apical half; scape shining dark grey with distinct white anterior line, dorsally with white basal streak and narrow white apical ring. Labial palpi white, segment 1 very short, segment 2 three-quarters the length of segment 3, with brownish grey subapical ring, segment 3 ventrally and apically dark grey. Thorax and tegulae shining reddish brown with greyish diffusion anteriorly. Legs dark grey; femora of mid- and hindlegs white with grey streaks; tibiae with subbasal, medial and subapical white rings, spurs white; tarsal segment 1 of forelegs with white subbasal and apical rings, segments 2 and 5 entirely white, tarsal segments of mid- and hindlegs strongly mixed whitish. Forewings reddish brown, with narrow grey costal streak from base to apex, costal half from base to 2/5 strongly suffused grey; indistinct, strong outward oblique, narrow whitish streak subcostally from 1/5 to fold, broadly edged dark grey on both sides or completely mottled grey; irregular dark grey spot on and below fold at or beyond half, not reaching dorsum, narrowly edged white inwards and broadly white outwards; apical half darkened in middle by indistinct and irregular grey streak, edged at apex with some whitish scales on both sides; cilia reddish brown around apex, ochreous grey towards dorsum, with distinct dark grey streak on costal side and similar, but more indistinct, streak at dorsal side of apex. Hindwings shining pale grey, tinged ochreous towards apex, cilia ochreous grey. Abdomen dorsally shining ochreous grey, segments posteriorly lighter banded, ventrally shining white; anal tuft ochreous dorsally, white ventrally.

Variation. The grey suffusion on the forewings is variable. In some cases the suffusion almost or completely covers the whitish oblique line in costal half.

MALE GENITALIA.– Tegumen almost square. Right brachium bent, gradually tapering distally and with small apical hook; left brachium rather stout, apex inwardly bent. Valvae long and narrow, with large basal lobe. Aedeagus rather slender, pointed; manica wide at base, with parallel-sided middle section, apex bent downwards. Left valvella straight and very thin, half as long as manica, right one absent. The very short and thin left valvella in combination with the bent apex of manica are characteristic.

FEMALE GENITALIA.– Apophyses posteriores slightly longer than apophyses anteriores. Ostium small with sclerotized ring. Ductus bursae narrow, widening towards ostium, about two-thirds the length of corpus bursae. Corpus bursae bulbous in upper half and with protrusion near entrance of ductus bursae, lower half very narrow; signa formed by two small rounded scaled plates.

DISTRIBUTION.– Spain, Morocco and Malta.

BIOLOGY.– Unknown. Adults have been collected from mid-April to mid-May and again from mid-July to mid-October, indicating two generations.

MATERIAL.– Holotype. ♂, 'Espana-Alicante, La Marina, 15/22.9.1999, Wolschrijn' 'Genitalia on slide JCK 4837' (ZMA).

Paratypes. 1 ♂, 'Spain, Almeria, Mini Hollywood, 230 m, 14.-15.x.1992, leg. M. Fibiger' (ZMUC); 1 ♂, 'Spain, Malaga, Marbella, Urb. El Mirador, 100 m, 21.ix.1987, Peder Skou leg.' 'Genitalia on slide JCK 4655' (ZMUC); 1 ♀, ditto, 21.ix.1987, Peder Skou leg.' (ZMUC); 1 ♀, 'Hispania, Andalusia, Marbella, El Mirador, 100 m, 10.4.1988, E. Traugott-Olsen'; 1 ♂ and 1 ♀, ditto, 1.11.1990; 1 ♂, ditto, 11.5.1991; 1 ♂, ditto, 19.7.1991; 1 ♀, ditto, 3.8.1991; 1 ♀, ditto, 5.8.1991 (ZMUC, ZIN); 1 ♂, ditto, but genitalia on slide JCK 4774 (coll. Koster); 1 ♀, 'Hispania, Andalusia, Marbella-Casa y Campo, ca 100 m., 14.4.1991, E. Traugott-Olsen' 'Genitalia on slide JCK 4838' (ZMUC); 1 ♂ and 1 ♀, ditto, 26.4.1991; 1 ♂, ditto, 8.5.1991 (ZMUC, ZIN); 7 ♂ and 16 ♀, 'Espana-Almeria, San Jose (Nihar), 6/20.ix.2000, J. Wolschrijn' (WLN); 2 ♀, 'Morocco, Rabat, 1.-15.5.1962, Schwarz leg.' (ZFMK).

ETYMOLOGY.– The species is named after Mr. Jacques Wolschrijn, The Netherlands, who collected most of the material.

89. Pyroderces apicinotella (Chrétien, 1915)

Lallia apicinotella Chrétien, 1915: 351.
Blastodacna cinnamomina Turati, 1930: 85.
Tenuipenna simplicella Amsel, 1959: 68.

DIAGNOSIS.– Adult. Wingspan 8-9 mm. Head yellowish white; antennae greyish brown, ciliate on both sides, with white line from base of scapus to 1/4. Thorax greyish brown, edged white posteriorly; tegulae greyish brown. Forewings yellowish white with strong greyish brown suffusion, strongest in basal area, interrupted by inwardly oblique irregular yellowish white fascia at 1/3 and by broad yellowish white fascia widening towards dorsum at 1/2; yellowish white geniculate fascia before apex followed by dark brown apical streak along costa prolonged into cilia; cilia pale ochreous. Hindwings shining pale grey, basal half suffused greyish brown, cilia on costa and basally on dorsum pale yellow.

MALE GENITALIA.– Tegumen long, slightly tapering distally. Both brachia strongly bent; right brachium apically hooked, about twice as long as left one. Valvae long and slender, distally with group of strongly bent setae. Aedeagus short and truncate; manica almost parallel-sided, distally tapering. Left valvella triangular, right one strongly reduced.

Similar species. Resembles *klimeschi*, but differs by the shape of the manica and the right valvella.

FEMALE GENITALIA.– Unknown.

DISTRIBUTION.– Tunisia, Libya, Iraq.

BIOLOGY.– Unknown. Adults have been collected in May.

Anatrachyntis Meyrick, 1915

Anatrachyntis Meyrick, 1915: 325.
Type species: *Gracillaria falcatella* Stainton, 1859, by original designation.
Lacciferophaga Zagulajev, 1959: 310.
Type species: *Lacciferophaga yunnanea* Zagulajev, 1959, by original designation.
Sathrobrota Hodges, 1962: 10 (key), 73.
Type species: *Batrachedra rileyi* Walsingham, 1882, by original designation.
Amneris Riedl, 1993: 113; homonym of *Amneris* Whitly, 1935.
Type species: *Amneris flexiloquella* Riedl, 1993, by original designation.
Euamneris Riedl, 1996b: 299; unnecessary substitute name for *Amneris* Riedl, 1993.
Type species: *Amneris flexiloquella* Riedl, 1993, by original designation for *Amneris*.

DIAGNOSIS.– Small moths with forewings narrow to very narrowly lanceolate, ground colour reddish brown with irregular fasciae and streaks, cilia lines prominent. Hindwings very narrow and sharply pointed.

MALE GENITALIA.– Tegumen long and narrow. Brachia of different length; right brachium 1 to 2 times as long as left one, pointed. Valvae symmetrical, spatulate or lobate. Valvellae strongly asymmetrical; right valvella well developed and free, left one strongly reduced. Aedeagus tubular or pointed distally.

FEMALE GENITALIA.– Apophyses posteriores slightly longer than apophyses anteriores. Ostium bursae in pouch-like membrane between sternites VII and VIII; sterigma weakly sclerotized, cylindrical or spherical. Ductus bursae narrow and membranous. Corpus bursae membranous, oval; signa absent.

DISTRIBUTION.– The genus has a pan-tropical distribution and is found in Europe only in the Mediterranean region and on the Canary Islands.

BIOLOGY.– The larvae live as scavengers on a variety of fruits and flowers and can be harmful on cotton and other crops (Sinev, 1994d).

90. *Anatrachyntis rileyi* (Walsingham, 1882)

Batrachedra rileyi Walsingham, 1882: 198.
Batrachedra stigmatophora Walsingham, 1897: 104.

DIAGNOSIS.– Adult. Wingspan 9-11 mm. Head pale reddish brown, laterally darker. Antennae blackish brown, annulated white to beyond 1/2, apical part with three dark brown sections of 5 segments each separated by a single white segment. Labial palpi white, segment 2 with dark brown basal and medial bands on outside and greyish brown medial spot on inside, segment 3 with dark brown, mixed reddish brown, subbasal, medial and subapical rings. Thorax and tegulae reddish brown. Forewings reddish brown with brown suffusion in costal half; indistinct and very irregular ochreous fascia at 1/4, at costa prolonged towards base and inwardly edged blackish brown; second fascia before middle, not reaching costa and dorsum and mottled with blackish brown scales; dark greyish brown subcostal streak from second fascia into termen; narrow longitudinal ochreous streak from tornus into dorsal cilia; blackish brown apical spot, lined on inside by ochreous costal spot; narrow blackish brown cilia line above and two below apical spot, latter as an upside down V; cilia reddish brown and blackish brown around apex, ochreous towards dorsum. Hindwings shining very dark grey with ochreous streak at base, cilia ochreous. Abdomen dorsally with segments I-III ochreous, remaining segments pale yellowish grey.

MALE GENITALIA.– Right brachium almost straight, sharply pointed; left brachium short and thick. Valvae spatulate, narrow at base, upper margin concave in basal half, lower margin straight, apex rounded. Left valvella as long as tegumen, slightly widening distally and with bulbous apex, right valvella very short. Aedeagus short, narrowing apically, apex pointed.

FEMALE GENITALIA.– Sterigma cylindrical, posteriorly rounded, slightly protruded from fold between sternites VII and VIII. Ductus bursae shorter than corpus bursae.

DISTRIBUTION.– Canary Islands (Tenerife).

BIOLOGY.– Scavenger. The larvae have been recorded from rotten cotton bolls, cotton seeds, corn, Milo maize, stems of corn, corn husks, flowers of castor bean, etc. (Meyrick, 1914c; Hodges, 1962; Sinev, 1994d). Adults were found in almost every month of the year. Three generations have been observed, and the larvae of the third generation hibernate (Douglas et al., 1962). In tropics there is probably no diapause.

91. *Anatrachyntis badia* (Hodges, 1962)

Sathrobota badia Hodges, 1962: 76.

DIAGNOSIS.– Adult. Wingspan 7-11 mm. Similar to *rileyi* but differs by the more pinkish brown ground colour of the forewings, the lack of the inner cilia line below apex, the shining grey hindwings and the abdomen with segments I-III dorsally pale yellow and the other segments shining grey.

MALE GENITALIA.– Resembles *rileyi* but differs by slender parallel-sided left brachium without bulbous apex and by the valvae gradually widened from base.

FEMALE GENITALIA.– Resembles *rileyi*, but sterigma long, narrowing distally and protruding much further out of fold between sternites VII and VIII. Ductus bursae slightly longer than length of corpus bursae.

DISTRIBUTION.– France (Varenne & Nel, 2001 as *rileyi*), Spain, Canary Islands (Tenerife); U.S.A. (Hodges, 1978). Accidentally imported in Holland (Koster, pers. obs.) and Great Britain (Langmaid & Young, 2002).

BIOLOGY.– Scavenger. The larvae have been found on a variety of materials like: pine cones infested by *Dioryctria* sp. (Pyralidae), rust infected cones of several pine-trees, limes, grapefruit, bananas, cabbage, blossoms of coconut, elm leaves, etc. (Hodges, 1962). There are two, in southern regions maybe three, generations, and adults can be found in most months of the year.

92. *Anatrachyntis coridophaga* (Meyrick, 1925)

Pyroderces coridophaga Meyrick, 1925: 211.
Heinemannia pygmaella Turati, 1927: 342.

DIAGNOSIS.– Adult. Wingspan 9 mm. Head pale reddish brown. Antennae blackish brown, annulated white to 1/3, followed by three dark brown and three white rings of approximately 4 segments each, apical part dark brown. Labial palpi white, segment 2 with brown basal spot on outside and brown subapical ring, segment 3 with dark brown subbasal, medial and subapical rings. Thorax and tegulae reddish brown, broadly edged white; thorax with white band beyond middle. Forewings reddish brown; two white fasciae, first at 1/4, bending towards base at costa, second just before 1/2, slightly outward oblique and not reaching costa, both inwardly edged blackish brown; white dorsal streak at base, similar streak from outside of second fascia along fold to dorsum, bending inward from dorsum towards apex; rather indistinct white subcostal streak from beyond second fascia towards apex, more ore less suffused by reddish brown scales; reddish brown area beyond second fascia mottled by blackish brown scales; two cilia lines; cilia reddish brown around apex, yellowish white towards dorsum. Hindwings shining yellowish white, cilia yellowish white.

Similar species. The shorter and broader forewings with two almost straight white fasciae distinguish *coridophaga* from other species.

MALE GENITALIA.– Brachia stout, left brachium just slightly shorter than right one and abruptly tapering to sharp point. Valvae spatulate, upper margin straight, lower margin straight convex, apex rounded. Left valvella straight, tapering distally, apex rounded; right valvella very small. Aedeagus long and cylindrical, slightly tapering distally. Differs from *rileyi* and *badia* by the thicker brachia and by the large, cylindrical aedeagus.

FEMALE GENITALIA.– Sterigma spherical with large ostium in middle. Ductus bursae longer than corpus bursae. Corpus bursae egg-shaped with elongated lobe ending in ductus seminalis.

DISTRIBUTION.– Egypt, Libya.

BIOLOGY.– Larvae have been found in the buds of *Hibiscus* sp. (Malvaceae). Adults have been recorded in April, July and October.

REMARKS.– *Heinemannia pygmaella* was described from Libya and then erroneously synonymized with *Sathrobrota simplex* by Riedl (1978). The designation of male lectotype deposited now in the BMNH (Sinev, 1999) has shown its identity with male lectotype of *Anatrachyntis coridophaga* described two years earlier from Egypt.

93. Anatrachyntis simplex (Walsingham, 1891)

Pyroderces simplex Walsingham, 1891a: 119.
Stagmatophora gossypiella Walsingham, 1906: 178.
Anatrachyntis hemizopha Meyrick, 1916a: 565.
Pyroderces repandatella Legrand, 1966: 61.
Amneris flexiloquella Riedl, 1993: 113.

DIAGNOSIS.– Adult. Wingspan 10 mm. Head creamy white, dorsally strongly mixed reddish brown; neck tuft creamy white, laterally edged reddish brown. Labial palpi reddish brown outside, white inside, segment 1 very short, segment 2 one-fifth shorter than segment 3, with white apical ring and ochreous subapical inner spot, segment 3 white with blackish medial inner spot and blackish subapical ring. Antennae dark brown, annulated white, apical part with three dark brown sections of 5 segments each; scapus dorsally ochreous with dark grey subapical ring, ventrally white. Thorax reddish brown with broad medial white streak; tegulae reddish brown, inwardly edged white mixed with dark brown. Forewings reddish brown, at apex ending in narrow streak to termen; white fascia at 1/4, bending towards base and narrowing on costa; white spot on dorsum at 1/2, outwardly edged by dark brown spot; several narrow longitudinal white streaks, more or less intermixed with dark brown, one on dorsum from base to fascia, one in fold ending in cilia, one from fascia just above fold, splitting in disc into costal and dorsal parts, one subcostal, starting opposite dorsal spot and ending in costal cilia, last one at costa from 1/3 towards apex; dark brown spot on tornus; blackish spot at apex at each side of narrow reddish brown streak; short blackish cilia line on termen around apex; cilia yellow-ochreous on costa and around termen, reddish brown on costa at extreme apex, more greyish towards dorsum. Hindwings shining grey, cilia ochreous grey. Abdomen dark grey, anal tuft creamy white.

MALE GENITALIA.– Sternite VII square with deep V–shaped excavation anteriorly. Right brachium long, strongly bent and with enlarged apex; left brachium straight, with length less than one third of right one. Valvae spatulate and strongly ciliate, with large lobe at base of lower margin. Left valvella short and bent, right one absent. Aedeagus very long, cylindrical, widened at base. Differs from other species of *Anatrachyntis* by the shape of the right brachium and the valvae.

FEMALE GENITALIA.– Posterior edge of sternite VII convex. Sterigma bucket-shaped. Ductus bursae widening towards ostium and slightly longer than corpus bursae.

DISTRIBUTION.– Spain, Portugal, Cyprus, Morocco, Egypt. Accidentally imported into Great Britain (Heckford, pers. comm.).

BIOLOGY.– In the tropics larvae feed on injured seeds, fibre and general debris found in cotton bolls attacked by other invertebrates (Fletcher, 1920). Also reported from African oil palm, *Elaeis guineensis* Jacq. (Arecaceae) and *Eriodendron anfractuosum* DC. (Bombacaceae) (Fletcher, 1933). The moths in Europe were collected in August and October. Several generations in the tropics.

REMARKS.– *Batrachedra coriacella* was described from Java by Snellen (1901) and treated (following Fletcher, 1928) as a junior synonym of *Anatrachyntis simplex* by the authors. However, the careful study of its female lectotype deposited in the RMNH has shown clear differencies in genital structures, and species has been restored from the synonymy (Sinev, 2002a).

Coccidiphila Danilevsky, 1950

Coccidiphila Danilevsky, 1950: 49.
Type species: *Coccidiphila gerasimovi* Danilevsky, 1950, by original designation.
Batrachedropsis Amsel, 1955: 15.
Type species: *Cosmopteryx ledereriella* Zeller, 1850, by monotypy.

DIAGNOSIS.– Small to very small moths. Head smoothly scaled; antennae three-quarters length of forewings; labial palpi cylindrical and porrect. Forewings narrowly lanceolate, ground colour pale yellow to reddish ochreous and more or less suffused by dark brown, pattern of broad dark brown costal streak and very outward oblique dark brown streak from dorsum to apex. Hindwings linear.

MALE GENITALIA.– Tegumen short and broad. Brachia of unequal length; right brachium long and often bent, left one short and cone-like. Valvae elongate, widening distally. Valvellae long and narrow, bent or hooked. Aedeagus short and wide; coecum penis with pair of protrusions.

FEMALE GENITALIA.– Apophyses posteriores of same length or slightly longer than apophyses anteriores. Sterigma large and sclerotized, rounded and often with lateral protrusions. Ductus bursae membranous, rather wide and about half length of corpus bursae. Corpus bursae egg-shaped, often covered with pectinations.

DISTRIBUTION.– Mainly restricted to the West-Palaearctic Region, but a few species are also known from Africa and South America.

BIOLOGY.– The larvae have been mentioned as egg-feeders in the colonies of scale insects (Homoptera: Coccoidea), but sometimes they also live on detritus and seed pods.

REMARKS.– *"Coccidiphila" charlierella*, described from a single male specimen (Real, 1989), has been recently synonymized with *Phyllonorycter millierella* (Staudinger) in the Gracillariidae (Réal, 1997).

94. Coccidiphila gerasimovi Danilevsky, 1950

Coccidiphila gerasimovi Danilevsky, 1950: 49.

DIAGNOSIS.– Adult. Wingspan 7-11 mm. Forewings yellowish ochreous, costa from base to beyond 2/3 dark grey; slightly bent and very outward oblique dark grey streak from middle of dorsum to apex, prolonged towards tornus; two dark brown cilia lines around apex; cilia dark grey at apex, ochreous grey towards dorsum. Hindwings shining grey.

Variation. Dark grey pattern of the forewings sometimes intensified and spread all over the wing surface, with only some traces of the ochreous coloration visible.

Similar species. Resembles *ledereriella*, but generally darker. Only examination of genitalia makes correct identification possible; for males it will be sufficient to unscale the anal tuft to reveal the peculiar right brachium.

MALE GENITALIA.– Right brachium porrect, tapering distally; left brachium cone-shaped. Valvae from narrowest part beyond middle gradually widening into triangular cucullus. Valvellae strongly bent near base, remaining part just slightly bent, widest in middle and gradually tapering distally. Aedeagus with apex rectangular, ventrally with long, tongue-shaped protrusion.

FEMALE GENITALIA.– Excavation of posterior margin of sternite VII very shallow. Ostium very wide; sterigma rounded anteriorly and flattened posteriorly, with large and oblique lateral lobes. Corpus bursae densely covered with crescent-shaped pectinations.

DISTRIBUTION.– Mediterranean from Spain to Asia Minor, Near East and western Caucasus; Canary Islands.

BIOLOGY.– Larvae feeds on the eggs of Homoptera, Coccoidea of the following species: *Pseudococcus maritimus* Ehrh., *Laccifer lacca* Kerr., *Leucanium prunastri* Fonsc., and *Aonidella citrina* (Danilevsky, 1950; Zagulajev, 1981; Zhang, 1994). Adults have been collected from April to October.

95. *Coccidiphila patriciae* J. Nel & A. Nel, 2000

Coccidiphila patriciae J. Nel & A. Nel, 2000: 381.

DIAGNOSIS.– Adult. Wingspan 9-10 mm. This species cannot be separated with certainty from *gerasimovi* by external characters, because of the overall similarity and the very limited material of *patriciae*.

MALE GENITALIA.– Unknown.

FEMALE GENITALIA.– Resemble those of *gerasimovi*.

DISTRIBUTION.– Canary Islands (Tenerife).

BIOLOGY.– Both known females of this species have been reared from seed pods of *Ixanthus viscosus* (Alt.) Griseb. (Gentianacaeae), of which the larva eats the seeds. When full-grown it makes a silken cocoon mixed with dry excrement. The cocoon inside the seed pod is prolonged with a short silken tube used as exit. Adults emerged in the first third of February (Nel & Nel, 2000).

REMARKS.– Figures of the adult and female genitalia are not given. The species has been described from two females, and the genitalia slide of the holotype is difficult to interpret due to distortion. Additional material, especially males, is needed to clarify the status of *patriciae*.

96. *Coccidiphila ledereriella* (Zeller, 1850)

Cosmopteryx ledereriella Zeller, 1850: 198.
Teleia? unedella Millière, 1876: 331.
Coccidiphila rungsella J. Nel & Brusseaux, 1997: 455. **Syn. n.**

DIAGNOSIS.– Adult. Wingspan 8-10 mm. Head shining white, tinged ochreous in middle; neck tuft strongly mottled with dark grey scales. Antennae basal two-thirds ochreous grey, annulated dark grey; apical third dark grey with 5 white rings. Thorax dark grey; tegulae dark grey, broadwhite edged laterally. Forewings yellow-ochreous, often more or less mottled with dark grey scales, especially in apical area; grey costal streak from base to beyond 2/3, widest in middle; irregular dark grey spot in fold at 1/3; dark grey streak, very outward oblique and bent, from middle of dorsum to apex; small grey dorsal streak at base. Cilia ochreous grey with two dark brown cilia lines around apex. Hindwings shining grey. Abdomen grey.

MALE GENITALIA.– Right brachium shorter than tegumen, moderately to strongly curved, laterally flattened and heavily sclerotized; left brachium short and rounded. Valvae rather compact, with cucullus almost circular. Valvellae straight, bent at right angle beyond middle; distal part tapering into sharp apex. Aedeagus obliquely cut at apex.

FEMALE GENITALIA.– Posterior margin of sternite VII with semicircular excavation and sharp and thorny lateral edges. Ostium circular; sterigma spherical with narrow sclerotized horizontal rim anteriorly and with short and wide lobe posteriorly. Corpus bursae densely covered with crescent-shaped pectinations.

DISTRIBUTION.– Mediterranean area. Old records from Asia Minor and northern Africa should be verified; most of them deal with *gerasimovi* and *danilevskyi*. Records from the Canary Islands can be accounted for the confusion with *riedli*.

BIOLOGY.– The larva lives on insect remains and debris from spider webs or nests of gregarious lepidopteran larvae (Millière, 1876), also found on dry rubbish of old seed-pods, seed heads and galls of numerous plants (Walsingham, 1891b, 1908; Suire, 1940; Lhomme, [1948-1949]; Rungs, 1979). Probably develops in several generations. Adults can be found from early April to the end of September, in Morocco also in January and March.

REMARKS.– There are no reliable differences between *ledereriella* and *rungsella* in both external characters and genitalia. The drawings of the paratype and holotype of *rungsella* in Nel & Brusseaux (1997) and Nel (2000) clearly shows the typical male genitalia of *ledereriella*.

97. Coccidiphila danilevskyi Sinev, 1997

Coccidiphila danilevskyi Sinev, 1997a: 205.
Coccidiphila ledereriella auctt., (*nec* Zeller, 1850); misidentification.

DIAGNOSIS.– Adult. Wingspan 7-10 mm. Very similar to *ledereriella*, but the ground colour more intensely ochreous or reddish ochreous. A safe identification is possible only by the genitalia characters.

MALE GENITALIA.– Very similar to *ledereriella* but differs by the right brachium which is longer than the tegumen, narrow valvae, and the less widened and more spatulate cucullus. Valvellae slightly bent upwards, distal third acutely bent. Aedeagus with triangular protrusion on left side of apex.

FEMALE GENITALIA.– Resembles very much *ledereriella* but differs by the smaller oval ostium and sterigma with the broader horizontal rim and the large and almost square posterior lobe. Corpus bursae less densely covered with smaller pectinations.

DISTRIBUTION.– Southern France (including Corsica), Spain, Morocco and Tunisia.

BIOLOGY.– Unknown. Adults have been collected from May to September. In North Africa also in February (Morocco) and September-October (Tunisia), indicating the possibility of two generations (Sinev, 1997a).

REMARKS.– The distribution of *danilevskyi* and *ledereriella* requires further investigation. These externally very similar species are probably allopatric and both restricted mainly to the western Mediterranean area. The validity of *danilevskyi* remains somewhat doubtful as many characters of wing pattern and genitalia are slightly varying geographically. It can be treated as a subspecies of *ledereriella*, but in every case, more material is necessary to make a final decision.

98. Coccidiphila kasypinkeri Traugott-Olsen, 1986

Coccidiphila kasypinkeri Traugott-Olsen, 1986: 40.

DIAGNOSIS.– Adult. Wingspan 6-7 mm. Forewings yellow ochreous; large triangular dark grey costal streak from base to 2/3, with ochreous yellow subcostal spot in middle; broad and very outward oblique dark grey streak extending around apex to tornus; cilia ochreous grey with two grey brown cilia lines at apex, pale ochreous towards dorsum. Hindwings shining pale ochreous.

Similar species. Differs from the other species by its small wingspan and by the yellow-ochreous spot on the dark grey costal streak of the forewings.

MALE GENITALIA.– Right brachium about as long as tegumen, strongly curved, apex rounded; left brachium short and triangular. Valvae narrow, cucullus ventrally enlarged in triangular extension. Valvellae very long and narrow, squarely bent in middle, distal part gradually tapering. Aedeagus with oblique apex.

FEMALE GENITALIA.– Posterior margin of sternite VII with shallow V–shaped excavation. Sterigma trapezoidal and strongly sclerotized, anterior margin concave. Corpus bursae without pectinations.

DISTRIBUTION.– Canary Islands (Lanzarote, Fuerteventura).

BIOLOGY.– Unknown. Adults have been collected in January and March.

99. Coccidiphila riedli Traugott-Olsen, 1986

Coccidiphila riedli Traugott-Olsen, 1986: 37.
Coccidiphila ledereriella auctt., (*nec* Zeller, 1850); misidentification.

DIAGNOSIS.– Adult. Wingspan 6-9 mm. Forewings yellow ochreous; costa grey from base to beyond 2/3, dark grey spot in fold; base dorsally ochreous yellow; very outward oblique dark grey streak broad.

Similar species. Differs from *gerasimovi*, *ledereriella* and *danilevskyi* by the lack of the dark grey scales on dorsum at base of the forewings and from *kasypinkeri* by the absence of the yellow spot in the grey costal area. For certain identification, see genitalia characters.

MALE GENITALIA.– Right brachium longer than tegumen, strongly curved and distally with fin-like protrusion; left brachium very small. Valvae very narrow, cucullus broadened, spatulate and bent upwards. Valvellae slightly bent and strongly widening distally, apex narrow and sharply and squarely bent. Aedeagus with slightly oblique apex.

FEMALE GENITALIA.– Posterior margin of sternite VII with large U–shaped excavation; lateral edges rounded and thorny. Sterigma with large triangular lateral lobes, concave anterior margin, and square posterior lobe. Corpus bursae densely covered with crescent-shaped pectinations.

DISTRIBUTION.– Canary Islands (Gran Canaria, Tenerife).

BIOLOGY.– The larvae have been found among the dead leaves of *Ceballosia* (= *Messerschmidia*) *fruticosa* (L. f.) G. Kunkel (Boraginaceae), but also *Plumbago capensis* Thunb. (Plumbaginaceae) might be a possibility (Klimesch, 1983). Adults have been collected from February to April.

Stagmatophora Herrich-Schäffer, 1853

Stagmatophora Herrich-Schäffer, 1853: 13 (key), 49.
Type species: *Oecophora heydeniella* Fischer von Röslerstamm, 1841: 256, by monotypy.

DIAGNOSIS.– Small moths. Forewings lanceolate, apex pointed, ground colour varying from reddish orange to dark brown with costal spots or fasciae and often with tubercular spots; cilia line present or absent. Hindwings very narrow.

MALE GENITALIA.– Tegumen well developed. Brachia strongly asymmetrical; right brachium much longer than left one. Valvae broad and rounded. Right valvella larger than left one. Aedeagus with bulbous base and long tubular distal part.

FEMALE GENITALIA.– Apophyses posteriores longer than apophyses anteriores. Sterigma rounded. Ductus bursae narrow, membranous. Corpus bursae elongate, membranous, with two inwardly projecting signa.

DISTRIBUTION.– Palaearctic.

BIOLOGY.– The only species in Europe feeds on Lamiaceae.

REMARKS.– In the past many European species were placed in *Stagmatophora* but subsequently they have been transferred to various genera, e.g. *Eteobalea*, *Vulcaniella*, *Isidiella*, *Hodgesiella* and *Tolliella*. At present in the western Palaearctic only *heydeniella* remains in this genus.

100. *Stagmatophora heydeniella* (Fischer von Röslerstamm, 1841)

Oecophora heydeniella Fischer von Röslerstamm, 1841: 256, pl. 88, figs 1a-d.

DIAGNOSIS.– Adult. Wingspan 7-9 mm. Head shining bronze grey; collar shining dark brown. Antennae shining dark brown with two white subapical rings. Thorax and tegulae shining bronze grey. Forewings bright orange in middle, shining dark brown in basal fourth; slight inwardly oblique silver metallic fascia at 1/5, not reaching dorsum, and dark brown-edged outwardly; orange section with five dark brown-edged, silver metallic spots, two costal, one subdorsal, one tornal and very small median, outer two with pale golden gloss; apical third shining dark brown; large silver metallic spot with purplish reflections around apex; cilia on costa dark brown, paler around apex towards dorsum. Hindwings greyish brown. Abdomen shining greyish brown.

MALE GENITALIA.– Tegumen long and rather narrow. Right brachium long, wide at base, narrowed in distal half, apex rounded and with hooked rim; left brachium less than twice length of right one, apex pointed. Valvae semicircular, inner margin with shallow hump. Right valvella tapering distally, apex blunt; left valvella fully attached to aedeagus and with row of small thorns. Aedeagus at base bulbous and almost circular, with large and wide coecum penis; tubular part long, bent before middle and widening apically.

FEMALE GENITALIA.– Segment VIII short, strongly tapering before posterior edge. Ostium with V–shaped sclerotization; sterigma egg-shaped. Ductus bursae narrow, about one-third the length of corpus bursae. Corpus bursae elongate, with two large, funnel-shaped, dentate signa of equal size.

DISTRIBUTION.– Central Europe southwards to the Balkans and eastwards to European Russia.

BIOLOGY.– Host plants: *Stachys (=Betonica) officinalis* (L.) Trev., *S. sylvatica* L. and

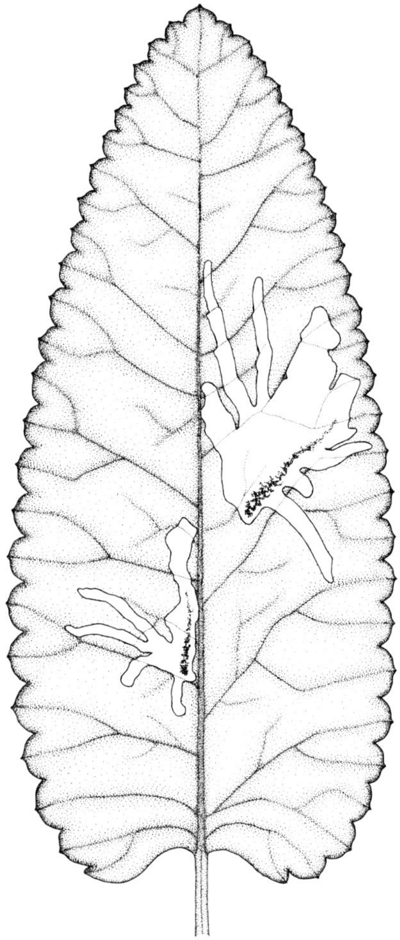

Text-fig. 47. *Stagmatophora heydeniella* (Fischer von Röslerstamm), larval mine in leaf of *Stachys officinalis* (L.) Trev.

S. alopecurus (L.) Bth. (Lamiaceae). The larva mines a leaf in August and September. The mine forms an irregular blotch with many short galleries; frequently several mines are placed in one leaf. Later the larva constructs a silk-lined gallery on the underside of a leaf, longitudinal to the midrib, serving as a shelter when it does not feed. One end of this shelter debouches into the mine. Frass is heaped in a string in the middle of the mine or partly removed. Pupation takes place in the shelter or in a fold on the edge of a leaf in a small white cocoon. The pupa hibernates (Fischer von Röslerstamm, 1841). Adults fly in May and June. See text-fig. 47.

Eteobalea Hodges, 1962

Eteobalea Hodges, 1962: 10 (key), 66.
Type species: *Gelechia sexnotella* Chambers, 1878, by original designation.
Parastagmatophora Riedl, 1965: 423.
Type species: *Oecophora serratella* Treitschke, 1833, by original designation.

DIAGNOSIS.– Small to medium sized moths. Forewings lanceolate, apex pointed, ground colour varying from yellow ochreous to dark blackish brown with white costal spots or fasciae and often with tubercular spots; cilia line present or absent. Hindwings shining white to grey.

MALE GENITALIA.– Tegumen well developed, quadrangular, upper part with deep excavation. Brachia strongly asymmetrical; right brachium 1.5-2 times as long as left one and often with broadened and sclerotized tip. Valvae symmetrical, broadly lobate, distally with dense setae. Right valvella well developed, left one strongly reduced. Aedeagus more or less bent with bulbous base and tubular distal end; coecum penis pointed, cornuti absent. Lobes of segment VIII rounded. The shape of the right brachium, the aedeagus and the right valvella are distinctive for the species.

FEMALE GENITALIA.– Apophyses posteriores longer than apophyses anteriores. Ostium bursae opens in membrane ventrally between sternites VII and VIII; sterigma weakly sclerotized, bowl-shaped or spherical. Ductus bursae long and narrow, membranous. Corpus bursae oval to elongate, membranous, with posterior extension ending in ductus seminalis; two inward projecting signa, sometimes absent. Sterigma shows important diagnostic characters.

DISTRIBUTION.– The genus has mainly Holarctic distribution. In Europe most species occur in the Mediterranean area.

BIOLOGY.– The host plants and life style of many species is insufficiently known or completely undescribed. The larvae feed on herbs of the families Scrophulariaceae, Globulariaceae, and Lamiaceae, living in the roots, stems or flowerheads, sometimes as gall-makers.

REMARKS.– Some of the species cannot be separated by their external appearance.

Key to species of *Eteobalea*

1.	Forewings ground colour yellow ochreous or golden brown	2
–	Forewings ground colour dark brown or blackish brown	4
2.	Forewings ground colour golden brown with 15-18 tubercular shining golden spots	***isabellella***
–	Forewings yellow ochreous with only 5 tubercular greyish brown spots with golden gloss	3
3.	Forewings with white dorsal patch at base, apical cilia white without cilia line	***dohrnii***
–	Forewings yellow ochreous at base, apical cilia white with two dark brown cilia lines	***siciliae***
4.	Forewings blackish brown with silky white markings	5
–	Forewings dark brown or blackish brown with silky white markings and/or silver, pale golden or golden brown markings	6
5.	Thorax with white medial line, dorsal spot in middle of forewings rounded and not reaching dorsum; antennae dorsally with white basal line	***albiapicella***
–	Thorax without white medial line, dorsal spot in middle of forewings elongate and reaching dorsum; apex of antennae with 5 white segments separated by 3 brown segments	***alypella***
6.	Forewings with three silvery fascia	***tririvella***

–	Forewings with tubercular pale golden or golden brown spots	7
7.	Forewings with tubercular golden brown spots with purplish gloss, dorsal half of forewings paler	***thaumatella***
–	Forewings with tubercular spots pale golden and lack purplish gloss	8
8.	Forewings with tubercular pale golden spot posteriorly connected to third white costal spot	9
–	Third white costal spot on forewings without tubercular pale golden spot	10
9.	Femur of hindleg dark brown	***anonymella***
–	Femur of hindleg white	***serratella***
10.	Underside of forewings shining yellowish grey; very large species	***klisieckii***
–	Underside of forewings shining brown	11
11.	Subdorsal retinaculum yellow ochreous	***teucrii***
–	Subdorsal retinaculum brown	12
12.	Third white costal spot with concave outer edge, thorax and tegulae with some posterior golden edging	***intermediella***
–	Third white costal spot square, thorax and tegulae with very sparse golden edging	***beata***
–	Third white costal spot rectangular and large, thorax and tegulae broadly edged golden	***sumptuosella***

External characters are usually insufficient for proper identification of *beata*, *intermediella* and *sumptuosella*. These species can be safely determined by genitalia only.

101. Eteobalea anonymella (Riedl, 1965)

Stagmatophora (Parastagmatophora) anonymella Riedl, 1965: 433.
Stagmatophora serratella auctt., (*nec* Treitschke, 1833); misidentification.

DIAGNOSIS.– Adult. Wingspan 8-15 mm. Head yellowish white, ochreous to light brown laterally and posteriorly; antennae dark brown with five white annulations apically. Thorax shining dark brown, posteriorly with some lighter spots; legs shining dark brown. Forewings shining dark brown; three white costal spots: first at 1/6, prolonged into tubercular pale golden, outward oblique, metallic streak to beyond fold, but not reaching dorsum, second before 1/2, changing into tubercular pale golden, slightly outward oblique, metallic streak, reaching to middle, third at 3/4, more than twice as large as inner costal spots, and with small tubercular pale golden metallic spot connected underneath; two tubercular pale golden metallic spots on dorsum, inner opposite second costal, outer almost perpendicular to dorsum and inward of third costal spot; tubercular pale golden metallic spot at apex; series of small tubercular pale golden metallic streaks at termen, mainly dorsally. Hindwings shining light grey, darker towards apex. Abdomen dorsally shining brown.

Similar species. Resembles *serratella*, but differs by the smaller wingspan, the more silvery tubercular markings of the forewings and the dark brown femora of hindlegs.

MALE GENITALIA.– Tegumen narrowing posteriorly, with V–shaped excavation. Right brachium almost straight, twice as long as left one, tip only slightly broadened; left brachium with bluntly pointed apex. Valvae pear-shaped. Right valvella slightly tapering distally, with rounded tip. Aedeagus slightly bent, bulbous part oval, tubular part narrowing distally, apex sharply pointed.

FEMALE GENITALIA.– Segment VIII about as long as wide. Ostium circular; sterigma spherical. Ductus bursae about as long as length of corpus bursae. Corpus bursae short and oval, abruptly narrowed posteriorly; two crescent-shaped signa of different size.

DISTRIBUTION.– Europe, except for northern Fennoscandia; northern Africa, Asia Minor, Near and Middle East, eastwards to Iran and Central Asia. In the south restricted to mountain regions.

BIOLOGY.– Unknown. Adults can be found, sometimes in plenty, from May to mid-August. Probably a second generation from late August to early September (Riedl, 1969).

102. *Eteobalea serratella* (Treitschke, 1833)

Oecophora serratella Treitschke, 1833: 168.
Stagmatophora sareptensis Walsingham, 1907b: 178.
Stagmatophora cinereocapitella Caradja, 1920: 148.

DIAGNOSIS.– Adult. Wingspan 13-19 mm. Similar to *anonymella*, but differs by the larger wingspan, the more prominent pale golden markings with a bronzy gloss on the thorax, tegulae and forewings, the white femora of the hindlegs, and by the ochreous first four segments of the abdomen.

MALE GENITALIA.– Tegumen rather narrow, almost parallel-sided, excavation U–shaped. Right brachium bent, about twice as long as left one, tip broadened and round, heavily sclerotized; left brachium with blunt point. Valvae strongly concave to boot-shaped, cucullus narrowly rounded. Right valvella large, thickest in middle and bluntly tapering. Aedeagus slightly bent, bulbous part slender, tubular part very wide, slightly tapering to sharp apex.

FEMALE GENITALIA.– Segment VIII longer than wide. Ostium circular; sterigma pouch-shaped with transverse sclerotized ridge in middle. Ductus bursae less than half as long as corpus bursae. Corpus bursae elongate, distinctly widened posteriorly; signa very small or absent.

DISTRIBUTION.– Central Europe from eastern Germany and Poland southwards to the Mediterranean coast, south-east of European Russia, Asia Minor, Near East, Central Asia.

BIOLOGY.– Host plants: *Linaria vulgaris* Mill. and *L. genistifolia* ssp. *euxina* Velen. (Scrophulariaceae). Also *Antirrhinum majus* L. has been mentioned (Joannis, 1921; Poutiers, 1921) but these records should be verified. Univoltine. Eggs are laid singly on the lower part of the stems from whence the larvae mine downwards to the central root. The final instar larva bores back into the lower parts of the stem. The larval development lasts about 11 months. Pupation takes place in the stem base or in the upper part of the root (Saner, 1991). Adults fly from the end of June to early September.

REMARKS.– The name *serratella* Treitschke, generally accepted for this species, was mentioned as a junior synonym of *Phalaena gronoviella* Scopoli, 1772 by Werneburg (1864) without any substantiation of this opinion. Such a synonymization was not accepted by contemporaries, and the name *gronoviella* Scopoli was not in use for more than one hundred years. Only very recently it was restored in the sense of Werneburg in the List of Iberian Lepidoptera (Vives Moreno, 1992). Unfortunately, the material of the Scopoli collection has been lost, and his description of *gronoviella* is too vague. It could refer to any of the dark species of the genus *Eteobalea*, but also to *Eulamprotes wilkella* (Linnaeus, 1758), belonging to Gelechiidae. Thus, *serratella* Treitschke remains the first valid name for the species, based on well preserved type specimens and repeatedly used in both faunistic and taxonomic publications during the last hundred and fifty years (Sinev, 1999).

103. *Eteobalea intermediella* (Riedl, 1966)

Stagmatophora intermediella Riedl, 1966a: 77.
Stagmatophora beata auctt., (*nec* Walsingham, 1907); misidentification.

Diagnosis.– Adult. Wingspan 11-17 mm. Forewings slightly narrowing towards apex, pattern in general like *serratella*, but without the pale golden spot underneath the third white costal spot.

Similar species. The species strongly resembles *beata* and *sumptuosella*, but it usually differs from these species by the third costal spot, which is smaller, more square and often with a convex inner edge and a concave outer edge. For certain identification genitalia should be examined.

Male genitalia.– Tegumen long, slightly narrowing posteriorly, excavation rounded. Right brachium slightly bent, about three times as long as left one, tip broadened, rounded and heavily sclerotized; left brachium with square apex. Valvae concave with rounded cucullus. Right valvella parallel-sided, gradually tapering to blunt apex. Aedeagus rather slender, bent, tubular part hardly tapering, apex sharp. Differs from *beata* and *sumptuosella* by the rounded, slightly broadened and sclerotized tip of the right brachium in combination with the slender aedeagus with wide tubular part and the large tapering right anellus arm.

Female genitalia.– Segment VIII about as long as wide. Ostium semi-circular or oval; sterigma pouched, narrowing distally. Ductus bursae slightly shorter than length of corpus bursae. Corpus bursae elongate, widest anteriorly, gradually tapering posteriorly; two crescent-shaped signa of different size.

Distribution.– Central and southern Europe, the Caucasus; Asia Minor, Near and Middle East to Iran and Central Asia.

Biology.– Host plants: *Linaria vulgaris* Mill., *L. pontica* L., *L. genistifolia* (L.) Mill., *L. dalmatica* (L.) Mill., and *Anarrhinum bellidifolium* (L.) Willd. (Scrophulariaceae). Bivoltine or univoltine (in the north and in the east of the distribution area). Eggs are laid in clusters on the lower part of the stems from whence the larvae mine downwards to the central root, often gregariously (Saner, 1991; Sinev, 1993a). Most of the development takes place in the root, where both the cortex and central parts are attacked. The final instar larvae bore back into the lower parts of the stem, where they may damage the base of several stems considerably. Pupation takes place in the stem base or in the upper part of the root, without distinct cocoon. The number of larvae per root varies with the root diameter. The immature stages are very similar to *serratella*, and differences have been found only in the chorion structure of the eggs (Rizza & Pecora, 1979). Adults fly from early May till early November.

104. *Eteobalea beata* (Walsingham, 1907)

Stagmatophora beata Walsingham, 1907b: 180.

DIAGNOSIS.– Adult. Wingspan 13-22 mm. Resembles *intermediella* and *sumptuosella*. Differs from *intermediella* by the third costal spot, which is more prolonged towards the apex. For separation from *sumptuosella*, q. v. Only genitalia examination yields safe identification.

MALE GENITALIA.– Tegumen parallel-sided, excavation of its posterior margin very large and deep, rounded. Right brachium slightly bent, about two to two and half times as long as left one, with large and flattened apex; left brachium sharply pointed. Valvae boot-shaped. Right valvella thick, club-shaped. Aedeagus almost straight, tubular part stout, gradually tapering. The large and flattened apex of the right brachium, the short and stout aedeagus and the large club-shaped right valvellae are characteristic.

FEMALE GENITALIA.– Segment VIII wider than long. Ostium bursae semi-circular; sterigma bowl-shaped with transverse sclerotized ridge. Ductus bursae less than twice length of corpus bursae. Corpus bursae elongate, posterior part widest, but then with bulge and long, gradually tapering extension; two small triangular signa.

DISTRIBUTION.– Western Mediterranean area (southern France, Italy, Iberian Peninsula, and northern Africa). All records of this species from more eastern regions belong to *intermediella* or *anonymella*.

BIOLOGY.– The species has been collected in Italy from *Antirrhinum* sp. (Scrophulariaceae) (Saner, 1991). The food plant and the larva life style are probably similar to those of *intermediella* and *serratella*. Adults can be found from May to August.

REMARKS.– The distributional data published before 1965 refer to *Stagmatophora beata* sensu lato (a complex of *beata*, *intermediella*, and *anonymella*).

105. *Eteobalea sumptuosella* (Lederer, 1855)

Lita admirandella Bruand, [1851]: 80; nomen nudum.
Elachista sumptuosella Lederer, 1855: 234.
Stagmatophora amseli Osthelder, 1940: 115.

DIAGNOSIS.– Adult. Wingspan 13-20 mm. Resembles closely *intermediella* and *beata*, but differs by the bronzy gloss of the pale golden markings of the forewings and the large pale golden posterior edging of thorax and tegulae. From *intermediella* it also differs by the third costal spot, which is larger and more or less considerably extended towards the apex. Safe identification only possible by examination of the genitalia.

MALE GENITALIA.– Tegumen broad, slightly narrowing posteriorly, with V–shaped excavation. Right brachium bent, two to two and half times as long as left one, tip broadened and heavily sclerotized; left brachium pointed. Valvae short, boot-shaped. Right valvella thickest in middle and tapering to rounded apex. Aedeagus strongly bent, bulbous part almost circular, tubular part strongly narrowing distally, apex very sharp.

Specimens from the western Mediterranean differ in the male genitalia by the less bent, almost parallel-sided and wider tubular part of the aedeagus and the less sclerotized apex of the right brachium. Specimens from former Yugoslavia are intermediate in these characters between those from the western and eastern part of the Mediterranean area. The large and rounded tip of the right brachium, the almost circular bulbous part of the aedeagus and the trapezoid right valvella distinguish this species from *beata* and *intermediella*.

FEMALE GENITALIA.– Segment VIII shorter than wide. Ostium oval; sterigma pouch-shaped. Ductus bursae two-thirds length of corpus bursae. Corpus bursae posteriorly with short extension; two crescent-shaped signa, left one often very small, sometimes absent.

Distribution.– Mediterranean area, eastwards to Asia Minor, the Caucasus, Near East and Central Asia.

Biology.– Unknown. Adults have been collected from early June to the end of August.

106. *Eteobalea teucrii* (Walsingham, 1907)

Stagmatophora teucrii Walsingham, 1907b: 179.

Diagnosis.– Adult. Wingspan 16-19 mm. Head white, ochreous to light brown laterally and posteriorly; antennae dark brown, lighter annulations in apical third very indistinct or absent. Forewings broad, similar to *sumptuosella*, but second white costal spot larger, pale golden metallic spot underneath very small, third white costal spot large and distinctly tapering apically, sometimes with pale golden metallic spot underneath, small white spot at apex, often with some pale golden scales. Underside of forewings shining brown with yellow ochreous patch in middle near base. Hindwings shining brownish grey, lighter at base; retinaculum yellow ochreous.

Similar species. Resembles *klisieckii*, but differs by the smaller size and the brown underside of the forewings.

Male genitalia.– Tegumen long, almost parallel-sided; excavation of posterior edge V–shaped. Right brachium straight, about three times as long as left one, flattened, widened distally, with large lobe at base on inner side; left brachium with rather blunt apex. Valvae concave with rounded cucullus. Right valvella narrow, gradually tapering to blunt apex, with small, very heavily sclerotized ridge on inner side before apex; similar ridge opposite on tubular part of aedeagus. Aedeagus rather slender, straight, tubular part long and narrow, slightly tapering, apex sharp.

Female genitalia.– Segment VIII short, wider than long. Ostium circular; sterigma large and sclerotized, spherical, with bent sclerotized ridges towards ostium. Ductus bursae about one half length of corpus bursae. Corpus bursae posteriorly narrowed and with long, gradually tapering extension; two large crescent-shaped signa of different length.

Distribution.– Southern Spain and Morocco.

Biology.– Host plant: *Teucrium fruticans* L. (Lamiaceae). The larva produces an elongate gall on the stem or branch and can be found in a period from February till August (Walsingham, 1907b). Adults fly from September till early October.

107. *Eteobalea klisieckii* (Riedl, 1966)

Stagmatophora klisieckii Riedl, 1966b: 104.

Diagnosis.– Adult. Wingspan 25 mm. Very large species, superficially similar to *teucrii*. Tegulae pale golden posteriorly. Forewings with white costal spots very large, almost reaching middle of wing, second and third nearly square; three tubercular pale golden metallic spots on dorsum, first small, near base, inward of first costal, second just opposite second costal and nearly touching it, third long and inward of third costal; tiny, white edged, pale golden metallic spot at apex. Underside of forewings shining yellowish grey. Hindwings shining whitish grey, light grey on costa.

Male genitalia.– Unknown.

FEMALE GENITALIA.– Segment VIII short, wider than length, dorsally with V–shaped sclerotization. Ostium semi-circular; sterigma pouched, narrowing distally. Ductus bursae slightly shorter than corpus bursae. Corpus bursae posteriorly with very long extension; two crescent-shaped signa, one of them very small.

DISTRIBUTION.– Morocco.

BIOLOGY.– Unknown. Adults have been collected in June and November.

108. Eteobalea tririvella (Staudinger, 1871)

Stagmatophora tririvella Staudinger, 1871: 322.
Stagmatophora kasyi Riedl, 1965: 436.
Eteobalea tririvella bernhardiella Kasy, 1973b: 276. Subspecies.

DIAGNOSIS.– Adult. Wingspan 11-15 mm. Head yellowish white, laterally dark brown; antennae shining dark brown in basal third, middle part annulated yellowish white, apical third yellowish white. Thorax shining dark brown with white posterior spot. Forewings shining dark blackish brown; three metallic silver fasciae with small white spot on costa, first at 1/6, outward oblique, reaching just beyond fold, second slightly before 1/2, perpendicular to costa or slightly outward oblique, reaching fold, third at 3/4, inward oblique and bending outwardly from fold to dorsum; silver metallic spot at apex. Hindwings shining grey. Abdomen shining greyish brown, shining grey laterally; anal tuft yellowish white.

Similar species. Differs from all western Palaearctic species of *Eteobalea* by the presence of silver metallic fasciae on the forewings; these fasciae are usually uninterrupted, but in ssp. *bernhardiella* described from Austria (Kasy, 1973b) the second and third ones are broadly divided in the middle.

MALE GENITALIA.– Tegumen parallel-sided, posterior edge with V–shaped excavation. Right brachium bent, about two and half times as long as left one, apically broadened; left brachium rounded at apex. Valvae pear-shaped. Right valvella parallel-sided, apex rounded with small heavily sclerotized hook-shaped tip; small heavily sclerotized patch opposite this tip on tubular part of aedeagus. Aedeagus slender, slightly bent, transition between bulbous and tubular part not well defined, tapering to sharp apex.

FEMALE GENITALIA.– Segment VIII narrow, twice as long as wide. Ostium semicircular; sterigma vase-shaped, narrowed distally. Ductus bursae slightly more than half as long as corpus bursae. Corpus bursae elongate, signa absent.

DISTRIBUTION.– Central and eastern Europe, southern Fennoscandia, eastwards to southern Siberia and Mongolia.

BIOLOGY.– Unknown. Adults fly from mid-June to the end of August.

109. Eteobalea albiapicella (Duponchel, [1843])

Lita albi-apicella Duponchel, in Godart, [1843]: 328.

DIAGNOSIS.– Adult. Wingspan 10-12 mm. Head with frons white, vertex shining bronze brown with narrow lateral white line; antennae shining dark brown with longitudinal white line from scape to the seventh segment. Thorax shining bronze brown with white medial line. Forewings shining dark brown with shining white markings, at one-sixth a very outward oblique fascia, slightly bent inwardly and reaching just beyond the fold, before one half an outward oblique costal streak towards the middle, at three-quarters a triangular costal streak, a small subdorsal spot, opposite inner costal streak, a second subdorsal spot at tornus, just opposite or slightly inward of outer costal streak, and a spot at apex. Hindwings shining grey. Abdomen shining greyish brown to dark brown, anal tuft white.

Similar species. The species resembles *alypella*, but differs by the bronze brown vertex, the white line at the base of the antennae, the white medial line on the thorax and the white subdorsal spot in the middle of the forewings.

MALE GENITALIA.– Tegumen slightly narrowing distally, with shallow V–shaped excavation of posterior edge. Right brachium bent, wide and flattened, about twice as long as left one, tip small and rounded; left brachium with blunt apex. Valvae pear-shaped, slightly concave. Right valvella thickest in middle, gradually tapering with blunt, heavily sclerotized, hook-shaped tip; opposite this tip small sclerotized patch on tubular part of aedeagus. Aedeagus rather thick and almost straight, gradually tapering, apex blunt.

FEMALE GENITALIA.– Segment VIII about as long as wide, apophyses anteriores stout and heavily sclerotized. Ostium circular; sterigma very shallowly pouch-shaped. Ductus bursae about one-third length of corpus bursae. Corpus bursae elongate, widest anteriorly, posteriorly with gradually tapering extension; signa absent.

DISTRIBUTION.– Central and southern Europe, eastwards to the south of European Russia, Mediterranean Islands; Asia Minor. The records from northern Africa (Morocco) probably refer to *alypella*.

BIOLOGY.– Host plants: *Globularia vulgaris* L., *G. nudicaulis* L. and *G. elongata* Hegetschw. (Globulariaceae). Bivoltine. The larva first lives in the flower-stem, later in the seed-head. It can be found from August to April, hibernating in the seedheads, and in June (Schütze, 1931; Lhomme, [1948-1949]; Klimesch, 1961). Adults fly in May-June and in August-September.

110. *Eteobalea alypella* (Klimesch, 1946)

Stagmatophora alypella Klimesch, 1946: 72.

DIAGNOSIS.– Adult. Wingspan 10-11 mm. Head white, vertex strongly mixed brown posteriorly; antennae shining dark brown, basal half annulated whitish, apical third with five white annulations. Thorax shining dark brown, posteriorly with scarce white scales. Forewings very similar to *albiapicella*, but fasciae and streaks narrower, and the white spot opposite the inner costal spot placed at dorsum. Hindwings shining grey. Abdomen dark grey, anal tuft white.

Similar species. Differs from all other species of the genus (except for *albiapicella*) by the absence of metallic scales on the forewings.

MALE GENITALIA.– Differs from *albiapicella* by the slender valvae with more pronounced cucullus, the long and gradually tapering right brachium, and the longer and more slender aedeagus.

FEMALE GENITALIA.– Unknown.

DISTRIBUTION.– Southern France, Spain, Italy, and Morocco.

BIOLOGY.– Host plants: *Globularia alypum* L. (Globulariaceae). The larva lives in the flowerheads; it is fully grown at the end of October and hibernates in its feeding place until spring (Klimesch, 1946). Adults fly in June.

111. *Eteobalea isabellella* (Costa, [1836])

Ornix isabellella Costa, [1836]: [299].
Ornix hisabellella Costa, [1836]: [299]; incorrect original (of a multiple) spelling of *Ornix isabellella* Costa, [1836].
Stagmatophora opulentella Herrich-Schäffer, 1853: 217.
Stagmatophora rutilella Chrétien, 1896: 105.

DIAGNOSIS.– Adult. Wingspan 9-16 mm. Head yellowish white, mixed goldish posteriorly; antennae at basal two-thirds shining dark brown, annulated whitish, apical third with five white annulations. Thorax shining golden brown, shining white-edged posteriorly. Forewings shining golden brown with yellowish streak on dorsum near base, with pattern of numerous tubercular golden metallic spots; three golden spots near base forming outward oblique row from just below costa to beyond fold; white costal spot at 1/5 (sometimes absent), outwardly connected to three golden spots from costa forming outward oblique row, slightly bent inwards to dorsum; second white costal spot at 2/5, outwardly connected to golden spot; third white costal spot at 3/4 with golden spot connected underneath; several golden spots, first subdorsal, slightly outward of second costal; second just above middle of wing between second and third costal; third at tornus, inward of third costal; fourth at apex; six to eight small golden streaks around apical area. Hindwings shining grey. Abdomen shining greyish brown with golden gloss.

Similar species. Differs from all other species of *Eteobalea* by the golden brown colour in combination with the numerous shining golden tubercular spots on the forewings.

MALE GENITALIA.– Tegumen slightly tapering distally, with U–shaped excavation of posterior edge. Right brachium bent, little less than twice as long as left one, strongly broadened at heavily sclerotized apex, with tiny lobe at base on inner side; left brachium bent, apex blunt. Valvae concave with short and rounded cucullus. Right valvella large, parallel-sided, almost as long as tubular part of aedeagus, distally tapering to rounded apex. Aedeagus slightly bent, bulbous part oval, tubular part rather wide, tapering to sharp apex.

FEMALE GENITALIA.– Segment VIII longer than wide. Ostium semicircular; sterigma pouch-shaped. Ductus bursae slightly shorter than corpus bursae. Corpus bursae oval, posteriorly with short, broad and abruptly tapering extension; two large scale-shaped signa of different size.

DISTRIBUTION.– Southern Europe, to Romania in the north; Asia Minor, northern Africa.

BIOLOGY.– Unknown. Adults can be collected from the end of May to early September.

112. *Eteobalea dohrnii* (Zeller, 1847)

Elachista dohrnii Zeller, 1847b: 890.
Stagmatophora dohrni auctt.; incorrect subsequent spelling of *Elachista dohrnii* Zeller, 1847.

DIAGNOSIS.– Adult. Wingspan 10-16 mm. Head yellowish white; antennae dark grey, apical third with five white annulations. Thorax shining ochreous. Forewings shining ochreous mottled with brown, especially in apical half; white patch on dorsum at base; three white costal spots, first at 1/6, long and outward oblique, towards fold, second at 2/5, third at 3/4, triangular; white spot at apex into cilia; five tubercular greyish brown spots with golden gloss, first small, connected outward of first costal, second just below fold, between first and second costal, third outwardly connected to second costal, fourth subdorsal, just outward of third, fifth and largest at tornus, often touching third costal spot; several white streaks between tornal and apical spot. Hindwings at base shining white, more yellowish grey towards apex. Abdomen ochreous yellow to dark grey.

Variation. The ground colour of the forewings can vary from pale ochreous yellow to brown.

Similar species. Resembles *siciliae*, but differs by the absence of cilia lines around the apex of the forewings.

MALE GENITALIA.– Tegumen long, slightly tapering distally, with V–shaped excavation of posterior edge. Right brachium bent, about twice as long as left one, strongly widened apically and heavily sclerotized; left brachium slightly bent, apex square. Valvae truncated boot-shaped. Right valvella narrow, gradually tapering and apically rounded. Aedeagus bent, bulbous part oval, tubular part parallel-sided, apex very sharp. For separation from *siciliae*, q. v.

FEMALE GENITALIA.– Segment VIII longer than wide. Ostium semicircular; sterigma bowl-shaped, ventral margin heavily sclerotized. Ductus bursae about one and a half times length of corpus bursae. Corpus bursae oval, posteriorly with long bulbous, gradually tapering extension; two large crescent-shaped signa of different size.

DISTRIBUTION.– Mediterranean area, from Portugal to Syria in the east.

BIOLOGY.– Host plant: *Plantago albicans* L. (Plantaginaceae). According to observations of Dumont, the larva lives in the root (Lhomme, [1948-1949]). Adults fly from May till early October, probably in two generations.

113. Eteobalea siciliae (Riedl, 1966)

Stagmatophora siciliae Riedl, 1966a: 79.

DIAGNOSIS.– Adult. Wingspan 13-14 mm. External appearance very much like *dohrnii*, but differs by the two distinct dark brown cilia lines around the apex of the forewings.

MALE GENITALIA.– Tegumen tapering distally, with U–shaped excavation of posterior edge. Right brachium slightly bent, about twice as long as left one, broadened apically and strongly sclerotized, with large lobe at base on inner side; left brachium almost straight, apex square. Valvae short and boot-shaped. Right valvella short, triangular and with sclerotized ridge at apex. Aedeagus bent, bulbous part almost circular, tubular part parallel-sided, apex sharp; large, strongly sclerotized patch on outside, opposite apex of right valvella.

Similar species. Differs from *dohrnii* by the large basal lobe on inner side of right brachium and the sclerotized apical ridge of the right valvella.

FEMALE GENITALIA.– Unknown.

DISTRIBUTION.– Mediterranean area (Spain, Sicily, Algeria, Turkey).

BIOLOGY.– Unknown. Adults have been collected in June and from August to the beginning of September.

114. Eteobalea thaumatella (Walsingham, 1907)

Stagmatophora thaumatella Walsingham, 1907a: 150.
Stagmatophora lactipunctella Turati, 1924: 181.

DIAGNOSIS.– Adult. Wingspan 8-15 mm. Head yellowish white, mixed ochreous posteriorly; antennae shining dark brown, apical third with five white annulations. Thorax dark brown, posteriorly edged ochreous brown. Forewings shining dark brown, dorsal and apical part more or less mixed ochreous brown; three very large white costal spots, first at 1/6, slightly outward oblique, to fold, underneath connected by tubercular golden brown metallic spot, second at 1/2, almost square, often with some golden brown metallic scales connected to posterior edge, third between second and apex, elongate; three tubercular golden brown metallic spots with purplish gloss, first opposite second costal spot subdorsally, second on dorsum at tornus, opposite or slightly inward of inner edge of third costal spot, similar spot, mixed white, at apex. Hindwings shining white. Abdomen pale brown; anal tuft white.

Similar species. The forewing pattern of tubercular shining golden brown spots with purplish gloss in combination with the dark brown and ochreous brown colour distinguishes this species from other species of the genus.

MALE GENITALIA.– Tegumen short, slightly tapering distally, with U–shaped excavation of posterior edge. Right brachium almost straight, less than twice as long as left one, widened apically and sclerotized, with large lobe at base on inner side; left brachium slightly bent, apex tapering to blunt point. Valvae truncated boot-shaped. Right valvella short, abruptly tapering, apex rounded. Aedeagus short and bent, bulbous part circular, tubular part tapering, shorter than bulbous part, apex sharp.

FEMALE GENITALIA.– Segment VIII longer than wide. Ostium semicircular; sterigma pouch-shaped with bent and heavily sclerotized transverse ridge. Ductus bursae slightly longer than corpus bursae. Corpus bursae oval, posteriorly with long gradually tapering extension; two large crescent-shaped signa of different size.

DISTRIBUTION.– Northern Africa, Canary Islands; eastwards to the Arabian Peninsula and Iran.

BIOLOGY.– Unknown. Adults have been collected from early March to early May, but also in October.

Isidiella Riedl, 1965

Isidiella Riedl, 1965: 437.
Type species: *Stagmatophora nickerlii* Nickerl, 1864, by original designation.

DIAGNOSIS.– Small moths. Forewings lanceolate, apex pointed; ground colour shining golden, with pattern of large white costal spots and pale golden metallic tubercular spots. Hindwings half width of forewings and acutely pointed.

MALE GENITALIA.– Tegumen wide at base, rounded distally. Right brachium stout, about one- fourth longer than left one, apex unwidened. Valvae shortly spatulate. Right valvella very large, reaching well beyond aedeagus, dorso-ventrally flattened and bent downwards; left valvella strongly reduced or absent. Aedeagus straight, basal half bulbous.

FEMALE GENITALIA.– Apophyses posteriores one-fourth longer than apophyses anteriores. Sterigma crescent-shaped, simple. Ductus bursae narrow, about as long as or shorter than corpus bursae. Corpus bursae oval to pear-shaped, membranous, posteriorly with elongate extension, ending in ductus seminalis; two bowl-shaped signa.

DISTRIBUTION.– Central Europe, Mediterranean area; South Africa.

BIOLOGY.– Larvae are leaf-miners or live in silky webs among the terminal leaves of Asteraceae.

115. Isidiella nickerlii (Nickerl, 1864)

Stagmatophora nickerlii Nickerl, 1864: 7.
Stagmatophora nickerli auctt.; incorrect subsequent spelling of *Stagmatophora nickerlii* Nickerl, 1864.

DIAGNOSIS.– Adult. Wingspan 8-11 mm. Head shining golden to golden ochreous, laterally with white line above eyes; antennae shining brown, basal two-thirds annulated white, in apical third five white segments, each separated by three dark brown segments. Thorax and tegulae shining golden; tegulae edged with silver posteriorly. Forewings shining golden, costa edged brownish; pale golden basal streak on dorsum; white costal streak, outwardly oblique and edged pale golden on costa at 1/6 to beyond the fold; two white costal spots, more or less edged dark brown, first at about 1/2, with tubercular pale golden metallic spot connected underneath, second at 3/4, outwardly edged with some pale golden scales; three dorsal pale golden metallic tubercular spots edged dark brown, first in fold between costal streak and first costal spot, second, and largest, at dorsum, between first and second costal spots, third on tornus; small tubercular pale golden metallic apical spot and some pale golden metallic and dark brown scales around apex; cilia greyish brown. Hindwings shining greyish brown. Abdomen grey, segments whitish banded posteriorly.
Similar species. For separation from *divitella*, q. v.

MALE GENITALIA.– Tegumen narrowing distally, with U–shaped excavation posteriorly. Right brachium tapering apically and bent, less than twice length of left one; left brachium just slightly curved, apex pointed. Valvae pear-shaped. Right valvella very large, much longer than tubular part of aedeagus and strongly bent downwards; left valvella reduced. Aedeagus with tubular part almost straight and basal part large and bulbous; diameter of entry of ductus ejaculatorius distinctly wider than one half of largest diameter of aedeagus.

FEMALE GENITALIA.– Segment VIII slightly tapering distally. Ostium with semicircular, broadly sclerotized anterior edge. Sternite VII posteriorly with distinct triangular sclerotization. Ductus bursae narrow, as long as corpus bursae. Corpus bursae oval, with two cup-shaped signa.

DISTRIBUTION.– Central and southern Europe, from the Iberian Peninsula to the Balkans in the east.

BIOLOGY.– Host plant: *Achillea millefolium* L. and *Helichrysum arenarium* (L.) Moench. (Asteraceae). The larva lives in a web on the young terminal leaves (Lhomme, [1948-1949]; Riedl, 1969). Adults have been collected from May till July.

116. Isidiella divitella (Constant, 1885)

Stagmatophora divitella Constant, 1885: 10.
Stagmatophora nickerli tyrrhenica Hartig & Amsel, 1939: 75.

DIAGNOSIS.– Adult. Wingspan 8.5-12.5 mm. Head shining white; antennae shining brown, apical half indistinctly annulated white. Thorax shining white, laterally shining golden; tegulae shining golden, posteriorly edged shining white. Forewings shining golden, shining golden brown at costa; indistinct golden brown basal streak between fold and dorsum; otherwise forewing pattern generally as in *nickerlii*, but white costal streak more pronounced and pale golden tubercular spots smaller, apical spot white.

Similar species. The most obvious differences with *nickerlii* are the white head and the broad white dorsal streak on the thorax.

MALE GENITALIA.– Very similar to those of *nickerlii*, but valvae more slender, right valvella shorter and wider, and aedeagus with diameter of entry of ductus ejaculatorius one half of the largest diameter in its bulbous part.

FEMALE GENITALIA.– Very similar to *nickerlii*, but can be distinguished by the very thinly sclerotized anterior edge of the ostium, the indistinct posterior sclerotization of sternite VII, and the ductus bursae, which is shorter than the corpus bursae.

DISTRIBUTION.– Western Mediterranean area (southern France, Corsica, Sardinia, Spain, Portugal).

BIOLOGY.– Host plants: *Helichrysum angustifolium* D.C., *H. serotinum* Boiss., *H. italicum* (Roth) Guss., *H. stoechas* (L.) DC., and *Santolina* sp. (Asteraceae). The larva mines several leaves, starting from the base. It constructs a globular case enwrapped with the silky hairs of the food plant and uses this as a shelter from which it mines the leaves. The case looks like a gall on the stem and is also used for pupation (Constant, 1885; Chrétien, 1902; Hartig & Amsel, 1951; Hering, 1957). Adults can be found from June till August.

Vulcaniella Riedl, 1965

Vulcaniella Riedl, 1965: 439.

Type species: *Elachista pomposella* Zeller, 1839, by original designation.

DIAGNOSIS.– Small moths. Head smoothly scaled; antennae three-quarters to four-fifths length of forewings; labial palpi cylindrical, porrect, segment 3 strongly angled upwards and often reaching well above head. Forewings lanceolate, in some species distinctly constricted subapically in females; ground colour dark to blackish brown, sometimes golden brown, with pattern of white and metallic silver or pale golden fasciae and spots. Hindwings very narrow.

MALE GENITALIA.– Tegumen well developed, quadrangular, posterior edge with deep excavation. Brachia asymmetrical; right brachium longer than left one and often with broadened tip. Valvae narrow and elongate, distally with sparse setae. Right valvella well developed, left one strongly reduced and fused with aedeagus. Aedeagus bulbously swollen in middle, distal part tapering, coecum penis short and tubular, cornuti absent. Lobes of segment VIII rounded. The shape of the right brachium, aedeagus and the right valvella are of diagnostic value.

FEMALE GENITALIA.– Apophyses posteriores about one and half times as long as apophyses anteriores. Ostium bursae in membrane ventrally between sternites VII and VIII; sterigma with characteristic sclerotized antevaginal plate opposite sclerotization at posterior edge of sternite VII. Ductus bursae long and narrow, membranous, sometimes with sclerotized patches distally. Corpus bursae oval to elongate; two funnel-shaped and inwardly projecting signa, often surrounded by strong circular wrinkles.

The shape of both the antevaginal plate and the sclerotization of the posterior edge of sternite VII show important diagnostic characters.

DISTRIBUTION.– The genus has a Palaearctic distribution. In Europe most species occur in the Mediterranean area.

BIOLOGY.– Larvae are leaf-miners on Lamiaceae and Asteraceae.

Key to species of *Vulcaniella* based on external features

1.	Forewings ground colour shining golden brownish	2
–	Forewings ground colour shining dark brown	3
2.	Thorax and tegulae shining golden brownish	***gielisi***
–	Thorax and tegulae shining white (form with golden brownish forewings)	***glaseri***
3.	Head shining white	4
	Head shining ochreous brown or dark golden brown	7
4.	Thorax and tegulae shining white (form with dark brown forewings)	***glaseri***
-	Thorax and tegulae shining dark brown	5
5.	Forewings with first dorsal spot more towards base, between basal fascia and first costal spot	6
–	Forewings with first dorsal spot in alignment with first costal spot; thorax without white markings	***fiordalisa***
6.	Thorax anteriorly and posteriorly with white medial spot	***anatolica***
–	Thorax without anterior white spot, posteriorly broadly edged white towards tegulae	***pontica***
7.	Forewings with first dorsal spot in alignment with first costal spot; thorax without white markings	8
–	Forewings with first dorsal spot more towards base, between basal fascia and first costal spot	10
8.	Thorax and tegulae without white edging	***rosmarinella***
–	Thorax and/or tegulae edged white	9
9.	Forewings without white spot at base	***grabowiella***
–	Forewings with small white spot at base; thorax with posterior white spot	***klimeschi***
10.	Cilia of forewings at apex greyish brown	11
–	Cilia of forewings at apex shining white	***extremella***
11.	Head dark golden brown, apex of antennae white, tegulae edged silver metallic	12
–	Head light ochreous, apex of antennae dark brown, tegulae posteriorly edged white	***caucasica***
12.	Forewings with small white basal spot and large white costal markings	***karadaghella***
–	Forewings without basal white spot, costal white markings small or partly absent	13
13.	Larger species, wingspan up to 13 mm; tegulae broadly edged silver metallic	***grandiferella***
–	Smaller species, wingspan less than 10 mm; tegulae with only few silver metallic scales posteriorly (identification on external characters not possible, examination of genitalia necessary)	***pomposella, cognatella***

117. Vulcaniella pomposella (Zeller, 1839)

Elachista pomposella Zeller, 1839: 212.

DIAGNOSIS.– Adult. Wingspan 8-10 mm. Head shining dark brown with golden reflection; antennae shining dark brown, basal three-quarters annulated white, apical fourth with five white segments, each divided by three dark brown segments. Thorax and tegulae shining dark brown with golden reflection, both posteriorly edged with silver metallic scales. Forewings shining dark brown with golden reflection; outward oblique, tubercular silver metallic fascia on costa at 1/6 to beyond fold; six silver metallic tubercular spots, two on costa, first before 1/2, second at 3/4 with white streak from costa to cilia, three on dorsum, first in fold before 1/2, second on dorsum, between two costal spots, third on tornus, just beyond third costal spot, one apical spot; cilia dark brown, grey around apex and towards base. Hindwings shining grey, more brownish towards apex. Abdomen dorsally dark grey.

Variation. Female with forewings distinctly narrowed before apex and abdomen ochreous brown.

MALE GENITALIA.– Right brachium inwardly bent, strongly tapering distally, apex slightly wider; left brachium swollen in middle, apex rounded. Valvae wider at base, distal half straight. Right valvella as long as valvae, slightly wider at base and with rounded tip, basal part ventrally bulbous; left valvella short, gradually tapering. Aedeagus straight, distal part gradually tapering.

FEMALE GENITALIA.– Segment VIII narrowing distally. Sterigma with semicircular antevaginal plate. Posterior sclerotization of sternite VII semi-circular, narrowing laterally. Ductus bursae slightly longer than corpus bursae. Corpus bursae weakly wrinkled and with two large signa.

DISTRIBUTION.– Central Europe, northwards to Latvia. All records from the Mediterranean area and eastern Europe should be verified; at least some of them refer to *cognatella* and *grandiferella*.

BIOLOGY.– Host plants: *Helichrysum arenarium* (L.) Moench, *Hieracium pilosella* L. (Asteraceae) (Hering, 1957). Also recorded from *Salvia officinalis* L. (Lamiaceae) (Klimesch, 1942), but this observation probably corresponds to *cognatella*. The specimens bred from *Salvia aethiopis* L. (Mitic-Mužina & Bogavac, 1971) belong to *grandiferella*. Probably bivoltine. The larva mines a leaf, hiding in silky galleries on its underside. Pupation takes place inside the mine. Adults can be found from early May till the beginning of July.

118. Vulcaniella grandiferella Sinev, 1986

Vulcaniella grandiferella Sinev, 1986: 62.
Vulcaniella pomposella auctt., (*nec* Zeller, 1839); misidentification.

DIAGNOSIS.– Adult. Wingspan 8-13 mm. Male and female externally very similar to *pomposella*, but differing by the larger wingspan, the more elongated and tapered forewings in the male, the broad silver metallic edging of the tegulae and by the paler abdomen. A safe identification is often only possible by examination of genitalia characters.

MALE GENITALIA.– Differs from *pomposella* by the apically wider right brachium, the broader valvae with narrowed base and slightly bent distal part, and the broader right valvella, which is distinctly shorter than the valvae.

Female genitalia.– Differs from *pomposella* by the longer and more slender apophyses, the broad and semicircular, band-shaped antevaginal plate, the large and triangular posterior sclerotization of sternite VII, and by the strongly wrinkled corpus bursae.

Distribution.– Eastern Mediterranean area, from the Balkans to the Caucasus.

Biology.– Host plants: *Salvia aethiopis* L. and *S. sclarea* L. (Lamiaceae) (Sinev, 1993a, 1994d). Bivoltine; according to the data erroneously attributed to *pomposella* (Mitic-Mužina & Bogavac, 1971), three or even four generations are sometimes possible. The larva mines the leaves, especially those lying on the ground. The irregularly shaped mine can usually be found between the main veins near the leaf base. The larva changes its mine several times and spends most of its life outside the mine, in silky galleries on the underside of a leaf. Frass is ejected from the mine and accumulated near its entrance. Pupation takes place in a light cocoon spun between the hairs on the surface of the leaf or the stem. In the Crimea the moths of the first generation fly from mid-May to early July, and those of the second generation in July-August.

119. Vulcaniella cognatella Riedl, 1990

Vulcaniella cognatella Riedl, 1990a: 68.
Vulcaniella pomposella auctt., (*nec* Zeller, 1839); misidentification.

Diagnosis.– Adult. Wingspan 7-9 mm. Externally very similar to *pomposella*; the silver metallic tubercular spots on dorsum often slightly pale golden and with some purplish gloss, abdomen dorsally with first three segments shining ochreous brown. Cannot be safely identified by external characters.

Male genitalia.– Similar to *pomposella*. Right brachium more strongly bent inwards, distally abruptly tapered, with knob-shaped apex; left brachium more straight. Valvae narrow and parallel-sided. Right valvella slender, bent laterally, as long as distal part of aedeagus and longer than valvae; left valvella short and rounded. Aedeagus slightly bent in distal part.

Female genitalia.– Similar to *pomposella* but differs by the broad and shallowly V–shaped antevaginal plate and by the posterior sclerotization of sternite VII, which is indistinct, triangular and has two inward oblique lateral ridges and a broad weak medial ridge.

Distribution.– Eastern Mediterranean area, from Italy to Asia Minor.

Biology.– Unknown. Adults have been collected in June and July.

120. Vulcaniella extremella (Wocke, 1871)

Stagmatophora extremella Wocke, 1871: 321.
Stagmatophora naviella Chrétien, 1907: 250.
Stagmatophora buhri M. Hering, 1935: 13.

Diagnosis.– Adult. Wingspan 8-11 mm. Forewings shining dark brown with golden reflection and outward oblique, tubercular silver metallic fascia in basal area; six silver metallic tubercular spots, three on dorsum, apical spot very small or absent; series of small silver metallic spots around apex; cilia dark brown, shining white at apex.

Variation. Female with the forewings distinctly narrowed before apex.

Similar species. Differs from all other species of the genus by the conspicuous white section in the cilia around apex.

MALE GENITALIA.– Right brachium with very broad apical part, having rounded excavation on outside; left brachium slender, bent and slightly swollen before apex. Valvae broadest at base and gradually tapering distally. Right valvella longer than distal part of aedeagus but shorter than valvae, slender, almost straight; left valvella rather long and gradually tapering. Aedeagus gradually tapering distally, apex slightly bent.

FEMALE GENITALIA.– Similar to *pomposella*, but the antevaginal plate semi-circular and sclerotization of the posterior edge of sternite VII triangular with rounded edges.

DISTRIBUTION.– Central and southern Europe from the Iberian Peninsula to northern Caucasus.

BIOLOGY.– Host plants: *Salvia pratensis* L. (Klimesch, 1943), *S. verticillata* L. (Kasy, 1965), *S. bertolonii* Viss. (Riedl, 1969), *S. austriaca* L. (Szöcs, 1981), *S. tesquicola* Klok. et Pobez. (Sinev, 1993a) and *Prunella grandiflora* (L.) Scholler (Lamiaceae) (Lhomme, [1948-1949]). Bivoltine, but also three generations are possible. The larva makes a leaf mine, which starts as a greenish gallery following the midrib of a lateral vein and later changes into an irregular contracted whitish blotch oriented to the leaf edge. When not feeding or if disturbed, the larva draws back into the gallery, which is lined with silk. Because the larva is not visible in its hiding place, the mine seems to be vacated. The shape of mine and larval life-style on various host plants and from generation to generation are slightly different. In spring, the larva feeds on the lower leaves, often changing mines; before entering a new leaf the larva makes a short whitish tube, usually in the angle between midrib and a lateral vein, from which the leaf is entered and the frass dispersed. In summer, the lower leaves are already dried, and the larvae feed on the leaves situated higher on the stem and does not change mine. Pupation takes place outside the mine in a whitish, paper-like cocoon in the spring generation, or inside the mine in a very light web in the summer generation. Adults can be found from mid-May to mid-September.

121. Vulcaniella caucasica Sinev, 1986

Vulcaniella caucasica Sinev, 1986: 64.

DIAGNOSIS.– Adult. Wingspan 8-10 mm. Head pale ochreous. Thorax and tegulae shining dark brown with golden gloss, both posteriorly edged white. Forewings shining dark brown with golden reflection; basal fascia and spots arranged as in *pomposella*, tubercular, dark golden; first and second costal spots preceded by small, and third costal spot by large white spot. Abdomen dorsally first three segments light grey, other segments dark grey.

Similar species. The ochreous head and neck tuft in combination with the dark golden metallic pattern of the forewings are characteristic.

MALE GENITALIA.– Right brachium distally very wide, apex truncate with outer margin slightly concave; left brachium with apex square. Valvae parallel-sided. Right valvella slightly shorter than distal part of aedeagus, broad at base with bulbous projection dorsally, tapering towards apex; left valvella small and very narrow. Aedeagus very slender, distal part gradually tapering.

FEMALE GENITALIA.– Segment VIII narrowing distally. Antevaginal plate with posterior margin concave, anterior margin convex. Sclerotization on posterior edge of sternite VII wedge-shaped, with two lateral, inwardly bent rims. Ductus bursae about half length of corpus bursae. Corpus bursae very strongly wrinkled; signa large and shallow.

DISTRIBUTION.– Asia Minor, the Caucasus (Armenia and Azerbaijan).

BIOLOGY.– Unknown. Adults have been collected in June and July.

122. *Vulcaniella grabowiella* (Staudinger, 1859)

Pancalia grabowiella Staudinger, 1859: 251.

DIAGNOSIS.– Adult. Wingspan 7-9 mm. Head ochreous brown. Thorax and tegulae shining brown with golden reflection, tegulae posteriorly edged white. Forewings shining dark brown with golden reflection; tubercular markings silver or pale golden, costal markings preceded by distinct white spots, first dorsal spot underneath first costal spot. Hindwings shining grey. Abdomen dorsally with first three segments shining greyish brown, others shining dark brown.

MALE GENITALIA.– Right brachium bent, narrowing distally; left brachium slightly curved, apex truncate. Valvae parallel-sided, narrow. Right valvella shorter than distal part of aedeagus, strongly bent upwards; left valvella very small, knob-shaped. Aedeagus with distal part gradually tapering, apex slightly bent.

FEMALE GENITALIA.– Segment VIII parallel-sided. Antevaginal plate semi-circular. Sclerotization on the posterior edge of sternite VII triangular, with weak medial ridge.

DISTRIBUTION.– Mediterranean area, from Iberian Peninsula to Asia Minor.

BIOLOGY.– Host plants: *Lavandula stoechas* L., *Thymus vulgaris* L., and *T. algeriensis numidicus* Poir. (Lamiaceae) (Staudinger, 1859; Chrétien, 1917; Hering, 1957). Bivoltine. In March and June the larva mines the leaves, which are often spun together. It mines from a silken case densely covered with blackish frass. The case resembles that of *Coleophora*, but can be distinguished by the more irregular shape and the frass covering. Pupation takes place outside the mine. Adults can be found from early June till early August.

123. *Vulcaniella rosmarinella* (Walsingham, 1891)

Stagmatophora rosmarinella Walsingham, 1891b: 139.

DIAGNOSIS.– Adult. Wingspan 7-8 mm. Head, thorax and tegulae shining dark brown with golden reflection.

Similar species. Forewings like *grabowiella*. Differs by the shining dark brown head being concolorous with the thorax, and the lack of white edging of the tegulae.

MALE GENITALIA.– Right brachium slightly bent and widened distally, apex truncated; left brachium slightly bent, apex rounded. Valvae parallel-sided. Right valvella shorter than distal part of aedeagus, stout, slightly bent, very wide at base, gradually tapering towards apex; left valvella long, gradually tapering. Aedeagus with distal part rather abruptly tapering, apex bent.

FEMALE GENITALIA.– Segment VIII narrowing posteriorly. Antevaginal plate V–shaped. Sclerotization on posterior edge of sternite VII pouch-shaped.

DISTRIBUTION.– Southern Europe, eastwards to Greece and Crete; northern Africa.

BIOLOGY.– Host plant: *Rosmarinus officinalis* L. (Lamiaceae). Probably bivoltine. In spring and summer the larva mines the leaves, which are often spun together. The mine starts as an irregular gallery, not very transparent and lined with silk inside, later occupying more than half of the topside of the leaf, contracting it and turning it brown. The mines are changed, and when entering a new leaf the larva gnaws a hole in the underside of the leaf, more or less in the middle of it. The frass is partly removed from the mine through the entrance hole. Pupation takes place inside the mine without cocoon. Adults can be found from late May to July. See text-fig. 48.

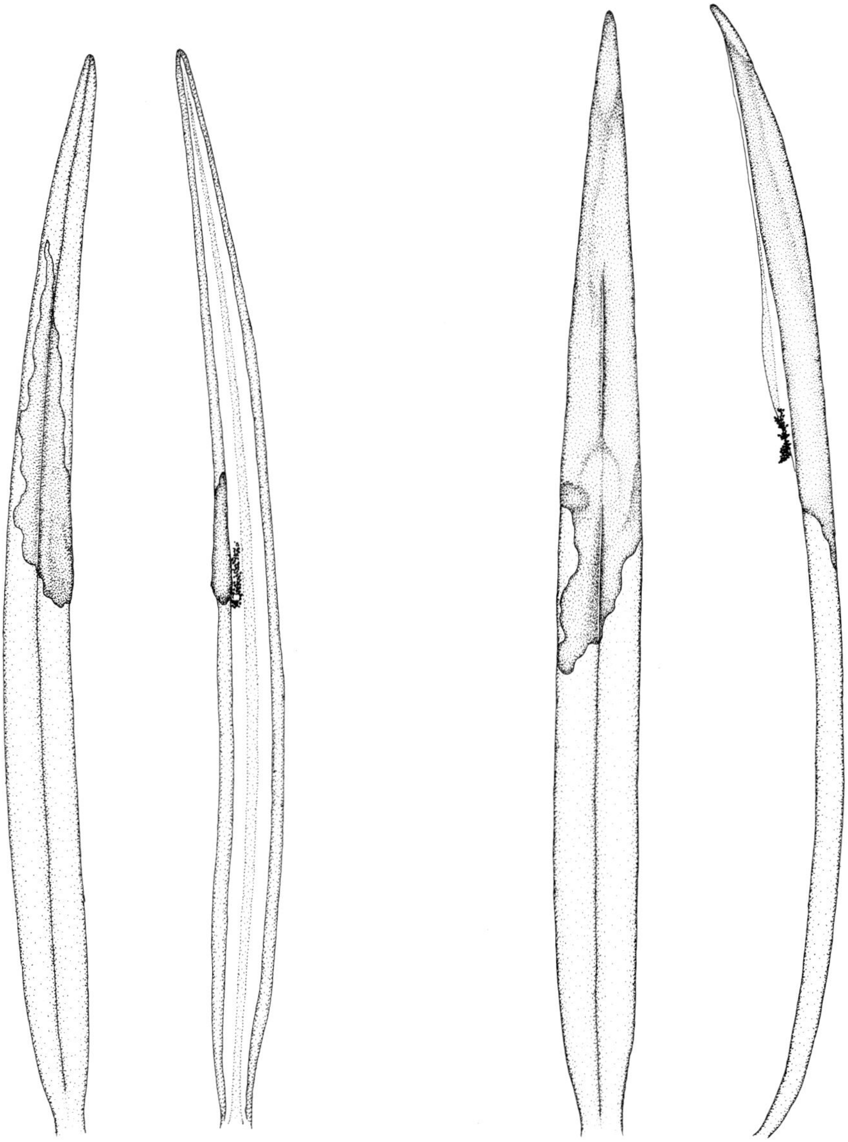

Text-fig. 48. *Vulcaniella rosmarinella* (Walsingham), larval mines in leaves of *Rosmarinus officinalis* L. The furthest right leaf contains a pupa.

124. Vulcaniella karadaghella Sinev, 1986

Vulcaniella karadaghella Sinev, 1986: 66.

DIAGNOSIS.– Adult. Wingspan 10-12 mm. Head shining dark golden brown; antennae shining dark brown, basal three-quarters annulated white, subapical dark brown section of three segments followed by white or almost white apical part. Thorax and tegulae shining golden brown; tegulae with silver metallic edging on outside.

Similar species. Forewings like *grandiferella*, but with a small white spot at the extreme base, the white spots on costa more pronounced and the apex often with some pale golden metallic streaks in the cilia. Hindwings shining brownish grey.

MALE GENITALIA.– Right brachium broad, slightly inwardly bent and truncated; left brachium broad, gradually tapering distally. Valvae parallel-sided, distally twisted. Right valvella shorter than distal part of aedeagus and of equal length with valvae, wide and dorso-ventrally flattened; left valvella long, gradually tapering. Aedeagus bulbous, distal part stout and rounded.

FEMALE GENITALIA.– Resembles *grandiferella*, but the antevaginal plate more irregular and ostium wider, posterior edge of sternite VII strongly concave and its sclerotization more narrowly triangular, signa apically rounded.

DISTRIBUTION.– The Crimea and Asia Minor.

BIOLOGY.– Host plant: *Salvia grandiflora* Ettl. (Lamiaceae). The larva mines the tip of a leaf. Very soon the apical portion of the leaf becomes brownish and rolled into a characteristic rosette, and the injured plants (usually in the shadow of shrubs) become well visible. Pupation takes place inside the leaf rosette. Adults fly from late June to early July.

125. Vulcaniella klimeschi (Riedl, 1966)

Stagmatophora klimeschi Riedl, 1966a: 80.

DIAGNOSIS.– Adult. Wingspan 7-9 mm. Head bronze brown. Thorax and tegulae shining dark golden brown, both posteriorly edged white.

Similar species. Forewings similar to *grabowiella*, but with a small white spot at base. External characters are usually insufficient for safe identification.

MALE GENITALIA.– Right brachium bent, narrowing distally; left brachium slightly curved, apex rounded. Valvae parallel-sided, rather broad. Right valvella slightly shorter than distal part of aedeagus, very stout, narrowing distally to rounded apex; left valvella elongate triangular. Aedeagus with distal part narrow and gradually tapering. Differs from *grabowiella* by the shape of brachia, valvae and both valvellae.

FEMALE GENITALIA.– Very similar to *grabowiella*, but antevaginal plate V–shaped, sclerotization on posterior edge of sternite VII pouch-shaped, and length of ductus bursae about half the length of corpus bursae.

DISTRIBUTION.– Macedonia, Greece and Crete.

BIOLOGY.– Host plant: *Salvia ringens* Sibith. & Sm. (Lamiaceae). The larva makes a bilateral blotch mine in a leaf. The mine starts as a somewhat darkened gallery on the midrib, then turns into a blotch; it lacks frass and is lined with silk. At the beginning of the mine the larva spins a silken case, covered with frass. Adults have been bred in the second half of July (Klimesch, 1968); in Greece and Crete the moths have been collected in the first half of June.

126. *Vulcaniella fiordalisa* (Petry, 1904)

Stagmatophora fiordalisa Petry, 1904: 250.
Stagmatophora gnaphaliella Chrétien, 1922: 120.

Diagnosis.– Adult. Wingspan 7-9 mm. Resembles *grabowiella* externally, but differs by the white head, the larger white costal markings on the forewings with the connected tubercular metallic spots often reduced to a few scales.

Male genitalia.– Right brachium parallel-sided, distally truncate; left brachium broad, slightly tapering, apex blunt. Valvae parallel-sided, tapering distally. Right valvella slightly shorter than distal part of aedeagus, slender, bent upwards; left valvella triangular. Aedeagus with distal part straight and gradually tapering.

Female genitalia.– Segment VIII slightly narrowing distally. Antevaginal plate shallowly V–shaped, broadened distally. Sclerotization on posterior edge of sternite VII pouch-shaped.

Distribution.– Mediterranean area, from Portugal and Morocco, eastwards to the Balkans and Lebanon.

Biology.– Host plants: *Helichrysum serotinum* Boiss., *H. angustifolium* D.C., *H. stoechas* (L.) DC. and *Phagnalon* sp. (Asteraceae) (Chrétien, 1922; Amsel, 1936). The larva mines both the old and young leaves and can be found from April to June. Pupation takes place inside the mine without a proper cocoon, just in a very light web. Adults have been collected from May to August.

127. *Vulcaniella glaseri* (Riedl, 1966)

Stagmatophora glaseri Riedl, 1966b: 108.

Diagnosis.– Adult. Wingspan 10 mm. Head, thorax and tegulae shining white. Forewings shining dark golden brown; shining white spot at base; white costal spots large and without tubercular metallic scales, dorsal and apical spots silver metallic and tubercular. Hindwings shining light grey.

Similar species. The white head and thorax in combination with the dark golden brown forewings are characteristic.

Male genitalia.– Right brachium strongly tapering distally, bent outwardly, apex broadened and rounded; left brachium slightly tapering distally and slightly bent before rounded apex. Valvae gradually narrowing distally but broadened again before apex. Right valvella shorter than distal part of aedeagus, slightly bent upwards, ventrally with large rounded basal lobe; left valvella short, stretched outwardly. Aedeagus with distal part gradually tapering and bent.

Female genitalia.– Sterigma with dish-shaped sclerotization. Posterior sclerotization of segment VII square, posterior edge concave. Length of ductus bursae slightly longer than half the length of corpus bursae. Corpus bursae wrinkled, both signa large.

Distribution.– Asia Minor.

Biology.– Unknown. Adults have been collected in June and August.

Remarks.– Some specimens from Turkey have the ground colour of the forewings brownish golden instead of dark golden brown. However, no differences could be found in the remaining external features and in the male genitalia of these striking specimens.

128. Vulcaniella gielisi sp. n.

DIAGNOSIS.– Adult. Wingspan 10-11 mm. Head with frons, vertex and neck tuft shining white, collar shining brownish golden. Labial palpi white, segment 2 with brown subapical spot, segment 3 brown, dorsally and ventrally with white longitudinal streak. Antennae shining dark brown, basal three-quarters annulated white, apical quarter with five white segments, each separated by three dark brown segments, apex brown; scapus dorsally shining dark brown with white anterior line and white apical spot, ventrally white. Thorax and tegulae shining brownish golden; thorax white-edged posteriorly. Legs shining dark brown, foretibiae with white basal and apical rings, two short longitudinal white lines and white spot medially; mid- and hindtibiae with white basal, medial and apical rings; tarsi with two white medial bands; spurs white, ochreous white apically. Forewings shining brownish golden; small white spot on costa at base changing into short silver metallic streak in fold; outward oblique fascia at 1/6, starting with white streak on costa and changing into tubercular silver metallic beyond fold; six silver metallic tubercular spots, two on costa, first before 1/2, connected to costa by white spot, second at 3/4, with white streak from costa into cilia, three on dorsum, first in fold opposite first costal, second on dorsum, between two costal spots, third on tornus, beyond third costal spot, one spot at apex; some silver metallic scales around apex; cilia greyish brown. Hindwings shining grey, more brownish towards apex. Abdomen dorsally with first three segments shining ochreous grey, others shining grey; all segments light grey-banded posteriorly, ventrally shining grey; anal tuft greyish white dorsally, white ventrally.

Similar species. Resembles the form of *glaseri* with the golden brownish forewings, but can easily be distinguished from this species by the golden brownish thorax and tegulae.

MALE GENITALIA.– Right brachium broad, almost parallel-sided, strongly bent inwards, apex square; left brachium broad, almost straight, apex truncate. Valvae broad at base, gradually tapering distally, tip of cucullus curled inwardly. Right valvella shorter than distal part of aedeagus, broad at base, narrowed in middle, apically rounded; left valvella rather long with broad rounded apex. Aedeagus with distal part gradually tapering, narrowed in middle.

FEMALE GENITALIA.– Unknown.

DISTRIBUTION.– Spain.

BIOLOGY.– Unknown. Adults have been collected in a mountainous region in a clearing of a pine forest, where *Salvia* spp. were abundant.

MATERIAL.– Holotype. ♂, 'Cuenca, Hisp., 14.vii.1985, leg. Gielis', 'Genitalia on slide JCK 3112' (RMNH).

Paratypes. 1 ♂, 'Spain, Cuenca, 7 km NE Cuenca, 18.vii.1992, leg. C. Gielis' (GLS); 1 ♂, 'Cuenca, Hisp., 29.vi.1982, leg Gielis.' (RMNH).

ETYMOLOGY.– The species is named after the Dutch lepidopterist Dr. Cees Gielis, who has collected the type series.

129. Vulcaniella anatolica sp. n.

DIAGNOSIS.– Adult. Wingspan 9 mm. Head shining white. Labial palpi variegated, segment 1 greyish brown, segment 2 white with greyish brown spot on outside at base and brown subapical ring, segment 3 white, spotted brown laterally. Antennae shining dark brown, basal two-thirds annulated white, apical third with two dark brown sections of three segments separated by one white segment, annulated by two double white and brown segments towards apex; scapus dorsally shining dark brown with white anterior line and white apical spot, ventrally white. Thorax shining dark golden brown, with white spot anteriorly, edged white posteriorly. Tegulae shining dark golden brown, edged white posteriorly. Legs shining dark brown, foretibiae with two basal spots and medial and apical rings; mid- and hindtibiae with white basal, medial and apical bands; foretarsi with white basal and apical ring on segment 1, segments 2 and 5 white; midtarsi with white basal and apical band on segment 1, segments 2, 3 and 5 with white apical ring; hindtarsi with segment 1 as at midtarsi, other segments white dorsally and ochreous ventrally; spurs ochreous white. Forewings shining dark golden brown; outward oblique fascia at 1/6, white from costa to fold, tubercular silver metallic beyond fold; six silver metallic tubercular spots, two costal spots, first before 1/2 with white edge on costa, second at 3/4 with white streak from costa into cilia, three dorsal spots, first in fold at 1/3, second on dorsum, equidistant to costal spots, third on tornus, beyond third costal spot, one apical spot; cilia greyish brown. Hindwings shining light grey, more brownish towards apex. Abdomen dorsally with first three segments ochreous grey, others shining grey, ventrally shining grey with pale golden gloss; anal tuft white.

Similar species. Resembles *fiordalisa* and *pontica*, but differs from the first species by the first dorsal tubercular spot, which is not situated under the first costal spot and from both species by the presence of a white anterior spot on the thorax.

MALE GENITALIA.– Right brachium tapering apically, bent inwardly, tip broadened and bent outwardly; left brachium straight, slightly tapering apically. Valvae long and narrow, cucullus twisted. Right valvella shorter than anterior part of aedeagus, upwardly bent, tip rounded, ventrally a large round lobe at base; left valvella short with a blunt point. Aedeagus rather slender, anterior part gradually tapering, narrowed in the middle. Resembles *glaseri*, but differs by the different apex of the right brachium, the longer, narrower and twisted valvae and by the more slender right valvella.

FEMALE GENITALIA.– Unknown.

DISTRIBUTION.– Asia Minor.

BIOLOGY.– Unknown.

MATERIAL.– Holotype ♂, 'Turkey, Mersin, 10 km SE Arslanköy, Taurus 1300 m, 11.vii.1987, leg. M. Fibiger', 'Genitalia in vial'. (ZMUC).

Paratypes. 1 ♂, ditto, 'Genitalia on slide JCK 3899 (RMNH); 1 ♂, ditto (ZMUC).

ETYMOLOGY.– The species name is derived from Anatolia (Asia Minor).

130. *Vulcaniella pontica* sp. n.

DIAGNOSIS.– Adult. Wingspan 11 mm. Head white. Labial palpi variegated, segment 1 shining ochreous-brown, segment 2 white with subapical greyish spot on outside, segment 3 white with interrupted dark lateral lines, apex white. Antennae shining dark brown, basal two-thirds annulated white, apical third with five dark brown sections of three segments separated by one white segment or by white and brown annulations; scapus dorsally shining dark brown with white anterior line and white apical spot, ventrally white. Thorax and tegulae shining dark golden brown, broadly edged white posteriorly; thorax with some white scales anteriorly. Legs shining dark brown, foretibiae with two white basal spots and white medial and apical rings; mid-

and hindtibiae with white basal, medial and apical rings; foretarsi with two medial bands, mid- and hindtarsi with white apical rings; spurs ochreous white. Forewings shining dark golden brown; outward oblique, tubercular silver metallic fascia at 1/6 from costa to beyond fold, white on costa and interrupted in fold; six silver metallic tubercular spots, two costal spots, first before 1/2 with white edge on costa, second at 3/4 with white streak from costa into cilia, three dorsal spots, first in fold beyond 1/3, second on dorsum, equidistant to costal spots, third on tornus, beyond third costal spot, one apical spot; cilia greyish brown. Hindwings shining light grey, more brownish towards apex. Abdomen dorsally shining light grey with ochreous tinge, first segment almost white, last segment narrow white-banded posteriorly, ventrally dark grey, broad shining white-banded posteriorly; anal tuft white.

Similar species. Resembles *anatolica*, but differs by the colour of the apical third of the antennae and the broad white posterior edging of the thorax.

MALE GENITALIA.– Right brachium broad, rounded towards apex and hooked; left brachium tapering apically, apex pointed. Valvae long, rather broad and wider at base. Right valvella shorter than distal part of aedeagus, broad at base, narrowed in middle, dorso-ventrally broad and flat, apex rounded; left valvella small and triangular, apex pointed. Aedeagus large and bulbous, distal part tapering, apex pointed.

FEMALE GENITALIA.– Unknown.

DISTRIBUTION.– Asia Minor.

BIOLOGY.– Unknown. The holotype was collected at the beginning of July.

MATERIAL.– Holotype. ♂, 'Turkey, Nevsehir, Göreme by Ürgüp, Love Vally, 1300 m., 2.vii.1987, leg. M. Fibiger', 'Genitalia on slide JCK 3897' (ZMUC).

ETYMOLOGY.– The species name is derived from Pontus, the ancient name of the Black Sea.

Ramphis Riedl, 1969

Ramphis Riedl, 1969: 786.
Type species: *Ramphis ibericus* Riedl, 1969, by original designation.

DIAGNOSIS.– Head smoothly scaled; antennae three-quarters length of forewings; labial palpi cylindrical, recurved above head. Forewings lanceolate, blackish brown with pattern of white costal streaks and metallic silver or pale golden spots. Hindwings narrow and acute, slightly more than half width of forewings. Matches representatives of the genus *Vulcaniella*.

MALE GENITALIA.– Brachia symmetrical, of equal length, slightly bent and tapered distally. Valvae narrow and elongate, distal half sparsely setose. Right valvella well defined, left one almost absent. Aedeagus long and stout, with bulbous basal section and long and tapering distal part.

FEMALE GENITALIA.– Apophyses posteriores about one and a half times as long as apophyses anteriores. Sterigma with semicircular antevaginal plate opposite to sclerotization at posterior edge of sternite VII. Ductus bursae membranous, narrow, about as long as or shorter than corpus bursae. Corpus bursae elongate, posteriorly with elongated extension ending in ductus seminalis, two large funnel- or bowl-shaped signa.

DISTRIBUTION.– A small genus with disjunct distribution (Iberian Peninsula and the easternmost part of the Mediterranean area).

BIOLOGY.– Larvae, as far as known, are leaf miners on Lamiaceae.

131. *Ramphis ibericus* Riedl, 1969

Ramphis ibericus Riedl, 1969: 787.

DIAGNOSIS.– Adult. Wingspan 7-11 mm. Head shining white; antennae shining dark brown, basal two-thirds annulated white, apical third with five white segments. Thorax and tegulae dark brown with golden reflection; thorax with broad shining white medial streak. Forewings shining dark brown with golden reflection; two white and very outward oblique costal streaks near base and before 1/2; costal spot at 3/4, outwardly edged with silver metallic scales; white dorsal line from base to 1/3, four silver metallic tubercular spots, first in fold, underneath second costal streak, second on dorsum, equidistant from second costal streak and costal spot, third on tornus, one spot at apex; several small silver metallic streaks around apex between costal and tornal spots; cilia brown. Hindwings shining light grey, more brownish towards apex; cilia greyish brown. Abdomen with first three segments ochreous grey, other segments shining grey.

Variation. In some specimens the tubercular silver metallic markings on the forewings have a pale golden gloss.

MALE GENITALIA.– Brachia gradually tapering distally, sharply pointed. Valvae parallel-sided and very narrow, shorter than aedeagus. Right valvella about two-thirds of length of distal part of aedeagus and apically rounded; left valvella strongly reduced. Aedeagus with bulbous basal part and prominent coecum penis, distal part slightly bent and narrowing to sharp apex.

FEMALE GENITALIA.– Segment VIII with band-shaped sclerotization. Antevaginal plate indistinct, pouch-shaped, with semicircular or bent, weakly sclerotized rims and lateral rows of spicules. Posterior edge of sternite VII concave with irregular triangular sclerotization. Ductus bursae narrow, shorter than corpus bursae. Corpus bursae elongate with two large funnel-shaped signa.

DISTRIBUTION.– Iberian Peninsula.

BIOLOGY.– Unknown. Adults have been collected in June and July.

132. *Ramphis libanoticus* Riedl, 1969

Ramphis libanoticus Riedl, 1969: 788.

DIAGNOSIS.– Adult. Wingspan 9-11 mm. Resembles *ibericus*, but thorax with a narrow white line and tegulae edged white posteriorly; on forewings the first costal streak is less oblique, dorsum without the white basal line and the tubercular spots are pale golden metallic.

MALE GENITALIA.– Similar to those of *ibericus*, but differs by the valvae, which are broader and as long as aedeagus and by the right valvella, which is less than half the length of the distal part of the aedeagus and apically pointed.

FEMALE GENITALIA.– Differs from those of *ibericus* by the more prominent and shorter sclerotized rims of the antevaginal plate, by the V–shaped posterior edge of sternite VII with smaller and more regular triangular sclerotization, and by the bowl-shaped signa.

DISTRIBUTION.– Islands of the eastern Mediterranean (Rhodos, Cyprus), Asia Minor, Near East (Lebanon, Syria).

BIOLOGY.– Host plant: *Salvia triloba cypria* Kotchy and *S. triloba libanotica* Boiss. & Faill (Riedl, 1986). The larva mines the leaves. Adults have been collected from May till July and also in October.

Hodgesiella Riedl, 1965

Hodgesiella Riedl, 1965: 440.
Type species: *Stagmatophora rebeli* Krone, 1905, by original designation.

DIAGNOSIS.– Small moths. Forewings lanceolate with pointed apex; ground colour dark brown, pattern of three, often incomplete or interrupted, white fasciae. Hindwings lanceolate and sharply pointed.

MALE GENITALIA.– Tegumen well developed, short but wide, with U–shaped excavation at posterior edge. Brachia of different length; right brachium about one and half times or twice as long as left one. Valvae lobate, cucullus with long and coarse setae. Right valvella free, left one partly jointed to aedeagus. Aedeagus bottle-shaped, distally oblique, apex sharp.

FEMALE GENITALIA.– Apophyses posteriores very long, about three times as long as apophyses anteriores. Sterigma elongated and heavily sclerotized. Ductus bursae membranous and very narrow, of about same length as corpus bursae. Corpus bursae oval, posteriorly with large elongated lobe ending in ductus seminalis; two signa, small and thorny or large and funnel-shaped.

DISTRIBUTION.– Mediterranean area and Central Asia; a few species are also known from the Oriental Region.

BIOLOGY.– The larvae, as far as known, are leaf miners on Convolvulaceae.

133. *Hodgesiella rebeli* (Krone, 1905)

Stagmatophora rebeli Krone, 1905: 83.

DIAGNOSIS.– Adult. Wingspan 9-11 mm. Head with frons shining white changing into shining grey on vertex, shining dark brown laterally and posteriorly; antennae shining dark brown, annulated white, apical one-fifth white. Thorax and tegulae shining dark brown. Forewings shining dark brown; three shining white fasciae, edged yellowish on both sides, basal and intermediate fasciae perpendicular on costa or slightly inward oblique, apical fascia outwardly bent at costa, both outer fasciae narrowing towards dorsum; cilia greyish brown. Hindwings shining light grey. Abdomen with first three segments almost shining white, other segments shining grey.

Similar species. For separation from *rhodorrhisella* and *christophi*, q. v.

MALE GENITALIA.– Right brachium parallel-sided with abruptly broadened and flat apex; left brachium straight, narrowing distally. Valvae lobate. Right valvella triangular and pointed. Aedeagus with tubular part straight, apex sharply pointed.

FEMALE GENITALIA.– Apophyses posteriores three times as long as apophyses anteriores. Sterigma parallel-sided, abruptly tapering anteriorly, heavily sclerotized. Ductus bursae as long as corpus bursae. Corpus bursae egg-shaped, with very wide elongated extension; signa of two small thorns surrounded by irregular sclerotized patch.

DISTRIBUTION.– The Balkans, Italy.

BIOLOGY.– Host plants: *Convolvulus cantabricus* L. and *C. althaeoides* L.. The larva in May produces a leaf mine, which starts as a gallery, but soon turns into a blotch, longitudinally contracted by the inner web. The frass is lumped in the gallery-part of the mine (Krone, 1905; Klimesch, 1942; Hering, 1957). Adults fly from May to July.

134. *Hodgesiella rhodorrhisella* Kasy, 1970

Hodgesiella rhodorrhisella Kasy, 1970: 195.

DIAGNOSIS.– Adult. Wingspan 9-12 mm. Head with frons silvery white, vertex shining grey, laterally and posteriorly shining dark brown. Antennae dark brown, annulated white to beyond one-half, apical part with two or three white rings in subapical part and four indistinct white rings in apical part. Thorax and tegulae shining dark brown. Forewings shining dark brown, more greyish brown on dorsum; two irregular white fasciae at 1/5 and 1/2, inner fascia inwardly oblique, slightly narrowing towards fold, very narrow beyond fold and not reaching dorsum, outer fascia perpendicular on costa, broadened in middle, then abruptly narrowing and not reaching dorsum, with brown spot in middle of widest part; white costal spot at 4/5; small white tornal spot, inward of costal spot; small white apical spot, sometimes reduced to several white scales; cilia grey around apex, greyish brown towards dorsum. Hindwings shining grey. Abdomen shining grey, in female shining dark brown.

Similar species. Resembles *rebeli* and *christophi*, but differs by the forewings with two incomplete and more irregular inner fasciae, lacking yellow or ochreous edging; in distinction from *rebeli*, the outer fascia is medially interrupted.

MALE GENITALIA.– Right brachium bent, apex strongly widened and hook-shaped; left brachium widening distally. Valvae triangularly widened distally, with coarse setae on basal half of widened section and fine setae on distal half. Right valvella slender and pointed. Aedeagus with tubular part straight and bent at extreme apex.

FEMALE GENITALIA.– Segment VIII shorter than in *rebeli* and *christophi*. Apophyses posteriores about two and half times as long as apophyses anteriores. Sterigma short, anteriorly rounded and with sclerotized rim. Ductus bursae very narrow. Corpus bursae egg-shaped, with narrow elongated extension; two large and heavily sclerotized, funnel-shaped signa.

DISTRIBUTION.– Canary Islands (Tenerife, Gran Canaria, La Palma). On stony places in the dry zone.

BIOLOGY.– The larva mines the leaves of *Convolvulus floridus* L. The mine starts regularly from the tip of the leaf or from the midrib and looks like a large, cloudy brownish blotch. Inside the mine the larva constructs a swollen and opaque shelter, which is also used to deposit frass. Adults have been reared from early December to the end of May, but also collected at light in February (Klimesch, 1983).

135. *Hodgesiella christophi* sp. n.

DIAGNOSIS.– Adult. Wingspan 8-10 mm. Head with frons shining white, vertex and neck tuft shining bronze brown. Antennae with basal two-thirds shining dark brown, annulated white, subapically dark brown with white ring, apical one-fifth white. Thorax and tegulae shining dark brown. Legs with fore- and midfemora shining grey, hind-femora shining greyish white; tibiae and tarsi shining dark brown, foretibiae with two white longitudinal lines, midtibiae with white medial and apical rings, hindtibiae with white basal, medial and apical rings; foretarsi with white line and white apical ring on segments 1 and 2, segment 5 white; mid- and hindtarsi with white apical rings; spurs white. Forewings shining dark brown; two shining white fasciae, first at 1/5, second at 1/2 and not reaching dorsum, both perpendicular to costa or slightly inward oblique and laterally edged ochreous; white costal spot at 3/4 and smaller white tornal spot, just inward of costal spot, both edged ochreous inwardly; cilia greyish brown, darker at apex. Hindwings shining ochreous grey, somewhat darker towards apex, cilia ochreous grey. Abdomen shining grey.

Similar species. Can be distinguished from *rebeli* and *rhodorrhisella* by the bronze brown head; from *rebeli* also by the forewings with the incomplete second fascia and the interrupted outer fascia; from *rhodorrhisella* also by the more regular inner fasciae.

MALE GENITALIA.– Very similar to those of *rebeli* but the right brachium slightly widened distally, the tubular part of the aedeagus bent distally, and the right valvella more rounded.

FEMALE GENITALIA.– Very similar to those of *rebeli* but sterigma more slender, gradually tapering apically and bent in middle, ductus bursae very narrow, and extension of the corpus bursae more elongated.

DISTRIBUTION.– Asia Minor.

BIOLOGY.– Unknown. Adults have been collected from mid to late June.

MATERIAL.– Holotype. ♂, 'Asia min., Turcia, Ivriz/Eregli, 1200 m, Bolkar daglari, am 24.6.1968, leg. M. u. W. Glaser' (LSNK).

Paratypes. 1 ♂, 'Turkey, Sivas, 10 km W Gürün, 1650m, 27.vii.1989, leg. Fibiger & Esser.' 'Genitalia on slide JCK 3985 (ZMUC); 6 ♀, ditto (RMNH, ZMUC); 1 ♀, 'Turkey, Nevsehir, 7 km SE Avanos, 1200 m, 2.viii.1989, leg. Fibiger & Esser (ZMUC); 1 ♀, 'Asia min., Turcia, Ivriz/Eregli, 1200 m, Bolkar daglari, am 14.6.1969, leg. F. Hahn' (LSNK).

ETYMOLOGY.– Named after the famous Russian entomologist Hugo Fedorovitsch Christoph.

Tolliella Riedl, 1969

Tolliella Riedl, 1969: 798.

Type species: *Stagmatophora fulguritella* Ragonot, 1895, by original designation.

DIAGNOSIS.– Very large species with a wing pattern similar to *Hodgesiella*.

MALE GENITALIA.– Shows similarity to *Eteobalea*, but right brachium very long and straight and valvae almost parallel-sided.

FEMALE GENITALIA.– Sterigma protruded and sclerotized. Ductus and corpus bursae membranous, signa absent.

DISTRIBUTION.– A monotypic genus known only from the eastern part of the Mediterranean area.

BIOLOGY.– Unknown.

136. Tolliella fulguritella (Ragonot, 1895)

Stagmatophora fulguritella Ragonot, 1895: 108.
Limnoecia copidobathra Meyrick, 1936: 30.

DIAGNOSIS.– Adult. Wingspan 17-24 mm. Head with frons shining ochreous brown, remaining parts dark brown. Antennae white. Thorax and tegulae shining dark brown. Forewings shining dark blackish brown with white basal streak; three white fasciae with irregular and dentate edges, first at 1/5, inwardly oblique, from costa to beyond fold, second at 1/2, slightly inward oblique, from costa to just above dorsum, third at 5/6, both outer fasciae zigzag-shaped; cilia brown. Hindwings shining brown, cilia ochreous brown. Abdomen with first three segments yellowish, others shining brown.

Similar species. Resembles *Eteobalea tririvella* and *Hodgesiella rebeli*, but differs by its much larger wingspan; from *tririvella* also differing by the brown head, the pure white fasciae and the lack of the apical spot on the forewings.

MALE GENITALIA.– Right brachium long and stout, more than three times as long as left one, narrowed in middle and hooked apically; left brachium triangular, apex pointed. Valvae long and weakly lobate. Right valvella triangular and pointed; left valvella short with two very long setae. Aedeagus bottle-shaped, coecum penis large and cylindrical, distal part narrow, setose and widened at apex.

FEMALE GENITALIA.– Apophyses posteriores slightly longer than apophyses anteriores. Ostium oval; sterigma funnel-shaped. Ductus bursae membranous, slightly shorter than corpus bursae. Corpus bursae egg-shaped with inception of ductus seminalis in posterior part; signa absent.

DISTRIBUTION.– Asia Minor and Near East (Turkey, Lebanon, Syria, Jordan, Iraq).

BIOLOGY.– Unknown. Adults have been collected in July.

Family CHRYSOPELEIIDAE

Chrysopeleiidae Mosher, 1916: 104. Type genus: *Chrysopeleia* Chambers, 1874.
Walshiidae Hodges, 1962: 7. Type genus: *Walshia* Clemens, 1864.

General part

MORPHOLOGY.– Wingspan 6 to 18 mm. Head (Text-fig. 49) covered with densely appressed scales. Antennae much shorter than forewings, scapus without bristled pecten. Eyes moderate, ocelli absent. Labial palpi rather long, curved upwards, without tufts of scales in front; segment 3 usually equal to segment 2, apically blunt. Forewings (Text-fig. 50) narrowly lanceolate, sometimes covered by numerous small tufts of raised scales; 2 or 3 median veins present, M1 sometimes stalked with R4+5. Hindwings lanceolate linear or linear; M1 usually stalked with Rs. Hindtibiae with middle pair of spurs before midpoint.

Abdominal tergites without specialized lanceolate scales (Text-fig. 51); longitudinal sclerotized struts of tergite II very short or absent. Male genitalia (Text-fig. 52): uncus digitate, fused at base with very short belt-like tegumen, or reduced; gnathos absent; vinculum with deep invaginations fused with base of aedeagus; valvae symmetrical or very slightly asymmetrical, unilobate or subdivided apically into 2 or 3 variously shaped lobes; anellus lobes absent; aedeagus sclerotized, sometimes long, without cornuti, fused basally with vinculum and saccus by a caulis. Sometimes very peculiar additional structures appear, like bubble-shaped appendices with a large sclerotized spine on the valvae, or different outgrowths on segment VIII. Female genitalia (Text-fig. 53): ovipositor moderately long; ostium often surrounded with peculiar corn-like structures or debouching into a protruded genital plate (sterigma); ductus seminalis merges into ductus bursae immediately in front of rather wide and usually sclerotized antrum; ductus bursae more or less curved, rough-walled or with longitudinal sclerotized belt; bursa copulatrix rounded or oval, with paired, drawing-pin or cap-shaped signa.

Larva rather short and broad ('fat'), spindle-shaped, slow-moving, without secondary setae, colourless, but becoming brownish before pupation. Pupa with rather hard cuticula and concealed labial palpi and forefemora; abdominal segments immobile due to wing cases that almost cover the complete ventral surface of the abdomen; apical cremaster not developed, only small curved crochets present at the apex of abdomen.

BIOLOGY.– Mainly distributed in arid and semiarid landscape (deserts, steppes and savannas); only very few species are found in woodlands of the temperate zone. Larvae predominately use plants of Fabaceae, Tamaricaceae, Polygonaceae and Rhamnaceae, boring into buds, twigs and branches, rarely living in a silky web between spun leaves or flowers; some species are able to form stem-galls. Adults are nocturnal and easily attracted to light. In resting position they press tightly to the substrate and start jumping when disturbed.

DISTRIBUTION.– The family is richly represented especially in Central and southern Asia, Africa and America. The considerable superficial similarity to some representatives of Momphidae and Cosmopterigidae was the reason that in the past many species from tropical regions were described as members of these two families. More than 50 species are known from the Palaearctic region (Riedl, 1996a), with 7 genera and 22 species in the region of study.

Text-fig. 49. Head of *Sorhagenia janiszewskae* Riedl.

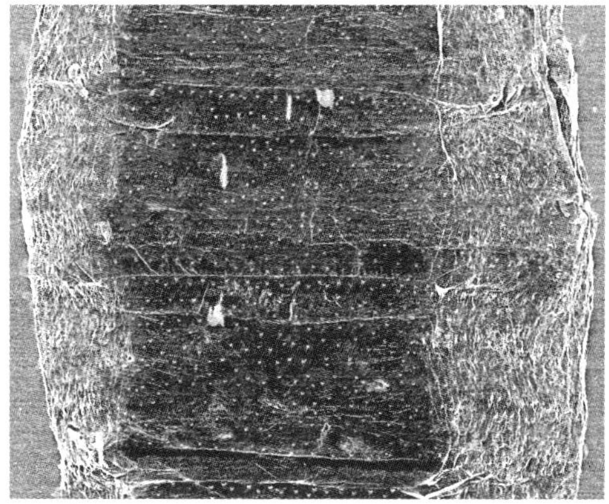

Text-fig. 51. Abdominal tergites of *Sorhagenia janiszewskae* Riedl.

SYSTEMATICS.– One of the most specialized families of gelechioid moths with very few morphological characters in a plesiomorphic state. The wide-winged forms are completely absent. In the male genitalia, a lot of apomorphies could be mentioned, i.e. reduction of tegumen, morphological integration of different parts (at least, a fusion between the valvae bases and aedeagus), modification of the segment VIII, etc. The same is true for the female. In many respects Chrysopeleiidae seem to be very similar with the most specialized forms of Cosmopterigidae and often they are treated as a subfamily of that family (Hodges, 1978, 1999). But a detailed analysis of morphological data has shown that most of the highly specialized characters in both families appeared independently or were erroneously regarded as homologous, i.e. the so-called ankylosed aedeagus (Sinev, 1992, 2002b). This is a reason why we consider Chrysopeleiidae a separate family.

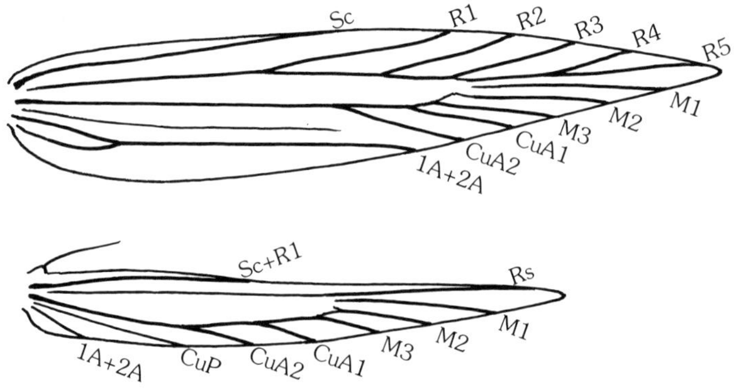

Text-fig. 50. Wing venation of *Sorhagenia rhamniella* (Zeller).

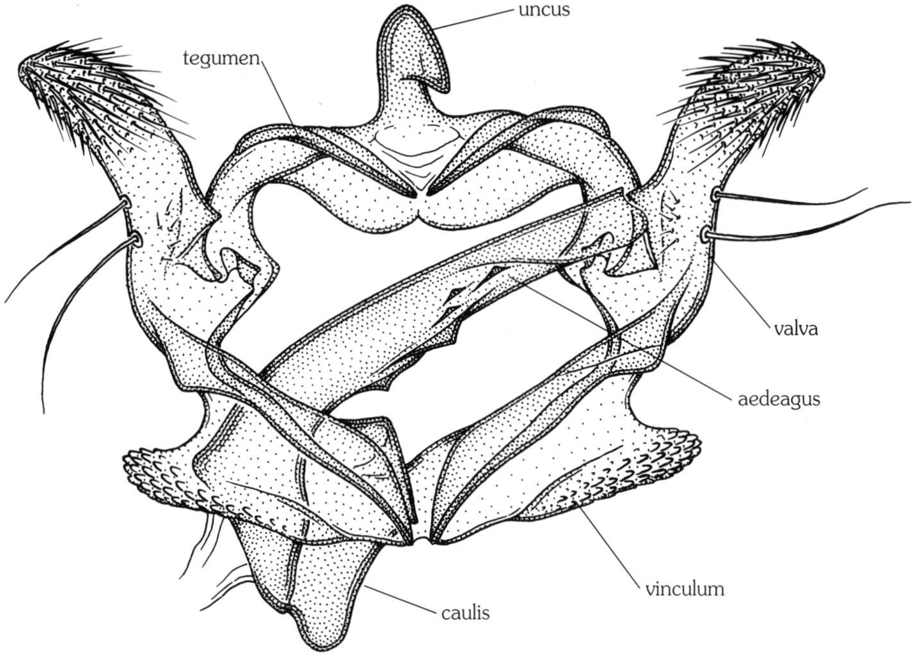

Text-fig. 52. Male genitalia of *Calycobathra calligoni* Sinev.

Systematic Treatment of the Genera and Species of Chrysopeleiidae in Europe, North Africa and the Near East

Sorhagenia Spuler, 1910

Sorhagenia Spuler, 1910: 384.
Type species: *Elachista rhamniella* Zeller, 1839, by monotypy.
Cystioecetes Braun, 1915: 195.
Type species: *Cystioecetes nimbosus* Braun, 1915, by original designation.
Sorhageniella Riedl, 1965: 464.
Type species: *Sorhagenia janiszewskae* Riedl, 1962, by original designation.

DIAGNOSIS.– Head smooth scaled. Labial palpi porrect, rather short. Forewings dull brownish grey, lanceolate, with four, often large, patches of raised scales. Hindwings narrower than forewings, lanceolate, pale to dark grey.

MALE GENITALIA.– Symmetrical. Uncus short, blunt or pointed. Tegumen band-shaped. Valvae with separate cucullus and sacculus and a large ampulla, costa often with a projection. Juxta triangular or trapezoidal, heavily sclerotized. Vinculum wide, saccus absent. Aedeagus short, coecum penis bulbous, distally tubular. Abdominal sternite VIII sclerotized and often with row of coarse setae posteriorly.

FEMALE GENITALIA.– Apophyses of equal length, apophyses anteriores often connected by sclerotized band. Ostium surrounded by a heavily sclerotized genital plate with triangular, conical or ribbon-like protrusions laterally. Antrum sclerotized and often bulbous anteriorly. Ductus bursae spiculate and wrinkled. Corpus bursae rounded or oval with two funnel-shaped signa with central thorn.

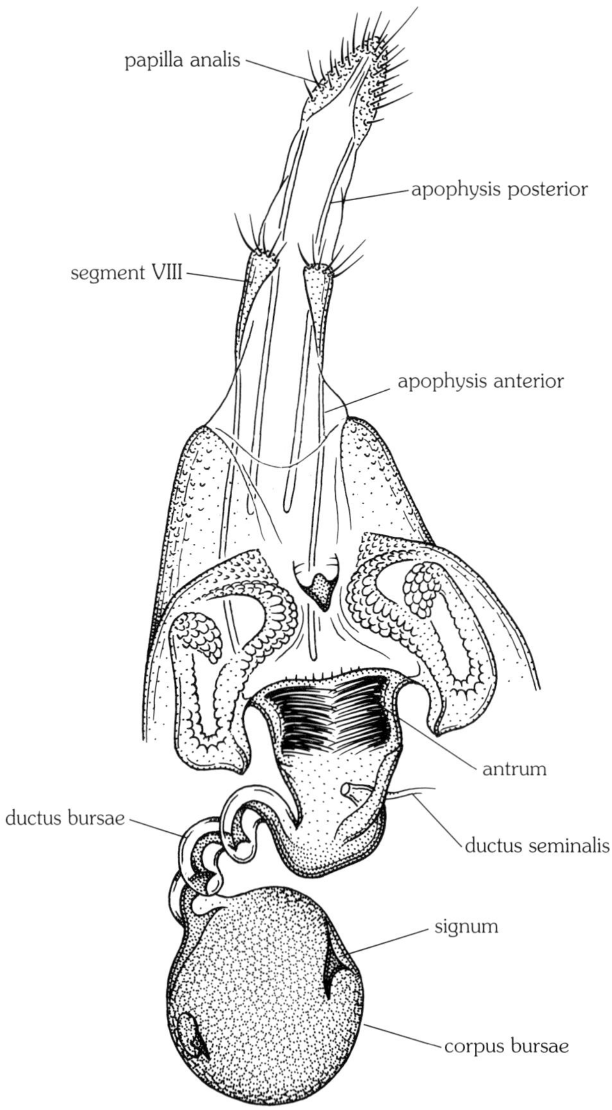

Text-fig. 53. Female genitalia of *Calycobathra calligoni* Sinev.

DISTRIBUTION.– Holarctic. Six species occur from Europe to Central Asia, five are described from the Russian Far East and five from North America.

BIOLOGY.– All species of *Sorhagenia* seem to be monophagous or oligophagous on Rhamnaceae.

REMARKS.– Most species of the genus *Sorhagenia* cannot be separated with certainty by external characters. For identification by examination of the genitalia it is often not necessary to dissect the specimen. In the male the scales of the anal tuft have to be removed to get a proper view of the cucullus, ampulla and sacculus to establish length and shape of these parts. In the female a removal of the scales on the posterior margin of sternite VI will reveal the upper part of the genital plate just enough for identification.

137. Sorhagenia rhamniella (Zeller, 1839)

Elachista rhamniella Zeller, 1839: 211.
Anacampsis asperipunctella Bruand, [1851]: 75.

DIAGNOSIS.– Adult. Wingspan 9-11 mm. Head brownish grey, mottled light grey. Thorax brownish grey, with two lateral scale tufts posteriorly. Forewings brownish grey, dark greyish brown at extreme base; four dark brown scale tufts, first between fold and dorsum at 1/4, second and largest above fold at 1/2, with grey-tipped scales or partly grey, third between fold and dorsum, just beyond the second, fourth above tornus at 3/4; paler area from 1/3 to second and third scale tufts; rather indistinct geniculate ochreous brown fascia at 3/4; large indistinct lighter grey patch in middle of wing from 1/3 to second and third scale tufts, not reaching costa; apex with series of small dark brown scale tufts. Hindwings dark grey.

Similar species. For separation from *reconditella*, *lophyrella*, *janiszewskae, fibigeri* and *taurensis*, q.v.

MALE GENITALIA.– Uncus short with blunt apex. Valvae with cucullus and ampulla of equal length, ampulla club-shaped, sacculus tapering distally and distinctly longer than cucullus and ampulla; almost square and strongly sclerotized projection on costa. Juxta trapezoidal. Vinculum square. Aedeagus short with wide coecum penis swollen and gradually tapering distally.

FEMALE GENITALIA.– Ostium with bent distal edge. Lateral protrusions of genital plate basally narrow, strongly widening distally and with rim on widest part. Antrum slightly widening towards ductus bursae. Ductus bursae parallel-sided and more than twice length of antrum. Corpus bursae circular, signa with a strong central thorn.

DISTRIBUTION.– Europe, Asia Minor and Central Asia (Sinev, 1993c).

BIOLOGY.– Host plants: *Frangula alnus* Mill., *Rhamnus cathartica* L., and *R. alpina* L. (Rhamnaceae). Univoltine. From early May till June the larva lives in a light web between flowers, eating both flowers and buds, rarely also young leaves (Malicky, 1966; Emmet, 1969; Malicky & Sobhian, 1971; Sinev, 1993c). Pupation takes place in a light yellowish cocoon on debris, sometimes amongst leaves. Adults can be found from mid-April to mid-August.

138. Sorhagenia reconditella Riedl, 1983

Sorhagenia reconditella Riedl, 1983: 314.

DIAGNOSIS.– Adult. Wingspan 8 mm. Very similar to *rhamniella* externally, but smaller; light grey patch on forewings not clearly defined, hindwings paler. Distinguishing characters only in the genitalia.

MALE GENITALIA.– Uncus very short. Valvae with cucullus twice as long as ampulla, ampulla slender and bent in middle, sacculus long and slender, almost twice as long as cucullus, slightly widening distally, apex rounded, two humps on costa. Juxta short and parallel-sided. Vinculum broad. Aedeagus very short. Sternite VIII square with rows of setae, which are slightly bent before middle.

FEMALE GENITALIA.– Distal edge of ostium rounded, with large V–shaped excavation inside. Lateral protrusions of genital plate short, narrowed at base, widening distally and with several sclerotized rims. Antrum slightly narrowed in middle. Ductus bursae twice as long as wide, slightly widening towards antrum. Corpus bursae oval, signa with bent thorn.

DISTRIBUTION.– Greece, Cyprus.

BIOLOGY.– Unknown. Adults have been found from mid-May to early July.

139. *Sorhagenia lophyrella* (Douglas, 1846)

Anacampsis lophyrella Douglas, 1846: 1270.
Sorhagenia tolli Riedl, 1962: 73.
Sorhagenia rhamniella auctt., (*nec* Zeller, 1839); misidentification.

DIAGNOSIS.– Adult. Wingspan 8-11 mm. Differs from *rhamniella* by the generally somewhat lighter forewings and the paler grey hindwings. Can only be distinguished with certainty by the genitalia.

MALE GENITALIA.– Uncus narrow with bundle of setae. Valvae with very large cucullus, slightly longer than ampulla; ampulla club-shaped and pointed apically; sacculus short, reaching to middle of cucullus, distally narrowing, subapically swollen, with apex pointed; costa with two small and pointed projections. Juxta triangular. Vinculum widely V–shaped. Aedeagus with small coecum penis.

FEMALE GENITALIA.– Distal edge of ostium straight. Lateral protrusions of genital plate long and gradually tapering into a thread-like apex. Antrum widening before entrance of ductus bursae. Ductus bursae widening towards corpus bursae, strongly wrinkled and about twice the length of antrum. Corpus bursae oval, slightly pectinate; signa funnel-shaped.

DISTRIBUTION.– Europe, Asia Minor and the Caucasus (Sinev, 1993c).

BIOLOGY.– Host plants: *Rhamnus cathartica* L., *R. saxatilis* L., and *R. pumila* L., rarely also *R. alpina* L. and *R. alaternus* L. (Rhamnaceae). Univoltine. The larva lives from early April to end of May in the buds and later between loosely spun terminal leaves (Malicky, 1966; Malicky & Sobhian, 1971). Pupation takes place in a light cocoon between debris on the ground. Adults fly from mid-May till early August and can be beaten from the food plant.

140. *Sorhagenia janiszewskae* Riedl, 1962

Sorhagenia janiszewskae Riedl, 1962: 70.
Sorhagenia rhamniella auctt., (*nec* Zeller, 1839); misidentification.

DIAGNOSIS.– Adult. Wingspan 10-13 mm. Very similar to *rhamniella* externally, but forewings darker and more elongate with length about 6 times width; lighter area in middle of wing not clearly defined. Safe identification only possible by examination of the genitalia.

MALE GENITALIA.– Uncus spatulate. Valvae with ampulla slender and pointed, about one-quarter longer than cucullus; sacculus short, broad at base and narrowing distally; costa with two protrusions, small and rounded near base followed by large square one. Juxta triangular. Vinculum broad, V–shaped. Aedeagus stout with large bulbous coecum penis.

FEMALE GENITALIA.– Ostium with two sharply pointed and strongly sclerotized lateral plates. Lateral protrusions of genital plate triangular. Antrum widening at entrance of ductus bursae. Ductus bursae slightly longer than antrum. Corpus bursae oval. Signa with a small central thorn.

DISTRIBUTION.– Europe, to southern Fennoscandia in the north; the Caucasus, western Transcaucasus.

BIOLOGY.– Host plants: *Frangula alnus* Mill., sometimes also on *Rhamnus cathartica* L., *R. alpina* L., and *R. fallax* L. (Rhamnaceae) (Malicky & Sobhian, 1971). Univoltine. The larva lives from May to June in a gallery in the young twigs eating the soft centre up to the wooden part. This activity causes the top leaves to wilt indicating the presence of the larva. The gallery has an opening on the side from where the excrements are partly removed. The opening is often situated at a leaf-axil and hard to find; only the removal excrements reveal its presence. Young larvae have also been found mining the leaf-stalks. Pupation takes place in a light cocoon among debris on the ground. Adults fly from the end of June to the end of August, but also have been found in late September and in March. Thus, it is possible that *janiszewskae* occasionally hibernates as an adult, in contrast to *rhamniella* and *lophyrella*.

141. Sorhagenia fibigeri sp. n.

DIAGNOSIS.– Adult. Wingspan 7-9 mm. Head with frons shining ochreous to greyish white; vertex, neck tuft and collar brownish grey. Antennae ochreous grey, indistinctly annulated dark greyish brown, apical third with 5 ochreous white segments. Labial palpi shining greyish white mottled ochreous on inside, segment 2 dark grey on outside, segment 3 strongly dark brown suffused and with dark grey basal spot and broad subapical streak on outside. Thorax and tegulae brownish grey, thorax with two lateral dark brown scale tufts posteriorly. Legs dark brownish grey, tibiae with ochreous white subbasal, medial and apical rings, tarsi ochreous white at joints. Forewings brownish grey, with four dark brownish grey scale tufts, first between fold and dorsum at 1/4, second and largest above fold before 1/2, with pale grey-tipped scales, third between fold and dorsum, just beyond the second, fourth above tornus at 3/4 almost in middle of wing; very indistinct geniculate ochreous brown fascia at 3/4, costa and dorsum in apical area with series of small dark brownish grey scale tufts; cilia brownish grey. Hindwings shining pale grey, darker along costa and dorsum; cilia brownish grey. Abdomen shining grey dorsally, sternites with broad yellowish white posterior bands; anal tuft pale grey dorsally, yellowish white ventrally.

Similar species. Externally very similar to a small-sized *rhamniella*, but can only be identified with certainty by examining of the genitalia.

MALE GENITALIA.– Uncus triangular with stout setae. Valvae with cucullus slightly longer than ampulla, ampulla club-shaped, bent at base; sacculus stout, more than twice as long as cucullus, widening distally and with sharp and bent apex; costa with two rounded humps. Juxta short and parallel-sided, widened at base. Vinculum V–shaped. Aedeagus rather slender. Abdominal sternite VIII with rows of stout setae, slightly bent at base.

FEMALE GENITALIA.– Ostium oval. Lateral protrusions of genital plate doubled, upper pair at ostium, short and pointed and followed by skirt-like protrusion, lower pair larger and situated at 1/2 of antrum. Antrum rather narrow, as long as ductus bursae. Ductus bursae parallel-sided. Corpus bursae egg-shaped, signa with short central spine.

DISTRIBUTION.– Asia Minor.

BIOLOGY.– Unknown. Adults were collected from early to mid-July at light.

MATERIAL.– Holotype. ♂, 'Turkey, Nevsehir, 3 km E Incesu, 1350 m, 3.vii.1987, leg. M. Fibiger' 'Genitalia on slide JCK 4145' (ZMUC).

Paratypes. 1 ♂, labelled as holotype, 'Genitalia on slide JCK 4060' (RMNH); 5 ♂ and 3 ♀, labelled as holotype (RMNH, ZMUC); 1 ♂, 'Turkey, Prov. Mersin, Güzeloluk NW Erdemil, 1400 m, 16.vii.1986, Taurus, leg. M. Fibiger' 'Genitalia on slide JCK 4141' (ZMUC); 1 ♀, ditto, but 'Genitalia on slide JCK 4144' (ZMUC); 1 ♂ and 1 ♀, ditto, (ZMUC); 1 ♂, 'Turkey, Mersin, 10 km SE Arslanköy, Taurus, 11.vii.1987, 1300 m, leg. M. Fibiger' 'Genitalia on slide JCK 4150' (RMNH).

ETYMOLOGY.– The name of the species is dedicated to its collector, Mr. Michael Fibiger from Denmark.

142. *Sorhagenia taurensis* sp. n.

DIAGNOSIS.– Adult. Wingspan 6-7 mm. Head with frons shining greyish white; vertex, neck tuft and collar grey, mottled light grey. Antennae ochreous grey, annulated dark brown, apical third with 5 ochreous white segments. Labial palpi white, outside and ventrally mottled dark grey, 3rd segment with dark grey ring in apical half. Thorax shining ochreous grey with two lateral dark brown scale tufts; tegulae shining ochreous grey. Legs dark brownish grey, mottled whitish, tibiae with mottled ochreous white basal, medial and apical rings. Forewings ochreous grey, dark grey at base, with four dark brownish grey scale tufts, first between fold and dorsum at 1/4, second and largest above fold, just beyond 1/2, third between fold and dorsum, just beyond the second, fourth above tornus at 3/4; two outer tufts half as large as inner ones; geniculate indistinct ochreous brown fascia at 3/4; costa and dorsum in apical area ochreous brown and with series of small dark brown scale tufts; cilia brownish grey. Hindwings shining grey; cilia brownish grey. Abdomen shining grey, segments with broad greyish white posterior bands ventrally; anal tuft yellowish white.

Similar species. The only external character that distinguishes *taurensis* from the other treated species is the small wingspan. Can only be identified with certainty by examination of the genitalia.

MALE GENITALIA.– Uncus triangular with strongly sclerotized hook at apex. Valvae with cucullus one and a half times as long as ampulla; ampulla bent at base, bulbous subapically and with pointed and bent apex; sacculus about twice as long as cucullus, broad at base, narrowing distally, apical part rounded and twisted. Juxta long and slender, slightly widened at base. Vinculum bent. Aedeagus short, acutely pointed, coecum penis almost circular. Abdominal sternite VIII tapering distally and with rows of broad, basally strongly bent, setae.

FEMALE GENITALIA.– Ostium elongate. Lateral protrusions narrow and very long, reaching beyond antrum. Antrum slender, slightly shorter and narrower than ductus bursae. Corpus bursae bulbous, signa shallow with small thorn.

DISTRIBUTION.– Asia Minor.

BIOLOGY.– Unknown. The few known specimens were collected in mid-July.

MATERIAL.– Holotype. ♂, 'Turkey, Mersin, 27 km N Anamur (Ermenek), Taurus 750 m, 13.vii.1987, leg. M. Fibiger' 'Genitalia on slide JCK 4129' (ZMUC).

Paratypes. 1 ♂, labelled as holotype, 'Genitalia on slide JCK 4060' (RMNH); 1 ♀, labelled as holotype, 'Genitalia on slide JCK 4065' (RMNH); 1 ♀, labelled as holotype (ZMUC).

ETYMOLOGY.– The species name is derived from Taurus, the type locality in Turkey.

Ascalenia Wocke, [1876]

Ascalenia Wocke, [1876]: 421.
Type species: *Laverna vanella* Frey, 1860, by monotypy.

DIAGNOSIS.– Head smoothly scaled. Antennae 3/4 to 4/5 length of forewings; scapus short and rather wide. Labial palpi cylindrical and porrect. Forewings lanceolate, somewhat tapering distally, greyish brown, often with darker spots and indistinct lighter spots or fascia. Hindwings very narrow, less than half of width of forewings.

MALE GENITALIA.– Uncus long, narrow and bent. Tegumen very short and wide. Valvae stout, sometimes with outgrowths, bearing bristles, long setae or spines. Aedeagus straight or bent, without cornuti.

FEMALE GENITALIA.– Apophyses of about equal length. Ostium at anterior part of slit-shaped excavation of sternite VII and under large fold of posterior margin of sternite VI, surrounded by complicated shields of rounded or knob-like sclerotizations with net-shaped or spiculate structures. Ductus bursae narrow to rather wide, with one or more loops and sclerotized longitudinal band. Corpus bursae rounded, sometimes covered with pectinations, with two small to large funnel or horn-shaped signa.

DISTRIBUTION.– Widely spread in the arid and semiarid regions of the Old World, especially richly represented in the deserts of northern Africa, Near and Middle East, and Central Asia.

BIOLOGY.– Only known for a few species. The larvae are bud- or twig-borers on Tamaricaceae, Polygonaceae, Mimosaceae and Fabaceae, sometimes producing galls.

143. Ascalenia vanella (Frey, 1860)

Laverna vanella Frey, 1860: 197.
Laverna seeboldiella Ragonot, 1882: 66.

DIAGNOSIS.– Adult. Wingspan 7-10 mm. Head, thorax and tegulae shining greyish brown, with tips of scales lighter. Antennae greyish brown. Forewings shining greyish brown, more or less mottled with light grey-tipped scales, forming very indistinct and irregular fascia at 1/3 and similar, slightly outward oblique, fascia at 2/3; dark oval spot in fold at 1/3; similar but smaller spot in middle, inwards of outer fascia; cilia greyish brown at apex, paler towards dorsum. Hindwings shining light grey, more brownish on costa and towards apex. Abdomen in male and female shining greyish brown with white anal tuft.

Similar species. Externally very similar to *vanelloides* and *echidnias*, genitalia have to be examined for certain identification.

MALE GENITALIA.– Uncus rather short, narrow, downwardly bent. Valvae slender, almost parallel-sided with rounded apex; rows of coarse setae at inner surface, hair-like setae and one very long curved seta in basal half at outer surface. Aedeagus almost straight and slightly tapering distally. – For separation from *vanelloides*, q. v.

FEMALE GENITALIA.– Slit-shaped excavation of sternite VII small and narrowly U–shaped; fold of sternite VI obtusely triangular. Lateral shields trapezoidal, sclerotized around ostium, with net-shaped structure at posterior edges connected on both sides to oval net-like patch on inner side of fold. Ductus bursae with two windings and narrow sclerotized longitudinal band. Corpus bursae oval, with two large, funnel-shaped signa with bent apex. – For separation from *vanelloides* and *antiqua*, q. v.

DISTRIBUTION.– Central (local, along banks of large rivers) and South Europe; Canary Islands, Asia Minor, the Caucasus.

BIOLOGY.– Host plants: *Myricaria germanica* Desv. and *Tamarix* L. (Tamaricaceae). Bivoltine. The larva bores in the bud or in the twig (Frey, 1860). Adults fly in June-July (first generation) and in September-October and after hibernation again in early spring (Kasy, 1969).

144. *Ascalenia vanelloides* Gerasimov, 1930

Ascalenia vanelloides Gerasimov, 1930: 38.
Ascalenia eremella Amsel, 1933: 48; nomen nudum.
Ascalenia eremella Amsel, 1935b: 270.

DIAGNOSIS.– Adult. Wingspan 7-8 mm. Head, thorax and tegulae shining greyish brown, with paler tipped scales. Forewings shining greyish brown, mottled by light grey-tipped scales, especially at 2/3, forming very indistinct fascia; dark spot in fold at 1/3 and similar spot in middle before 2/3. Hindwings shining light grey. Abdomen shining greyish brown with greyish white anal tuft in male, shining light grey, with darker subapical portion in female.

Similar species. Very similar to *vanella* and *echidnias*, and certain determination only possible by examination of genitalia.

MALE GENITALIA.– Uncus long and stout, with bent tip. Valvae slender, slightly bent and tapering distally; rows of coarse setae at inner surface, hair-like setae and two very long setae near base at outer surface. Aedeagus short, bent and strongly tapering distally. – The species can be distinguished from *vanella* by the apically narrowing valvae, the long uncus and the shape of the aedeagus.

FEMALE GENITALIA.– Slit-shaped excavation of sternite VII large and U–shaped, at each side prolonged to a bent spool-shaped protrusion with net-shaped structure, its medial wall spiculate; fold of sternite VI large with deep indentation in middle. Ostium surrounded by spiculated hump extended into a bent and widened protrusion. Ductus bursae with four loops; longitudinal sclerotized band narrow near ostium, widening towards corpus bursae. Corpus bursae rounded, with two small double-pointed signa. – The large slit-shaped excavation with long protrusions and central indentation of fold distinguishes this species from *vanella*.

DISTRIBUTION.– Asia Minor, Middle East, Central Asia eastwards to Afghanistan.

BIOLOGY.– Unknown. Kasy (1969) suggests that *Prosopis farcta* (Banks & Sol.)J.F.Macbr. (Fabaceae) could be the host plant. The reason is that the species was very abundant at light near Teheran (Iran) in a locality were, besides *P. farcta*, hardly anything else grew except *Alhagi* sp. (Fabaceae). Adults fly from April to August, probably in more than one generation.

REMARKS.– The binomen *Ascalenia eremella* was first published without description (Amsel, 1933); it was made nomenclaturally available later (Amsel, 1935b).

145. Ascalenia antiqua Meyrick, 1925

Ascalenia antiqua Meyrick, 1925: 213.

DIAGNOSIS.– Adult. Wingspan 10 mm. Head brownish grey, mottled with light grey-tipped scales; antennae shining grey with indistinct darker annulations. Thorax and tegula grey, postero-laterally ochreous grey. Forewings dark grey, mottled with light grey-tipped scales with very indistinct pattern, at 1/3, above and on fold an indistinct dark spot, lower further from base, outwardly edged greyish white, similar spot in middle at 2/3, edged greyish white on both sides, small greyish white costal spot at 3/4 and similar, but larger spot on tornus; cilia brownish grey, strongly intermixed with greyish white around costa, ochreous grey towards dorsum. Hindwings shining light grey, cilia ochreous grey. Abdomen dorsally brownish grey, segments paler banded posteriorly, ventrally brownish grey with broad ochreous white longitudinal streak; anal tuft ochreous white.

Similar species. The ochreous grey spots on the thorax and tegulae in combination with the indistinct markings on the forewings are distinctive characters.

MALE GENITALIA.– Uncus short and pointed. Valvae large, broad, outer edge convex, rows of course setae at inner surface, two long and bent setae in middle. Aedeagus stout, rather short and straight, tapering distally. – The bulky valvae are a distinguishing character.

FEMALE GENITALIA.– Resembles those of *vanella*, but the slit-shaped excavation of sternite VII wider and much longer; fold of sternite VI distinctly triangular. Lateral shields square, with oval net-like patches on inner side of fold larger.

DISTRIBUTION.– Israel and Egypt.

BIOLOGY.– Host plant: *Tamarix aphylla* (L.) Karst. (Tamaricaceae). The larva lives in the twig (Meyrick, 1925); according to Kasy (1974), it has also been found in galls caused by *Eriophyes* (Acarina). The few known specimens have been reared from September to January.

146. Ascalenia acaciella Chrétien, 1915

Ascalenia acaciella Chrétien, 1915: 352.
Scythris tergipunctella Turati, 1924: 33.
Scythris maculatella Lucas, 1937: 128.
Tischeria noviciata Gozmány, 1960: 420.

DIAGNOSIS.– Adult. Wingspan 6-8 mm. Head, thorax and tegulae shining greyish brown, with tips of scales lighter; antennae uniformly grey. Forewings without pattern, dark grey, mottled whitish by light tips of scales; cilia grey. Hindwings grey.

Similar species. Can be distinguished from *vanella* and *vanelloides* by the unicolorous forewings.

MALE GENITALIA.– Uncus long, strongly bent and very narrow. Valvae bulbous in basal 2/3, distal part parallel-sided with square apex; patch of rather long setae at inner surface; short and slightly bent setae around apex; single large and bent seta near base. Aedeagus almost straight, slightly narrowing distally, apex obliquely flattened. – Resembles *antiqua*, but differs by the long and narrow uncus and the square apex of the valvae.

FEMALE GENITALIA.– Slit-shaped excavation of sternite VII large, with narrow band posteriorly; fold of sternite VI large, triangular with short groove in middle. Ostium surrounded by large semicircular lateral shields with net-shaped structure and patch of spicules in middle; two smaller semicircular fields alongside excavation of sternite VII. Ductus bursae long, wide at ostium, then narrowing, and again gradually widening towards corpus bursae; longitudinal sclerotized band narrow, widening towards corpus bursae. Corpus bursae small, egg-shaped, with net-shaped structure; signa large, horn-like and bent.

DISTRIBUTION.– Canary Islands (Gomera), Malta, northern Africa, Near and Middle East, eastwards to Afghanistan and Pakistan.

BIOLOGY.– Host plants: *Acacia farnesiana* (L.) Willd., *A. horrida* Willd., and *A. tortilis* (Forsk.) Hayne (Mimosaceae) (Chrétien, 1915). The larva lives in the flowerheads. Pupation takes place among flowers in a transparent cocoon, covered with frass. Adults can be found in almost every month of the year, probably in several successive generations (Kasy, 1969).

147. *Ascalenia jerichoella* Amsel, 1935

Ascalenia jerichoella Amsel, 1933: 48; nomen nudum.
Ascalenia jerichoella Amsel, 1935a: 309.

DIAGNOSIS.– Adult. Wingspan 6 mm. Head bronze brown, vertex laterally with lighter tips of scales; antennae shining brown, apical half more greyish and slightly serrate. Thorax and tegulae shining dark brown with lighter tipped scales, thorax lighter posteriorly. Forewings shining dark brown, with irregular incomplete whitish fascia at 1/3; second and complete fascia at 2/3, widened in middle with some brown scales in widest part; cilia brown at apex, brownish grey towards dorsum. Hindwings shining light grey, more brownish on costa and towards apex; cilia pale ochreous grey.

Similar species. The small wingspan and the shining dark brown forewings with distinct white markings are characteristic.

MALE GENITALIA.– Unknown.

FEMALE GENITALIA.– Slit-shaped excavation of sternite VII long and very narrow. Fold rather shallow, posterior edge slightly bent. Sterigma in shape of curved sclerotized plate around ostium, sterigma and excavation enclosed by two large oval fields with net-shaped structure. Ductus bursae long, wide at ostium, narrow towards corpus bursae, with one and a half loops and narrow sclerotized band. Corpus bursae egg-shaped with two horn-shaped signa. – The narrow excavation in combination with the shape of the sterigma is characteristic.

DISTRIBUTION.– Israel (Wadi el Kelt).

BIOLOGY.– Unknown. The holotype was collected at light in May.

REMARKS.– The binomen *Ascalenia jerichoella* was first published without description (Amsel, 1933); it was made nomenclaturally available later (Amsel, 1935a).

148. *Ascalenia viviparella* Kasy, 1969

Ascalenia viviparella Kasy, 1969: 365.

DIAGNOSIS.– Adult. Wingspan 7-10 mm. Head, thorax and tegulae shining greyish brown, with tips of scales lighter; antennae shining grey. Forewings greyish brown, mottled with light grey-tipped scales; irregular and indistinct light grey subcostal streak from base to 1/3, partly bending towards fold, crossing it before 1/2 but not reaching dorsum; indistinct subcostal streak at 1/2; irregular light grey zigzag fascia at 3/4 with extensions towards base and apex; apical area more ochreous grey; cilia shining light grey. Hindwings shining light grey. In female underside of forewings with some large dark grey scales at tornus, hindwings with row of large dark grey scales in middle of dorsal area. Abdomen shining light grey, in female abdominal segments VI and VII blackish laterally and ventrally.

Variation. The light mottling of the ground colour, the shape and distinctness of the light grey streaks and the fasciae of the forewings can vary considerably, hence confusion with other similar species is possible.

MALE GENITALIA.– Uncus long and stout, slightly bent, apex blunt and hooked. Valvae long and slightly tapering in distal half; apex with large indentation edged by strong sclerotized rim on outer surface, and rows of course setae of different length on inner surface; two long and strongly bent setae at base. Aedeagus long and slender, bent before middle, with two short thorns of different length dorso-laterally in middle and smaller one at apex. Thorns of aedeagus are characteristic. – For separation from *echidnias*, q. v.

FEMALE GENITALIA.– Slit-shaped excavation of sternite VII U–shaped, with anterior half sclerotized; fold of sternite 6 large, trapezoidal. Ostium wide and circular, surrounded by two large spool-shaped lateral shields with net-shaped structure; shields narrowed in middle and sharply bent outwards to fold. Ductus bursae short and wide, with several loops; longitudinal sclerotized band wide, especially towards ostium. Corpus bursae egg-shaped with two small, often double-pointed signa. – For separation from *echidnias*, q. v.

DISTRIBUTION.– The most south-eastern part of Europe (the Crimea, southern Ural); Asia Minor, Transcaucasus, Central Asia eastwards to Mongolia (Sinev, 1984d). Very common in desert and semi-desert habitats.

BIOLOGY.– Host plants: *Calligonum junceum* (Fisch. Et Mey.) Litv., *C. setosum* (Litv.) Litv., and *C. leucocladum* (Schrenk) Bunge (Polygonaceae). Bivoltine. The larvae live in late spring and the second half of summer in the shoots and fruits (Falkovitsh, 1969). Adults can be found throughout the year and hibernate in the burrows of various rodents.

149. *Ascalenia echidnias* (Meyrick, 1891)

Elachista echidnias Meyrick, 1891: 61.
Ascalenia acaciella var. *signatella* Chrétien, 1915: 352.
Ascalenia oranella Lucas, 1939: 209.
Ascalenia satellita Gozmány, 1960: 420.

DIAGNOSIS.– Adult. Wingspan 7-10 mm. Head, thorax, and tegulae greyish brown, mottled with light grey-tipped scales. Forewings greyish brown, mottled with light grey-tipped scales; two indistinct short dark brown streaks, in fold at 1/3 and in middle at 2/3; white inwardly oblique costal streak and white outwardly oblique tornal streak forming indistinct chevron at 3/4; some white scales in apical area; cilia greyish brown at apex, grey towards dorsum. Hindwings shining light grey, more greyish brownish on costa and towards apex; cilia light grey. Abdomen dorsally shining pale brown, segments shining grey-banded posteriorly, anal tuft white; in female segment VII shining blackish ventrally.

Similar species. Superficially very similar to *vanella* and *vanelloides*, and cannot be separated from these species without examination of the genitalia.

MALE GENITALIA.– Uncus moderate, bent, with blunt apex. Valvae narrow at base, gradually widening to middle; distal part rather abruptly narrowing to rounded and slightly bent apex; triangular protrusion with rounded tip dorso-laterally on widest part; several coarse and slightly bent setae at inner surface; apex and protrusion ventrally with long, hair-like setae; one long and bent seta near base. Aedeagus stout and slightly bent. – Differs from *viviparella* by presence of a distal protrusion on the valvae and the lack of thorns on the aedeagus.

FEMALE GENITALIA.– Slit-shaped excavation of sternite VII small, U–shaped; fold of sternite VI broadly trapezoidal. Ostium surrounded by large spool-shaped lateral shields with net-shaped structure, narrowed in middle and sharply bent towards fold. Ductus bursae short and wide with several loops; longitudinal sclerotized band wide, especially towards ostium. Corpus bursae oval, with net-shaped surface; signa large, often double-pointed. – Differs from *viviparella* by the more shallow fold, the shorter, but stouter lateral shields, and the somewhat larger signa.

DISTRIBUTION.– Madeira, northern Africa, Asia Minor.

BIOLOGY.– Unknown. Adults have been collected from April to June and in September-October.

150. *Ascalenia bifasciella* Chrétien, 1915

Ascalenia bifasciella Chrétien, 1915: 352.
Elachista bizonatella Turati, 1930: 89, **syn.n.**

DIAGNOSIS.– Adult. Wingspan 7-10 mm. Head dark brownish grey, mottled with light grey-tipped scales. Thorax and tegulae dark brownish grey anteriorly, pale greyish ochreous posteriorly. Forewings dark brownish grey, mottled with light grey-tipped scales; broad irregular greyish white fascia at 1/3, not reaching costa and dorsum; similar fascia running from costa to dorsum at 2/3; cilia brownish grey around apex, ochreous grey towards dorsum. Hindwings shining light grey, more ochreous at apex; cilia pale ochreous grey. Abdomen shining pale grey dorsally, shining brownish grey ventrally; in female segments VI-VII and anal tuft dark brownish grey.

MALE GENITALIA.– Uncus very narrow, bent and distally widened. Valvae with large rounded basal part and narrow and bent caudal part; very large and coarse seta at base, single long seta at outer surface and two similar setae at edge of distal part, row of short and bent setae on inner surface. Aedeagus long and slender, tapering distally. – The large bulbous valvae with a narrow distal part and the very large and strong setae are characteristic.

FEMALE GENITALIA.– Slit-shaped excavation of sternite VII wide; fold of sternite VI rather shallow, posterior edge almost straight with small indentation in middle. Ostium surrounded by two sclerotized and bent lateral rims and paired small shields with net-like surface latero-caudally. Ductus rather long with several loops; longitudinal sclerotized band narrow, widening towards corpus bursae. Corpus bursae egg-shaped with two large, funnel-shaped signa. – The almost straight posterior edge of fold and rather simple surrounding of ostium are characteristic.

DISTRIBUTION.– Northern Africa.

BIOLOGY.– Unknown. Adults have been collected from April to June.

REMARKS.– The type of *Elachista bizonatella* (female) has been found very recently in the collection of the BMNH by the second author. In spite of the lack of abdomen, this specimen is evidently conspecific with *Ascalenia bifasciella*, having remarkable whitish fasciae on the forewings.

Calanesia Sinev, 1990

Calanesia Sinev, 1990: 141.
Type species: *Calanesia karsholti* Sinev, 1990, by original designation.

DIAGNOSIS.– Head smoothly scaled. Antennae 3/4 length of forewings; scapus long and wide. Labial palpi cylindrical and recurved above head. Forewings broadly lanceolate, tapering distally, with distinct pattern of longitudinal lines. Hindwings lanceolate, half width of forewings.

MALE GENITALIA.– Uncus very long and stout. Tegumen short and wide. Valvae rather short, stout, with patch of long and coarse setae. Aedeagus long and slender; cornuti absent.

FEMALE GENITALIA.– Apophyses of about equal length; apophyses anteriores connected by broad V–shaped transverse band. Ostium at anterior part of sternite VII, surrounded by large genital plate. Ductus bursae long and membranous. Corpus bursae small, with two signa.

DISTRIBUTION.– Northern Africa.

BIOLOGY.– Unknown.

151. Calanesia karsholti Sinev, 1990

Calanesia karsholti Sinev, 1990: 142.

DIAGNOSIS.– Adult. Wingspan 13-16 mm. Head grey. Thorax grey, laterally and posteriorly dark grey; tegulae dark grey, edged light grey. Forewings greyish white, with several narrow dark grey lines along veins in distal part; broad dark grey streak from base to 1/2, narrowing towards base and mottled light grey; narrow dark grey dorsal band from base towards apex, irrorated light grey; elongated dark grey spot in fold at 1/3, similar spot above tornus at 2/3; series of dark grey streaks to termen around apex; cilia shining grey. Hindwings shining light grey. Abdomen ochreous grey dorsally, shining light grey laterally and ventrally.

MALE GENITALIA.– Uncus strongly bent in middle and heavily sclerotized, apex pointed. Valvae rather narrow at base, gradually widening to sparsely setose bulbous part beyond middle, with blunt apex; long and stout, slightly bent spine on inner surface, surrounded by 5-6 setae of equal length. Aedeagus weakly bent, gradually tapering distally, apex blunt.

FEMALE GENITALIA.– Genital plate broad trapezoidal, lateral edges rounded, posterior edge with large indentation bearing two heavily sclerotized oval patches allied to ostium; ostium anteriorly margined by semicircular sclerotized band connected to oval patches. Ductus bursae narrow at ostium, slightly and gradually widening towards corpus bursae. Corpus bursae egg-shaped, with two tooth-like signa.

DISTRIBUTION.– Northern Africa: Tunisia (Nefta).

Biology.– Unknown. Adults have been collected with light in March.

Calycobathra Meyrick, 1891

Calycobathra Meyrick, 1891: 59.
Type species: *Calycobathra acarpa* Meyrick, 1891, by monotypy.

Diagnosis.– Head smoothly scaled. Antennae 3/4 length of forewings; scapus very broad and flattened, especially at base. Labial palpi cylindrical and recurved. Forewings broad lanceolate, dull ochreous or brownish with very indistinct pattern. Hindwings lanceolate, less than half width of forewings.

Male genitalia.– Uncus short, stout and hook-shaped. Tegumen very short, band-shaped. Valvae long and slender, finger-shaped, with setae at inner surface distally; few very long setae in middle ventrally. Vinculum large with two broad inner lobes. Aedeagus long to very long, slender, often with teeth or thorns.

Female genitalia.– Apophyses of about equal length. Posterior margin of sternite VI with large fold, covering ostium; sternite VII with short and wide slit-shaped excavation. Ostium on projecting ring at anterior edge of sternite VII, between large and often very complicated lateral shields with net-shaped structures and thorns. Ductus bursae with several loops and longitudinal sclerotized band; antrum often sclerotized and with dense inner hairs. Corpus bursae rounded, sometimes covered with pectinations; two small thorny or horn-shaped signa.

Distribution.– Northern Africa, Near and Middle East, the Caucasus, Central Asia eastwards to Mongolia.

Biology.– Known for only one species; larvae live in shoots and fruits of *Calligonum* L. (Polygonaceae).

152. *Calycobathra acarpa* Meyrick, 1891

Calycobathra acarpa Meyrick, 1891: 60.
Calycobathra acarpa var. *pinguescentella* Chrétien, 1915: 351.
Blastobasis sublineatella Lucas, 1933: 198, **syn.n.**

Diagnosis.– Adult. Wingspan 11-12 mm. Head pale ochreous. Thorax and tegulae pale ochreous, anteriorly darker. Forewings pale ochreous to greyish ochreous; small dark brown streak in fold at 1/3, similar streak in middle at 3/5; indistinct longitudinal brownish streaks in course of veins, more pronounced in apical area; cilia ochreous grey. Hindwings grey; cilia ochreous grey. Abdomen shining grey, anal tuft pale ochreous; in female segment VII entirely blackish ventrally.

Similar species. Superficially similar to *variapennella*, but thorax and forewings lighter, and brownish longitudinal streaks narrower. For differences with *calligoni*, q. v.

Male genitalia.– Uncus short and strongly bent. Valvae slender and inwardly bent distally, gradually narrowing towards apex, but apex widened, spatulate and densely covered with coarse setae at inner surface; three long and bent dorsal setae from base to 1/2; setae on outer surface hair-like. Lobes of vinculum triangular. Aedeagus long and slender, slightly S–shaped; apex with one very long and one short tooth. – The long and slender aedeagus with a long apical tooth is characteristic.

FEMALE GENITALIA.– Slit-shaped excavation of sternite VII bottle-shaped, with sclerotization anteriorly; fold of sternite VI widely rounded. Ostium surrounded by annular reinforcement with sclerotized rims connected to a large complicated circular genital plate, triangular in centre; lateral parts of plate with net-shaped structure turning into large protrusion of similar structure, which joins posteriorly at top of triangular plate and partly surrounds the slit-shaped excavation. Ductus bursae with two loops and longitudinal sclerotized band; antrum long and straight, supplied with two dense lateral inner rows of trap-like hairs inside. Corpus bursae rounded with net-shaped surface; two small signa with circular base and small central spine, sometimes accompanied by another tiny spine. – Differs from *calligoni* and *variapennella* by the circular genital plate and the long and narrow antrum.

DISTRIBUTION.– Northern Africa, Near East (Israel).

BIOLOGY.– Unknown. Adults fly from April till June and in September-October, probably in two generations (Kasy, 1968a).

REMARKS.– The type of *Blastobasis sublineatella* (female) have been found very recently in the collection of the MNHN by the second author. According to the genitalia, it is undoubtedly conspecific with *Calycobathra acarpa*.

153. Calycobathra calligoni Sinev, 1979

Calycobathra calligoni Sinev, 1979b: 85.

DIAGNOSIS.– Adult. Wingspan 10-12 mm. Head ochreous. Thorax and tegulae pale ochreous, mottled with ochreous. Forewings pale ochreous, mottled with ochreous grey, especially in basal area and along costa; two narrow dark grey streaks, first in fold at 1/3, second in middle at 3/5; cilia pale ochreous grey. Hindwings shining light grey; cilia pale ochreous grey. Abdomen dorsally shining brownish grey with segments whitish banded posteriorly; last two segments white, anal tuft yellowish white.

Similar species. Differs from *acarpa* and *variapennella* by lighter and almost unicolorous pale ochreous forewings with two rather distinct dark grey streaks.

MALE GENITALIA.– Uncus short, stout, sickle-shaped. Valvae slender and inwardly bent distally, with notch at base; caudal part parallel-sided, with rows of coarse setae at inner surface, apex rounded; three long and bent setae dorsally from near base to 1/2; setae on outer surface short and hair-like. Lobes of vinculum triangular, with caudal parts rounded and dentate, especially at margins. Aedeagus slightly bent and laterally flattened, widening to rounded apex with semicircular indention at the tip; two or three small subapical teeth on ventral edge. – The dentate lobes of the vinculum and the enlarged apex of the aedeagus are characteristic.

FEMALE GENITALIA.– Slit-shaped excavation of sternite VII anteriorly sclerotized, U–shaped; fold wide with rather straight margin. Ostium surrounded by almost square reinforcement connected to more or less trapezoidal genital plate with two pairs of lateral protrusions of net-shaped structure; inner protrusion bending before slit-shaped excavation, outer one almost straight, apically widened and hook-like bent inwardly. Ductus bursae with many loops and longitudinal sclerotized band. Antrum widely cup-shaped, densely covered with hairs in posterior half inside. Corpus bursae oval; two small horn-shaped signa on circular bases. – Differs from *acarpa* and *variapennella* by cup-shaped antrum.

DISTRIBUTION.– Extreme south-east of European Russia (eastern Caucasus, southern

Ural); eastern Transcaucasus, Central Asia. Very common in desert and semi-desert habitats, together with *Ascalenia viviparella*.

Biology.– Host plants: *Calligonum junceum* (Fisch. et Mey.) Litv., *C. eriopodum* Bunge, *C. microcarpum* Borszcz., and *C. leucocladum* (Schrenk) Bunge. (Polygonaceae). Bivoltine. In late spring and the second half of summer the larvae live in the shoots and fruits (Falkovitsh, 1969). Adults can be found throughout the year and hibernate in the burrows of various rodents.

154. *Calycobathra variapennella* Sinev, 1984

Calycobathra variapennella Sinev, 1984d: 331.

Diagnosis.– Adult. Wingspan 10-11 mm. Head greyish brown. Thorax and tegulae greyish brown. Forewings ochreous grey to greyish brown, with indistinct and irregular brown dots and streaks along veins, especially in apical area; cilia ochreous grey. Hindwings pale grey, darker on costa; cilia ochreous grey. Abdomen shining pale ochreous grey.

Similar species. Resembles *acarpa*, but darker, and forewings with more coarse longitudinal streaks. For differences with *calligoni*, q. v.

Male genitalia.– Uncus short and sickle-shaped. Valvae slender, inwardly bent distally, gradually tapering towards spatulate apex; inner surface with sharp tooth at base, densely covered with coarse setae; three long and bent setae dorsally, from near base to one-half; outer surface with short and hair-like setae. Lobes of vinculum rounded. Aedeagus bent, strongly flattened laterally, widened distally; from one to three triangular teeth on dorsal side and one tooth on ventral side. – Easily distinguished from other species of the genus by the laterally flattened aedeagus with strong apical teeth.

Female genitalia.– Slit-shaped excavation of sternite VII triangular and sclerotized; fold of sternite VI wide, with rather straight margin. Ostium in form of large oblique slit, situated slightly to the left; genital plate more or less narrowly trapezoidal, with two large V–shaped lateral shields, partly covered with net-shaped structures; inner arms of these lobes with long protrusions towards slit-shaped excavation. Ductus bursae with several loops and longitudinal sclerotized band; antrum very wide, pouch-shaped, densely covered with hairs inside. Corpus bursae oval, with two small spiny signa on circular bases.– The asymmetrical position of the very wide, pouch-shaped antrum is characteristic.

Distribution.– Turkey, north-eastern Caucasus, Transcaucasus, Central Asia.

Biology.– Unknown. Adults have been collected from May till July (Sinev, 1984d).

Bifascia Amsel, 1961

Bifascia Amsel, 1961: 52.

Type species: *Ascalenia nigralbella* Chrétien, 1915, by original designation.

Diagnosis.– Head smoothly scaled. Antennae 3/4 length of forewings; scapus short and broad. Labial palpi cylindrical and recurved. Forewings lanceolate, blackish brown with two broad white fasciae. Hindwings narrow lanceolate, about half of width of forewings.

Male genitalia.– Rather simple. Uncus very short. Tegumen large and widened basally.

Valvae completely divided into two similar lobes. Vinculum broad. Aedeagus long and slender, without cornuti.

FEMALE GENITALIA.– Apophyses posteriores slightly longer than apophyses anteriores. Ostium rounded, placed in middle of anterior edge of sternite VII. Posterior margin of sternite VI with very shallow fold, not covering ostium. Ductus bursae narrow, widening towards corpus bursae. Corpus bursae membranous, and with two large signa.

DISTRIBUTION.– Southwestern Europe (Spain), northern Africa, Near and Middle East, and Central Asia.

BIOLOGY.– Published data on the single known species are inconsistent.

155. *Bifascia nigralbella* (Chrétien, 1915)

Ascalenia nigralbella Chrétien, 1915: 351.
Limnoecia asteroleuca Meyrick, 1917: 43.

DIAGNOSIS.– Adult. Wingspan 7-9 mm. Head shining dark brown with reddish gloss; frons shining ochreous white; antennae shining greyish brown. Thorax and tegulae shining dark brown with slight reddish gloss. Forewings shining dark brown; two very distinct and broad white fasciae, first at 1/4, abruptly narrowing at costa, second at 2/3, narrowing towards costa and dorsum; cilia grey. Hindwings shining silvery white, more ochreous towards apex. Abdomen ochreous grey, brownish grey with white anal tuft in female.

Similar species. For separation from *Bifascioides leucomelanellus*, q. v.

MALE GENITALIA.– Uncus very short and rounded, tuberculate. Tegumen broad, narrowing distally, membranous, only broad dorsal stripe and narrow lateral margins sclerotized. Valvae consist of two narrow lobes of equal length, both sparsely covered with short hairs and few very long setae; dorsal lobe slightly spatulate, ventral one tapering distally. Vinculum rounded. Aedeagus long, bent in middle and tapering apically.

FEMALE GENITALIA.– Ostium circular and surrounded by sclerotized ring with two short lateral protrusions. Ductus bursae long, very narrow near ostium, gradually widening towards corpus bursae; inner surface weakly spiculate. Corpus bursae egg-shaped with two large and bent horn-like signa.

DISTRIBUTION.– Spain (Almeria), northern Africa, Near and Middle East eastwards to Pakistan and western India, Central Asia eastwards to Mongolia.

BIOLOGY.– Host plant: *Acacia* Mill. (Mimosaceae) or *Tamarix* L. (Tamaricaceae). The larva has been found feeding on the flowers of acacia (Dumont, 1932), but on another occasion one specimen has been reared from a dry stem of *Tamarix* sp. (Halperin & Sauter, 1992). Adults fly in May-June and also in September, probably in two generations.

Bifascioides Kasy, 1968

Bifascioides Kasy, 1968b: 509.

Type species: *Elachista leucomelanella* Rebel, in Rebel & Zerny, 1917, by original designation.

DIAGNOSIS.– Head smoothly scaled. Antennae 3/4 length of forewings; scapus rather short. Labial palpi cylindrical and recurved. Forewings lanceolate, blackish brown with two more or less distinct broad white fasciae. Hindwings lanceolate, very narrow, about 1/3 of width of forewings.

MALE GENITALIA.– Uncus absent. Tegumen narrow, band-shaped. Valvae slightly asymmetrical, with broad basal part, both forked distally into large ventral and small dorsal lobe. Aedeagus long, slender and strongly bent; cornuti absent.

FEMALE GENITALIA.– Papillae anales fused and heavily sclerotized. Apophyses long and stout, of equal length. Ostium very large and shifted to the left of anterior edge of sternite VII. Ductus bursae short, membranous, with more or less pronounced antrum. Corpus bursae membranous, with one signum.

DISTRIBUTION.– Northern and eastern Africa, Near East (Arabia).

BIOLOGY.– Unknown.

156. *Bifascioides leucomelanellus* (Rebel, 1917)

Elachista leucomelanellus Rebel, in Rebel & Zerny, 1917: 445.
Ascalenia pirastica Meyrick, 1936: 80.

DIAGNOSIS.– Adult. Wingspan 6-7 mm. Head shining dark brown with frons shining silvery white. Thorax and tegulae shining dark brown, posterior part of thorax yellowish white. Forewings shining dark brown with two broad yellowish white fasciae; first fascia near base, broad at dorsum and abruptly narrowing at costa, second fascia at 2/3, narrowing towards dorsum; cilia ochreous grey. Hindwings shining silvery white; cilia ochreous grey. Abdomen shining brownish grey, anal tuft white.

Similar species. Differs from *Bifascia nigralbella* by the yellowish white posterior end of the thorax, the broader and more basally placed inner fascia on the forewings, and the narrower hindwings.

MALE GENITALIA.– Tegumen membranous, sclerotized laterally, narrowing distally. Valvae broad in basal half and strongly narrowed in distal half, and covered with long and stout setae; right valva with large basal outgrowth bearing patch of short and thick setae and double row of long setae, its distal part with small hairy lobe before middle of dorsal edge; left valva with same small lobe but at 3/4 of basal half. Aedeagus long and bent at right angle, tapering distally.

FEMALE GENITALIA.– Ostium large, bag-shaped. Ductus bursae very narrow before ostium, rapidly widening towards corpus bursae; antrum bent, widened in middle and slightly sclerotized. Corpus bursae rounded; signum large, elongated and bifurcate.

DISTRIBUTION.– Northern Africa (Libya and Egypt).

BIOLOGY.– Unknown. Adults have been collected in February and in May-June.

Gisilia Kasy, 1968

Gisilia Kasy, 1968b: 516.
Type species: *Ascalenia stereodoxa* Meyrick, 1925, by original designation.
Melnea Clarke, 1986: 180 (key), 303.
Type species: *Ascalenia armigera* Meyrick, 1923, by original designation.

DIAGNOSIS.– Head smoothly scaled. Antennae 3/4 length of forewings; scapus slightly widening distally. Labial palpi cylindrical and recurved. Forewings lanceolate, with two sometimes incomplete fasciae and two opposite spots before apex. Hindwings very narrow, about 1/3 width of forewing.

MALE GENITALIA.– Uncus absent. Tegumen very short, narrowly triangular. Valvae asymmetrical, divided into two lobes of different shape covered with coarse setae distally. Vinculum band-shaped. Aedeagus long and sclerotized; cornuti absent.

FEMALE GENITALIA.– Papillae anales fused and sclerotized. Apophyses long and stout, both pairs of equal length. Ostium large and shifted to the left of anterior edge of sternite VII. Genital plate complicated, with several arched and strongly sclerotized folds around ostium. Ductus bursae membranous, sometimes with sclerotized inner band. Corpus bursae membranous or with fine net-like structure; one dentate signum.

DISTRIBUTION.– Mediterranean area, Africa, and from the Near East to Pakistan and India.

BIOLOGY.– Larvae of one species have been found on *Acacia* sp. (Mimosaceae).

157. Gisilia stereodoxa (Meyrick, 1925)

Ascalenia stereodoxa Meyrick, 1925: 212.
Ascalenia evitans Meyrick, 1925: 212.
Stagmatophora alfieriella Rebel, 1926: 190.

DIAGNOSIS.– Adult. Wingspan 7-9 mm. Head shining dark brown with frons shining greyish bronze; antennae shining dark brown, apical third yellowish white. Thorax and tegulae shining dark brown with purplish gloss, posterior part of thorax white. Forewings shining dark brown with slight purplish gloss and two white fasciae: first and widest fascia near base, narrowing towards costa, edged greyish brown, especially at inner side; second fascia beyond 1/2, narrowing at both ends and not reaching dorsum; white costal spot at 3/4, smaller spot opposite on dorsum; cilia dark brown around apex, greyish brown towards dorsum. Hindwings shining greyish brown in male, shining light grey in female. Abdomen shining brownish grey with whitish anal tuft in male, shining greyish white dorsally and shining yellowish white with segment VII greyish ventrally in female.

Variation. In some specimens the first fascia is completely darkened with greyish brown; the second fascia and subapical spots are sometimes reduced.

MALE GENITALIA.– Tegumen triangular. Valvae rather sharply asymmetrical, bilobate; right valva rather compact, dorsal lobe widely spatulate with hump in middle of costa, ventral lobe half length and narrow, both with distal patch of short and coarse setae; left valva slender and longer, dorsal lobe long and narrow, distally widened and with row of short spines, running from base to apex, ventral lobe half the length of dorsal, slender and slightly bent, with patch of coarse setae distally. Aedeagus acutely bent at distal third, apically pointed and with several sclerotized strips.

FEMALE GENITALIA.– Genital plate strongly sclerotized and arched, partly protruding and with irregular ridges. Ductus bursae rather wide, with narrow sclerotized zigzag inner band. Corpus bursae elongate, with one rectangular, strongly dentate signum.

DISTRIBUTION.– Southern Europe (Mediterranean coast), Sardinia, Malta; Egypt, Iran, India.

BIOLOGY.– Host plants: *Acacia nilotica* (L.) Dell. (Mimosaceae). The larva was recorded from the inflorescences in October (Meyrick, 1925; Rebel, 1926). Adults have been collected in January-February, April-May, July and August-October.

158. Gisilia lerautella Gibeaux, 1986

Gisilia lerautella Gibeaux, 1986: 168.

DIAGNOSIS.– Adult. Wingspan 6-8 mm. Differs from *stereodoxa* by the shorter white apical section of the antennae and the wider white fasciae without pronounced greyish brown edging.

MALE GENITALIA.– Resembles *stereodoxa*, but right valva with straight dorsal margin of dorsal lobe and pointed apex of longer and wider ventral lobe; left valva with setae on edge of spatulate apex of wider dorsal lobe and straight ventral lobe; aedeagus gradually bent and tapered.

FEMALE GENITALIA.– Can be distinguished from *sterodoxa* by the spiculate genital plate, the lack of a sclerotized zigzag inner band of ductus bursae, the weakly granulate corpus bursae, and the more elongated signum.

DISTRIBUTION.– France: Seine et Marne, fóret de Fontainebleau (type locality only). Most likely introduced.

BIOLOGY.– Unknown. Seven adults have been collected on trunks of oak trees from late April to early May.

Taxa formerly placed in Momphidae sensu lato

Neomariania Mariani, 1943

Neomariania Mariani, 1943: 175; objective replacement name for *Mariania* Rebel, 1937.
Type species: *Mariania partinicensis* Rebel, 1937 (of *Mariania* Rebel, 1937).
Mariania Rebel, 1937: 46; junior homonym of *Mariania* Airaghi, 1901.
Megaceraea Rebel, 1940: 40; nomen nudum.
Megaceraea Nye & Fletcher, 1991: 184. **Syn.n.**
Type species: *Megaceraea oecophorella* Rebel, 1940, by original designation.

The genus *Neomariania* (as *Mariania*) was originally placed by Rebel (1937) in Momphidae.This placement was firstly supported by Riedl (1969), but later he transferred *Neomariania* to Oecophoridae (Riedl, 1986) and then to Stathmopodidae (Riedl, 1990b). In most recent catalogues the genus is tentatively placed in Stathmopodidae (Riedl, 1997a; Vives Moreno, 1996). Its systematic position remains uncertain.

The generic name *Megaceraea* was proposed after 1930 (Rebel, 1940) for three nominal species and was not accompanied by the fixation of a type species. Thus, the name is technically unavailable. It was made nomenclaturally available only much later by Nye & Fletcher (1991). According to the genital characters, all species described in *Megaceraea* are clearly congeneric with *Neomariania partinicensis* (Rebel, 1937), the type species of the genus *Neomariania*.

159. *Neomariania partinicensis* (Rebel, 1937)

Mariania partinicensis Rebel, 1937: 46.

DIAGNOSIS.– Wingspan 10-12 mm. Head pale greyish ochreous irrorated brownish; antennae shining dark brownish grey, apical half little lighter. Thorax and tegulae pale greyish ochreous mixed brown. Forewings pale greyish ochreous, mottled with greyish brown, especially on costa at base and in apical area; broad, irregular and indistinct fascia in middle and indistinct spot at tornus; cilia ochreous grey. Hindwings grey.

Similar species. Resembles *rebeli*, but differs by the more greyish ochreous colour and by the length of the antennae, which reach beyond the length of the forewings.

MALE GENITALIA.– Valvae with large and arched costa, sacculus with inner side of perpendicular projection almost straight. Aedeagus cylindrical. – The valvae with the larger and flatter costa and the straight inner side of the perpendicular projection of the sacculi distinguish it from *rebeli*.

FEMALE GENITALIA.– Lamellae antevaginalis distally rounded. Ductus bursae about as long as corpus bursae. Signa no more than half the length of the corpus bursae. – The shape of the lamellae antevaginalis and the length of the ductus bursae are characteristic.

DISTRIBUTION.– Southern France, Spain, Sicily, Malta, Greece, Crete.

BIOLOGY.– Unknown. Adults have been collected from the end of June till mid-September.

160. *Neomariania rebeli* (Walsingham, 1894)

Laverna rebeli Walsingham, 1894: 554.

DIAGNOSIS.– Wingspan 8-14 mm. Head pale ochreous mottled with brownish; antennae shining brownish grey with some indistinct annulations basally. Thorax and tegulae pale ochreous mixed brown. Forewings pale ochreous, mottled with greyish brown, especially on costa at base; indistinct fascia of greyish brown scales at tornus, narrowed towards costa, similar fascia between tornus and apex, and indistinct spot in apical area; cilia ochreous-grey. Hindwings grey in male, dark grey in female. Abdomen shining grey, anal tuft grey dorsally, white ventrally.

Variation. There is some variation in the density of the dark brown scales on the forewings.

Similar species. For differences with *partinicensis*, q. v.

MALE GENITALIA.– Valvae with costa as semi-circular hump, sacculus with inner side of perpendicular projection strongly convex. Aedeagus slightly narrowing distally. – The valvae with the short and semi-circular costa and the convex inner side of the perpendicular projection of the sacculi distinguish it from *partinicensis*.

FEMALE GENITALIA.– Lamellae antevaginalis distally angular. Ductus bursae almost twice as long as corpus bursae. Signa more than half the length of corpus bursae. – For differences with *partinicensis*, q. v.

DISTRIBUTION.– Portugal (including Madeira), Spain (Canary Islands).

BIOLOGY.– Unknown. Adults have been collected from the beginning of July till mid-October.

161. Neomariania incertella (Rebel, 1940) comb.n.

Megaceraea incertella Rebel, 1940: 41.

ORIGINAL DESCRIPTION.– Rebel gave the following description (freely translated from German): "A single, very worn ♀ from Flores, Vales, June (Storå) looks like, except for the long antenna (which surpasses in length the apex of the forewing) and the dark hindwing, *Blast[obasis] flavescentella* Rbl. from Madeira. The forewings yellowish sand-coloured with a grey central spot and single grey scales at the edge. The hindwings dust grey with yellowish cilia. Forewing length 6 mm, wingspan 13 mm."

DISTRIBUTION.– Azores.

BIOLOGY.– Unknown.

REMARKS.– Probably only a female of *Neomariania oecophorella*.

162. Neomariania oecophorella (Rebel, 1940) comb.n.

Megaceraea oecophorella Rebel, 1940: 41.

ORIGINAL DESCRIPTION.– Rebel gave the following description (freely translated from German): "Three fresh specimens and one damaged large ♂ from San Miguel, Furnas, 19.–24. May (Storå) and Flores, Sta Cruz and Vales, 1–15. June (Storå) belong to a dark species, that somewhat recalls *Borkhausenia (Oecophora) fuscescens* Hw.

The antennae, strongly thickened by scales and with serrate apex, surpass by approx. 1/6 of their length the apex of the forewing. They are, as the universal colouring, umber brown. The short, little curved, labial palpi lighter on the inside. Their short apical segment hardly 1/3 of the length of the middle segment. Head umber brown with somewhat paler frons; the rough conical abdomen of the same colour, with 3/4 of its length surpasses the hindwing. The short anal tuft is somewhat brighter brownish, with white-spotted tarsal segments.

The forewings almost parallel-sided, with blunt rounded apex and slightly oblique (widening by the cilia) termen, brighter or darker umber brown with a blackish, not always distinctive spot at 1/2 of the fold and similar opposite spots beyond. The latter are faded and separated from the dark apical area by a somewhat brighter costal spot. Cilia unicoloured dark. The hindwings a little more than 1/2 the width of the forewings, very narrow, umber brown or blackish with unicoloured cilia. Underside of the forewings blackish, that of the hindwings somewhat brighter. Forewing length 5 mm, wingspan 11 mm.

A damaged ♂ from San Miguel, Furnas 19.–24. May (Storå) is considerably larger, forewing more than 6 mm, wingspan almost 14 mm, more blackish coloured, but would belong to the same species. It has only one (thickened) stump of the antenna."

DISTRIBUTION.– Azores.

BIOLOGY.– Unknown.

163. Neomariania scriptella (Rebel, 1940) comb.n.

Megaceraea scriptella Rebel, 1940: 42.

ORIGINAL DESCRIPTION.– Rebel gave the following description (freely translated from German): "A single, well preserved ♂ from Terceira, Praia da Victoria, 8.–10. June (Frey) differs much from the previous species by the bright grey, black-marked forewings.

Antenna less thickened than in the previous species, slightly surpasses the apex of the forewing, bright grey as the universal colouring. Likewise the diverged, curved labial palpi, of which middle segment is blackish on the outside. Frons yellowish, vertex and thorax bright grey. Legs bright brownish grey, darkened on the outside, with ends of segments lighter. Hindtibiae with yellowish brown hairs. The abdomen bright brownish with yellowish brown clipped anal tuft, surpasses beyond the end of the very broad cilia of the hindwings.

Forewings narrowly lanceolate, with the apex sharper than in previous species and very oblique termen. Ground colour bright grey with a very striking coal black, narrow dorsal mark at 1/2, that reaches to 3/4 of width of wing. In the apex of the right wing are also two oblique black streaks, on the left wing only indicated by some black scales. Cilia bright grey. The very pointed hindwings dust grey, with very broad bright grey cilia. Underside of the forewings blackish grey, that of the hindwings bright grey. Forewing length 7 mm, wingspan 14 mm."

DISTRIBUTION.– Azores.

BIOLOGY.– Unknown.

Urodeta Stainton, 1869

Urodeta Stainton, 1869: 226; described as *Urodela*, an incorrect original spelling.
Type species: *Urodeta cisticolella* Stainton, 1869, by monotypy.

The genus *Urodeta* was described by Stainton (1869) within the large family-group Tineina, and then was placed in Elachistidae (Elachistinae) by Rebel (Staudinger & Rebel, 1901). Later it was included in the "Lyonetiadae" by Fletcher (1929). In some modern

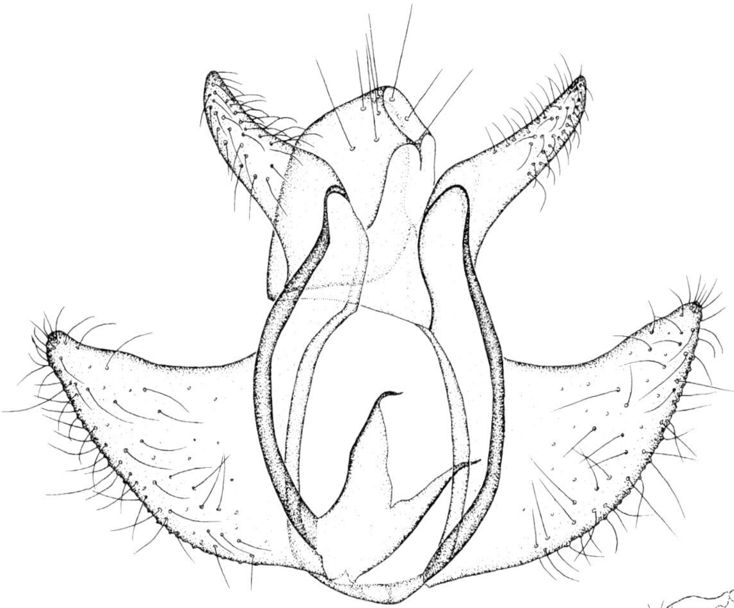

54

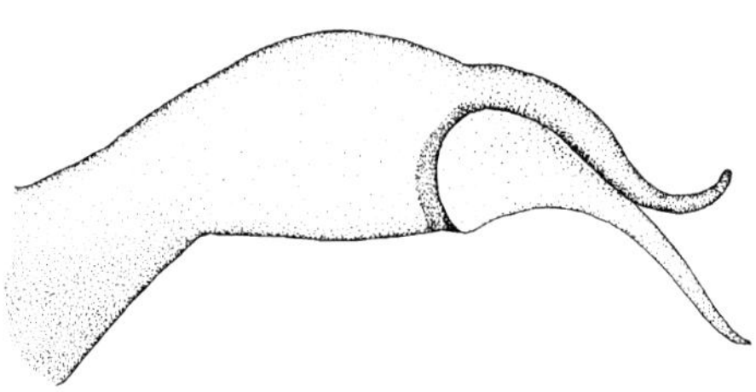

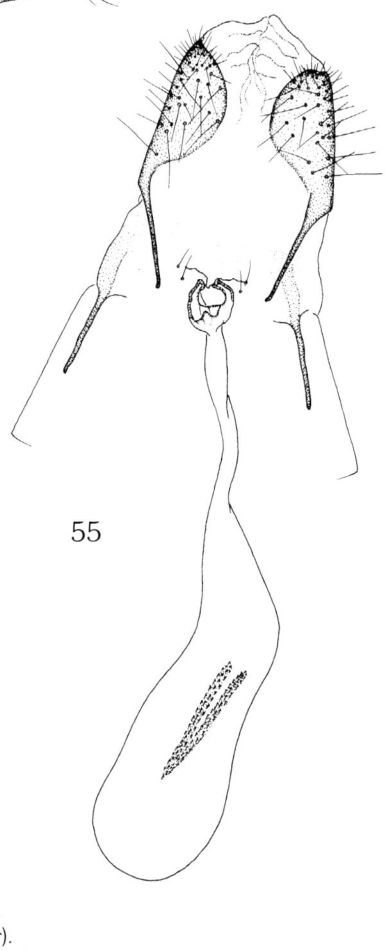

55

Text-fig. 54. Male genitalia of *Urodeta hibernella* (Staudinger).
Text-fig. 55. Female genitalia of *Urodeta hibernella* (Staudinger).

check lists the genus is placed in Momphidae (Leraut, 1980; Riedl, 1997b), but it does not belong to this family. The adult and the wing venation were figured by Bengtsson (1997), who included *Urodeta hibernella* as: 'Taxa misplaced in the Scythrididae'. Leraut (1997), however, again placed *Urodeta* in Elachistidae, stating that 'this genus is a truly primitive member of the Elachistinae related to the genus *Perittia* (Leraut & Minet, in prep.).' Kaila (pers. comm.) agrees that *hibernella* belongs to Elachistidae in the broad sense, but not to Elachistinae.

Urodeta hibernella (Staudinger, 1859)

Butalis hibernella Staudinger, 1859: 250.
Urodeta cisticolella Stainton, 1869: 235.

ORIGINAL DESCRIPTION.– Staudinger gave the following description (freely translated from German): "Forewings dark ash grey; hindwings very narrow, grey. Labial palpi very small, through which this species essentially differs from *B[utalis] incongruella*, which is also much larger. One ♂ on 26 February near Chiclana."

DISTRIBUTION.– Spain, Portugal, southern France, Morocco, Tunisia.

BIOLOGY.– Host plant: *Cistus monspeliensis* (Cistaceae). The larva mines leaves, making an elongate blotch, which is slightly puckered. The upper end of the mine is free from excrement, which seems all collected at the lower end. The larva moves from leaf to leaf; and when full-fed it quits the mine, and, attaching itself to the edge of a leaf, changes there to a naked pupa, which is only attached by the tail (Stainton, 1869). Adults have been collected in April and May.

REMARKS.– In Bengtsson (1997) the adult is pictured in watercolours and also a line drawing of the wing venation is given. To complete the above mentioned description, figures of the male and female genitalia are provided here (text-fig. 54, 55).

Taxa incertae sedis

Stagmatophora gronoviella (Scopoli, 1772)

Phalaena gronoviella Scopoli, 1772: 119; nomen dubium.
Stagmatophora gronoviella: Werneburg, 1864: 172.

ORIGINAL DESCRIPTION.– "Diagn. Tinea; alis nigris; punctis aureoargenteis 8-9: marginalibus tribus majoribus. Alae anticae lin. 4. logae. Caput, ani region aureo-argentea. Pedes nigro et aurato colore varii."

DISTRIBUTION.– Slovenia (Schemnitz).

REMARKS.– The material of the Scopoli collection has been lost, and the description of *gronoviella* can refer to any of the dark species of the genus *Eteobalea*, but also to *Eulamprotes wilkella* (Linnaeus, 1758), belonging to Gelechiidae. See also under *Eteobalea serratella*.

Distribution catalogue

The distribution catalogue mainly follows Karsholt & Razowski (1996).

Additional information has been obtained from the following sources: Ireland (Emmet & Langmaid, 1997); France (Leraut, 1997); Lebanon (Ellison & Wiltshire, 1939); Finland (Kerppola et al., 1995); Madeira Islands (Passos de Carvalho, 1995); Albania (Utech, 1962); North Africa and Middle-East (Amsel, 1933; Chrétien, 1917; Rungs, 1979); Portugal (Passos de Carvalho & Corley, 1995); Azores (Vieira, 1997); Russia and the Caucasus (Sinev, 1985a; 1986; 1993b; 1997b);

Furthermore unpublished records of various collections are included.

Acronyms of countries

PR	Portugal
ES	Spain
AD	Andorra
FR	France
CO	Corsica
SA	Sardinia
SI	Sicily
IT	Italy
MA	Malta
IR	Ireland
GB	Great Britain
BL	Belgium
NL	The Netherlands
LX	Luxemburg
DT	Germany
CH	Switzerland
FL	Principality of Liechtenstein
AU	Austria
PL	Poland
CZ	Czech Republic
SK	Slovakia
HG	Hungary
RO	Romania
SL	Slovenia
HR	Croatia
YU	Yugoslavia
BH	Bosnia-Herzegovina
MK	Macedonia
AL	Albania
BG	Bulgaria
GR	Greece
CR	Crete
CY	Cyprus
TR	Turkey
IC	Iceland
DK	Denmark
NR	Norway
SW	Sweden
SF	Finland
EE	Estonia
LV	Latvia
LT	Lithuania
RU	Russia (European part)
BR	Byelorussia
UI	Ukraine
MO	Moldavia
GO	Georgia
AM	Armenia
AZ	Azerbaijan
SY	Syria
LE	Lebanon
IS	Israel
JO	Jordan
EG	Egypt
LY	Libya
TU	Tunisia
AL	Algeria
MR	Morocco
CI	Canary Islands
MD	Madeira
AR	Azores

	PR	ES	AD	FR	CO	SA	SI	IT	MA	IR	GB	BL	NL	LX	DT	CH	FL
1. conturbatella				FR				IT			GB	BL	NL	LX	DT	CH	
2. ochraceella		ES		FR				IT		IR	GB	BL	NL		DT	CH	
3. lacteella		ES		FR				IT		IR	GB	BL			DT	CH	
4. propinquella		ES		FR				IT		IR	GB	BL	NL	LX	DT	CH	
5. divisella	?			FR	CO	SA		IT			GB	BL	NL	LX	DT	CH	
6. jurassicella	PR	ES		FR							GB					CH	
7. bradleyi											GB		NL		DT		
8. confusella																	
9. subbistrigella		ES		FR			SI	IT		IR	GB	BL	NL	LX	DT	CH	
10. sturnipennella				FR			SI	IT			GB	BL	NL	LX	DT	CH	
11. meridionella																	
12. epilobiella		ES		FR			SI	IT		IR	GB	BL	NL	LX	DT	CH	
13. langiella		ES		FR				IT			GB	BL	NL		DT	CH	
14. idaei				FR				IT				BL	NL	LX	DT	CH	
15. miscella	PR	ES	AD	FR	CO	SA	SI	IT			GB	BL	NL		DT	CH	
16. locupletella		ES		FR				IT		IR	GB	BL	NL		DT	CH	
17. maculata																	
18. terminella		ES		FR						IR	GB	BL	NL		DT	CH	
19. raschkiella				FR				IT		IR	GB	BL	NL	LX	DT	CH	
20. sexstrigella																	
21. praeangusta		ES		FR				IT		IR	GB	BL	NL		DT	CH	
22. pinicolella		ES		FR				IT			GB	BL	NL		DT	CH	
23. parvulipunctella	PR	ES		FR		SA	SI	IT	MA		GB						
24. amydraula																	
25. pedella		ES		FR		SA	SI	IT			GB	BL	NL		DT	CH	FL
26. auriferella																	
27. flavella		ES															
28. trigonella																	
29. charadritis																	
30. pallidella																	
31. graeca																	
32. hellerella	PR	ES		FR		SA	SI	IT			GB	BL	NL		DT	CH	
33. rossica																	
34. atra		ES		FR		SA	SI	IT		IR	GB	BL	NL		DT	CH	
35. vinolentella				FR									NL		DT	CH?	
36. georgiella																	
37. libanotica																	
38. flavicaput				FR				IT		IR	GB	BL	NL		DT	CH	
39. laspeyrella								IT							DT		
40. festivella		ES	AD	FR		SA?	SI	IT					NL		DT		
41. albidorsella		ES		FR	CO	SA		IT									
42. stephensi		ES		FR				IT			GB				DT	CH	
43. gelechiella																	
44. ochraceella		ES		FR			SI	IT							DT		
45. albanica	PR							IT									
46. buvati	PR	ES	AD	FR													
47. theae																	
48. linneella		ES		FR		SA		IT			GB	BL	NL	LX	DT	CH	
49. lathamella				FR							GB		NL	LX	DT		
50. splendida				FR								BL			DT		

	AU	PL	CZ	SK	HG	RO	SL	HR	YU	BH	MK	AL	BG	GR	CR	CY	TR
1. conturbatella	AU	PL	CZ	SK	HG				YU?					GR			
2. ochraceella	AU	PL	CZ	SK	HG	RO							BG	GR			TR
3. lacteella	AU	PL	CZ	SK	HG	RO											TR
4. propinquella	AU	PL	CZ	SK	HG	RO					MK						
5. divisella	AU	PL	CZ	SK	HG	RO		HR									
6. jurassicella	AU																
7. bradleyi	AU	PL	CZ	SK	HG	RO											
8. confusella	AU			SK	HG												
9. subbistrigella	AU	PL	CZ	SK	HG	RO		HR	YU?					GR	CR		TR
10. sturnipennella	AU	PL	CZ	SK	HG	RO											
11. meridionella														GR			
12. epilobiella	AU	PL	CZ	SK	HG	RO		HR	YU				BG	GR			
13. langiella	AU	PL	CZ	SK	HG	RO		HR				AL	BG				TR
14. idaei	AU	PL	CZ	SK	HG	RO							BG				
15. miscella	AU	PL	CZ		HG	RO	SL	HR	YU		MK			GR	CR	CY	TR
16. locupletella	AU	PL	CZ	SK	HG	RO							BG				
17. maculata																	
18. terminella	AU	PL	CZ	SK	HG	RO											
19. raschkiella	AU	PL	CZ	SK	HG	RO											
20. sexstrigella																	
21. praeangusta	AU	PL	CZ	SK	HG	RO											
22. pinicolella	AU	PL	CZ	SK	HG	RO											
23. parvulipunctella														GR	CR		
24. amydraula																	
25. pedella	AU	PL	CZ	SK	HG	RO											
26. auriferella																	
27. flavella																	
28. trigonella																	
29. charadritis																	
30. pallidella																	
31. graeca														GR	CR	CY	TR
32. hellerella	AU	PL	CZ	SK	HG	RO		HR					BG	GR			
33. rossica																	
34. atra	AU	PL	CZ	SK	HG	RO		HR			MK	AL	BG	GR			TR
35. vinolentella		PL	CZ	SK										GR			
36. georgiella																	
37. libanotica																	TR
38. flavicaput	AU	PL	CZ	SK	HG	RO	SL	HR			MK						TR
39. laspeyrella	AU	PL	CZ	SK	HG	RO					MK	AL	BG				
40. festivella	AU	PL	CZ	SK	HG	RO		HR		BH	MK	AL	BG	GR			TR
41. albidorsella																	
42. stephensi	AU	PL	CZ	SK	HG	RO		HR			MK	AL		GR			
43. gelechiella														GR			
44. ochraceella	AU		CZ	SK	HG	RO		HR	YU		MK	AL					TR
45. albanica	AU			SK	HG						MK	AL		GR	CR		
46. buvati																	
47. theae																	
48. linneella	AU	PL	CZ	SK	HG	RO											TR
49. lathamella		PL	CZ	SK													
50. splendida	AU	PL	CZ	SK	HG	RO					MK		BG				

	IC	DK	NR	SW	SF	EE	LV	LT	RU	BR	UI	MO	GO	AM	AZ	SY	LE
1. conturbatella	IC	DK	NR	SW	SF	EE	LV	LT	RU		UI						
2. ochraceella		DK		SW			LV		RU		UI			AM		SY	
3. lacteella		DK	NR	SW	SF	EE	LV		RU		UI				AZ		
4. propinquella		DK		SW	SF	EE	LV		RU		UI						
5. divisella		DK	NR	SW	SF	EE	LV	LT	RU				GO				
6. jurassicella											UI?						
7. bradleyi														AM			
8. confusella											UI			AM	AZ		
9. subbistrigella		DK	NR	SW	SF	EE	LV	LT	RU	BR	UI				AZ		
10. sturnipennella		DK	NR	SW	SF		LV	LT	RU								
11. meridionella									RU								
12. epilobiella		DK	NR	SW	SF	EE	LV	LT	RU		UI						
13. langiella		DK	NR	SW	SF		LV	LT	RU				GO				
14. idaei		DK	NR	SW	SF	EE	LV	LT	RU	BR	UI						
15. miscella		DK		SW	SF		LV		RU		UI			AM			
16. locupletella		DK	NR	SW	SF	EE	LV	LT	RU	BR	UI						
17. maculata													GO	AM	AZ		
18. terminella		DK		SW	SF	EE	LV	LT	RU								
19. raschkiella		DK	NR	SW	SF	EE	LV	LT	RU	BR	UI						
20. sexstrigella				SW	SF				RU								
21. praeangusta		DK	NR	SW	SF	EE	LV	LT	RU	BR	UI						
22. pinicolella		DK	NR	SW	SF	EE	LV	LT	RU	BR	UI						
23. parvulipunctella																	
24. amydraula																	
25. pedella		DK	NR	SW	SF	EE	LV	LT	RU	BR	UI		GO		AZ		
26. auriferella																	
27. flavella																SY	
28. trigonella																	
29. charadritis																	
30. pallidella																	
31. graeca																	
32. hellerella		DK	NR	SW		EE?	?		RU				GO	AM	AZ		
33. rossica									RU		UI						
34. atra		DK	NR	SW	SF	EE	LV	LT	RU	BR	UI						
35. vinolentella																	
36. georgiella													GO				
37. libanotica																	LE
38. flavicaput		DK		SW					RU?		UI						
39. laspeyrella			NR	SW	SF	EE	LV	LT	RU	BR	UI						
40. festivella				SW					RU		UI						LE
41. albidorsella																	
42. stephensi		DK		SW							UI		GO				
43. gelechiella																	
44. ochraceella																	
45. albanica																	
46. buvati																	
47. theae													GO		AZ		
48. linneella		DK	NR?	SW	SF	EE	LV	LT	RU	BR	UI						
49. lathamella		DK		SW	SF		LV		RU		UI						
50. splendida				SW?													

	IS	JO	EG	LY	TU	AL	MR	CI	MD	AR
1. conturbatella										
2. ochraceella										
3. lacteella										
4. propinquella										
5. divisella										
6. jurassicella										
7. bradleyi										
8. confusella										
9. subbistrigella										
10. sturnipennella										
11. meridionella										
12. epilobiella										
13. langiella										
14. idaei							MR			
15. miscella							MR			
16. locupletella										
17. maculata										
18. terminella										
19. raschkiella										
20. sexstrigella										
21. praeangusta										
22. pinicolella										
23. parvulipunctella					TU		MR			
24. amydraula	IS		EG	LY	TU	AL				
25. pedella										
26. auriferella	IS		EG	LY	TU					
27. flavella	IS				TU	AL	MR			
28. trigonella							MR			
29. charadritis	IS		EG	LY						
30. pallidella	IS		EG							
31. graeca										
32. hellerella										
33. rossica										
34. atra										
35. vinolentella										
36. georgiella										
37. libanotica										
38. flavicaput										
39. laspeyrella										
40. festivella										
41. albidorsella										
42. stephensi										
43. gelechiella										
44. ochraceella										
45. albanica										
46. buvati										
47. theae										
48. linneella										
49. lathamella										
50. splendida										

51. zagulajevi																	
52. abchasica																	
53. leuwenhoekella	PR	ES	AD	FR	CO		SI	IT		IR	GB	BL	NL	LX	DT	CH	
54. schwarzella	PR	ES		FR				IT		IR	GB	BL	NL		DT	CH	
55. baldizzonella								IT									
56. nodosella		ES		FR				IT					NL		DT		
57.woodiella											GB						
58. gemmatella																	
59. recisella	PR	ES			CO		SI										
60. haifensis																	
61. delozona																	
62. brachygrapta																	
63. deprinsi																	
64. francoeuriae																	
65. bellaqueifontis				FR													
66. reliqua				FR													
67. phragmitella	PR	ES		FR		SA		IT		IR	GB	BL	NL		DT	CH	
68. autophaea																	
69. zieglerella	PR	ES		FR				IT			GB	BL	NL		DT	CH	
70. schmidiella		ES		FR							GB				DT	CH	
71. orichalcea		ES		FR				IT		IR	GB				DT	CH	
72. sibirica																	
73. scribaiella		ES		FR				IT			GB	BL	NL		DT	CH	
74. pulchrimella	PR	ES		FR	CO	SA	SI	IT	MA		GB					CH	
75. attenuatella																	
76. crassicervicella	PR	ES					SI										
77. turbidella																	
78. pararufella		ES			CO												
79. coryphaea		ES					SI	IT	MA								
80. lienigiella		ES		FR						IR	GB		NL		DT	CH	
81. salahinella																	
82. semicoccinea													NL				
83. argyrogrammos	PR	ES		FR	CO	SA	SI	IT	MA						DT	CH	
84. caesaris		ES				SA	SI	IT									
85. klimeschi		ES						IT									
86. brosi								IT									
87. tethysella		ES															
88. wolschrijni		ES							MA								
89. apicinotella																	
90. rileyi		ES															
91. badia		ES									GB		NL				
92. coridophaga		ES															
93. simplex	PR	ES									GB						
94. gerasimovi	PR	ES		FR	CO			IT	MA								
95. patriciae																	
96. ledereriella	PR?	ES		FR	CO	SA	SI?	IT	MA								
97. danilevskyi	PR	ES		FR	CO												
98. kasypinkeri																	
99. riedli																	
100. heydeniella				FR				IT							DT	CH	
101. anonymella		ES		FR			SI	IT							DT	CH	

51. zagulajevi																	
52. abchasica																	
53. leuwenhoekella	AU	PL	CZ	SK	HG	RO	SL	HR	YU	BH	MK	AL	BG	GR			TR
54. schwarzella	AU	PL	CZ	SK	HG	RO		HR	YU	BH	MK	AL	BG	GR			TR
55. baldizzonella																	
56. nodosella	AU	PL		SK			SL	HR	YU	BH		AL	BG	GR			
57.woodiella																	
58. gemmatella																	
59. recisella														GR		CY	
60. haifensis																	
61. delozona																	
62. brachygrapta																	
63. deprinsi																	
64. francoeuriae																	
65. bellaqueifontis																	
66. reliqua																	
67. phragmitella	AU	PL	CZ	SK	HG	RO		HR	YU		MK		BG	GR			
68. autophaea																	TR
69. zieglerella	AU	PL	CZ	SK	HG	RO		HR	YU				BG				TR
70. schmidiella	AU	PL	CZ	SK		RO											
71. orichalcea	AU	PL	CZ	SK	HG	RO		HR	YU?				BG				
72. sibirica																	
73. scribaiella	AU	PL	CZ	SK	HG	RO		HR									TR
74. pulchrimella								HR				AL		GR	CR		TR
75. attenuatella																	
76. crassicervicella														GR	CR		TR
77. turbidella								HR						GR	CR	CY	
78. pararufella														GR		CY	
79. coryphaea								HR						GR	CR	CY	
80. lienigiella	AU	PL			HG								BG	GR			
81. salahinella																	
82. semicoccinea																	
83. argyrogrammos	AU		CZ	SK	HG	RO	SL	HR	YU		MK	AL	BG	GR	CR	CY	TR
84. caesaris						RO		HR					BG	GR			TR
85. klimeschi	AU	PL		SK	HG	RO											
86. brosi																	
87. tethysella																	
88. wolschrijni																	
89. apicinotella																	
90. rileyi																	
91. badia														GR			
92. coridophaga																	
93. simplex																CY	
94. gerasimovi														GR			TR
95. patriciae																	
96. ledereriella								HR?						GR			TR?
97. danilevskyi																	
98. kasypinkeri																	
99. riedli																	
100. heydeniella	AU	PL	CZ	SK	HG	RO					MK						
101. anonymella	AU	PL	CZ	SK	HG	RO	SL	HR	YU				BG	GR			TR

51. zagulajevi													GO				
52. abchasica													GO				
53. leuwenhoekella		DK	?	SW		EE	LV	LT	RU	BR	UI		GO	AM			
54. schwarzella		DK	NR	SW	SF	EE	LV	LT	RU	BR	UI						
55. baldizzonella																	
56. nodosella							LV		RU		UI						
57.woodiella																	
58. gemmatella																	
59. recisella																	
60. haifensis																	
61. delozona																	
62. brachygrapta																	
63. deprinsi																	
64. francoeuriae																	
65. bellaqueifontis																	
66. reliqua																	
67. phragmitella		DK	NR	SW	SF	EE	LV	LT	RU	BR	UI						
68. autophaea																	
69. zieglerella		DK		SW	SF?	EE	LV		RU	BR	UI	MO	GO				
70. schmidiella		DK							RU								
71. orichalcea	IC	DK	NR	SW	SF	EE	LV	LT	RU	BR	UI						
72. sibirica																	
73. scribaiella		DK		SW			LV	LT			UI						
74. pulchrimella													GO				
75. attenuatella																	
76. crassicervicella													GO		AZ		
77. turbidella																	
78. pararufella																	
79. coryphaea																	
80. lienigiella		DK		SW	SF	EE	LV	LT	RU	BR							
81. salahinella																	
82. semicoccinea													GO				
83. argyrogrammos									RU		UI	MO		AM	AZ	SY	LE
84. caesaris									RU?		UI	MO					LE
85. klimeschi																	
86. brosi																	
87. tethysella																	
88. wolschrijni																	
89. apicinotella																	
90. rileyi																	
91. badia																	
92. coridophaga																	
93. simplex																	
94. gerasimovi									RU				GO		AZ		
95. patriciae																	
96. ledereriella																	
97. danilevskyi																	
98. kasypinkeri																	
99. riedli																	
100. heydeniella									RU		UI						
101. anonymella				SW	SF	EE	LV	LT	RU	BR	UI		GO	AM	AZ		LE

51. zagulajevi										
52. abchasica										
53. leuwenhoekella										
54. schwarzella										
55. baldizzonella										
56. nodosella										
57.woodiella										
58. gemmatella					TU					
59. recisella	IS				TU	AL				
60. haifensis	IS									
61. delozona			EG							
62. brachygrapta			EG							
63. deprinsi										
64. francoeuriae					TU	AL	MR			
65. bellaqueifontis										
66. reliqua										
67. phragmitella						AL	MR			
68. autophaea										
69. zieglerella										
70. schmidiella										
71. orichalcea										
72. sibirica										
73. scribaiella							MR			
74. pulchrimella								CI	MD	AR
75. attenuatella							MR	CI	MD	
76. crassicervicella	IS	JO	EG	LY	TU	AL	MR			
77. turbidella								CI		
78. pararufella			EG		TU					
79. coryphaea	IS				TU			CI		
80. lienigiella										
81. salahinella	IS			LY	TU					
82. semicoccinea										
83. argyrogrammos	IS	JO	EG		TU	AL	MR	CI	MD	AR
84. caesaris						AL	MR			
85. klimeschi										
86. brosi										
87. tethysella					TU					
88. wolschrijni							MR			
89. apicinotella				LY	TU					
90. rileyi										
91. badia										
92. coridophaga										
93. simplex			EG				MR			
94. gerasimovi	IS					AL	MR	CI		
95. patriciae								CI		
96. ledereriella							MR	CI?		
97. danilevskyi					TU		MR			
98. kasypinkeri								CI		
99. riedli								CI		
100. heydeniella										
101. anonymella										

102. serratella	PR	ES		FR	CO		SI	IT	MA						DT	CH	
103 intermediella	PR	ES		FR		SA	SI	IT	MA								
104. beata	PR	ES		FR			SI	IT									
105. sumptuosella		ES		FR	CO	SA		IT									
106. teucrii		ES															
107. klisieckii																	
108. tririvella				FR				IT									
109. albiapicella		ES		FR	CO			IT							DT	CH	
110. alypella		ES		FR				IT									
111. isabellella		ES		FR	CO		SI	IT									
112. dohrnii	PR	ES		FR	CO	SA	SI	IT									
113. siciliae		ES					SI										
114. thaumatella																	
115. nickerlii	PR	ES		FR				IT								CH	
116. divitella	PR	ES		FR	CO	SA											
117. pomposella		ES		FR	CO		SI	IT							DT	CH	
118. grandiferella																	
119. cognatella						SA		IT									
120. extremella		ES		FR				IT									
121. caucasica																	
122. grabowiella	PR	ES	AD	FR	CO	SA		IT									
123. rosmarinella		ES		FR	CO												
124. karadaghella																	
125. klimeschi																	
126. fiordalisa	PR	ES		FR	CO	SA		IT									
127. glaseri																	
128. gielisi		ES															
129. anatolica																	
130. pontica																	
131. ibericus		ES															
132. libanoticus																	
133. rebeli								IT									
134. rhodorrhisella																	
135. christophi																	
136. fulguritella																	
137. rhamniella	PR	ES		FR				IT			GB	BL	NL		DT	CH	
138. reconditella																	
139. lophyrella		ES		FR				IT			GB				DT	CH	
140. janiszewskae				FR							GB		NL		DT	CH	
141. fibigeri																	
142. taurensis																	
143. vanella		ES		FR			SI	IT							DT	CH	
144. vanelloides							SI										
145. antiqua																	
146. acaciella									MA								
147. jerichoella																	
148. viviparella																	
149. echidnias																	
150. bifasciella																	
151. karsholti																	
152. acarpa																	

102. serratella	AU	PL	CZ	SK	HG		SL	HR	YU		MK	AL	BG	GR	CR		TR
103 intermediella	AU		CZ	SK	HG	RO	SL	HR	YU		MK		BG	GR	CR		TR
104. beata					HG	RO					MK		BG	GR	CR		TR
105. sumptuosella								HR			MK	AL	BG	GR	CR		TR
106. teucrii																	
107. klisieckii																	
108. tririvella	AU		CZ	SK	HG	RO		HR									TR
109. albiapicella	AU		CZ	SK	HG	RO	SL	HR			MK			GR			TR
110. alypella																	
111. isabellella					HG	RO		HR	YU	BH	MK	AL	BG	GR			TR
112. dohrnii								HR						GR	CR	CY	TR
113. siciliae																	TR
114. thaumatella																	
115. nickerlii	AU		CZ		HG	RO					MK						
116. divitella																	
117. pomposella	AU	PL	CZ	SK	HG	RO	SL	HR	YU?		MK			GR			TR
118. grandiferella									YU		MK			GR			TR
119. cognatella								HR			MK			GR			TR
120. extremella	AU	PL	CZ	SK	HG			HR									TR
121. caucasica																	TR
122. grabowiella								HR			MK			GR			TR
123. rosmarinella												AL		GR	CR		
124. karadaghella																	TR
125. klimeschi											MK			GR	CR		
126. fiordalisa								HR									
127. glaseri																	TR
128. gielisi																	
129. anatolica																	TR
130. pontica																	TR
131. ibericus																	
132. libanoticus														GR	CR	CY	TR
133. rebeli					HG	RO		HR			MK	AL		GR			
134. rhodorrhisella																	
135. christophi																	TR
136. fulguritella																	TR
137. rhamniella	AU	PL	CZ	SK	HG	RO		HR									TR
138. reconditella														GR		CY	
139. lophyrella	AU	PL	CZ	SK	HG	RO		HR	YU?		MK	AL	BG	GR			TR
140. janiszewskae	AU	PL	CZ	SK	HG	RO		HR	YU?								
141. fibigeri																	TR
142. taurensis																	TR
143. vanella	AU			SK	HG									GR	CR		TR
144. vanelloides																CY	TR
145. antiqua																	
146. acaciella																	
147. jerichoella																	
148. viviparella																	TR
149. echidnias																	TR
150. bifasciella																	
151. karsholti																	
152. acarpa																	

102. serratella			NR						RU		UI			AM	AZ	SY	
103 intermediella									RU		UI	MO	GO		AZ	SY	LE
104. beata																	
105. sumptuosella									RU		UI				AZ	SY	LE
106. teucrii																	
107. klisieckii																	
108. tririvella				SW	SF	EE	LV		RU		UI						
109. albiapicella									RU							SY	
110. alypella																	
111. isabellella																	
112. dohrnii																SY	
113. siciliae																	
114. thaumatella																	
115. nickerlii																	
116. divitella																	
117. pomposella							LV		RU		UI				AZ	SY	
118. grandiferella									RU		UI				AZ		
119. cognatella																	
120. extremella									RU		UI						
121. caucasica														AM	AZ		
122. grabowiella																	
123. rosmarinella																	
124. karadaghella											UI						
125. klimeschi																	
126. fiordalisa																	LE
127. glaseri																	
128. gielisi																	
129. anatolica																	
130. pontica																	
131. ibericus																	
132. libanoticus																SY	LE
133. rebeli																	
134. rhodorrhisella																	
135. christophi																	
136. fulguritella																SY	LE
137. rhamniella		DK	NR	SW		EE	LV	LT	RU	BR	UI	MO		AM	AZ		
138. reconditella																	
139. lophyrella		DK	NR	SW	SF	EE	LV		RU		UI			AM	AZ		
140. janiszewskae		DK	NR	SW	SF		LV	LT	RU		UI		GO				
141. fibigeri																	
142. taurensis																	
143. vanella									RU		UI			AM			
144. vanelloides															AZ		
145. antiqua																	
146. acaciella																	LE
147. jerichoella																	
148. viviparella									RU		UI			AM			
149. echidnias																	
150. bifasciella																	
151. karsholti																	
152. acarpa																	

102. serratella							MR			
103 intermediella					TU	AL	MR			
104. beata						AL	MR			
105. sumptuosella	IS				TU		MR			
106. teucrii							MR			
107. klisieckii							MR			
108. tririvella										
109. albiapicella							MR			
110. alypella							MR			
111. isabellella					TU	AL	MR			
112. dohrnii	IS	JO		LY	TU	AL	MR			
113. siciliae					TU	AL				
114. thaumatella	IS	JO		LY	TU	AL	MR	CI		
115. nickerlii										
116. divitella										
117. pomposella										
118. grandiferella										
119. cognatella										
120. extremella										
121. caucasica										
122. grabowiella					TU					
123. rosmarinella							MR			
124. karadaghella										
125. klimeschi										
126. fiordalisa							MR			
127. glaseri										
128. gielisi										
129. anatolica										
130. pontica										
131. ibericus										
132. libanoticus										
133. rebeli										
134. rhodorrhisella								CI		
135. christophi										
136. fulguritella		JO								
137. rhamniella										
138. reconditella										
139. lophyrella										
140. janiszewskae										
141. fibigeri										
142. taurensis										
143. vanella							MR	CI		
144. vanelloides	IS	JO								
145. antiqua	IS		EG							
146. acaciella			EG	LY	TU	AL		CI		
147. jerichoella	IS									
148. viviparella										
149. echidnias			EG	LY	TU	AL			MD	
150. bifasciella			EG		TU	AL				
151. karsholti					TU					
152. acarpa	IS				TU	AL	MR			

153. calligoni																	
154. variapennella																	
155. nigralbella		ES															
156. leucomelanellus																	
157. stereodoxa	PR	ES				SA			MA								
158. lerautella				FR													
159. partinicensis		ES		FR			SI		MA								
160. rebeli	PR																
161. scriptella																	
162. oecophorella																	
163. incertella																	

153. calligoni																	
154. variapennella																	TR
155. nigralbella																	
156. leucomelanellus																	
157. stereodoxa											MK						
158. lerautella																	
159. partinicensis																CY	
160. rebeli																	
161. scriptella																	
162. oecophorella																	
163. incertella																	

153. calligoni									RU								
154. variapennella									RU					AM			
155. nigralbella																	
156. leucomelanellus																	
157. stereodoxa																	
158. lerautella																	
159. partinicensis																	
160. rebeli																	
161. scriptella																	
162. oecophorella																	
163. incertella																	

153. calligoni										
154. variapennella										
155. nigralbella	IS				TU	AL				
156. leucomelanellus			EG	LY			MR			
157. stereodoxa			EG							
158. lerautella										
159. partinicensis										
160. rebeli								CI	MD	
161. scriptella										AR
162. oecophorella										AR
163. incertella										AR

Colour plates

All specimens are illustrated in 8 x natural size

Adults. Species Nos 1-11

1. *Mompha conturbatella* (Hübner) – ♂, Netherlands, De Lutte, Kampbrug, la. 22.v.1991, em. 11.vi.1991, leg. Koster (KTR).

2. *Mompha ochraceella* (Curtis) – ♂, Netherlands, Callantsoog, Zwanenwater, la. 29.iv.1992, em. 20.v.1992, leg. Koster (KTR).

3. *Mompha lacteella* (Stephens) – ♂, Switzerland, Wallis, Fiescherahal, 1200 m, 19.vii.1982, leg. Van der Wolf (KTR).

4. *Mompha propinquella* (Stainton) – ♂, Netherlands, Kerkrade, 19.vi.1989, leg. Schreurs (KTR).

5. *Mompha divisella* Herrich-Schäffer – ♂, Netherlands, Amerongen, 29.vii.1949, coll. Bentinck (RMNH).

6. *Mompha jurassicella* (Frey) – ♂, England, Oxford, la. 28.viii.1988, em. 21.ix.1988, leg. Langmaid (KTR).

7. *Mompha bradleyi* Riedl – ♂, Netherlands, Kerkrade, Erenstein, la. 4.viii.1993, em. 1.ix.1993, leg. Schreurs (KTR).

8. *Mompha confusella* Koster & Sinev – ♂, Hungary, Almar, Bukk-hegys, 4.iv.1951, leg. Reskowits (HNHM).

9. *Mompha subbistrigella* (Haworth) – ♂, Netherlands, Twello, ex l., 28.vii.1992, leg. Wolschrijn (KTR).

10. *Mompha sturnipennella* (Treitschke) – ♂, Netherlands, Limburg, Venlo, 29.viii.1982, leg. Van Aartsen (KTR).

11. *Mompha meridionella* sp. n. – ♀, paratype, Greece, Smolikas, 10 km W Paraskevi by Konitea, 600 m, 23.vii.1990, leg. Fibiger (ZMUC).

1

2

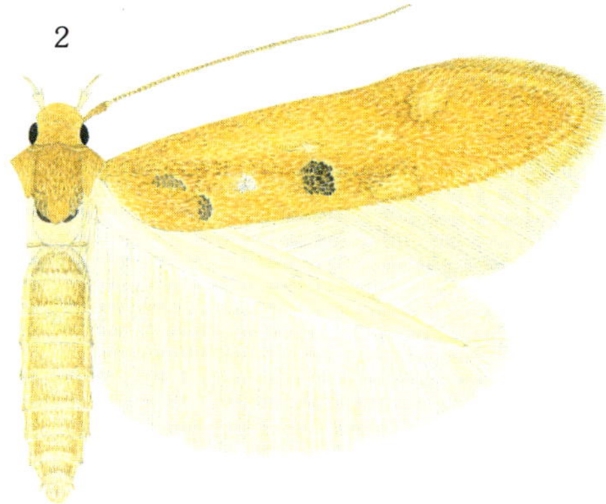

3

4

5

6

7

8

9

10

11

12. *Mompha epilobiella* ([Denis & Schiffermüller]) – ♂, Netherlands, Callantsoog, ex p., 1.viii.1992, leg. Koster (KTR).

13. *Mompha langiella* (Hübner) – ♂, England, Hampshire, Ashhurst, la. 12.vii.1993, em. 6.viii.1993, leg. Langmaid (KTR).

14. *Mompha idaei* (Zeller) – ♂, Belgium, Ardennen, Achouffe nr. Houffalise, 28.vi.-4.vii.1968, leg. Diakonoff (RMNH).

15. *Mompha miscella* ([Denis & Schiffermüller]) – ♂, Spain, Görliz-Visc., 1.-16.v.1986, leg. Wolschrijn (KTR).

16. *Mompha locupletella* ([Denis & Schiffermüller]) – ♂, Netherlands, Losser, Poortbulten, la. 2.v.1988, em. 15.v.1988, leg. Koster (KTR).

17. *Mompha maculata* (Sinev, 1986) – ♂, paratype, Georgia, Adzharia, Gogadzewi, 1300 m, 20.vi.1969, leg. Zagulajev (ZIN).

18. *Mompha terminella* (Humphreys & Westwood) – ♂, Netherlands, De Lutte, Egheria, la. 25.viii.1992, em. 12.v.1993, leg. Koster (KTR).

19. *Mompha raschkiella* (Zeller) – ♂, Netherlands, Callantsoog, Zwanenwater, la. 16.vii.1987, em. 8.viii.1987, leg. Koster (KTR).

20. *Mompha sexstrigella* (Braun) – ♂, Russia, Murmansk area, Luostari, 15.vi.1980, leg. Sinev (ZIN).

21. *Batrachedra praeangusta* (Haworth) – ♀, Netherlands, Losser, Poortbulten, 4.vii.1990, leg. Koster (KTR).

22. *Batrachedra pinicolella* (Zeller) – ♀, Netherlands, Losser, Duivelshof, 17.vi.1993, leg. Koster (KTR).

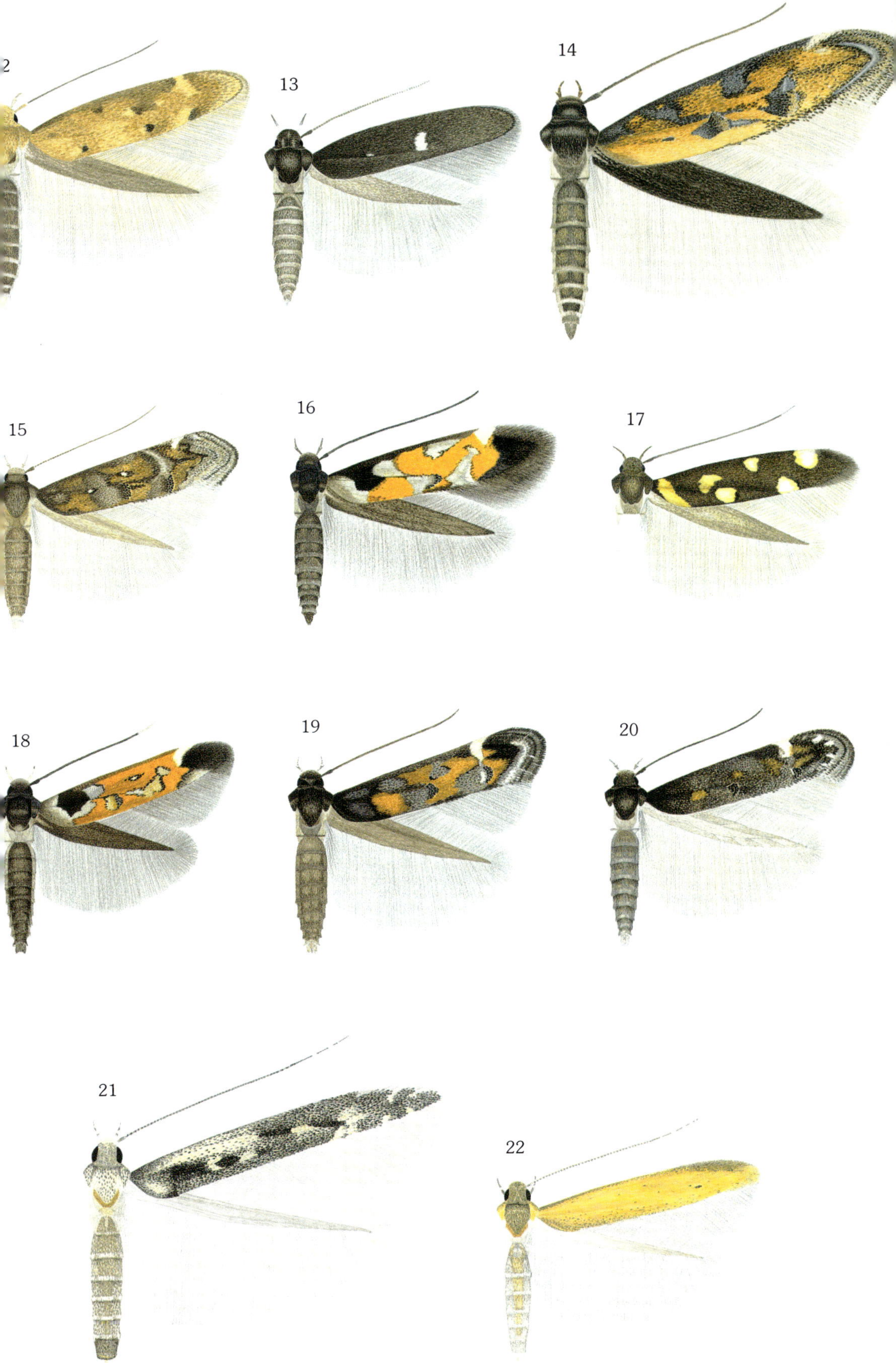
2
13
14
15
16
17
18
19
20
21
22

23. *Batrachedra parvulipunctella* Chrétien – ♂, France, Drôme, Le Penne S.Ouvèze, 23.vii.1980, leg. Van der Wolf (VDW).

24. *Batrachedra amydraula* Meyrick – ♀, Iraq, Baghdad, 100 ft., 21.v.1956, leg. Wiltshire (LSNK).

25. *Stathmopoda pedella* (Linnaeus) – ♂, Netherlands, Callantsoog, Zwanenwater, 12.vii.1988, leg. Koster (KTR).

26. *Stathmopoda auriferella* (Walker) – ♂, Israel, Tiberias, 7.iii.1968, leg. Bleszynski (ZFMK).

27. *Tortilia flavella* Chrétien – ♀, India, Kashmir, Srinagar, 5200 ft., 9.vi.1923, coll. Fletcher (BMNH).

28. *Tortilia trigonella* (Zerny) – ♀, Morocco, Hout-Atlas, Asni, 1200 m, 6.vii.1984, leg. De Prins (ZMAN).

29. *Tortilia charadritis* (Meyrick) – ♂, paratype, Egypt, Aswan (BMNH).

30. *Tortilia pallidella* Kasy – ♂, paratype, S-Iran, 8 km östl. Bandar-Abbas, 11.iv.1972, Exped. Mus. Vind. (NHMW).

31. *Tortilia graeca* Kasy – ♂, Greece, Peloponnisos, 6 km S of Monemvasia, 8.vii.1991, leg. Schouten (KTR).

32. *Blastodacna hellerella* (Duponchel) – ♂, Germany, Hessen, Dachnau bei Mörfelden, 20.v.1966, leg. Groß (LMAD).

23

24

25

26

27

28

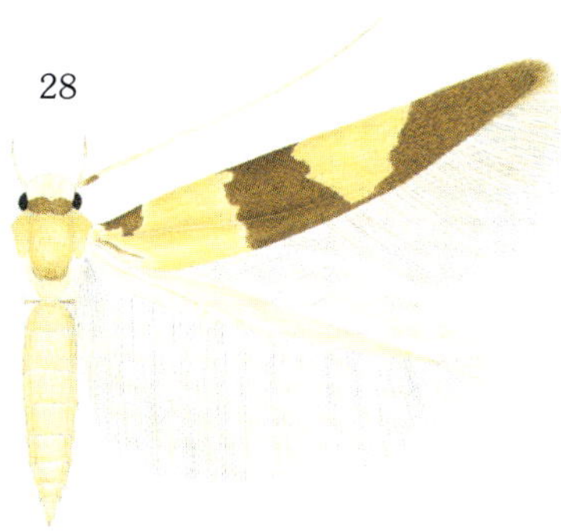

29

30

31

32

33. *Blastodacna rossica* Sinev – ♀, holotype, Ukraine, Krym, Karadagh, 12.vii.1987, leg. Sinev (ZIN).

34. *Blastodacna atra* (Haworth) – ♂, Germany, Köln-Longerich, 17.vi.1957, leg. Groß (LMAD).

35. *Blastodacna vinolentella* (Herrich-Schäffer) – ♀, Netherlands, Meerssen, 9.vi.1948, coll. Bentinck (RMNH).

36. *Blastodacna georgiella* Sinev – ♀, paratype, Georgia, Adzharia, Batumi, 19.vi.1977, leg. Sinev (ZIN).

37. *Blastodacna libanotica* Diakonoff – ♀, Turkey, Asia min., Gebse, 50 km östl. Istanbul, 17.v.1969, leg. Glaser (TLMF).

38. *Spuleria flavicaput* (Haworth) – ♀, Netherlands, De Lutte, Beverborgsbrug, 22.v.1991, leg. Koster (KTR).

39. *Heinemannia laspeyrella* (Hübner) – ♀, Russia, Novosibirsk, Akademgorodok, 21.-22.vi.1984, leg. Sinev (ZIN).

40. *Heinemannia festivella* ([Denis & Schiffermüller]) – ♂, France, Alps Hte Prov, La Condamine Chatelard, 1300-1450 m, 16.-27.vii.1991, leg. Gielis (GLS).

33

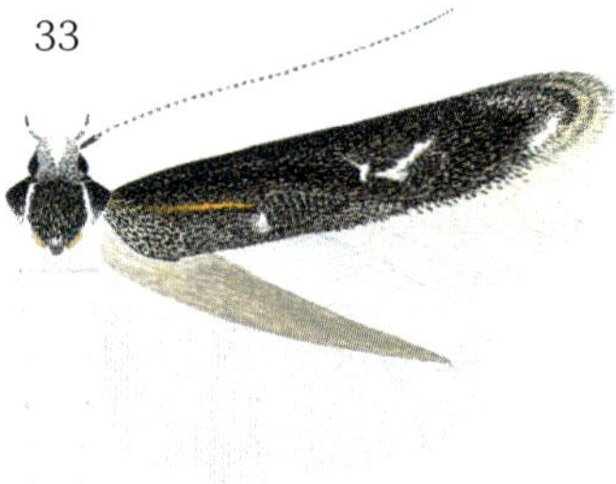

34

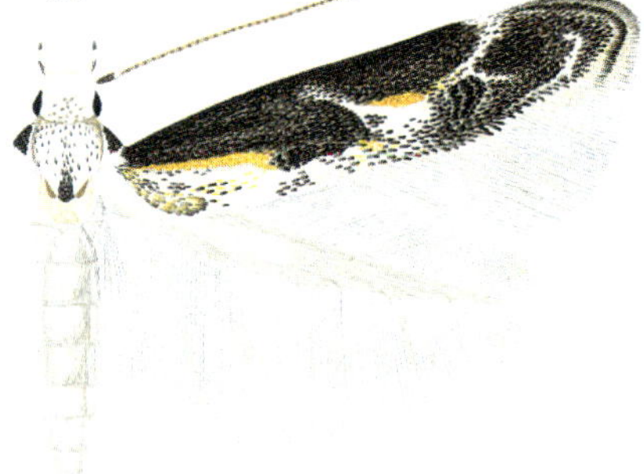

35

36

37

38

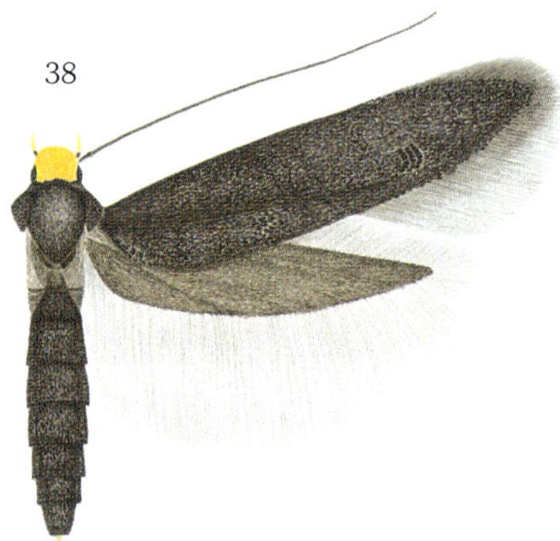

39

40

41. *Heinemannia albidorsella* (Staudinger) – ♀, France, Corsica, corse du Sud, Ste Lucie de Porto, 11.v.1996, leg. Huisman (HSN).

42. *Dystebenna stephensi* (Stainton) – ♂, Poland, Silesia, coll. Staudinger, 11.vii.1894 (RMNH).

43. *Haplochrois gelechiella* (Rebel) – ♀, Greece, S-Mt. Olympos, NW of Sikaminea, 12.vii.1999, leg. Schouten (KTR).

44. *Haplochrois ochraceella* (Rebel) – ♀, Slovakia, Zadiel, 18.-20.vii.1990, leg. Van der Wolf (VDW).

45. *Haplochrois albanica* (Rebel & Zerny) – ♂, Portugal, Galegos, Portalegre, Alentejo, 3.vi.1996, leg. Corley (KTR).

46. *Haplochrois buvati* (Baldizzone) – ♂, Portugal, N. of Alportel, Algarve, 14.ix.1995, leg. Corley (KTR).

47. *Haplochrois theae* (Kusnezov) – ♂, Georgia, Adzharia, Gonio, 3.viii.1976, leg. Sinev (ZIN).

48. *Chrysoclista linneella* (Clerck) – ♂, Germany, Dortmund, 3.vii.1944, coll. Grabe (LMAD).

49. *Chrysoclista lathamella* (Fletcher) – ♂, Netherlands, Amsterdam O, 3.vi.1972, leg. Lourens (ZMAN).

41

42

43

44

45

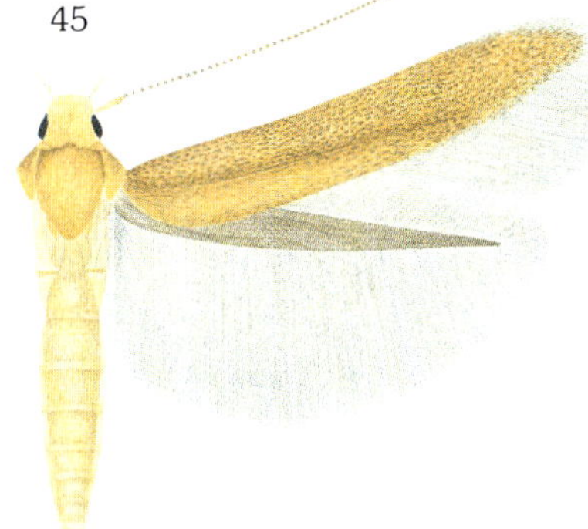

46

47

48

49

50. *Chrysoclista splendida* Karsholt – ♀, Austria, coll. Staudinger (RMNH).

51. *Chrysoclista zagulajevi* Sinev – ♂, holotype, Georgia, Adzharia, Tchchutuneti, 19.vii.1976, leg. Sinev (ZIN).

52. *Chrysoclista abchasica* (Sinev) – ♂, holotype, Georgia, Abkhazia, Gumistinskij nature reserve, 500 m, 8.vii.1978, leg. Zagulajev (ZIN).

53. *Pancalia leuwenhoekella* (Linnaeus) – ♂, Germany, Rheinland Eifel, Gilsdorf bei Münstereifel, 2.v.1974, leg. Groß (LMAD).

54. *Pancalia schwarzella* (Fabricius) –
a) ♂, Germany, Wahner Heide, Köln, 1.v.1955, leg. Groß (LMAD).
b) ♀, Netherlands, Halfweg, 5.vi.1910, coll. Van den Beek (RMNH).

55. *Pancalia baldizzonella* Riedl –
a) ♂, holotype, Italy, Mt. Pollino, 1500 m, Lucania, Piano di Ruggia, 9.vii.1991, leg. Baldizzone (BDZ).
b) ♀, paratype, Italy, Mt. Pollino, 1500 m, Lucania, Piano di Ruggia, 9.vii.1991, leg. Baldizzone (BDZ).

56. *Pancalia nodosella* (Bruand) –
a) ♂, Netherlands, Callantsoog, Zwanenwater, 14.v.1994, leg. Koster (RMNH).
b) ♀, Netherlands, Callantsoog, Zwanenwater, 4.v.1995, leg. Koster (RMNH).

50

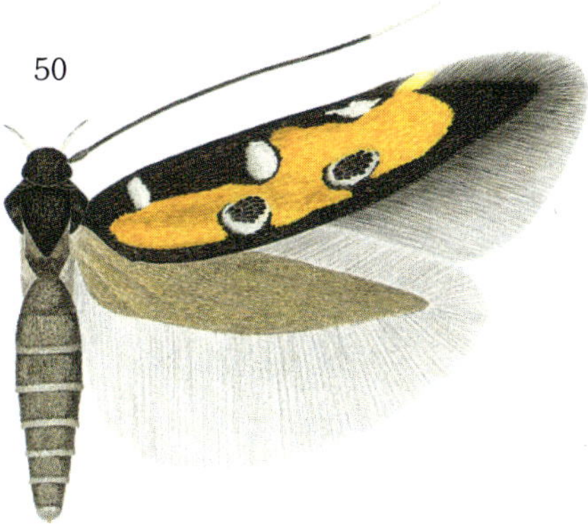

51

2

53

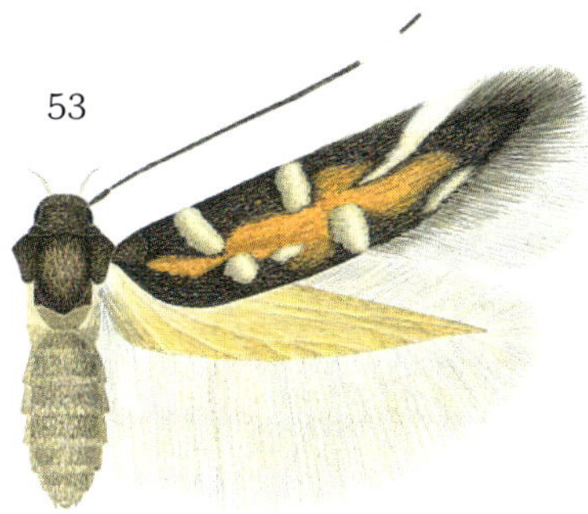

54a

4b

55a

55b

56a

56b

57. *Euclemensia woodiella* (Curtis) – ♂, holotype, England, Kersall-moor, Manchester, Lancashire, vi.1829 (MMA).

58. *Phosphaticola gemmatella* Viette – ♀, Tunisia, Metlasui, 3.ii.1930 (MNHN).

59. *Alloclita recisella* Staudinger – ♂, Spain, Huelva, Torre de la Higuera, 16.vii.1990, leg. Hull (KTR).

60. *Alloclita haifensis* Rebel – ♂, holotype, Israel, Haïfa, leg. Kalchberg (NHMW).

61. *Alloclita delozona* Meyrick – ♂, Oman, Northern Region, Jabal Shams, 19 km NW Al Hamra, 1100 m, 7.i.1993, leg. Skule (ZMUC).

62. *Alloclita brachygrapta* Meyrick – ♂, syntype, Egypt, Romani, AWB., ix.1916 (BMNH).

63. *Alloclita deprinsi* sp. n. – ♂, holotype, Turkey, St. 2388, Icel 10 m, 8 km SSE Kazanli, 10.viii.1997, leg. De Prins (ZMAN).

64. *Alloclita francoeuriae* Walsingham –
    a) ♂, Algeria, Hammam-es-Salahin, 6.iv.1904, Wlsm. 97943 (BMNH).

57

58

59

60

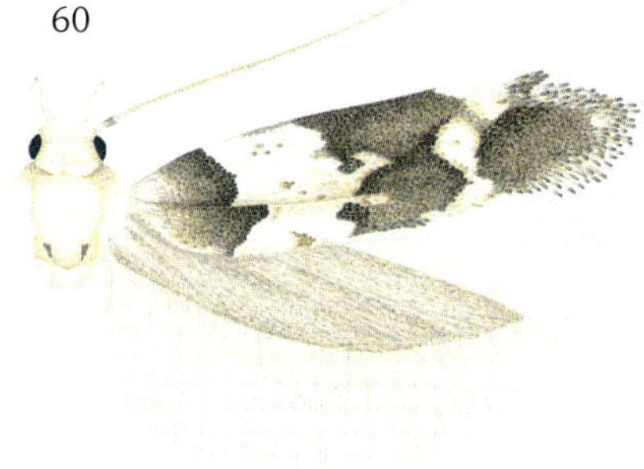

61

62

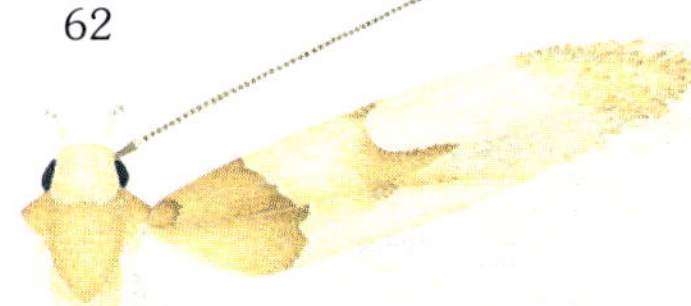

63

64a

64. *Alloclita francoeuriae* Walsingham –
b) ♂, Algeria, Hammam-es-Salahin, 12.iii.1904, Wlsm. 97937 (BMNH).

65. *Gibeauxiella bellaqueifontis* (Gibeaux, 1986) – ♂, holotype, France, Seine-et-Marne, forêt de Fontainebleau, La Tiliae, 8.v.1985, leg. Gibeaux (GBX).

66. *Gibeauxiella reliqua* (Gibeaux, 1986) – ♂, holotype, France, Seine-et-Marne, forêt de Fontainebleau, La Tiliae, 8.v.1985, leg. Gibeaux (GBX).

67. *Limnaecia phragmitella* Stainton – ♂, Netherlands, Callantsoog, Zwanenwater, ex l., 6.vii.1981, leg. Koster (KTR).

68. *Allotalanta autophaea* Meyrick – ♂, paralectotype, Turkey, Taurus Mts., Asia Minor, Joannis 1906 (BMNH).

69. *Cosmopterix zieglerella* (Hübner) – ♂, Netherlands, Losser, Zandbergen, 5.vii.1990, leg. Koster (KTR).

70. *Cosmopterix schmidiella* Frey – ♂, Switzerland, Zurich, coll. Wocke (ZIN).

71. *Cosmopterix orichalcea* Stainton – ♂, Germany, Bavaria, Regensburg (RMNH).

72. *Cosmopterix sibirica* Sinev, 1985 – ♂, Russia, Gornyj Altaj, Ongudai, 18.vi.1989, leg. Ustjuzhanin (KTR).

73. *Cosmopterix scribaiella* Zeller – ♂, Netherlands, Het Woold, Wooldse veen, 12.vi.1993, leg. Koster (KTR).

74. *Cosmopterix pulchrimella* Chambers – ♂, Spain, prov. Valencia, Pto. De Sagunto, 7.v.1990, leg. Koster (KTR).

75. *Cosmopterix attenuatella* (Walker) – ♂, Portugal, Madeira, Machico, 20.-25.v.1989, leg. Cox (COX).

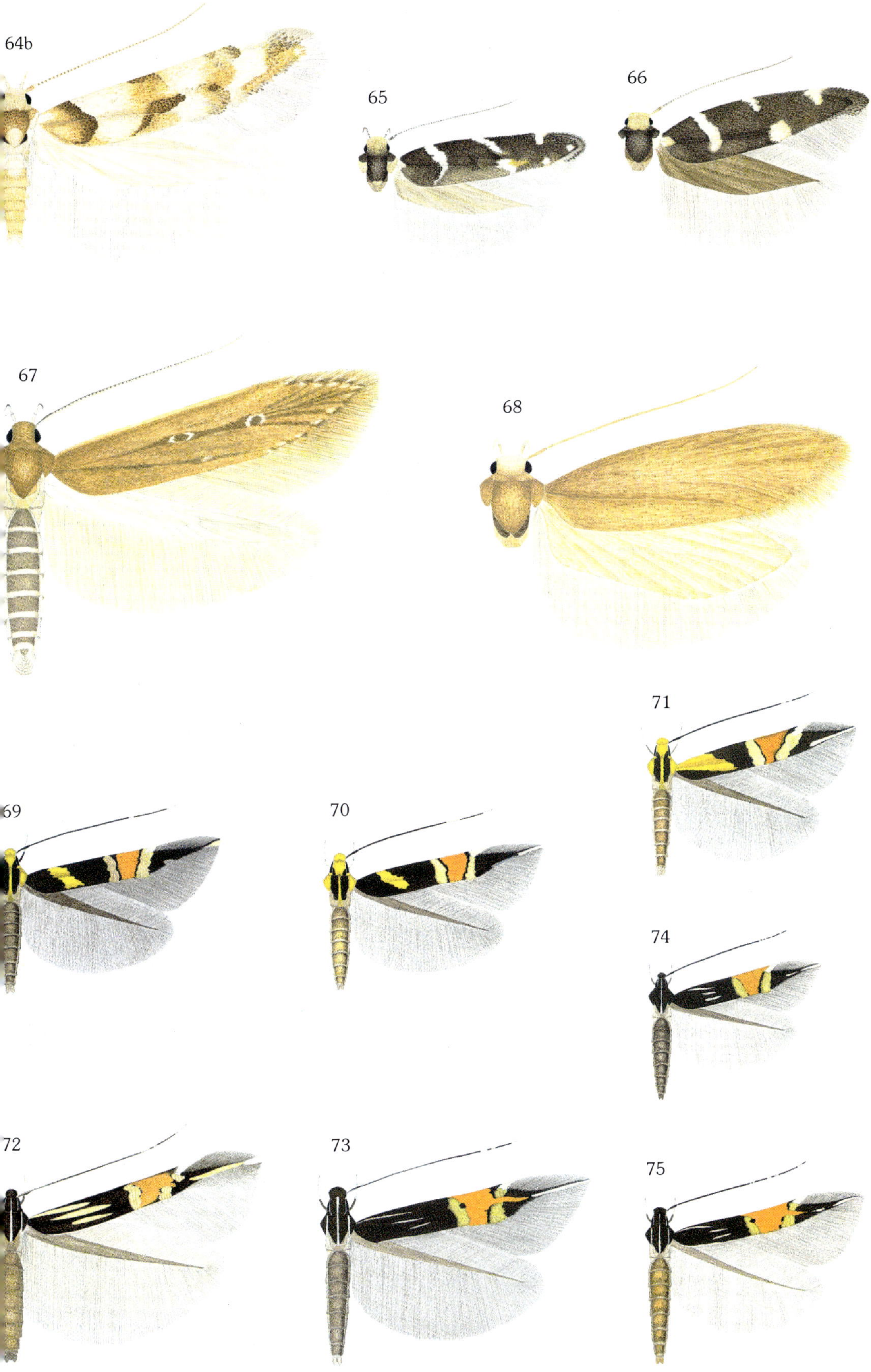
64b
65
66
67
68
71
69
70
74
72
73
75

76. *Cosmopterix crassicervicella* Chrétien – ♂, Spain, Castellon, Vinaroz, 26.v.1997, leg. Cox (COX).

77. *Cosmopterix turbidella* Rebel – ♂, Spain, Isl. Canarias, la Gomera, Hermigua, 123 m, 3.viii.1996, leg. Van den Berg (KTR).

78. *Cosmopterix pararufella* Riedl – ♂, Spain, Almeria, 6 km SW Tabernas, Mini Hollywood, 400 m., 30.ix.1993, leg. Van der Wolf (VDW).

79. *Cosmopterix coryphaea* Walsingham – ♂, Greece, Peloponnisos, Gorge 12 km SE of Kardhamili, 29.vii.1991, leg. Schouten (KTR).

80. *Cosmopterix lienigiella* Lienig & Zeller – ♂, Netherlands, Lauwersmeer, Sennerplaat, ex l. 2.vi.1996, leg. Jansen (KTR).

81. *Cosmopterix salahinella* Chrétien – ♀, Israel, Tel-Aviv, ex l., 21.iv.-28.v.1930, leg. Amsel (LSNK).

82. *Labdia semicoccinea* (Stainton) – ♂, Import Sri Lanka (PD Aalsmeer), ex l. 12.v.1986, leg. Van Frankenhuyzen (PDW).

83. *Pyroderces argyrogrammos* (Zeller) – ♂, Austria, Illmitz, 17.vii.1993 (KTR).

84. *Pyroderces caesaris* Gozmány – ♂, SW-Turkey, El-Antalya, Termessos, 7-800 m, Bey Daglari, 30.v.-1.vi.1974, leg. Groß (KTR).

85. *Pyroderces klimeschi* Rebel – ♀, Hungary, Kis Balaton p. Vörs, 21.-29.vi.1936, leg. Klimesch (LSNK).

76

77

78

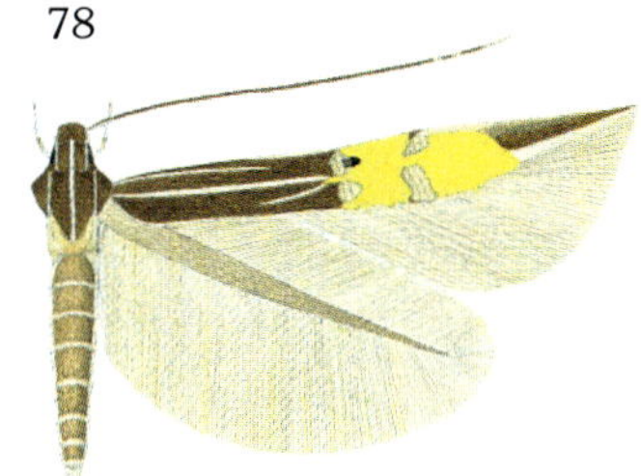

79

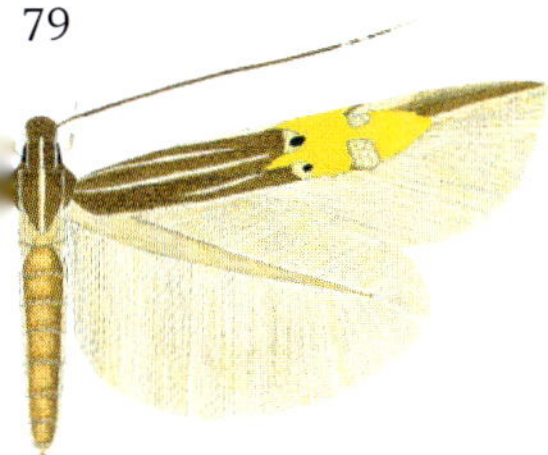

80

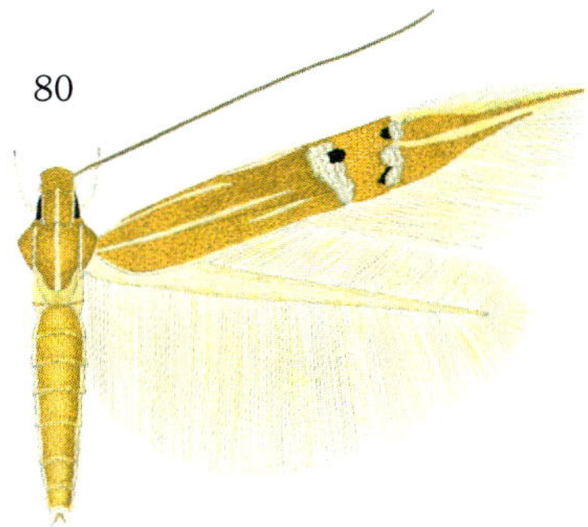

81

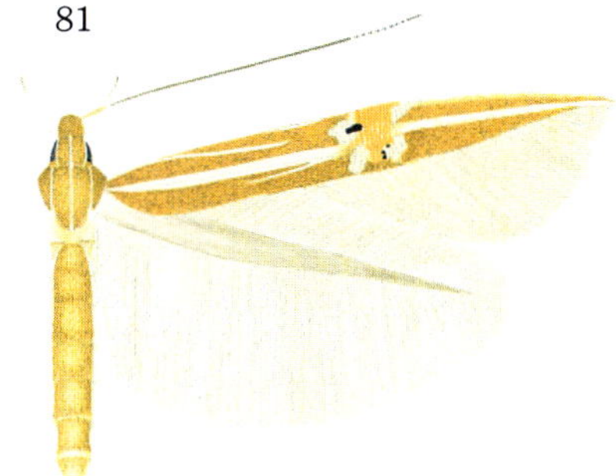

82

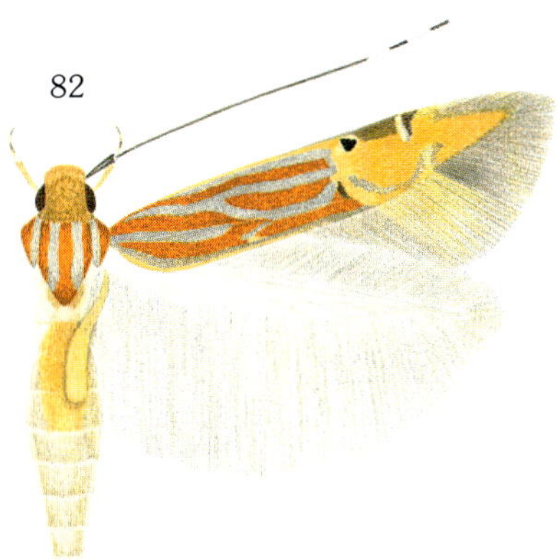

83

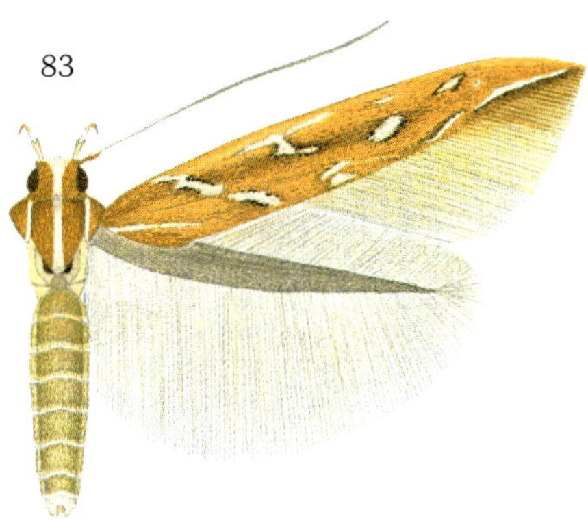

84

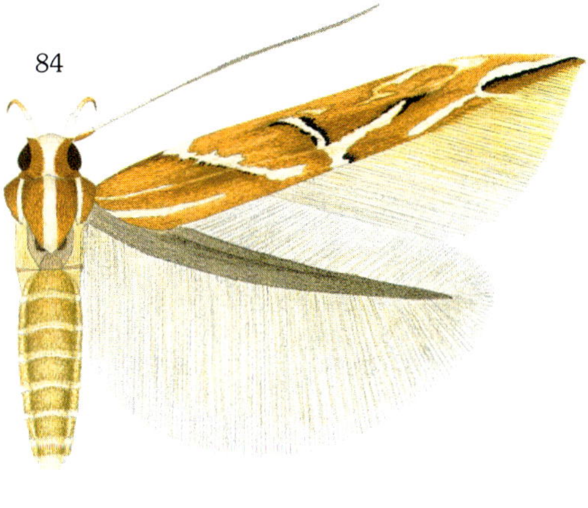

85

86. *Pyroderces brosi* Riedl – ♂, holotype, Italy, Fano (Marche), 9.vi.1953, coll. Toll (ISAK).

87. *Pyroderces tethysella* sp. n. – ♂, holotype, southern Tadzhikistan, Tigrovaya Balka, env., Dzhilikul, 19.viii.1990, leg. Noreika (ZIN).

88. *Pyroderces wolschrijni* sp. n. – ♀, paratype, Spain, Almeria, San Jose (Nihar), 6.-20.ix.2000, leg. Wolschrijn (WLN).

89. *Pyroderces apicinotella* (Chrétien) – ♂, Tunesia, Degache,14.iv.1977, leg. Embey-Isztin et Gozmány (TMAB).

90. *Anatrachyntis rileyi* (Walsingham) – ♂, Spain, Isl. Canarias, Tenerife, Los Christianos, 10.-15.i.2000, leg. Wolschrijn (WLN).

91. *Anatrachyntis badia* (Hodges) – ♂, Spain, Alicante, La Marina, 15.-22.ix.1999, leg. Wolschrijn (WLN).

92. *Anatrachyntis coridophaga* (Meyrick) – ♂, Egypt, Mariout-El-Agami, 1.-4.x.1974, leg. Zouhar (ZFMK).

93. *Anatrachyntis simplex* (Walsingham) – ♀, Egypt, Giza, 2.vi.1974, leg. Zouhar (ZFMK).

94. *Coccidiphila gerasimovi* Danilevsky – ♂, Spain, Huelva, Marismas del Odiel, 6.iv.1994, leg. Van der Wolf (VDW).

95. Not figured.

96. *Coccidiphila ledereriella* (Zeller) – ♂, France, Cannes, ex l. 7.vi.1883, coll. Heylaerts (RMNH).

97. *Coccidiphila danilevskyi* Sinev – ♀, Spain, 5 km W of Nerja, 1.v.1995, 250 m, leg. Schouten (KTR).

98. *Coccidiphila kasypinkeri* Traugott-Olsen – ♂, Spain, Lancerote, Haria, leg. Pinker (LSNK).

99. *Coccidiphila riedli* Traugott-Olsen – ♀, Spain, Islas Canarias, Tenerife, Chio, 900 m, 5 km SE road C-823, 12.iv.1993, leg. Van der Wolf (VDW).

87
88
90
91
94
93
92
96
97
99
98

100. *Stagmatophora heydeniella* (Fischer von Röslerstamm) – ♂, Germany, Wiesbaden, leg. Rössler (RMNH).

101. *Eteobalea anonymella* Riedl – ♂, Slovakia, Jablonov near Turnou, 19.vii.1990, leg. Van der Wolf (KTR).

102. *Eteobalea serratella* (Treitschke) – ♂, Slovakia, Zadiel, 18.-20.vii.1990, leg. Van der Wolf (KTR).

103. *Eteobalea intermediella* Riedl – ♂, Czech Republic, Moravia, Pavlovské Tabulová, 12.vii.1990, leg. Van der Wolf (KTR).

104. *Eteobalea beata* (Walsingham) – ♂, France, Alpes de Ht Provence, Les Mees, 11.-22.vii.1982, leg. Langohr (KTR).

105. *Eteobalea sumptuosella* (Lederer) – ♂, Greece, Fokís, Itea, 20.vii.1987, leg. Van der Wolf (KTR).

106. *Eteobalea teucrii* (Walsingham) – ♂, Spain, Andalucia, Camino d. Rhonda, Urb Madrenal, Loma de Colmenas, 500 m, 10.ix.1988, leg. Traugott-Olsen (ZMUC).

107. *Eteobalea klisieckii* Riedl – ♀, Maroc Saharien, Maader Sellam, 7.xi.1967, coll. Rungs (MNHN).

108. *Eteobalea tririvella* (Staudinger) – ♂, Slovakia mer., Sahy, 15.vii.1990, leg. Van der Wolf (KTR).

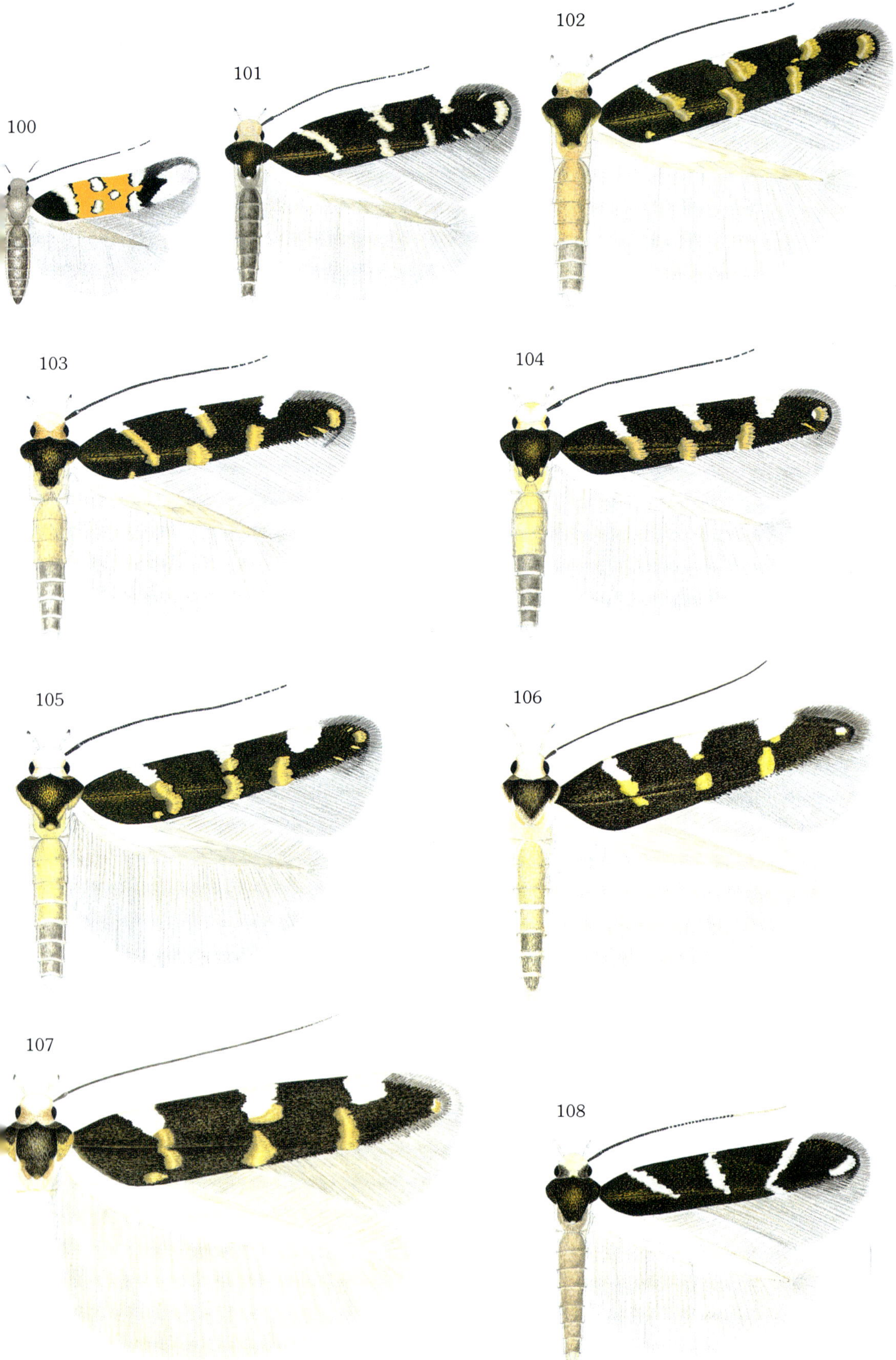
100
101
102
103
104
105
106
107
108

109. *Eteobalea albiapicella* (Duponchel) – ♂, Croatia, Dalmatien, Senj-Alan, 200 m, 3.v.1973, leg. Groß (LMAD).

110. *Eteobalea alypella* (Klimesch) – ♂, Spain, Castellon, 5 km E Cuevas de Vinroma, 200 m, 13.vii.1992, leg. Fibiger (ZMUC).

111. *Eteobalea isabellella* (Costa) – ♂, Greece, Peloponnisos, 6 km NE of Dhimitsana, 800 m, 20.vi.1991, leg. Schouten (KTR).

112. *Eteobalea dohrnii* (Zeller) – ♂, Spain, Teruel, Albarracin, Valdevecar, 1200 m, 25.vii.1991, leg. Van der Wolf (KTR).

113. *Eteobalea siciliae* (Riedl) – ♂, Turkey, St. 2386, Hatay, 2 km NW Yayladagi, 600 m, 8.viii.1997, leg. De Prins (KTR).

114. *Eteobalea thaumatella* (Walsingham) – ♂, Libya, Gharian, Wadi El Hira, 25.iii.1983, leg. Seneca (ZMUC).

115. *Isidiella nickerlii* (Nickerl) – ♂, Italy, Südtirol, Vintschau, Sonnenberg, Latsch - St. Martin, 650 m, 22.vii.1968, leg. Groß (LMAD).

116. *Isidiella divitella* (Constant) – ♂, Italy, Aritzo-Sardinia, 1000 m, vii.1938, leg. Noack (LMAD).

117. *Vulcaniella pomposella* (Zeller) –
   a) ♂, Greece, Peloponnes, Zachlorou 600 m, bei Kalávryta, 1.-14.vii.1959, leg. Noack (LMAD).
   b) ♀, Poland, Breslau, 23.vi.1865, coll. Wocke (ZIN).

118. *Vulcaniella grandiferella* Sinev –
   a) ♂, Ukraine, Krym, Karadagh, 12.vi.1987, leg. Sinev (ZIN).
   b) ♀, Ukraine, Krym, Karadagh, ex l. 15.vii.1987, leg. Sinev (ZIN).

119. *Vulcaniella cognatella* Riedl – ♂, Greece, Peloponnisos, Gorge, 12 km SE of Kardhamili, 29.vi.1991, leg. Schouten (KTR).

120. *Vulcaniella extremella* (Wocke) – ♂, Spain, Teruel, Albarracin, Valdevecar, 1200 m, 24.vii.1991, leg. Van der Wolf (KTR).

121. *Vulcaniella caucasica* Sinev – ♂, Turkey, Agri, 23 km W Dogubayazit, 1800 m, 5.ix.1993, leg. Fibiger (ZMUC).

122. *Vulcaniella grabowiella* (Staudinger) – ♂, France, La Penne-'s Ouvèze, Drôme, 24.vii.1984, leg. Van der Wolf (KTR).

123. *Vulcaniella rosmarinella* (Walsingham) – ♂, Spain, prov. Cuenca, 8 km N of Cuenca, 1000 m, la. 22.iv.1993, em. 14.vi.1993, leg. Van Nieukerken & Koster (KTR).

109
110
111
112
113
114
116
118a
118b
15
117a
117b
119
20
121
122
123

124. *Vulcaniella karadaghella* Sinev – ♂, Ukraine, Krym, Echkidagh, ex l., 21.vi.1991, leg. Budashkin (ZIN).

125. *Vulcaniella klimeschi* (Riedl) – ♀, paratype, Macedonia, Treska Schlucht, ex l., 20.-30.vii.1963, leg. J. Klimesch (LSNK).

126. *Vulcaniella fiordalisa* (Petry) – ♂, Spain, Cuenca, Valdecabras, 25-30.ix.1995, leg. Wolschrijn (WLN).

127. *Vulcaniella glaseri* (Riedl) –
a) ♂, Turkey, Sivas, 10 km W Gürün, 1650 m, 7.vii.1989, leg. Fibiger & Esser (ZMUC).
b) ♂, paratype, Turkey, Ankara, vi.1934, leg. Noack (NHMW).

128. *Vulcaniella gielisi* sp. n. – ♂, holotype, Spain, prov. Cuenca, 7 km NE Cuenca, 18.vii.1992, leg. Gielis (RMNH).

129. *Vulcaniella anatolica* sp. n. – ♂, holotype, Turkey, Prov. Konya, 20 km SE Hadim, Taurus, 1800 m, 14.vii.1986, leg. Fibiger (ZMUC).

130. *Vulcaniella pontica* sp. n. – ♂, holotype, Turkey, Mersin, 10 km SE Arslanköy, Taurus, 1300 m, 11.vii.1987, leg. Fibiger (ZMUC).

131. *Ramphis ibericus* Riedl – ♂, Spain, Teruel, Albarracin, 1170 m, 8.-10.vi.1994, leg. Cox (COX).

132. *Ramphis libanoticus* Riedl – ♂, Turkey, Elazig, Hazar See, NW Ufer, 1250 m, 11.-12.vi.1974, leg. Groß (LMAD).

133. *Hodgesiella rebeli* (Krone) – ♂, Greece, Fokís, Itea, 20.vii.1987, leg. Van der Wolf (VDW).

134. *Hodgesiella rhodorrhisella* Kasy – ♀, Spain, Canary Isl., Tenerife, Bco de Ruiz, ex l. 6.vi.1971, leg. Klimesch (ZMUC).

135. *Hodgesiella christophi* sp. n – ♀, paratype, Turkey, Sivas, 10 km W Gürün, 1650 m, 27.vii.1989, leg. Fibiger & Esser (ZMUC).

136. *Tolliella fulguritella* (Ragonot) – ♂, Turkey, Urfa, 10 km N Halfeti, 500 m, 9.vii.1987, leg. Fibiger (ZMUC).

124

125

126

27a

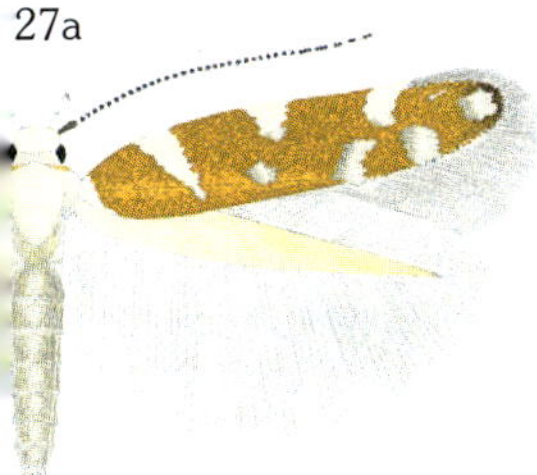

127b

128

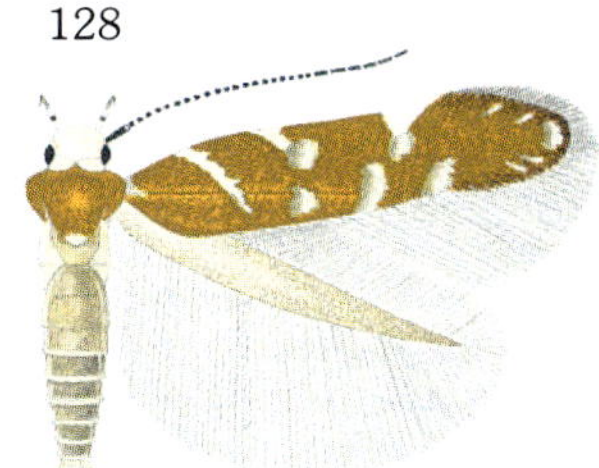

29

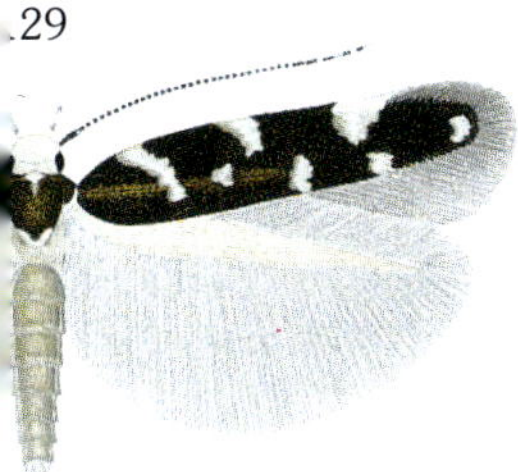

130

131

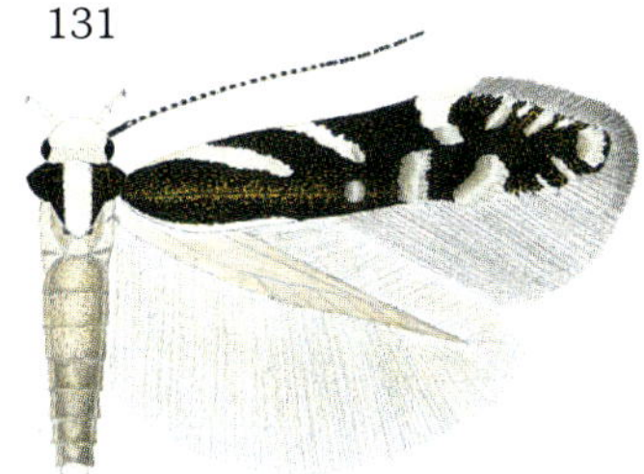

132

133

134

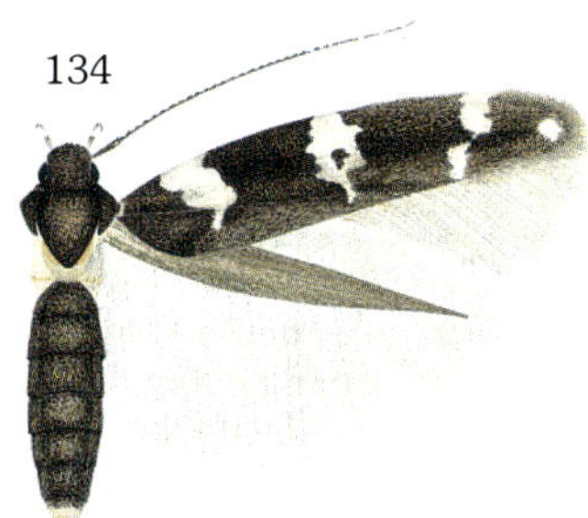

136

135

137. *Sorhagenia rhamniella* (Zeller) – ♂, Netherlands, Oisterwijk, De Rosep, 26.vii.1978, leg. Koster (KTR).

138. *Sorhagenia reconditella* Riedl – ♀, Greece, Peloponnisos, 2 km SW of Anoyia, 600 m, 3.vii.1991, leg. Schouten (KTR).

139. *Sorhagenia lophyrella* (Douglas) – ♂, Austria, Burgenland, Hackelsberg bei Jois, 25.v.1969, leg. Malicky (RMNH).

140. *Sorhagenia janiszewskae* Riedl – ♀, Netherlands, Losser, Duivelshof, 31.vii.1990, leg. Koster (KTR).

141. *Sorhagenia fibigeri* sp. n. – ♂, holotype, Turkey, Nevsehir, 3 km E Incesu, 1350 m, 3.vii.1987, leg. Fibiger (ZMUC).

142. *Sorhagenia taurensis* sp. n. – ♂, holotype, Turkey, Mersin, 27 km N Anamur (Ermenek), Taurus, 750 m, 13.vii.1987, leg. Fibiger (ZMUC).

143. *Ascalenia vanella* (Frey) – ♂, Italy, Adria-Küste, Tronto-Mündung bei S.Benedetto d. Tronto, 13.viii.1970, leg. Groß (LMAD).

144. *Ascalenia vanelloides* Gerasimov – ♂, Turkey, Prov. Icel, Taurus, 600 m, 15.vii.1986, leg. Fibiger (ZMUC).

145. *Ascalenia antiqua* Meyrick – ♂, Israel, Ein-el-Turaba, 27 km S Jericho, 12.i.1973, coll. Grinberg (LSNK).

146. *Ascalenia acaciella* Chrétien – ♀, Algeria, Biskra, 30.v. (MNHN).

147. *Ascalenia jerichoella* Amsel – ♀, holotype, Israel, Georgskloster, Wadi el Kelt, 21.v.1930, leg. Amsel (LSNK).

148. *Ascalenia viviparella* Kasy – ♀, Uzbekistan, Zhamansai, 140 km NW Shafrikan, 10.ix.1968, leg. Falkovitsh (ZIN).

149. *Ascalenia echidnias* (Meyrick) – ♂, Portugal, Madeira, Ponta de S. Lourenço, 100 m, 15.ix.1977, leg. Lomholdt & Wolff (ZMUC).

150. *Ascalenia bifasciella* Chrétien – ♀, 1920-1932, coll. De Joannis (MNHN).

137
138
139
140
141
142
143
144
145
146
147
148
149
150

151. *Calanesia karsholti* Sinev – ♂, holotype, Tunisia, Nefta, 14.-16.iii.1986, leg. Karsholt (ZMUC).

152. *Calycobathra acarpa* Meyrick – ♂, Morocco, Erfoud/Rissani area, 900 m, 13.-14.iv.1989, leg. Karsholt (ZMUC).

153. *Calycobathra calligoni* Sinev – ♂, Uzbekistan, Ispas, 70 km NW Chardzhou, 2.vi.1965, leg. Falkovitsh (ZIN).

154. *Calycobathra variapennella* Sinev – ♂, Kazakhstan, Lake Aral, Barsa-Kel'mes I., 30.vi.1975, leg. Piryulin (ZIN).

155. *Bifascia nigralbella* (Chrétien) – ♂, Tunisia, Nefta area, 1.-4.v.1988, leg. Karsholt (ZMUC).

156. *Bifascioides leucomelanellus* (Rebel) – ♂, S-Iran, Abad-Geno, 40 km nördl. Bandar-Abbas, 23.iv.1974, Exped. Mus. Vind. (NHMW).

157. *Gisilia stereodoxa* (Meyrick) – ♂, S-Iran, 8 km östl. Bandar Abbas, 23.iv.1974, Exped. Mus. Vind. (NHMW).

158. *Gisilia lerautella* Gibeaux – ♂, holotype, France, Seine-et-Marne, fôret de Fontainebleau, La Tiliae, 8.v.1985, leg. Gibeaux (GBX).

159. *Neomariania partinicensis* (Rebel) – ♂, Cyprus, J. Zakynthos, s. Vasilikos, 20 m, 7.ix.1997, leg. Sutter (STR).

160. *Neomariania rebeli* (Walsingham) – ♂, Portugal, Madeira, Fajã da Nogueira, 1000 m, 23.viii.1974, leg. Traugott-Olsen (ZMUC).

151

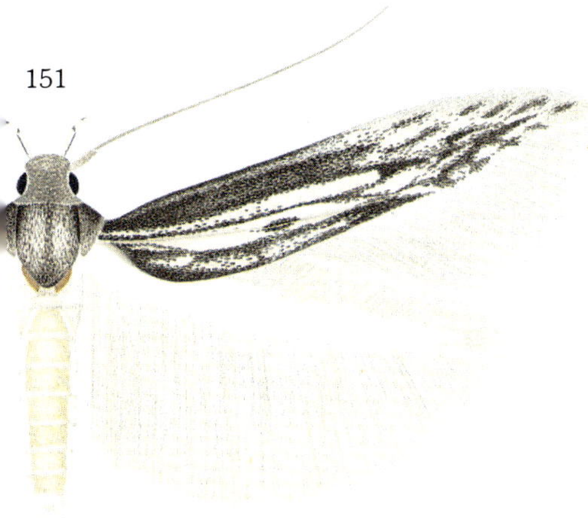

152

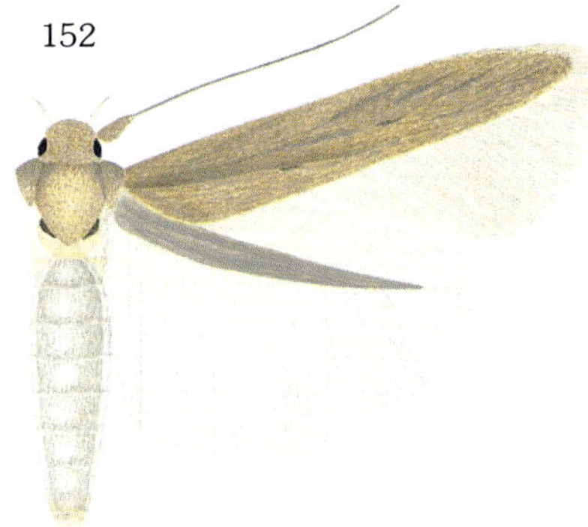

153

154

155

156

157

158

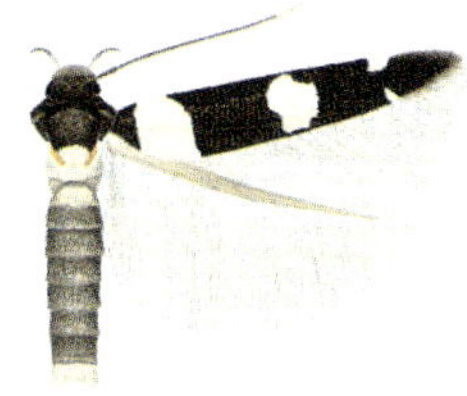

159

160

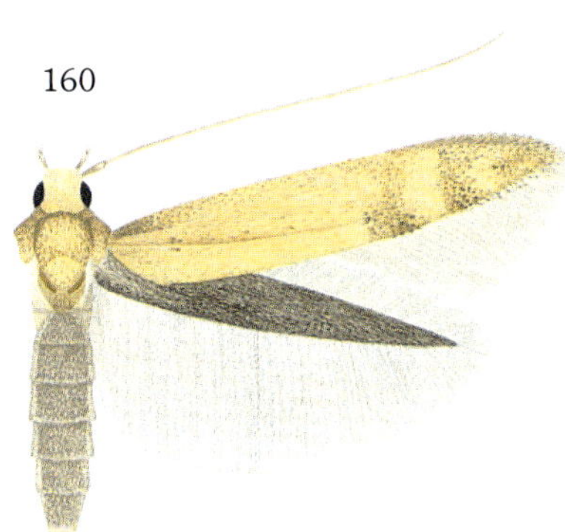

1

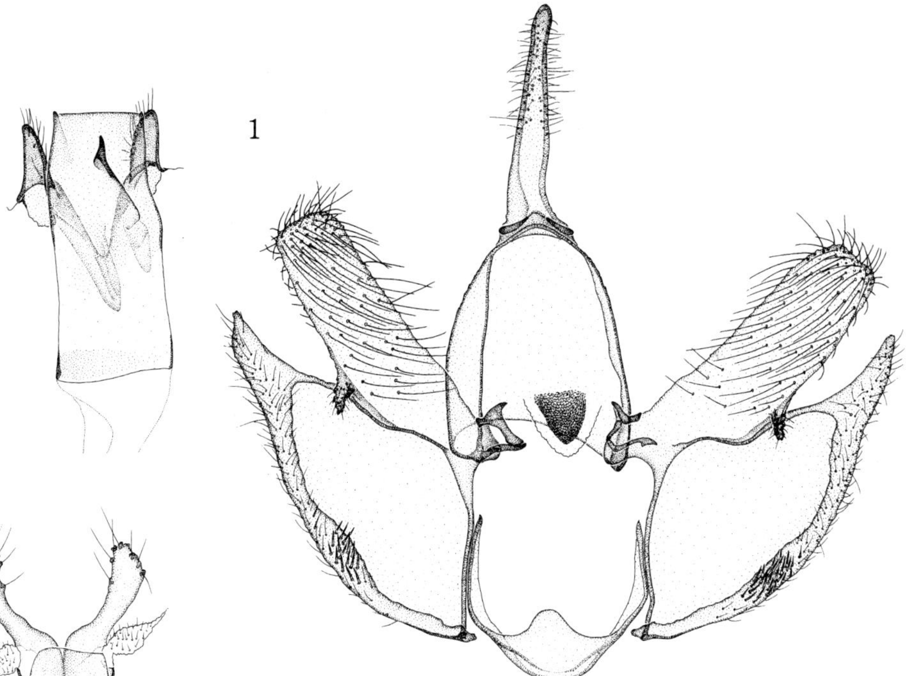

2

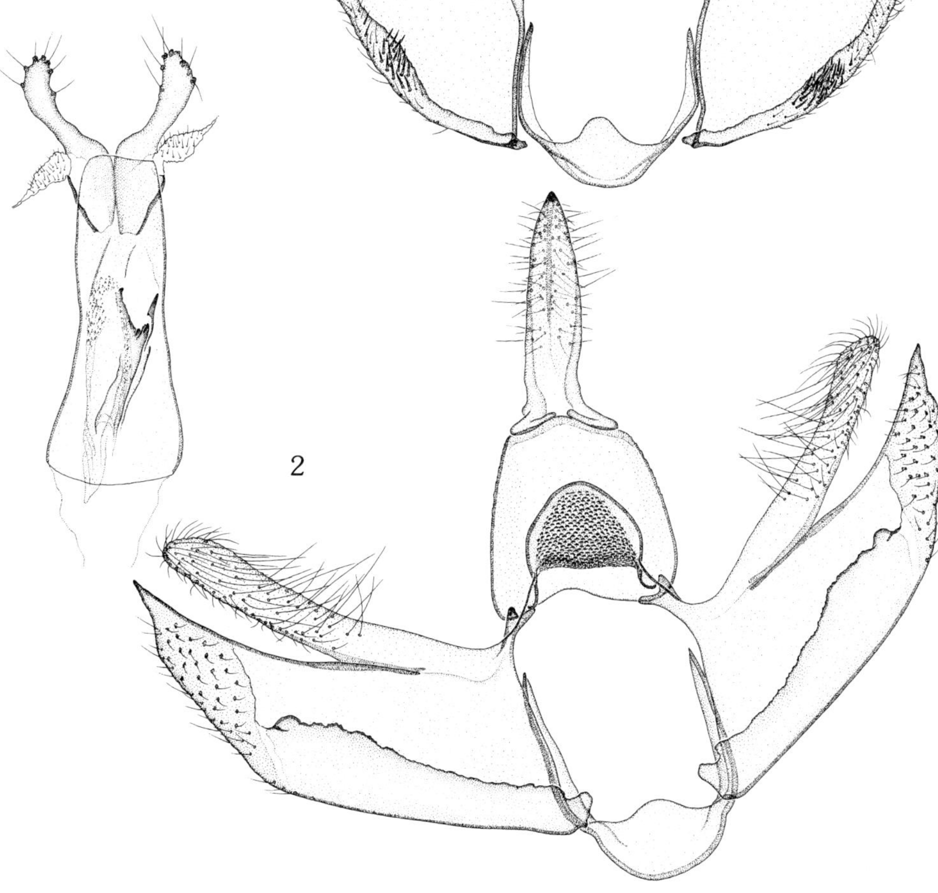

Male Genitalia

1. *Mompha conturbatella* (Hübner) – Denmark, gen. slide JCK 3259 (ZMUC).
2. *Mompha ochraceella* (Curtis) – Netherlands, gen. slide JCK 2394 (KTR).

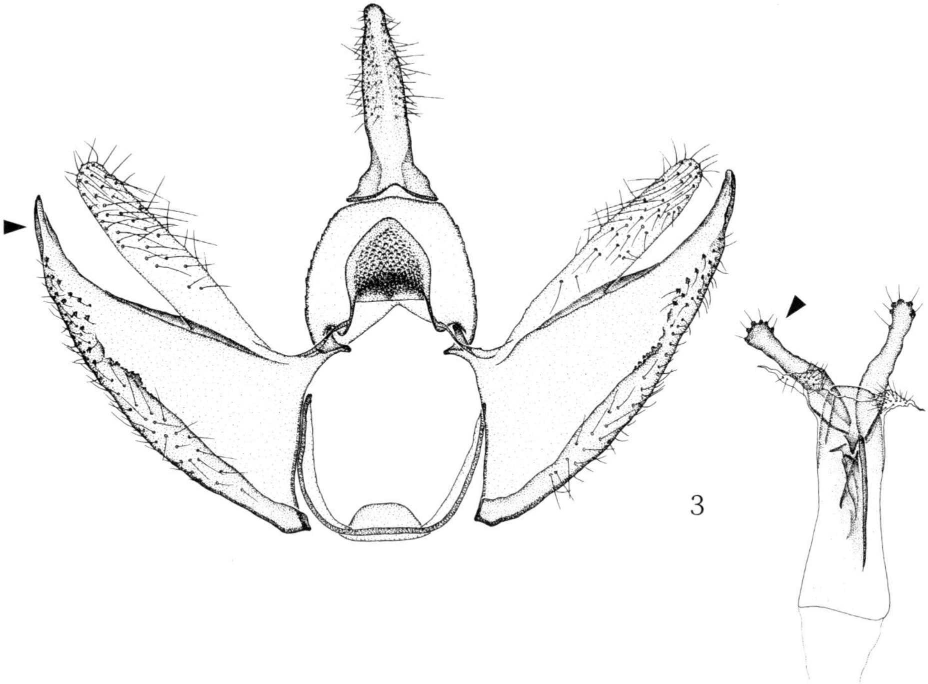

3

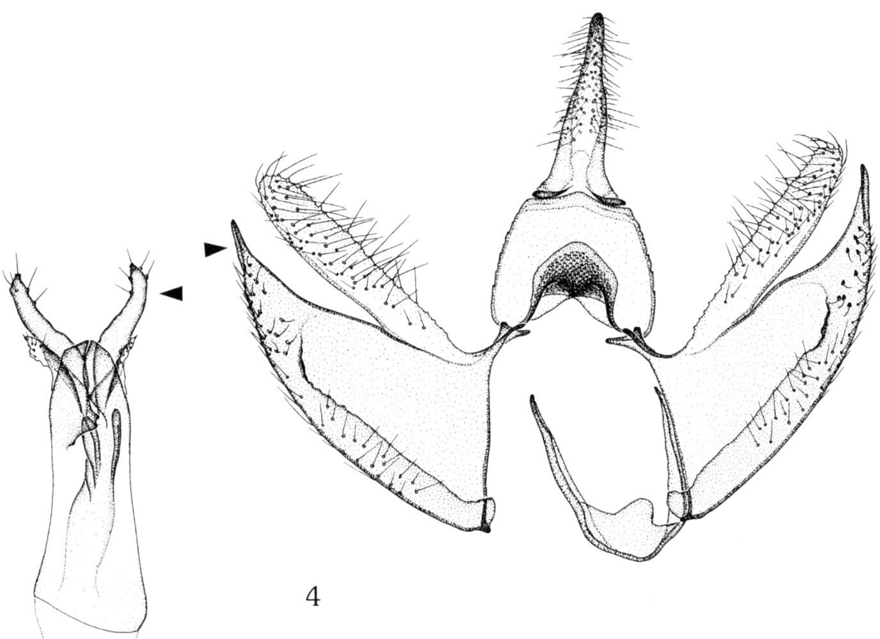

4

Male Genitalia

3. *Mompha lacteella* (Stephens) – Sweden, gen. slide JCK 2077 (KTR).
4. *Mompha propinquella* (Stainton) – Netherlands, gen. slide JCK 2318 (KTR).

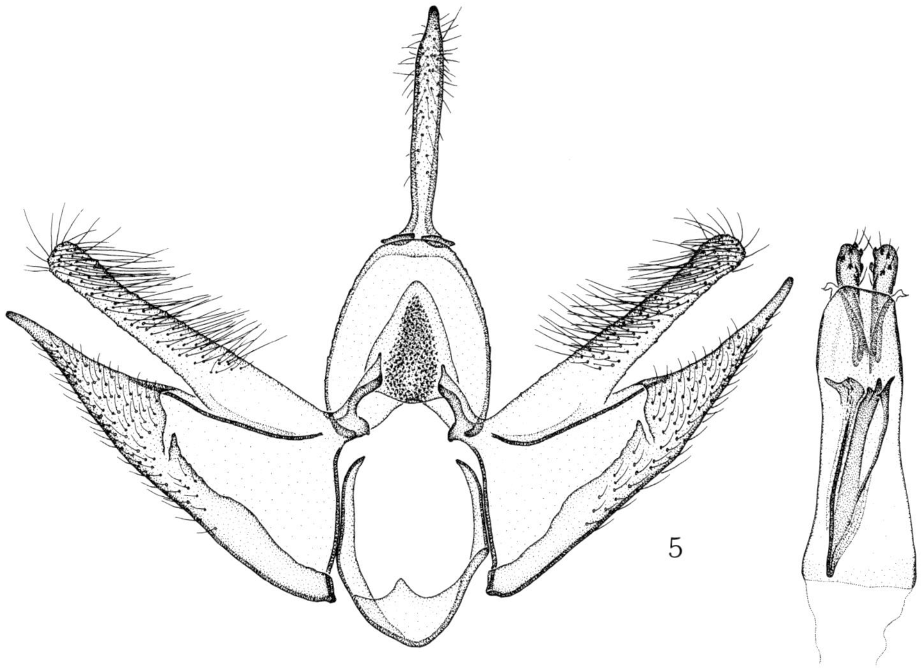

5

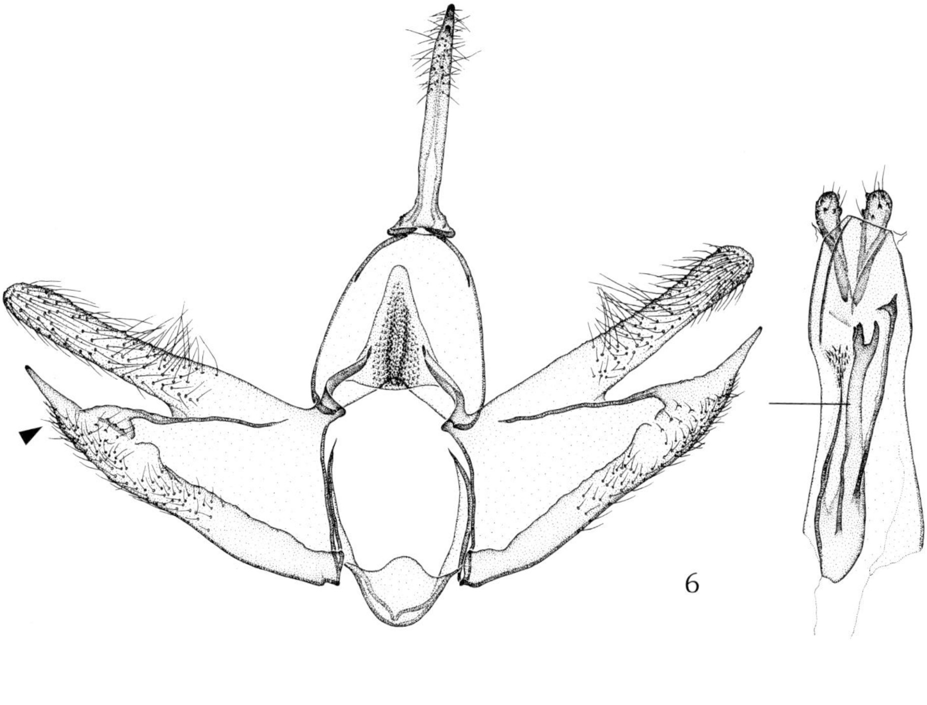

6

MALE GENITALIA

5. *Mompha divisella* Herrich-Schäffer – Denmark, gen. slide JCK 3710 (DPR).
6. *Mompha jurassicella* (Frey) – England, gen. slide JCK 2076 (KTR).

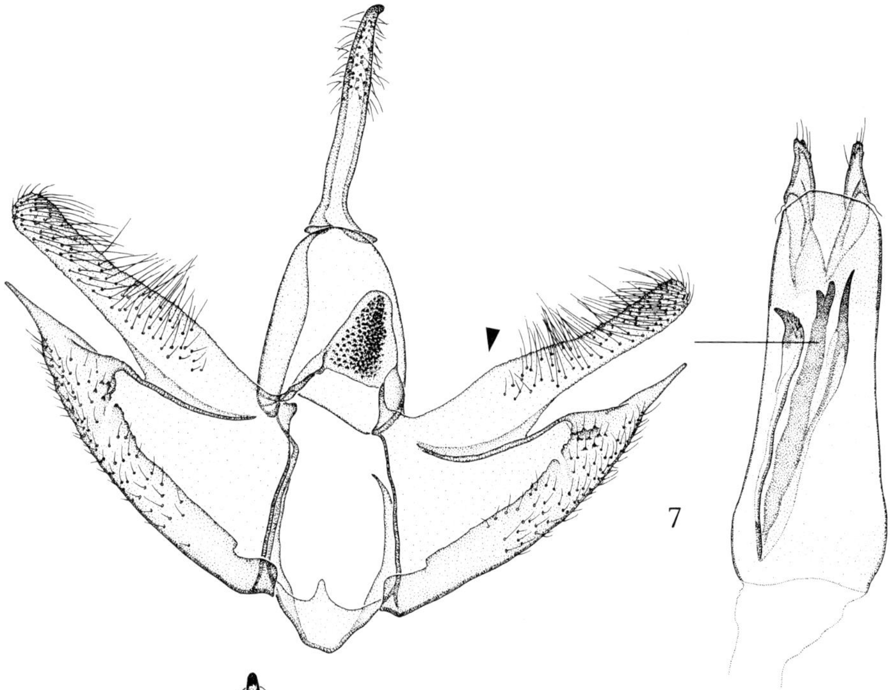

7

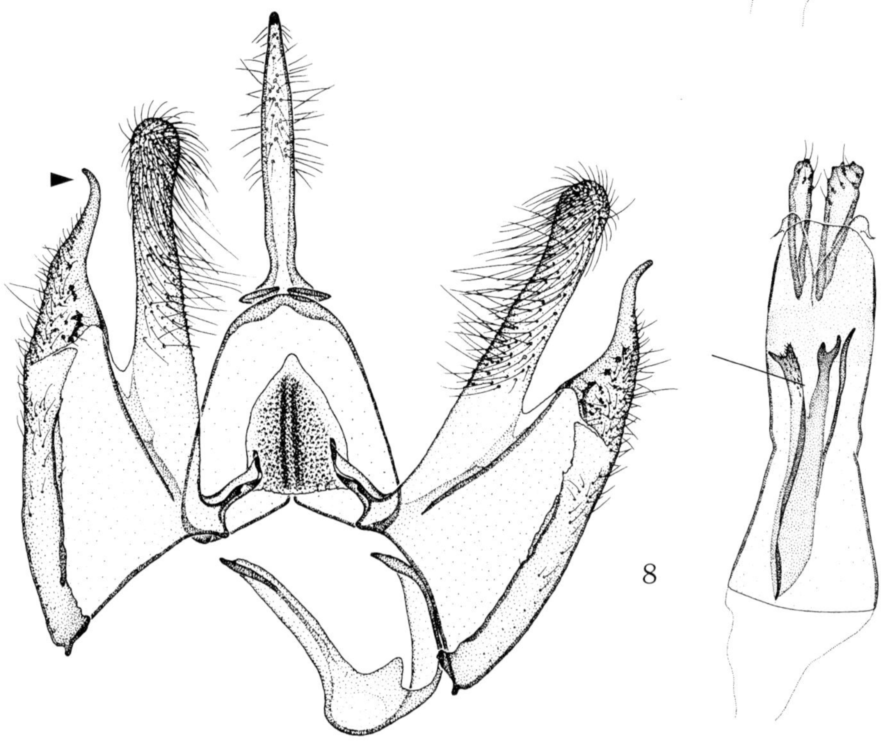

8

Male Genitalia

7. *Mompha bradleyi* Riedl – Netherlands, gen. slide JCK 2129 (KTR).
8. *Mompha confusella* Koster & Sinev – holotype, Azerbaijan, gen. slide JCK 3928 (ZIN).

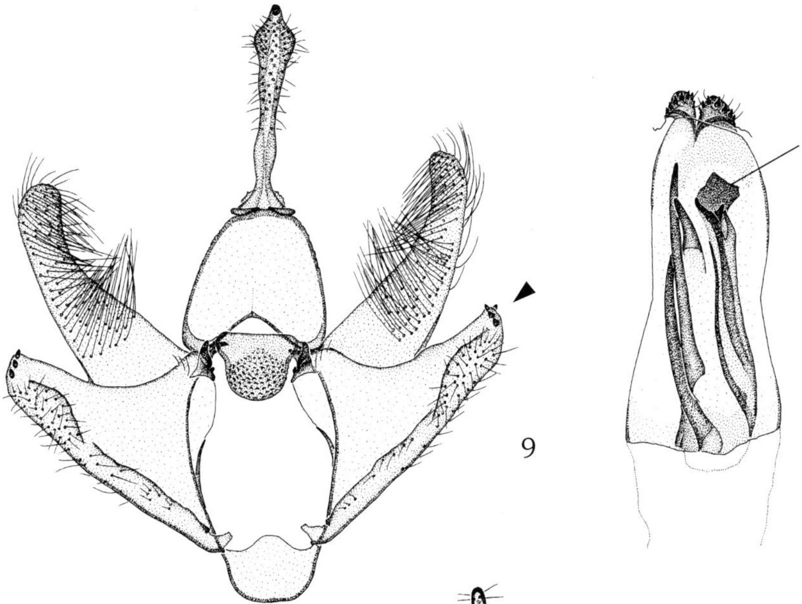

9

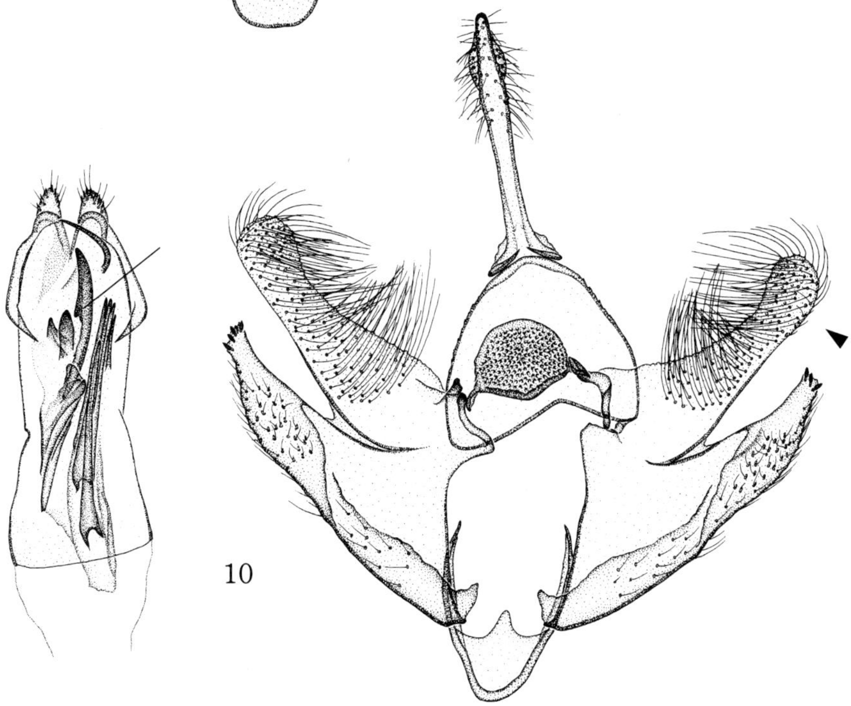

10

Male Genitalia

9. *Mompha subbistrigella* (Haworth) – Germany, gen. slide JCK 3645 (BSB).
10. *Mompha sturnipennella* (Treitschke) – Netherlands, gen. slide JCK 2319 (KTR).

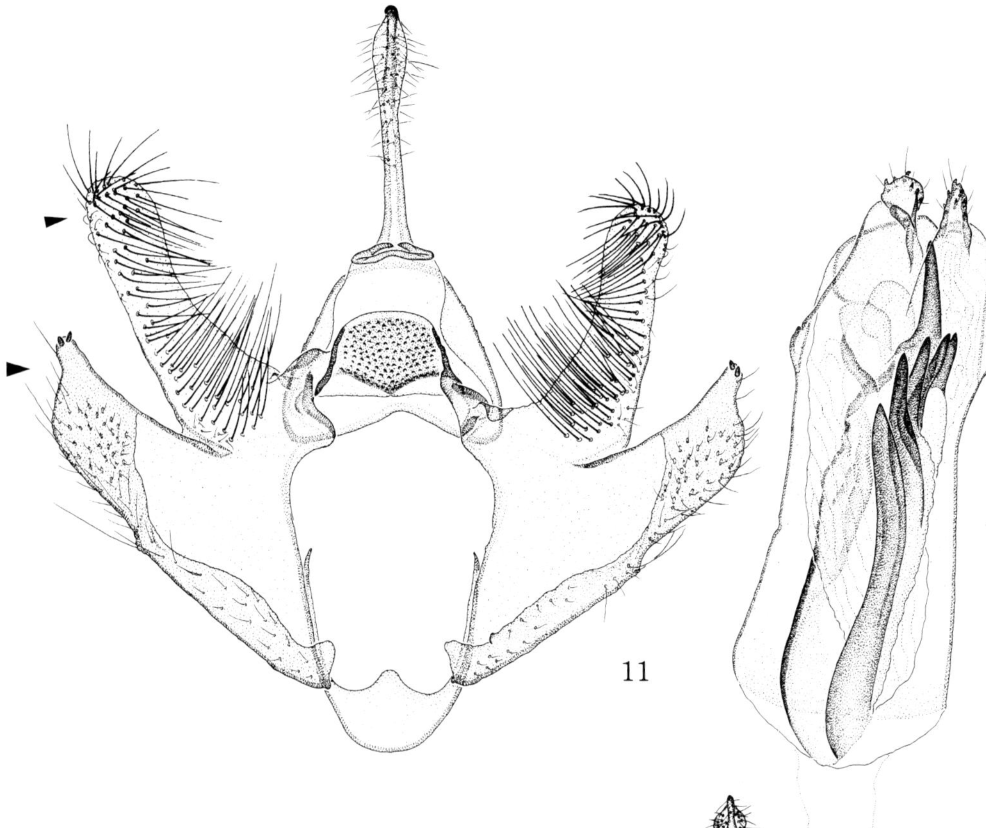

11

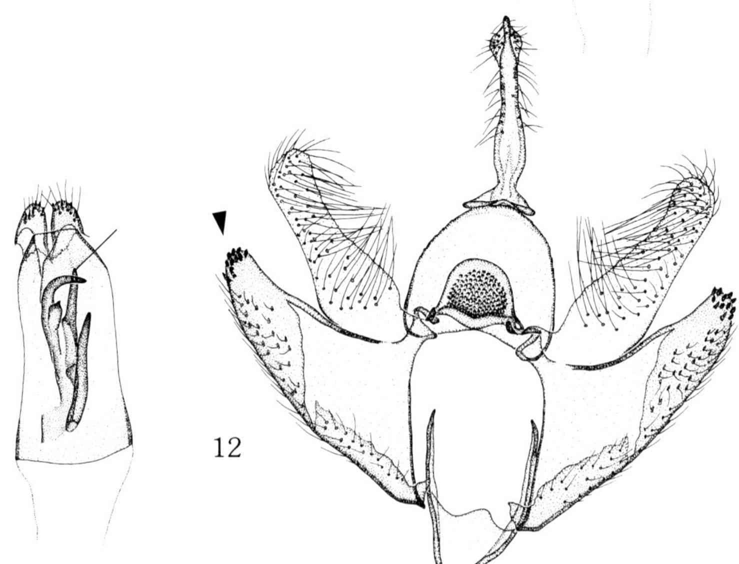

12

Male Genitalia

11. *Mompha meridionella* sp. n. – holotype, Russia, gen. in vial (ZIN).

12. *Mompha epilobiella* ([Denis & Schiffermüller]) – France, gen. slide JCK 3718 (VHN).

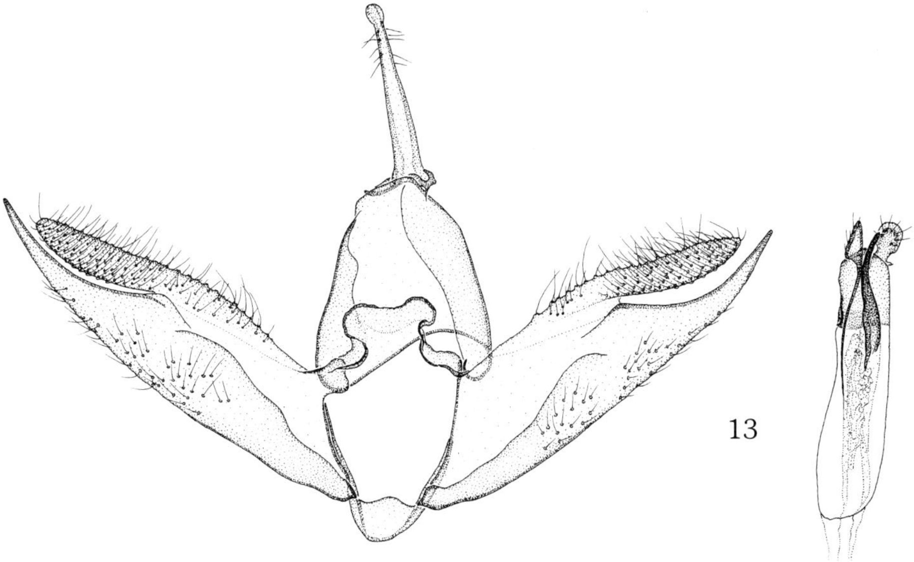

13

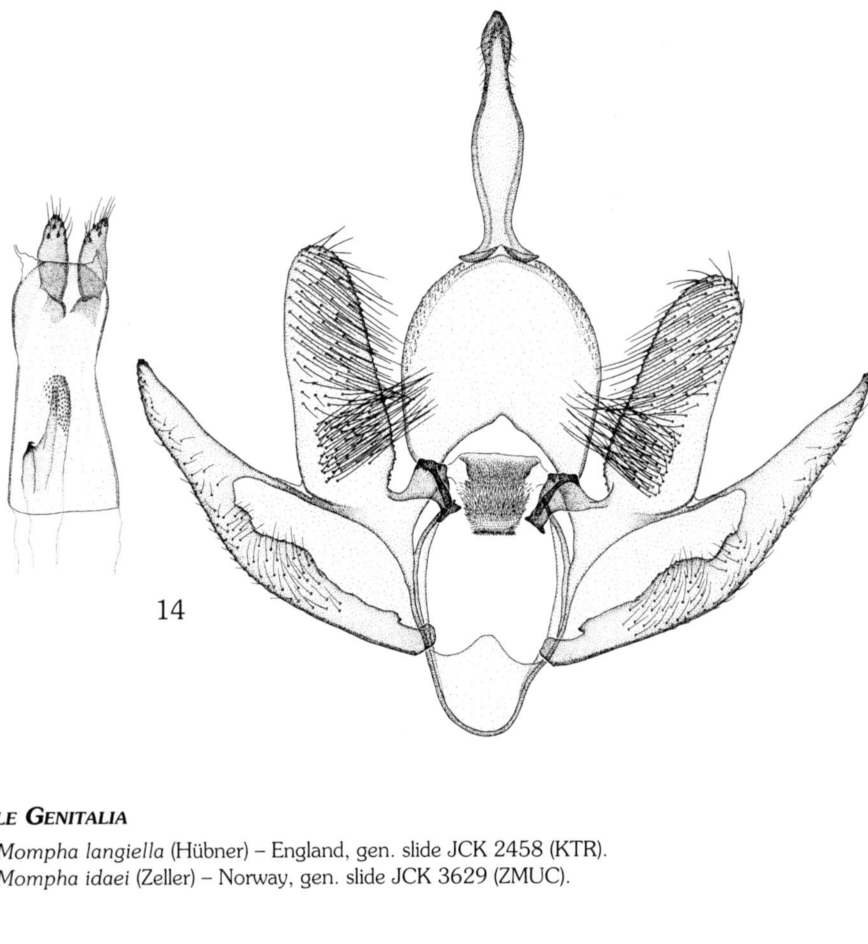

14

Male Genitalia

13. *Mompha langiella* (Hübner) – England, gen. slide JCK 2458 (KTR).
14. *Mompha idaei* (Zeller) – Norway, gen. slide JCK 3629 (ZMUC).

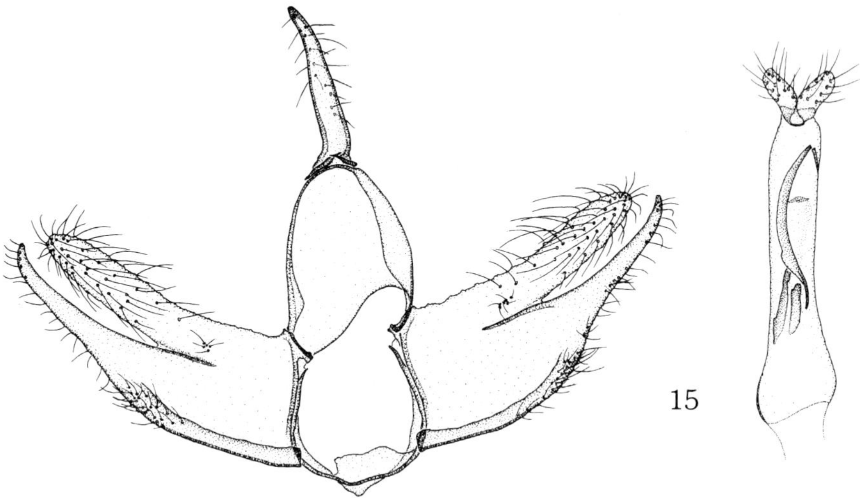

15

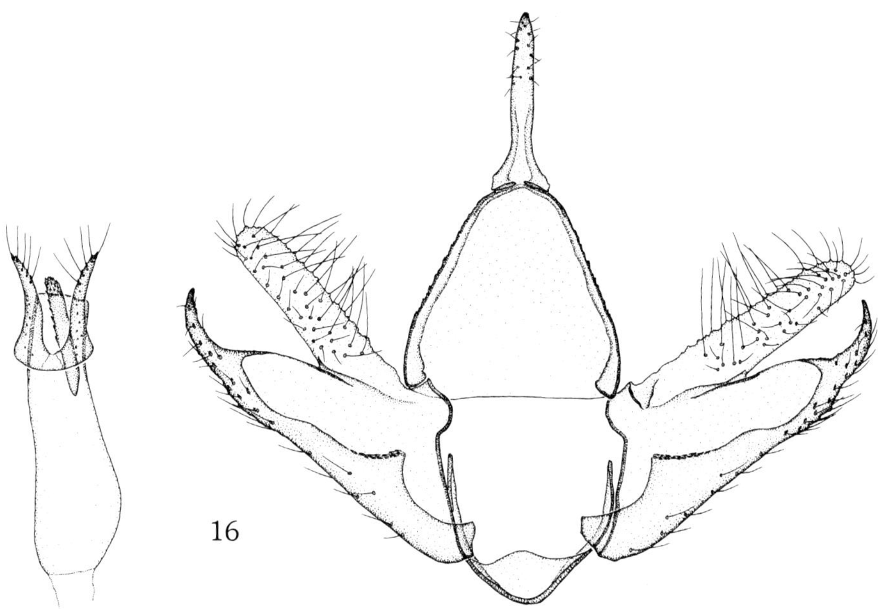

16

MALE GENITALIA

15. *Mompha miscella* ([Denis & Schiffermüller]) – Turkey, gen. slide JCK 3709 (DPR).
16. *Mompha locupletella* ([Denis & Schiffermüller]) – Netherlands, gen. slide JCK 2316 (KTR).

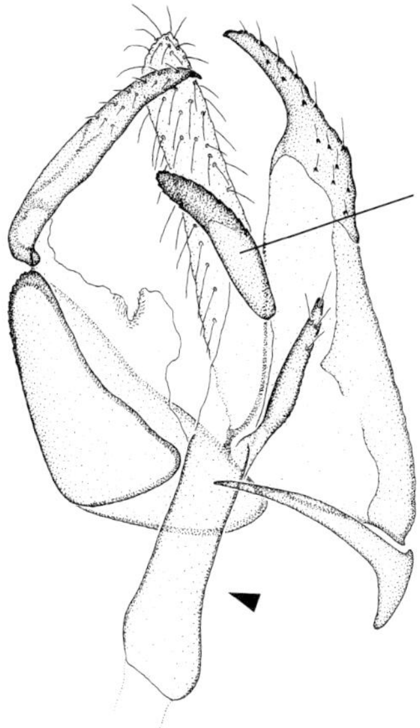

17

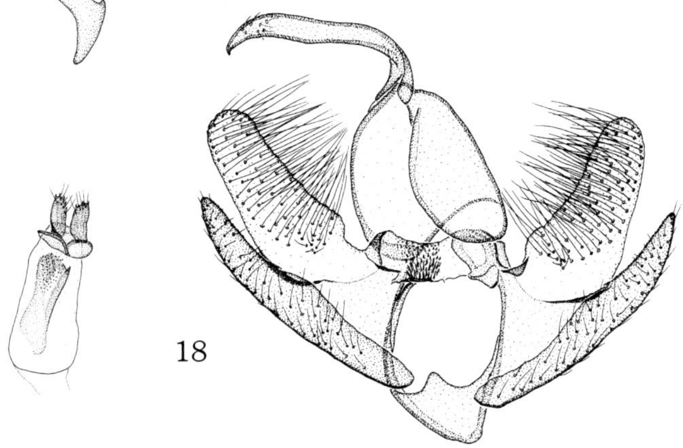

18

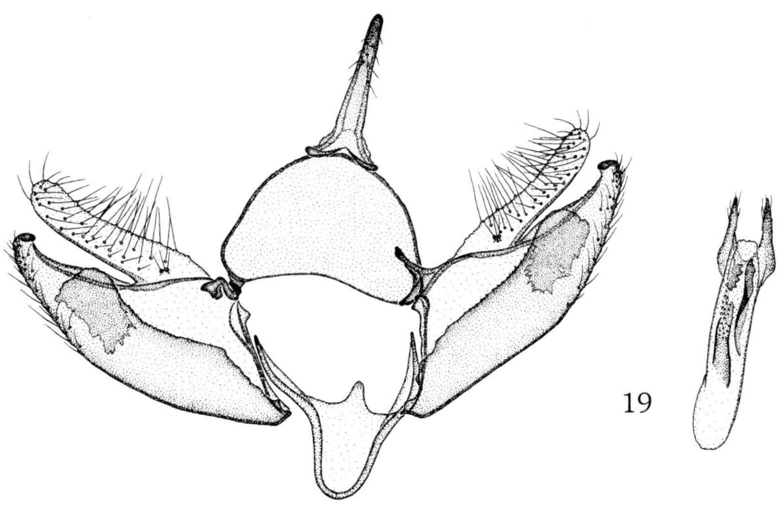

19

Male Genitalia

17. *Mompha maculata* (Sinev) – Georgia, gen. in vial (ZIN).
18. *Mompha terminella* (Humphreys & Westwood) – Netherlands, gen. slide JCK 2463 (KTR).
19. *Mompha raschkiella* (Zeller) – Netherlands, gen. slide JCK 2464 (KTR).

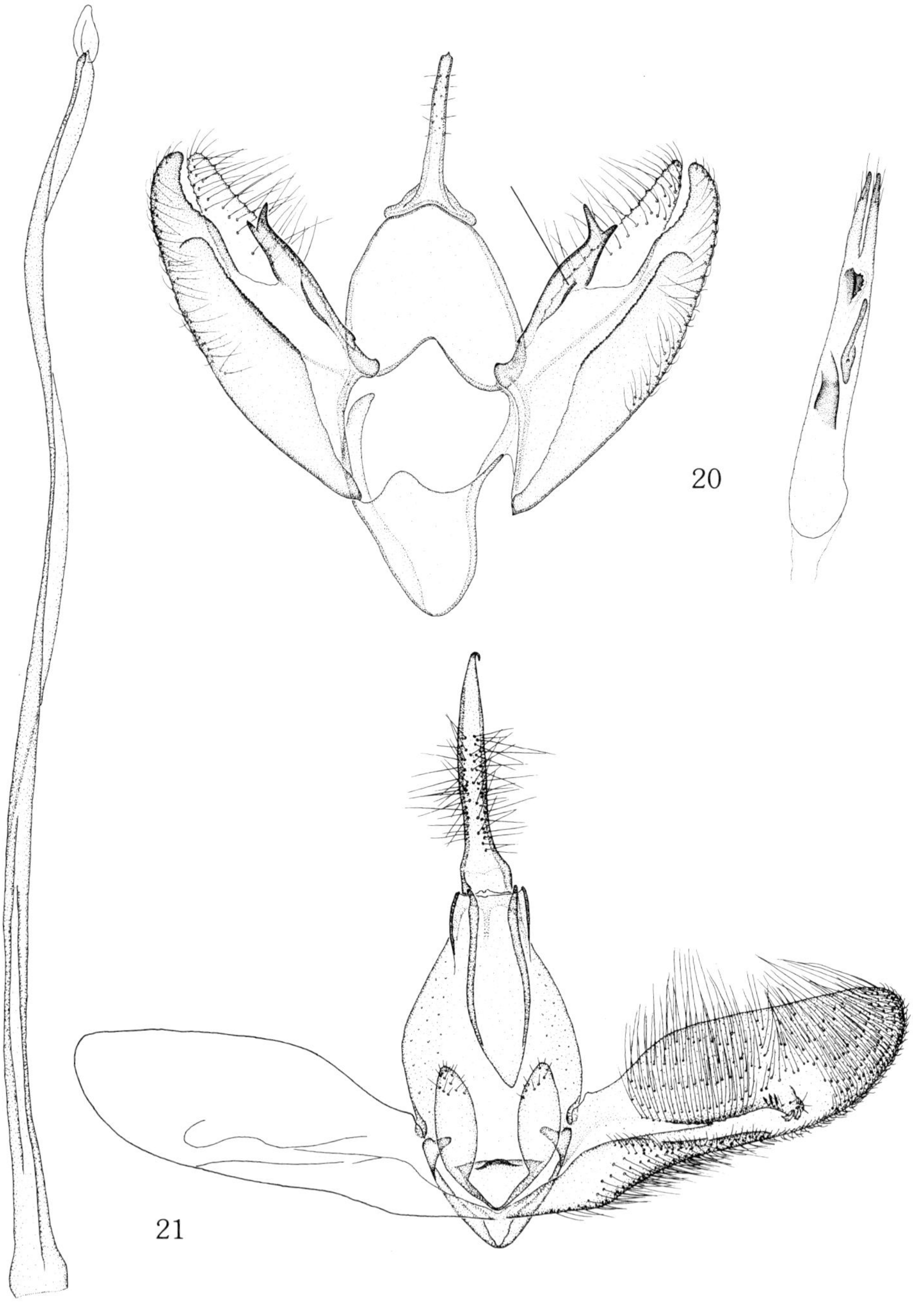

MALE GENITALIA

20. *Mompha sexstrigella* (Braun) – Finland, gen. slide JCK 2460 (KTR).
21. *Batrachedra praeangusta* (Haworth) – Germany, gen. slide JCK 3375 (LMAD).

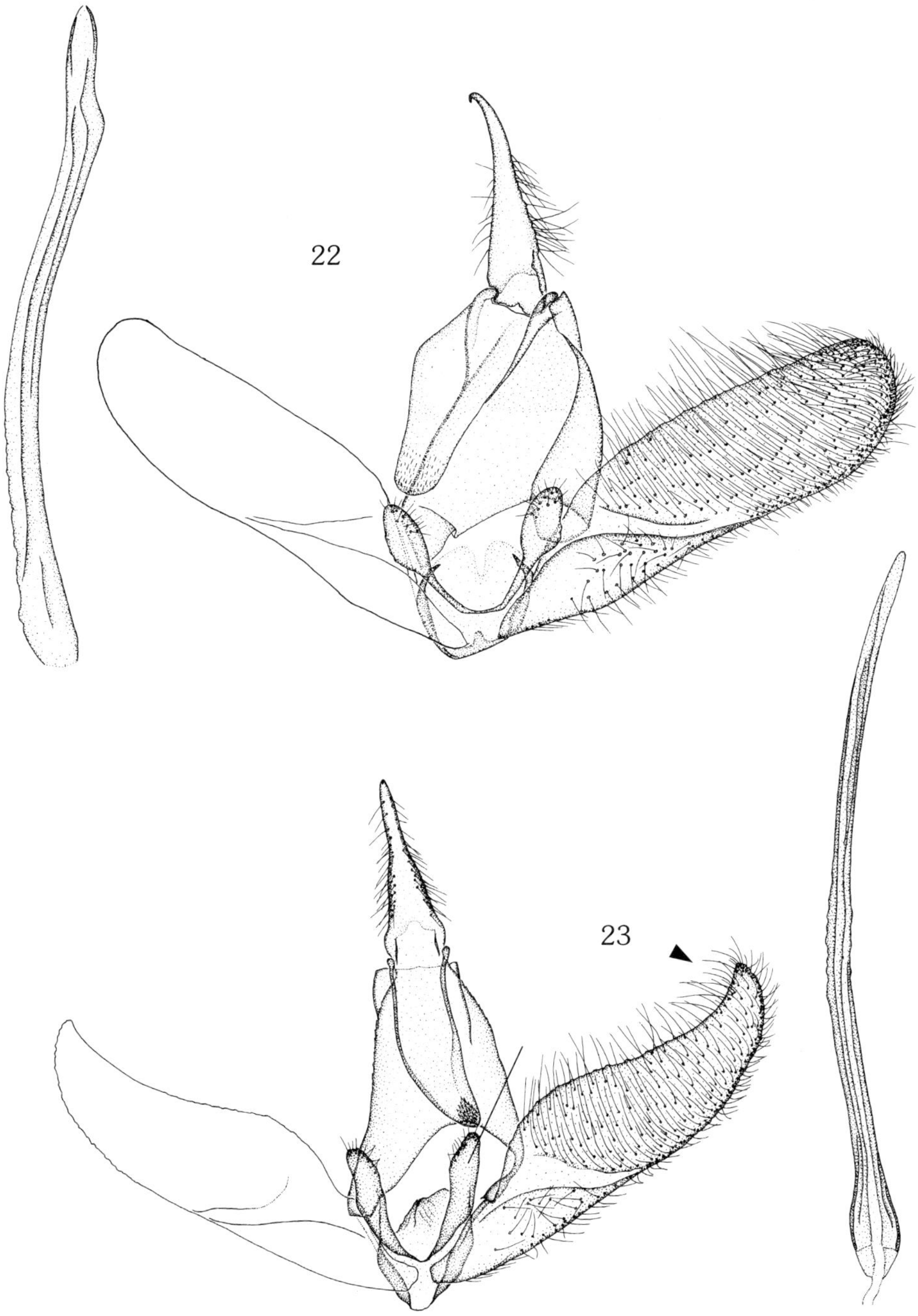

MALE GENITALIA

22. *Batrachedra pinicolella* (Zeller) – Germany, gen. slide JCK 3429 (LMAD).
23. *Batrachedra parvulipunctella* Chrétien – Spain, gen. slide JCK 3846 (GLS).

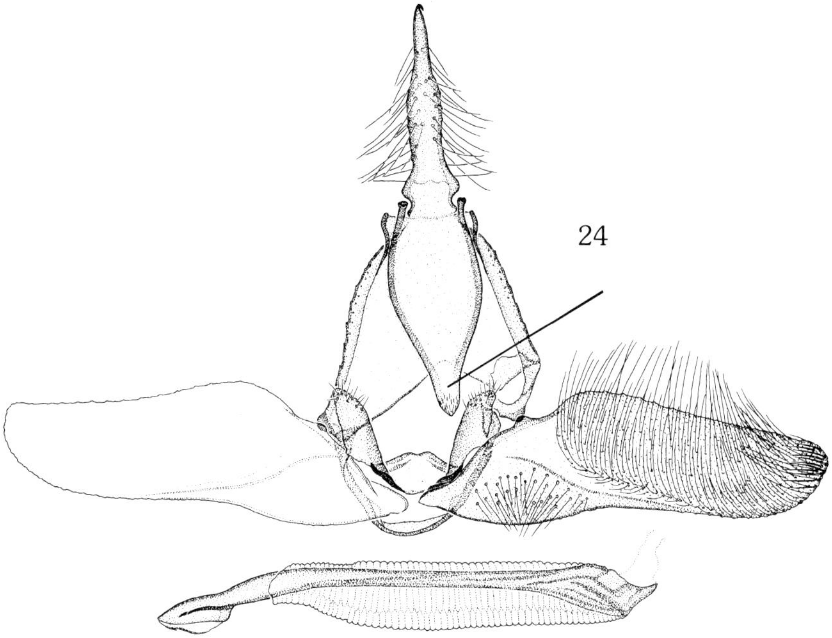

24

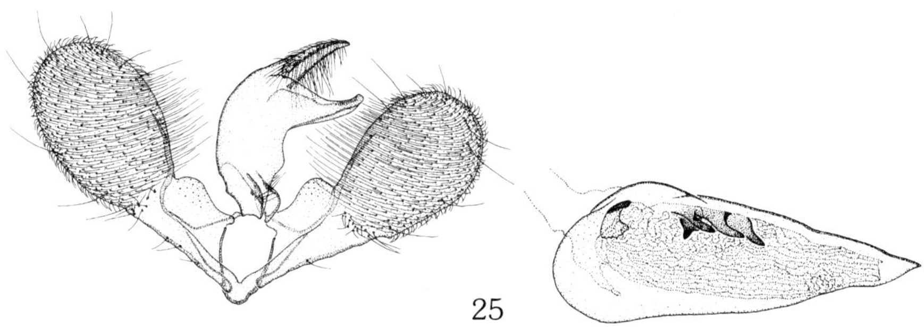

25

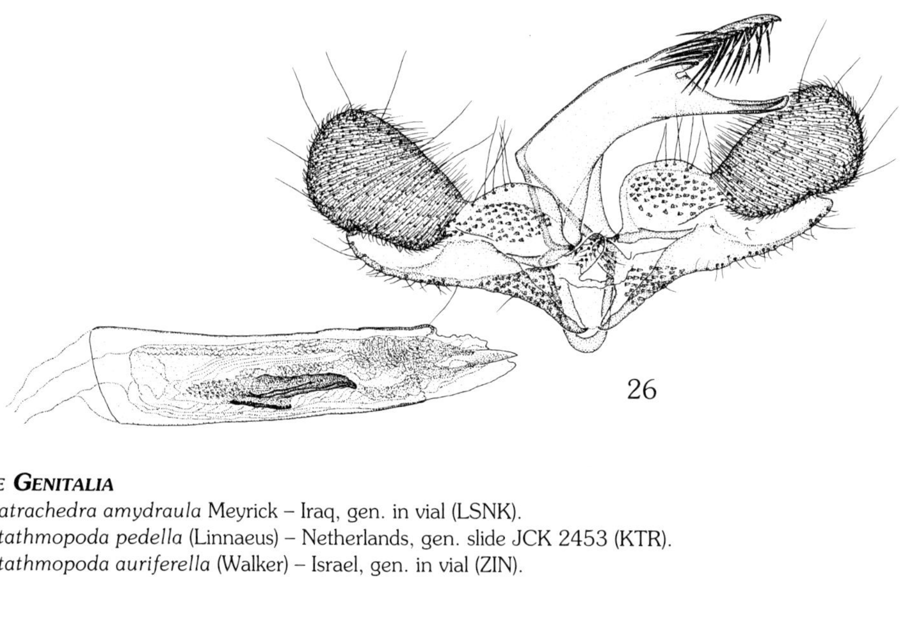

26

MALE GENITALIA

24. *Batrachedra amydraula* Meyrick – Iraq, gen. in vial (LSNK).
25. *Stathmopoda pedella* (Linnaeus) – Netherlands, gen. slide JCK 2453 (KTR).
26. *Stathmopoda auriferella* (Walker) – Israel, gen. in vial (ZIN).

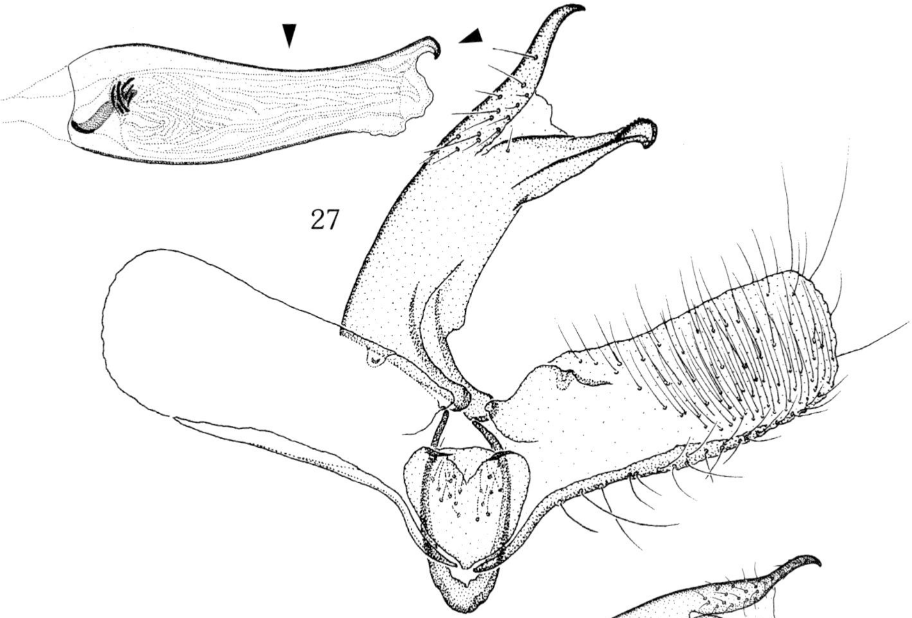

27

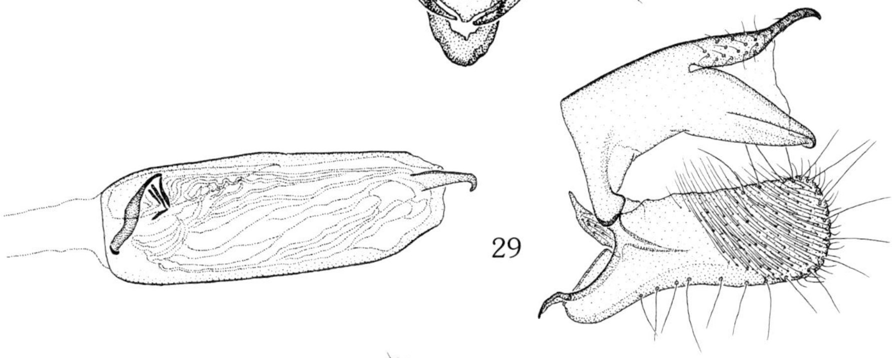

29

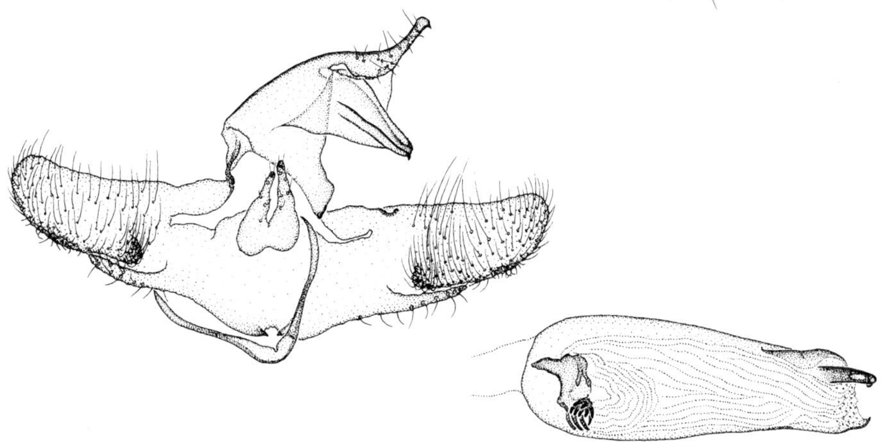

31

Male Genitalia

27. *Tortilia flavella* Chrétien – Algeria, gen. slide BMNH 4055 (BMNH).
28. *Tortilia trigonella* (Zerny) – not available.
29. *Tortilia charadritis* (Meyrick) – Egypt, gen. slide 15146 (BMNH).
31. *Tortilia pallidella* Kasy – Iran, gen. slide Mus. Vind. 3920 (NHMW).

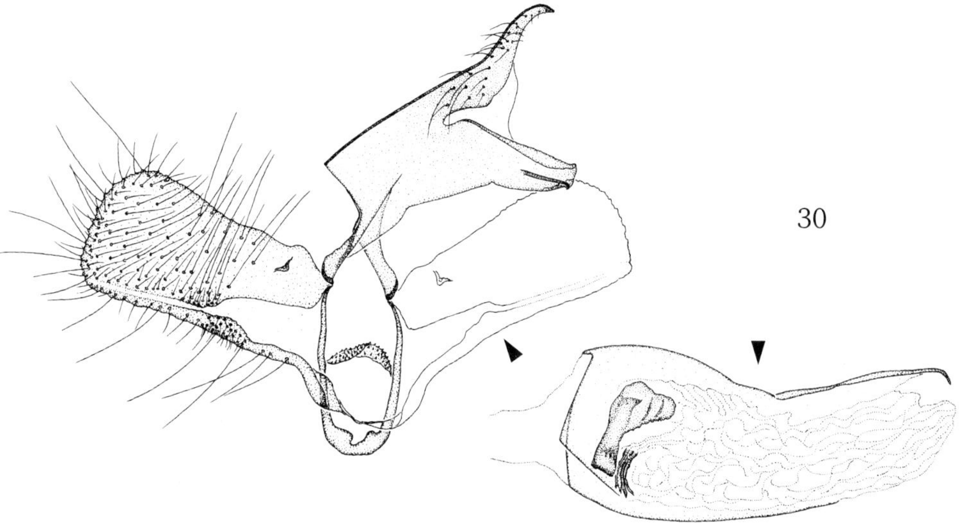

30

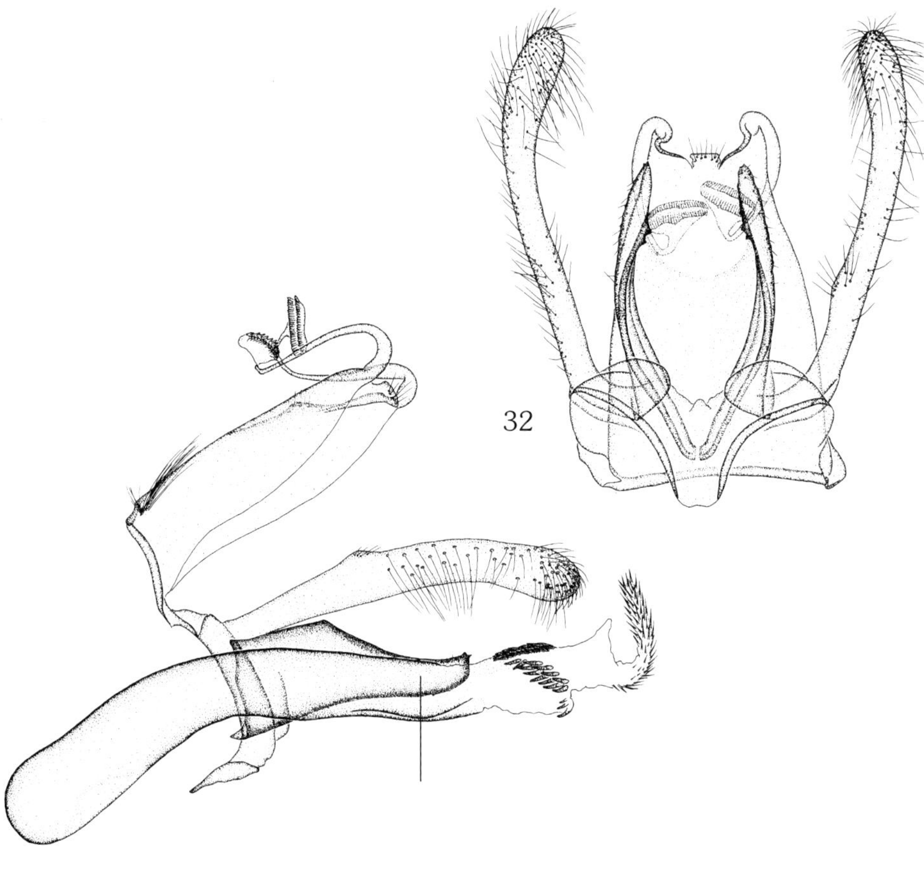

32

Male Genitalia

30. *Tortilia graeca* Kasy – Greece, gen. slide JCK 3660 (KTR).
32. *Blastodacna hellerella* (Duponchel) – Germany, gen. slide JCK 3363 (LMAD).

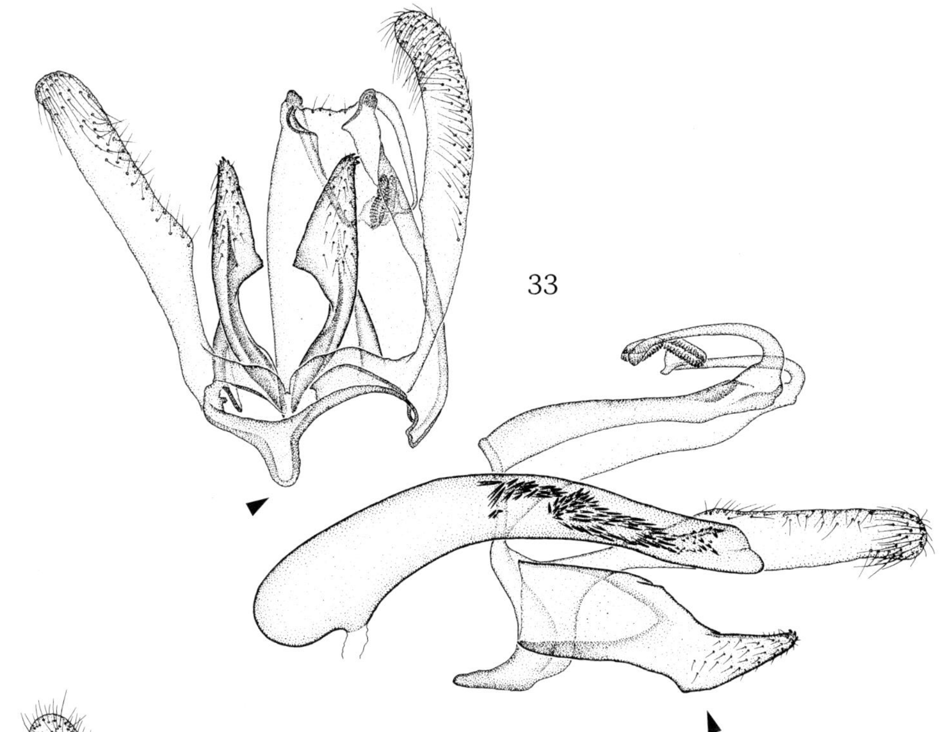

33

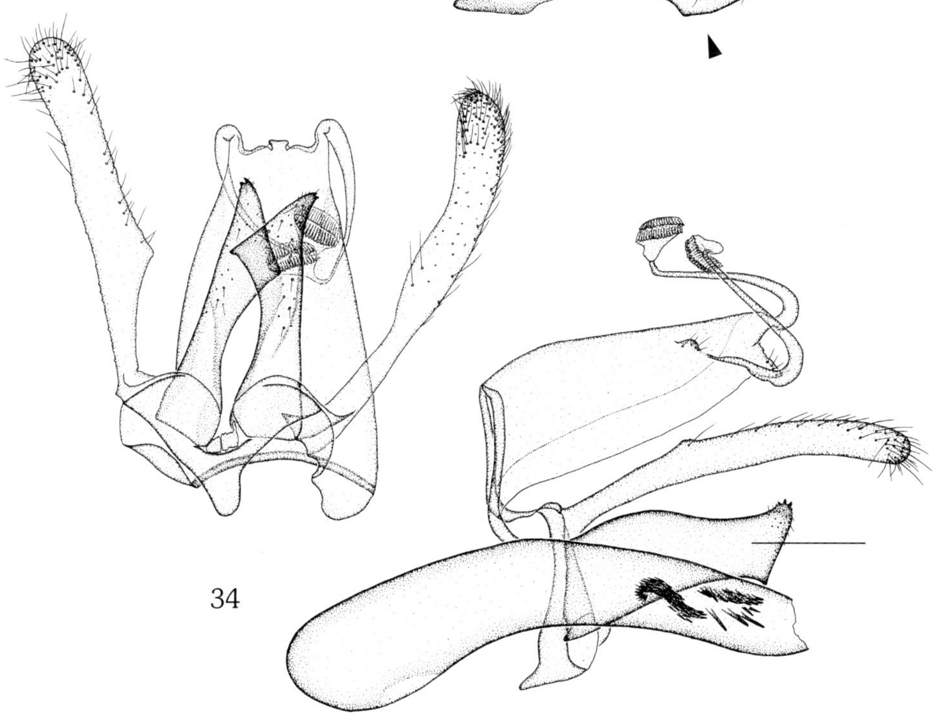

34

Male Genitalia

33. *Blastodacna rossica* Sinev – paratype, Ukraine (Crimea), gen. in vial (ZIN).
34. *Blastodacna atra* (Haworth) – Germany, gen. slide JCK 3435 (LMAD).

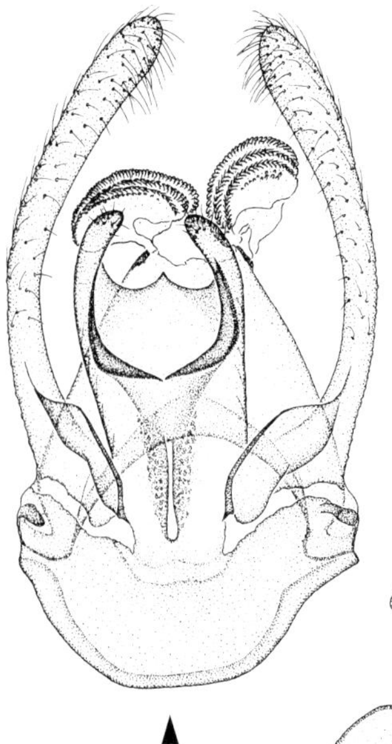

35

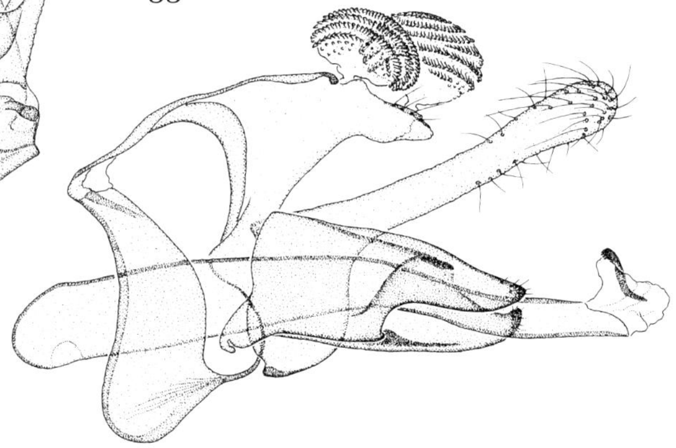

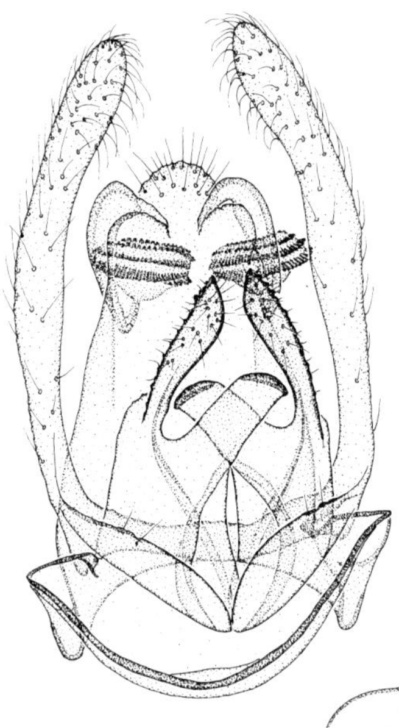

37

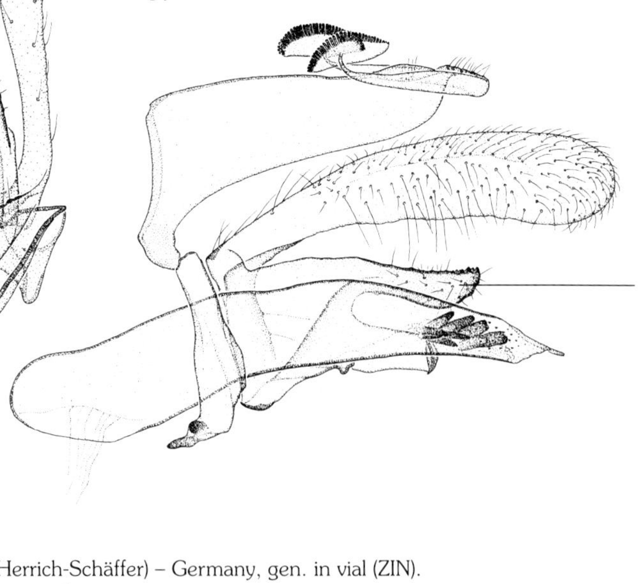

Male Genitalia

35. *Blastodacna vinolentella* (Herrich-Schäffer) – Germany, gen. in vial (ZIN).
36. *Blastodacna georgiella* Sinev – unknown.
37. *Blastodacna libanotica* Diakonoff – Turkey, gen. slide JCK 4278 (TLMF).

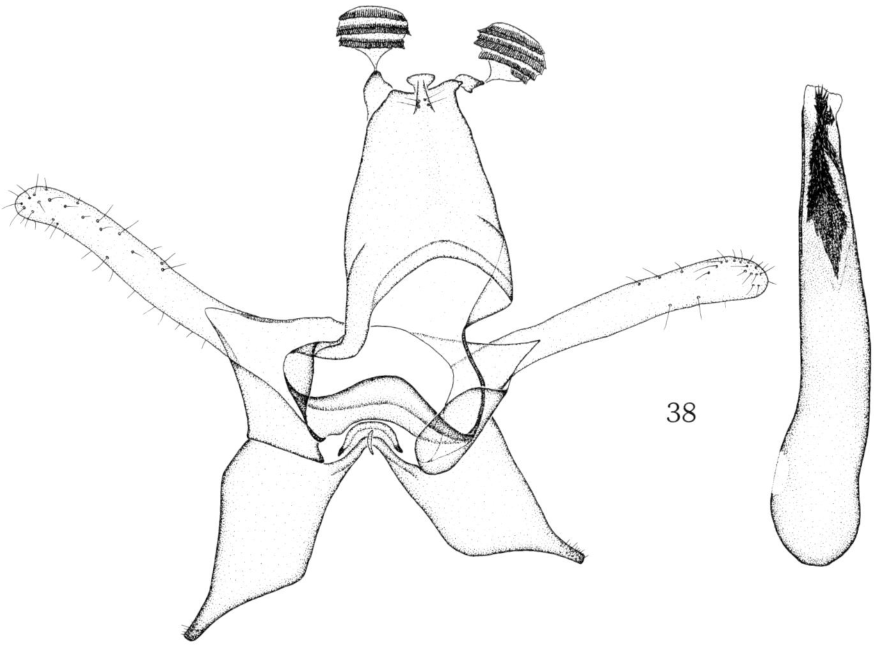

38

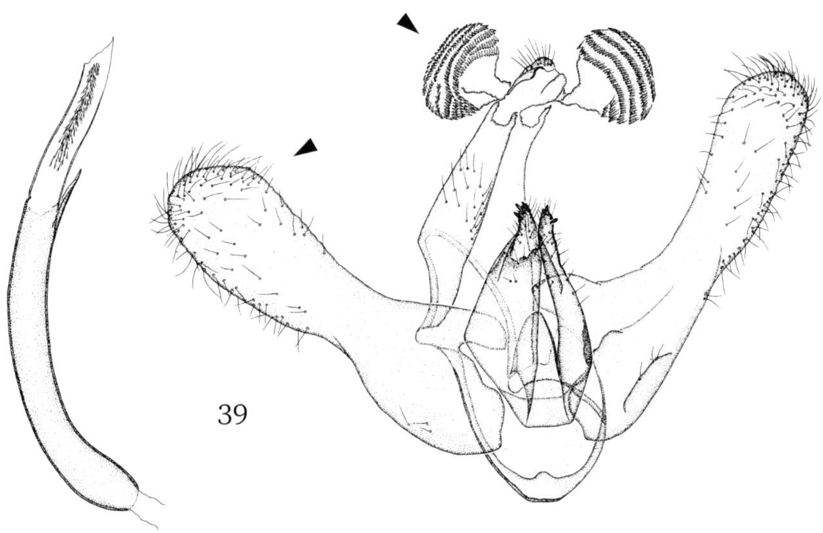

39

Male Genitalia

38. *Spuleria flavicaput* (Haworth) – Germany, gen. slide JCK 3371 (LMAD).
39. *Heinemannia laspeyrella* (Hübner) – Country unknown, gen. slide JCK 3758 (RMNH).

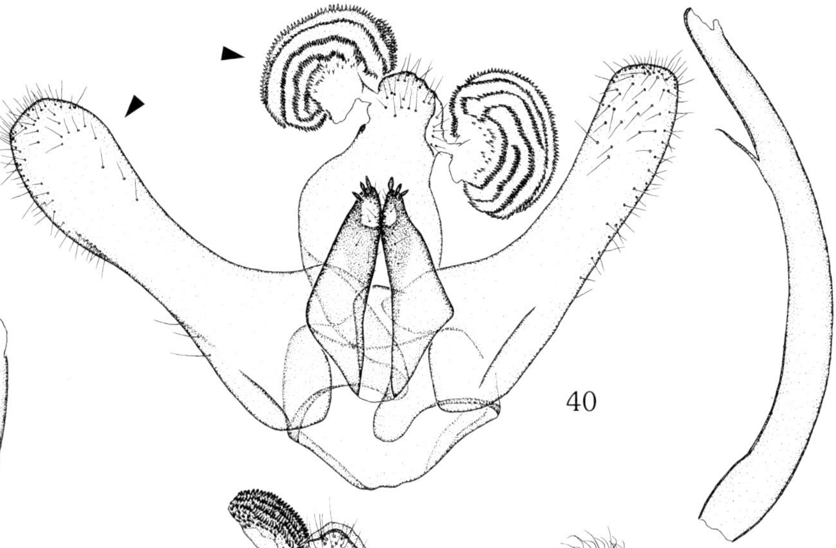

40

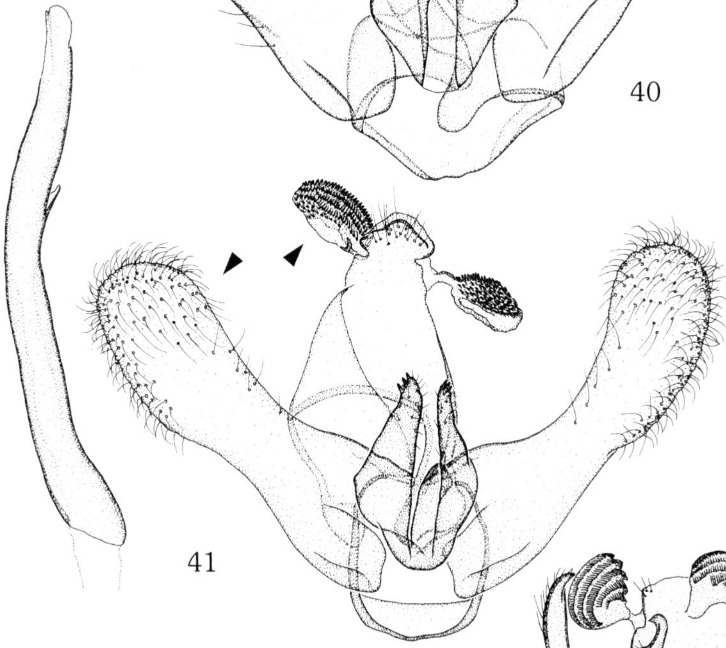

41

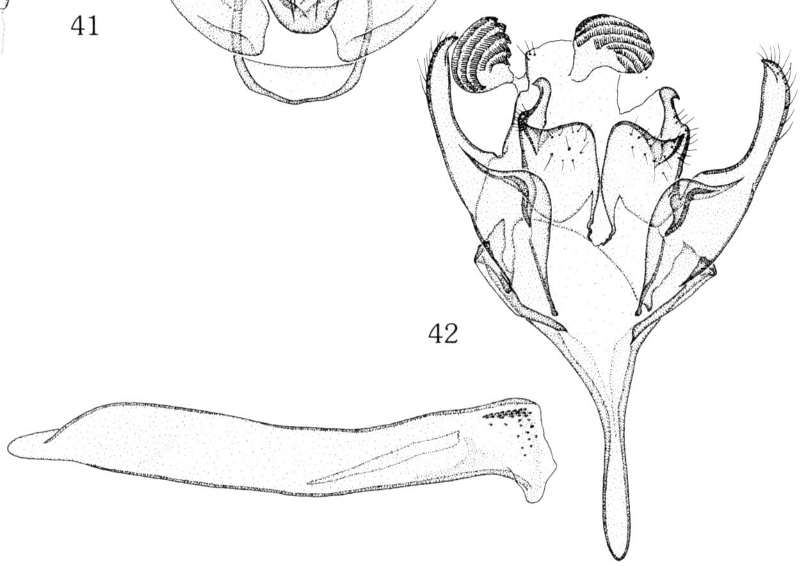

42

Male Genitalia

40. *Heinemannia festivella* ([Denis & Schiffermüller]) – Turkey, gen. slide JCK 3499 (LMAD).
41. *Heinemannia albidorsella* (Staudinger) – France, gen. slide JCK 4126 (HSN).
42. *Dystebenna stephensi* (Stainton) – Poland, gen. slide JCK 3865 (RMNH).

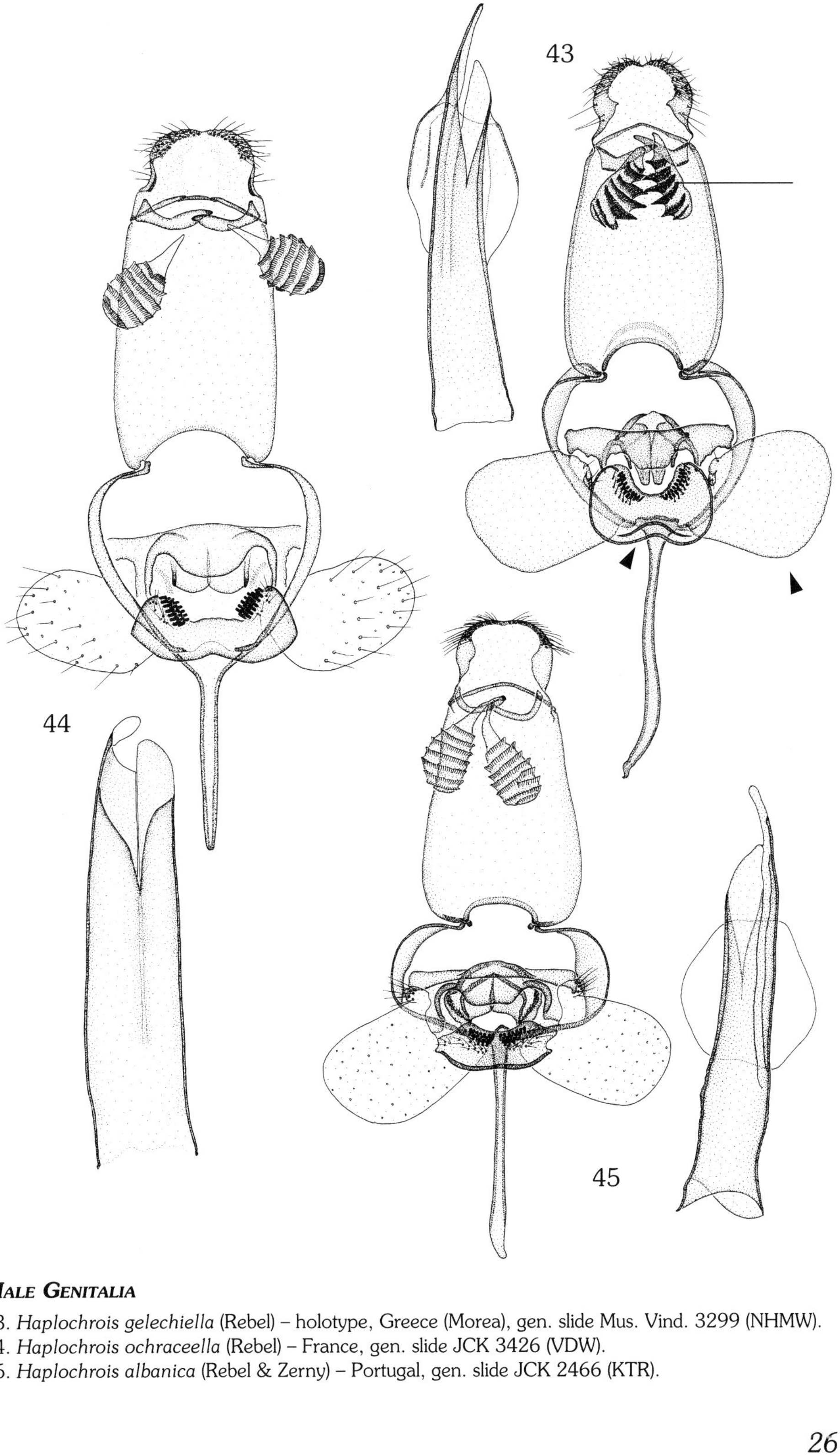

Male Genitalia

43. *Haplochrois gelechiella* (Rebel) – holotype, Greece (Morea), gen. slide Mus. Vind. 3299 (NHMW).
44. *Haplochrois ochraceella* (Rebel) – France, gen. slide JCK 3426 (VDW).
45. *Haplochrois albanica* (Rebel & Zerny) – Portugal, gen. slide JCK 2466 (KTR).

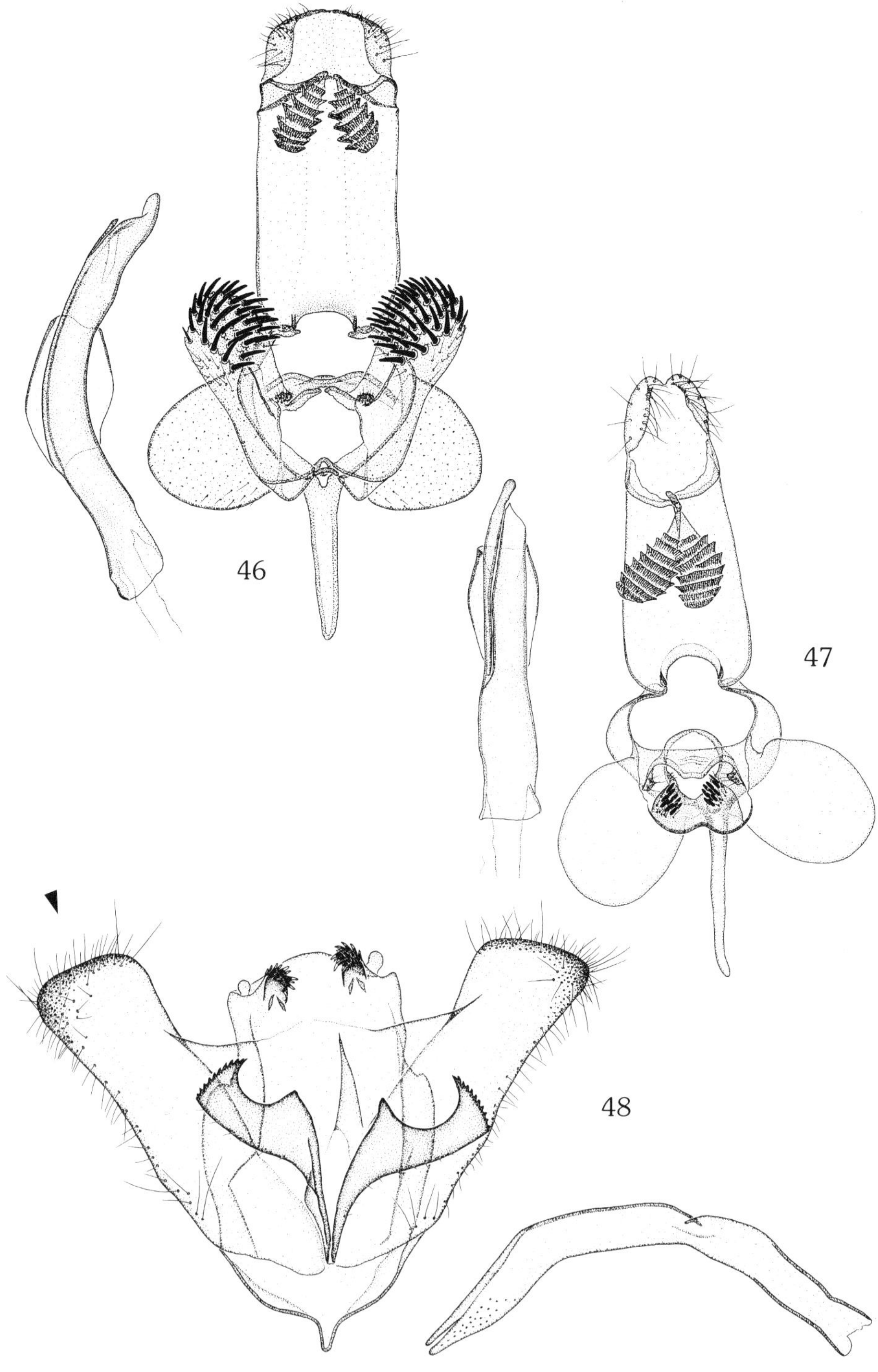

Male Genitalia

46. *Haplochrois buvati* (Baldizzone) – Spain, gen. slide WF 6860 (VDW).
47. *Haplochrois theae* (Kusnezov) – Georgia, gen. in vial (ZIN).
48. *Chrysoclista linneella* (Clerck) – Germany, gen. slide JCK 2097 (KTR).

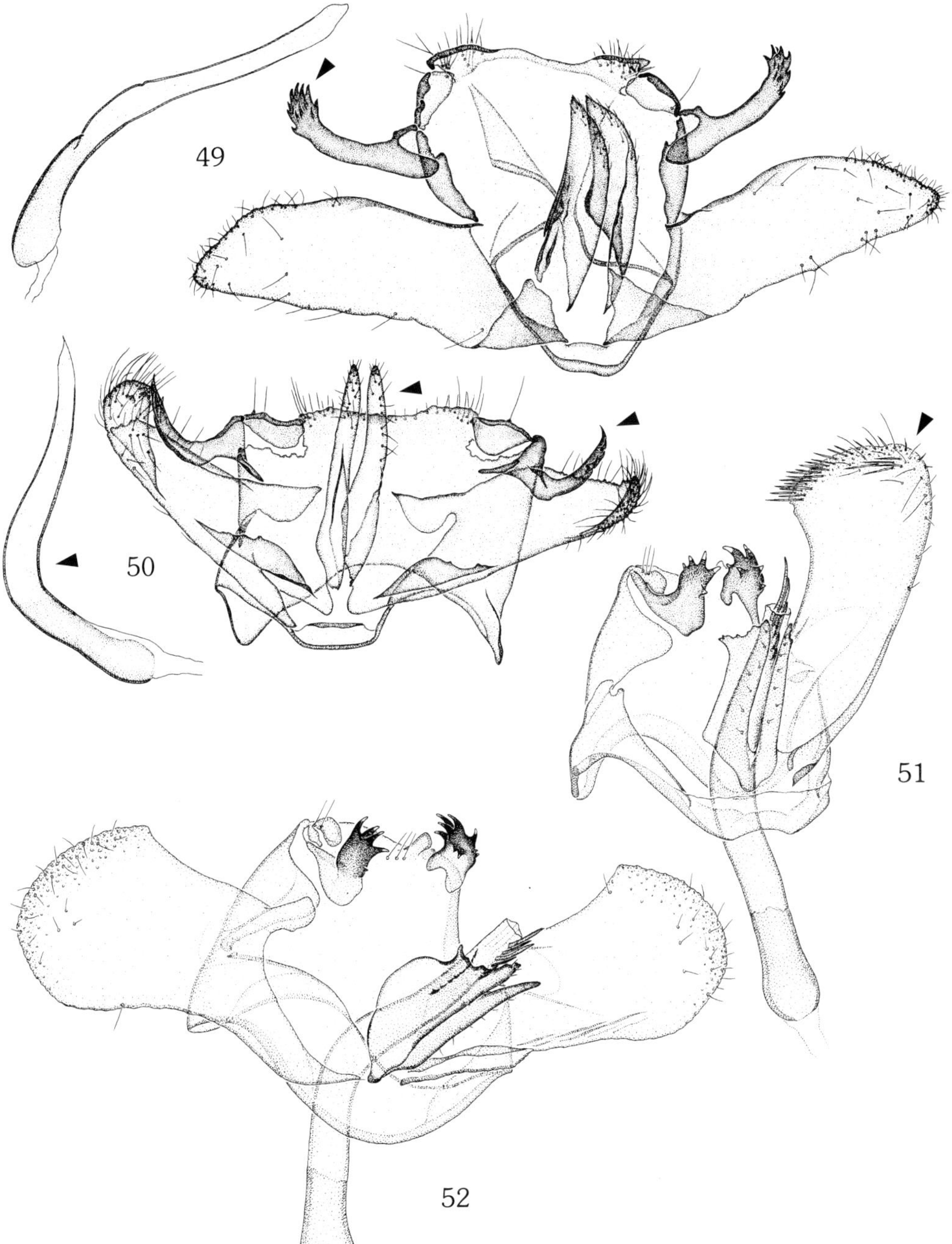

Male Genitalia

49. *Chrysoclista lathamella* (Fletcher) – Netherlands, gen. slide JCK 3755 (WLN).
50. *Chrysoclista splendida* Karsholt – Austria, gen. slide JCK 3756 (RMNH).
51. *Chrysoclista zagulajevi* Sinev – holotype, Georgia (Adzharia), gen. in vial (ZIN).
52. *Chrysoclista abchasica* (Sinev) – holotype, Georgia (Abchasia), gen. in vial (ZIN).

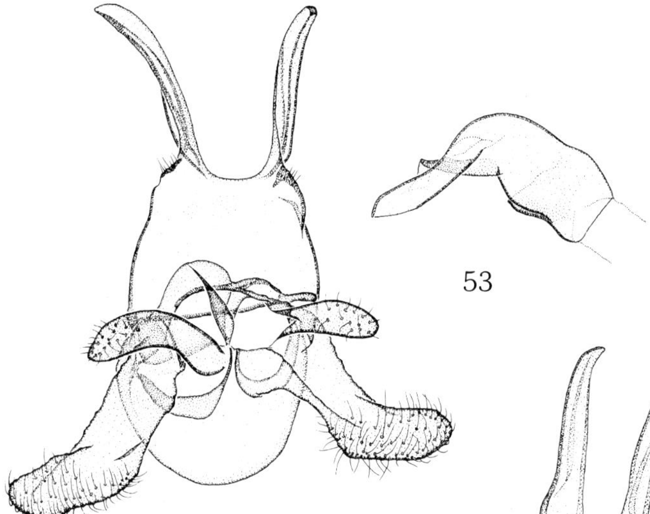

53

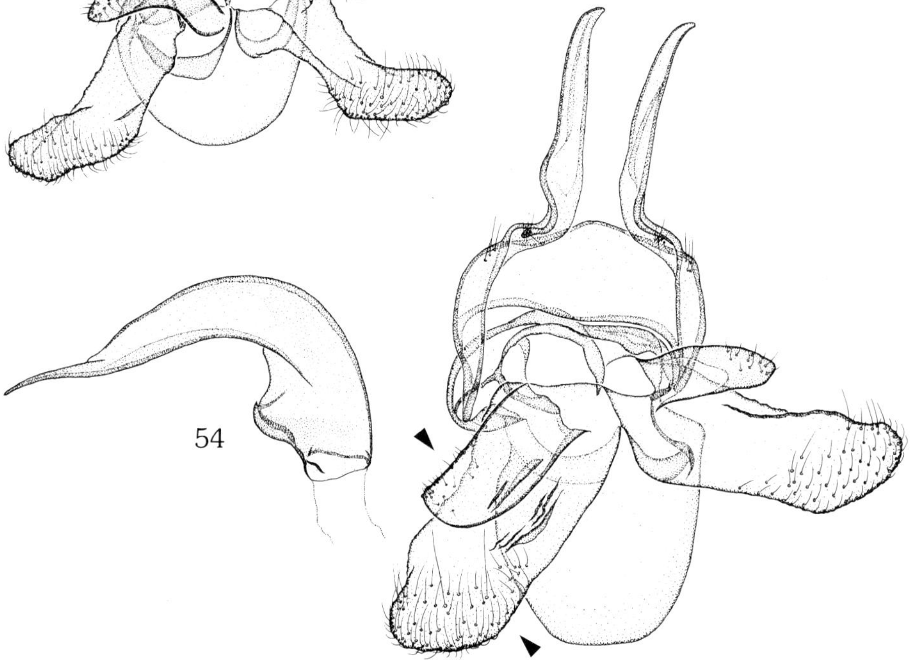

54

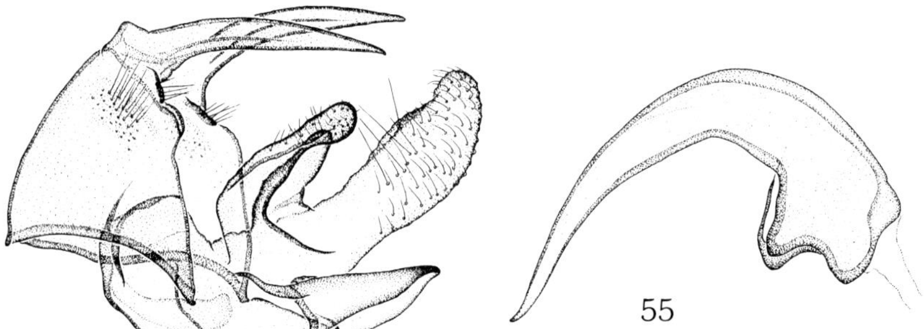

55

Male Genitalia

53. *Pancalia leuwenhoekella* (Linnaeus) – Italy, gen. slide JCK 3072 (GLS).
54. *Pancalia schwarzella* (Fabricius) – Netherlands, gen. slide JCK 2505 (KTR).
55. *Pancalia baldizzonella* Riedl – holotype, Italy, gen. slide 1415/R (BDZ).

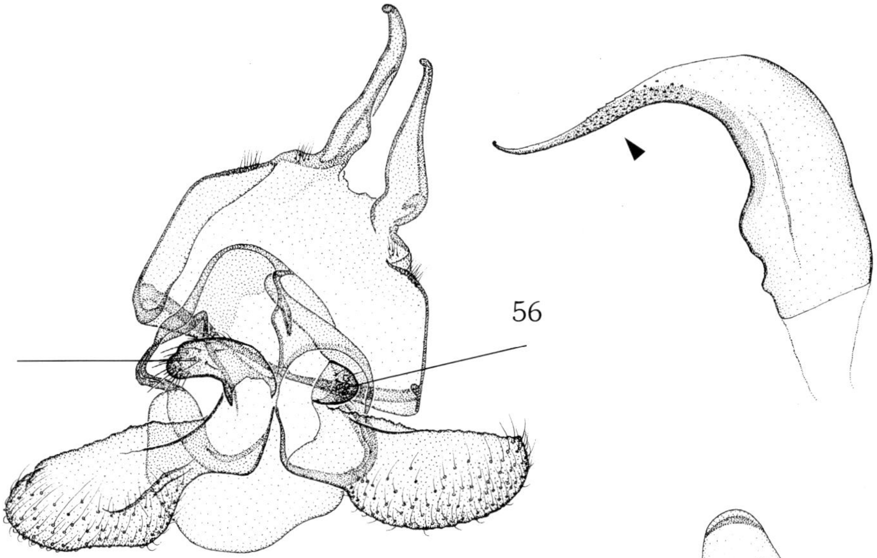

56

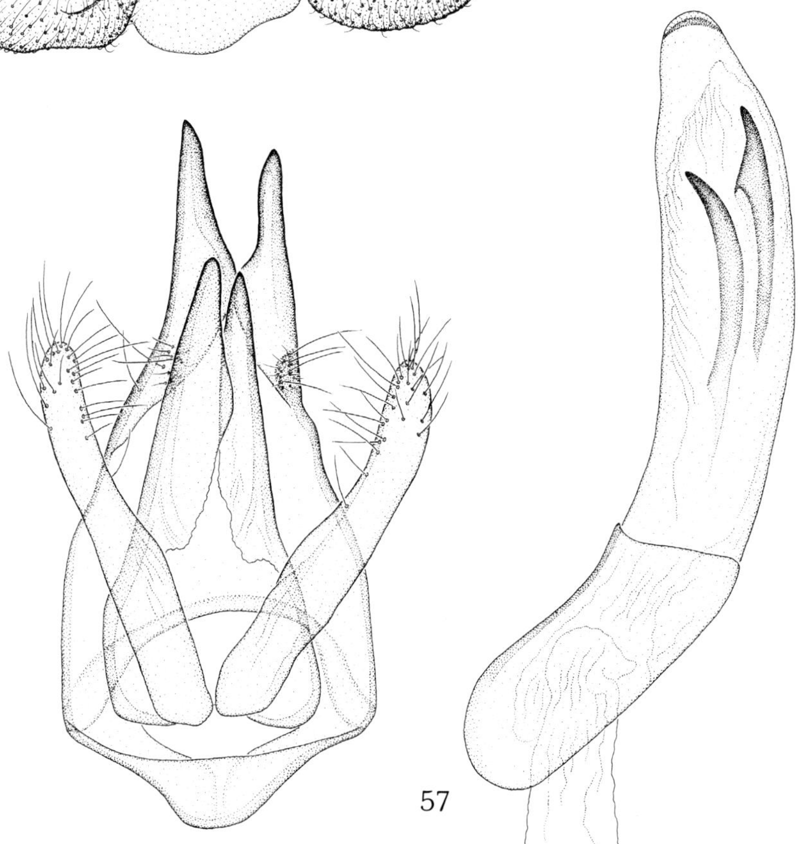

57

MALE GENITALIA

56. *Pancalia nodosella* (Bruand) – Netherlands, gen. slide JCK 1689 (KTR).
57. *Euclemensia woodiella* (Curtis) – holotype, England, gen. Slide JFL 1478 (MMA).

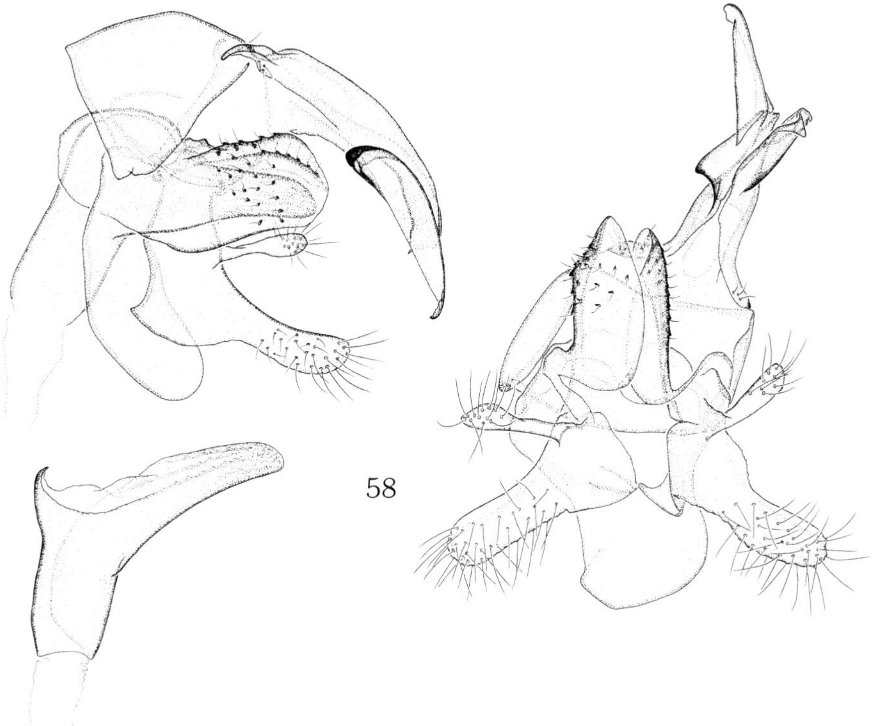

58

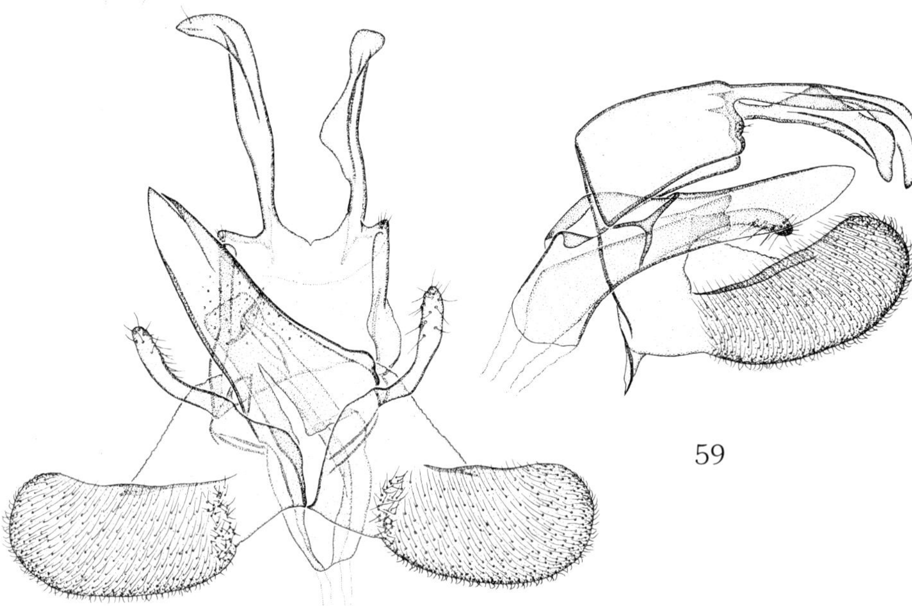

59

MALE GENITALIA

58. *Phosphaticola gemmatella* Viette – Tunisia, gen. in vial (MNHN).
59. *Alloclita recisella* Staudinger – Spain, gen. slide JCK 5256 (WLN).

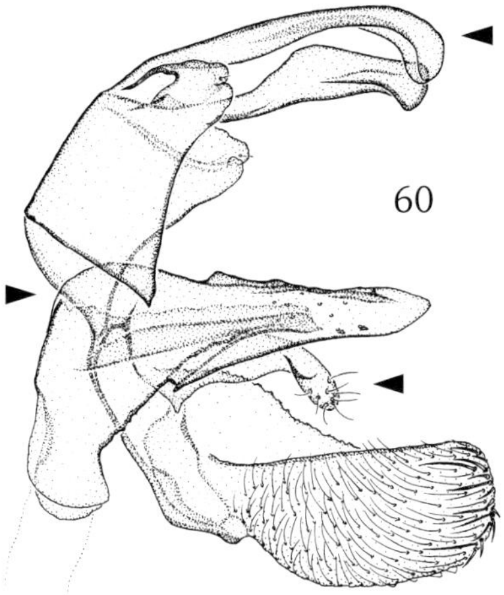

60

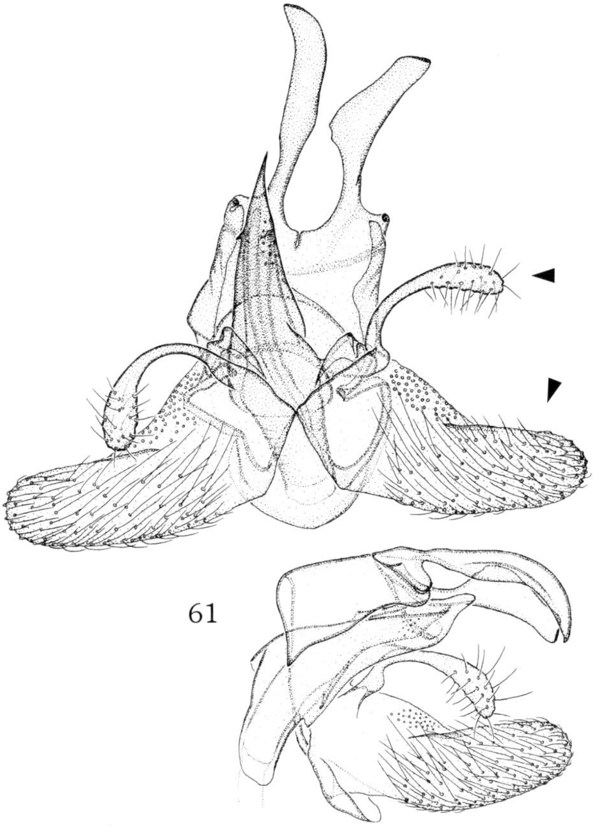

61

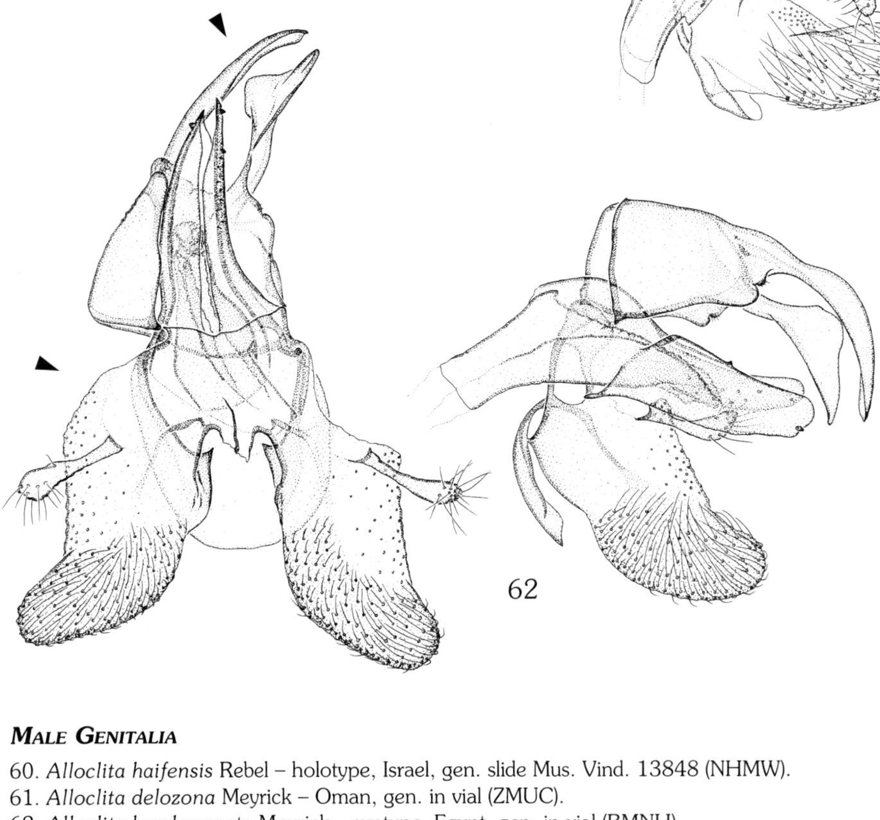

62

Male Genitalia

60. *Alloclita haifensis* Rebel – holotype, Israel, gen. slide Mus. Vind. 13848 (NHMW).
61. *Alloclita delozona* Meyrick – Oman, gen. in vial (ZMUC).
62. *Alloclita brachygrapta* Meyrick – syntype, Egypt, gen. in vial (BMNH).

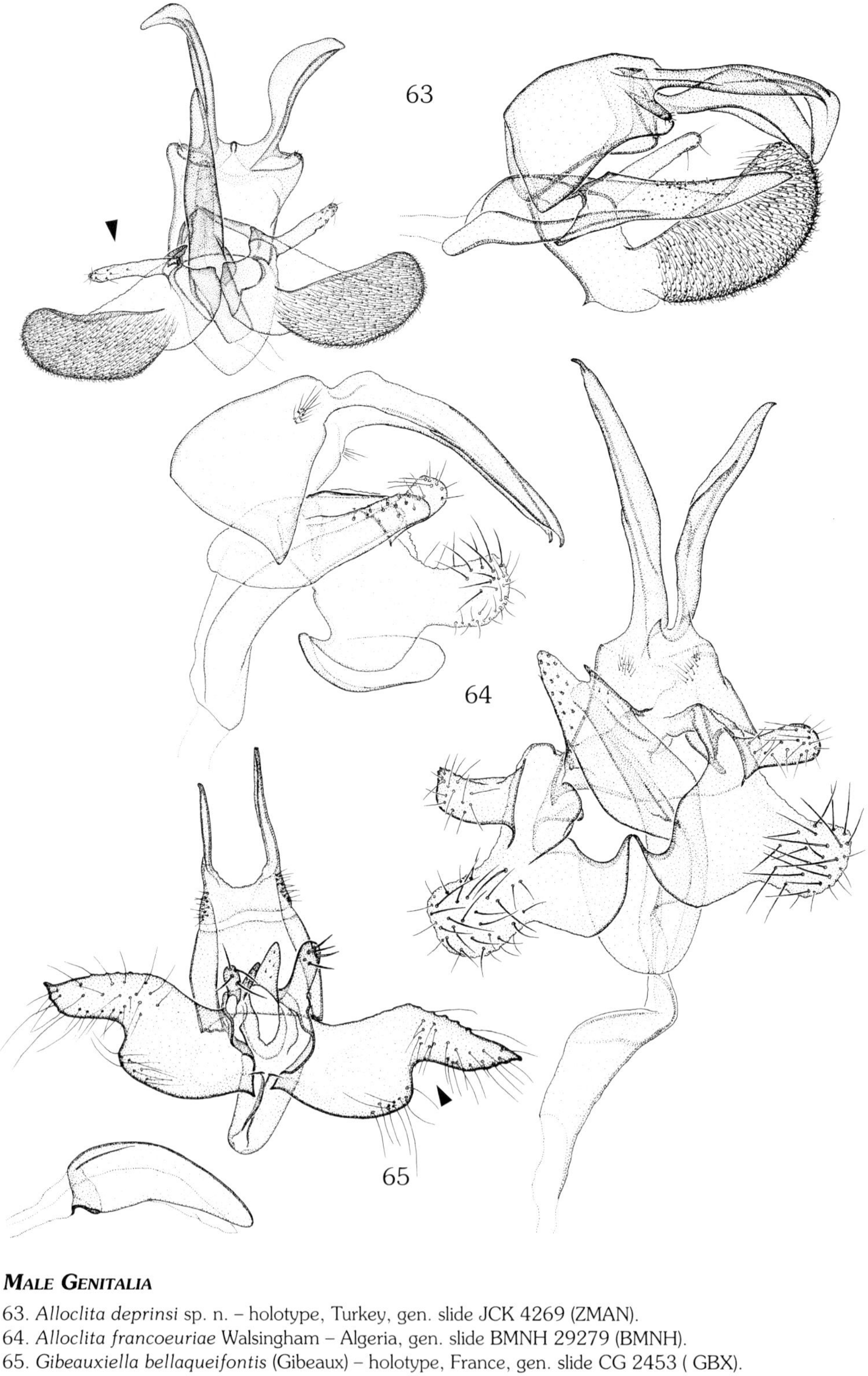

Male Genitalia

63. *Alloclita deprinsi* sp. n. – holotype, Turkey, gen. slide JCK 4269 (ZMAN).
64. *Alloclita francoeuriae* Walsingham – Algeria, gen. slide BMNH 29279 (BMNH).
65. *Gibeauxiella bellaqueifontis* (Gibeaux) – holotype, France, gen. slide CG 2453 (GBX).

67

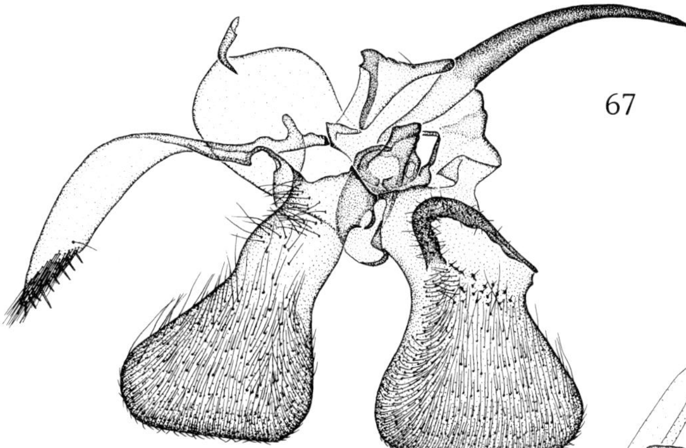

68

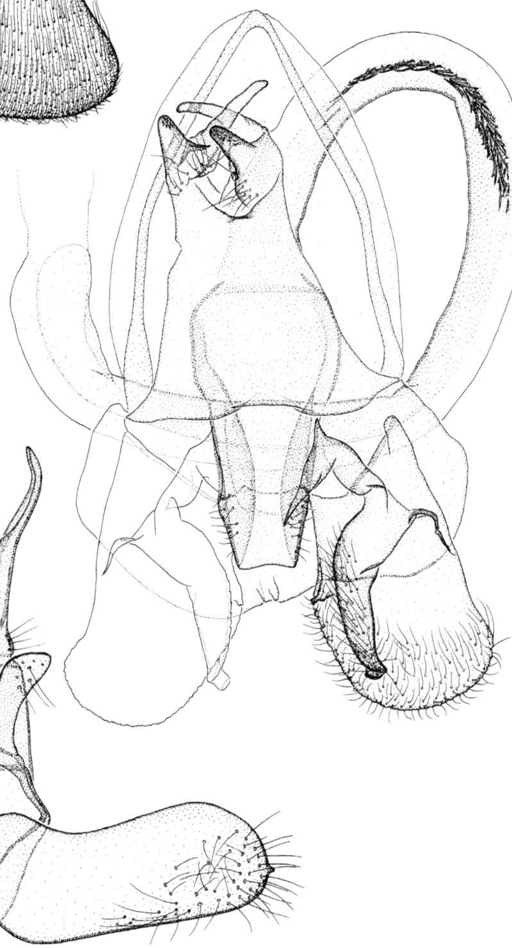

66

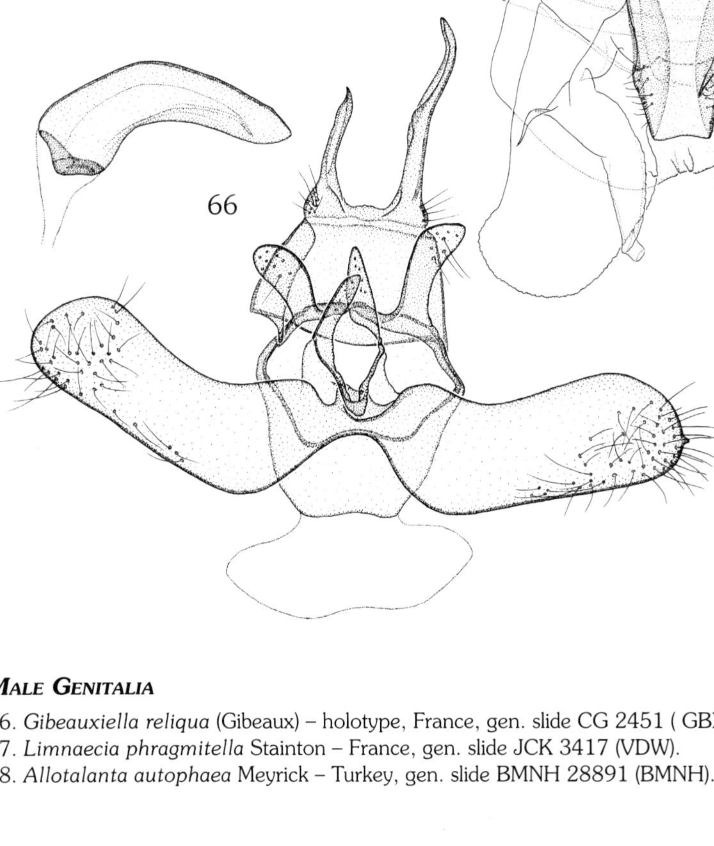

Male Genitalia

66. *Gibeauxiella reliqua* (Gibeaux) – holotype, France, gen. slide CG 2451 (GBX).
67. *Limnaecia phragmitella* Stainton – France, gen. slide JCK 3417 (VDW).
68. *Allotalanta autophaea* Meyrick – Turkey, gen. slide BMNH 28891 (BMNH).

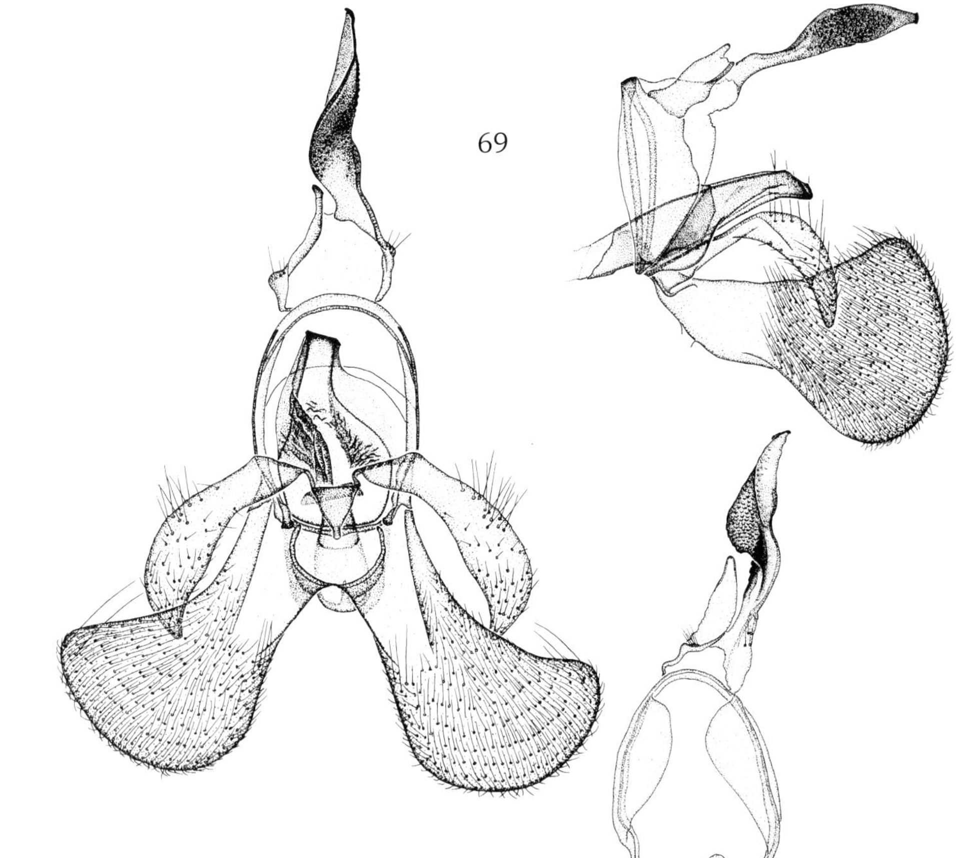

MALE GENITALIA

69. *Cosmopterix zieglerella* (Hübner) – France, gen. slide JCK 3719 (VHN)
70. *Cosmopterix schmidiella* Frey – Switzerland, gen. slide JCK 3981 (ZIN).

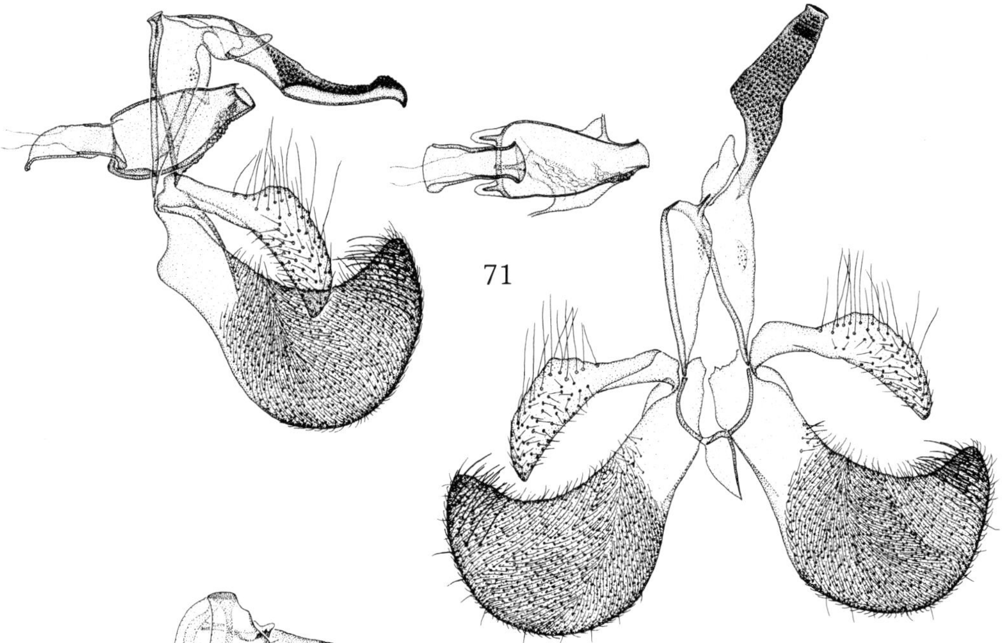

71

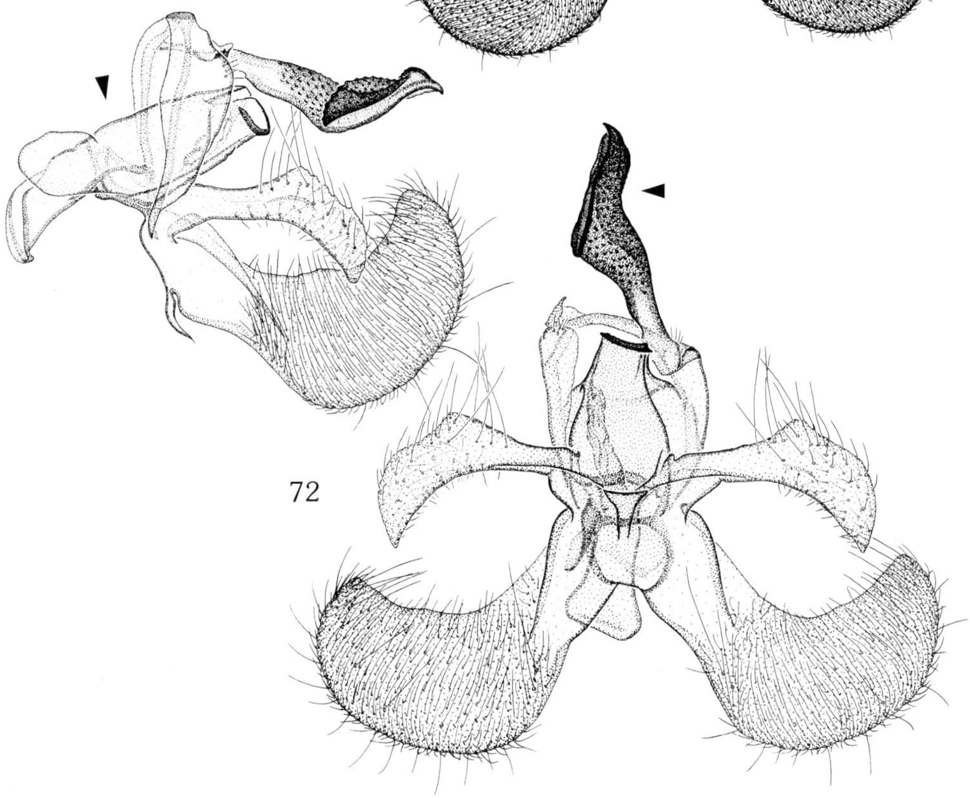

72

MALE GENITALIA

71. *Cosmopterix orichalcea* Stainton – Spain, gen. slide JCK 3732 (WLN).
72. *Cosmopterix sibirica* Sinev – Russia, gen. slide JCK 2594 (KTR).

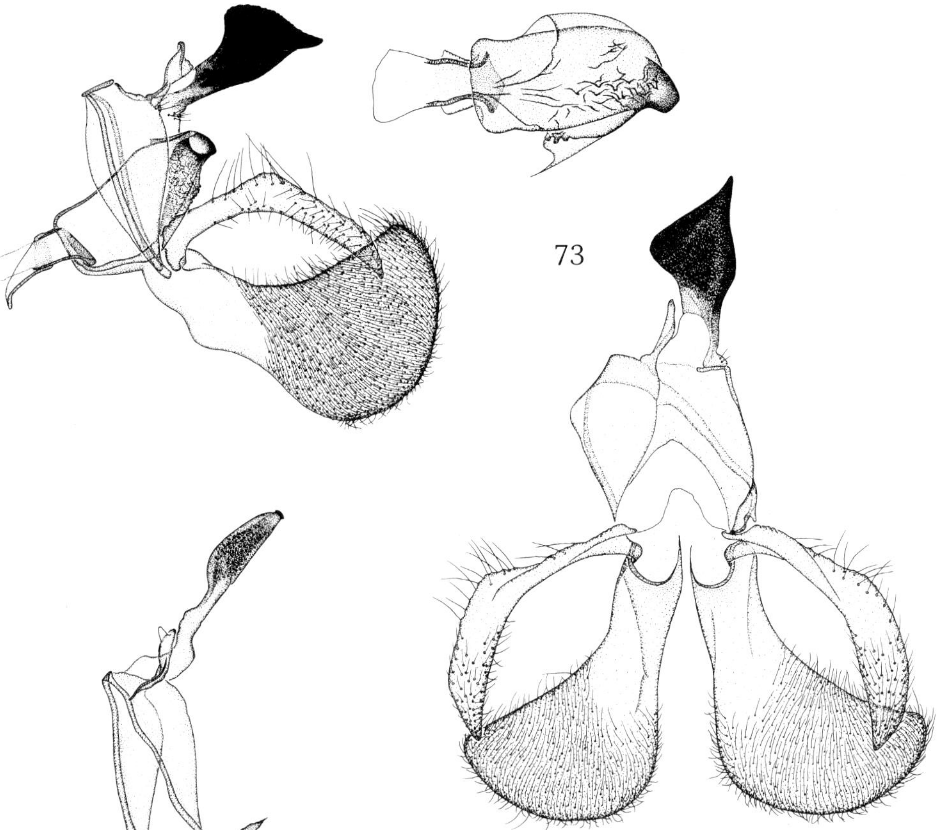

73

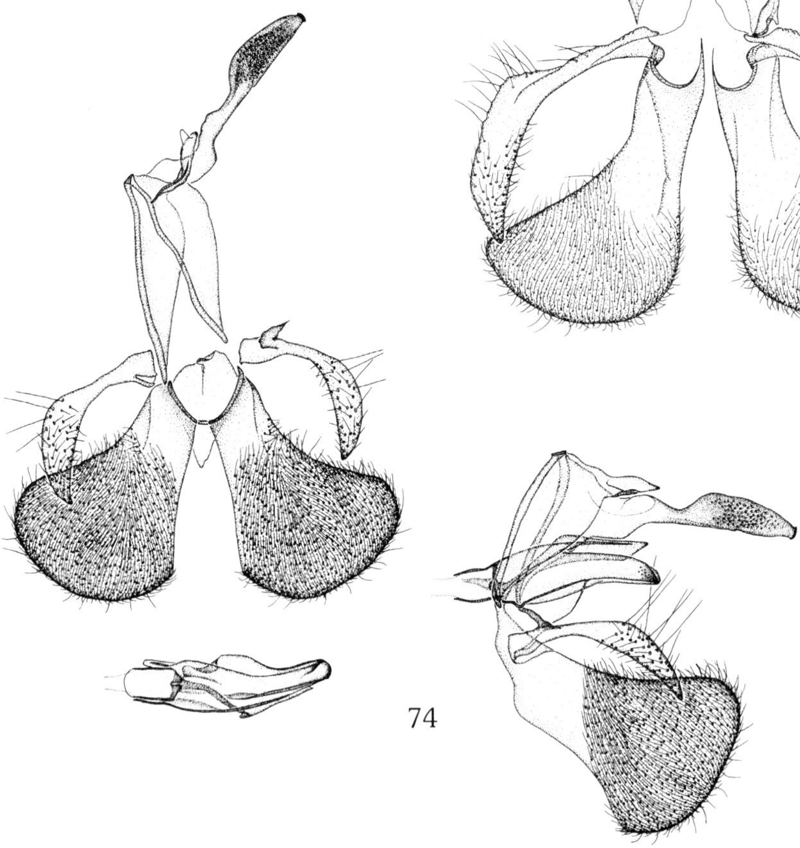

74

Male Genitalia

73. *Cosmopterix scribaiella* Zeller – Netherlands, gen. slide JCK 3752 (WLN).
74. *Cosmopterix pulchrimella* Chambers – Spain, gen. slide JCK 3750 (WLN).

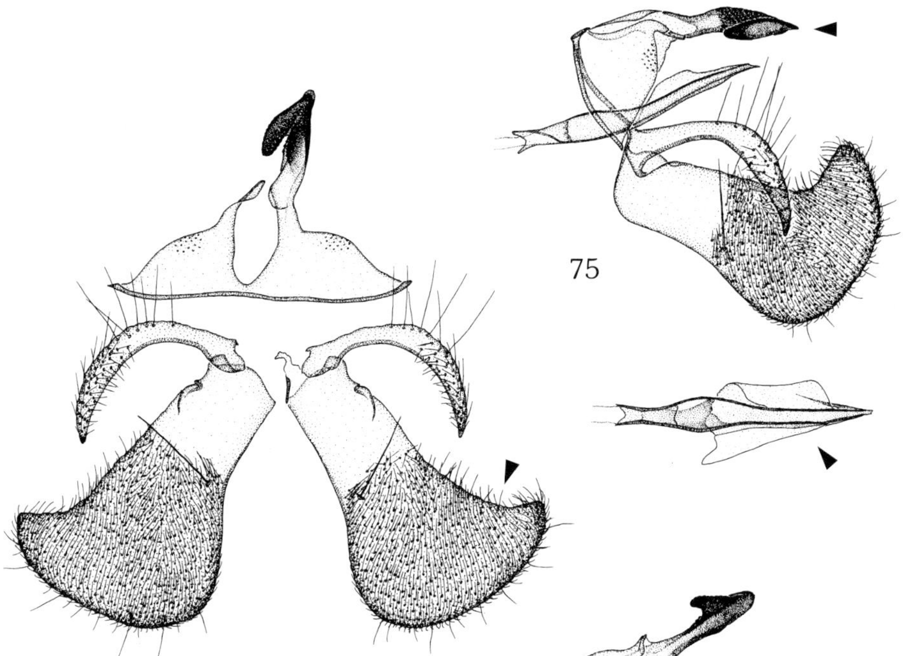

75

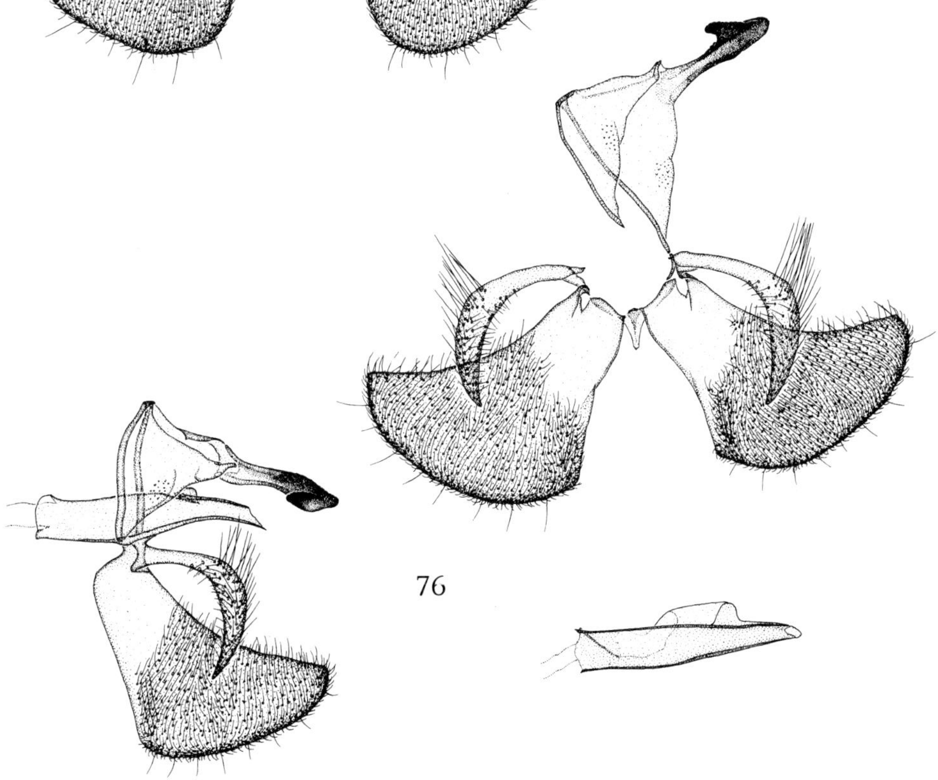

76

Male Genitalia

75. *Cosmopterix attenuatella* (Walker) – Madeira, gen. slide JCK 3728 (COX).
76. *Cosmopterix crassicervicella* Chrétien – Algeria, gen. slide JCK 3729 (STR).

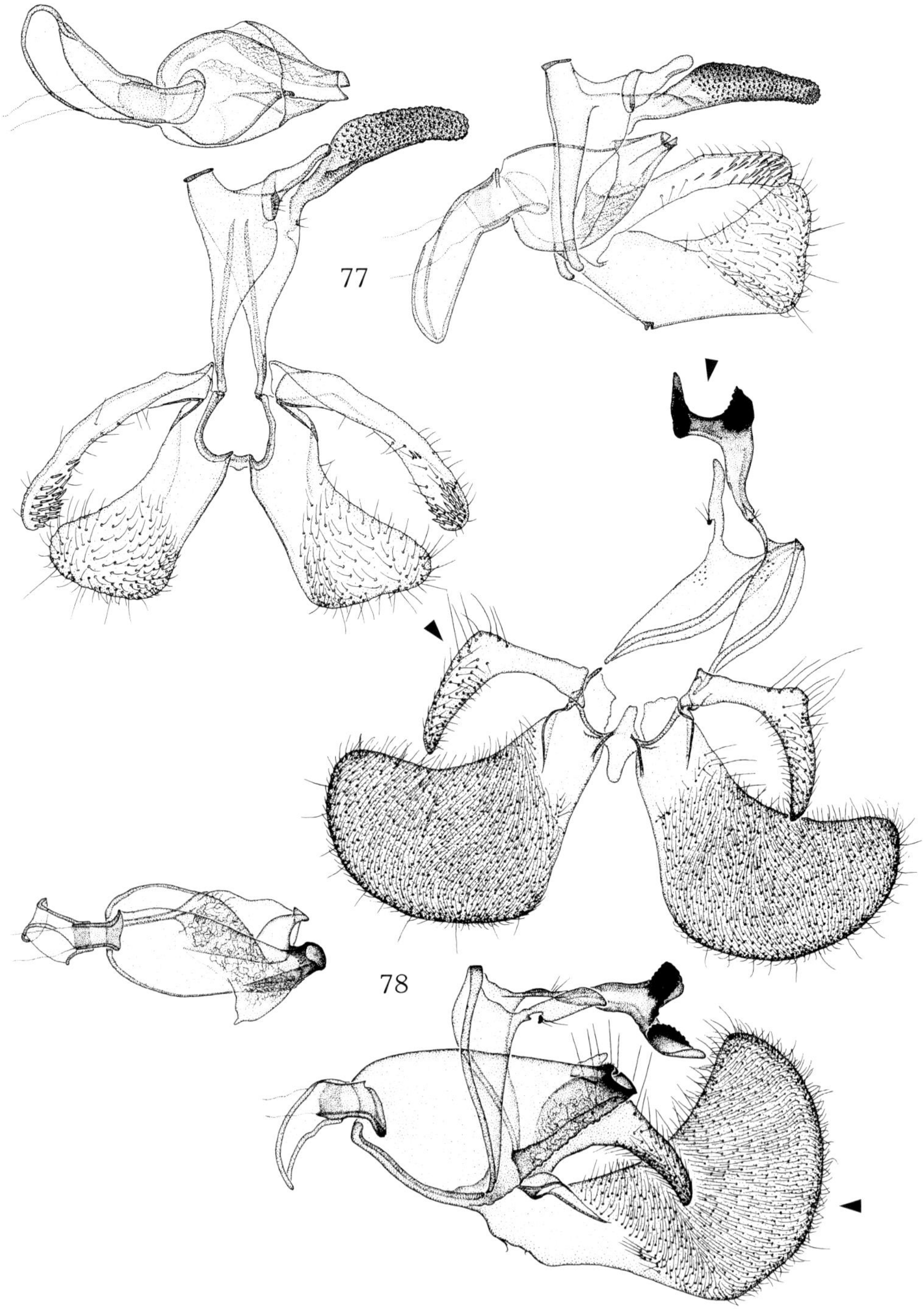

Male Genitalia

77. *Cosmopterix turbidella* Rebel – Canary Islands, gen. in vial (ZIN).
78. *Cosmopterix pararufella* Riedl – Spain, gen. slide JCK 3731 (VDW).

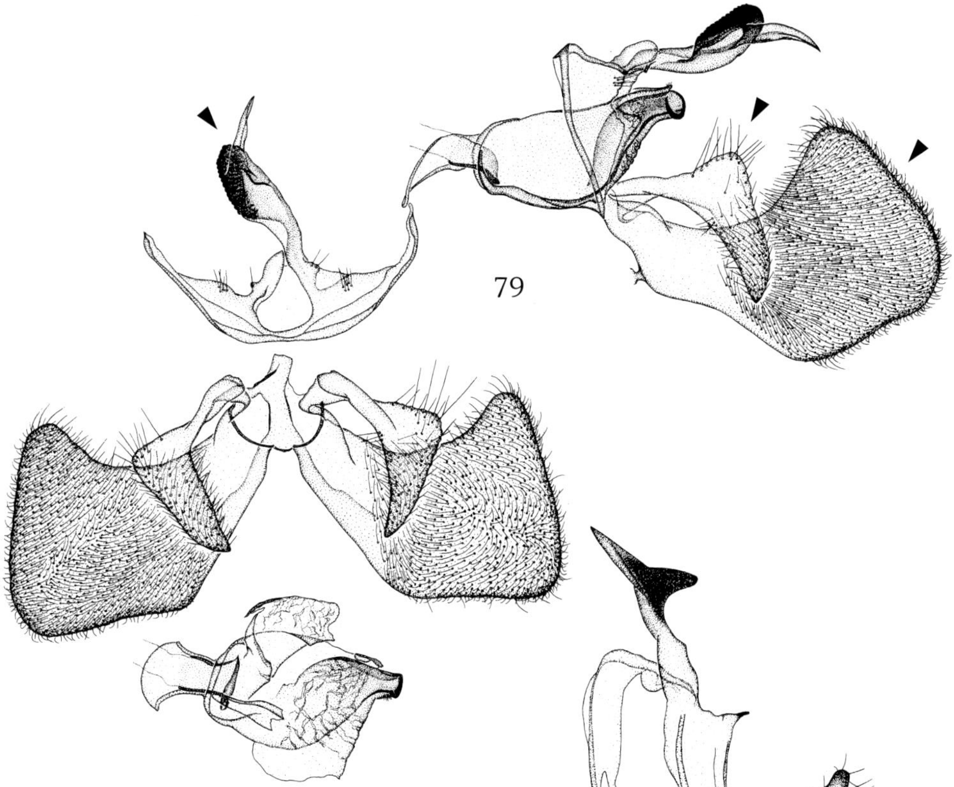

79

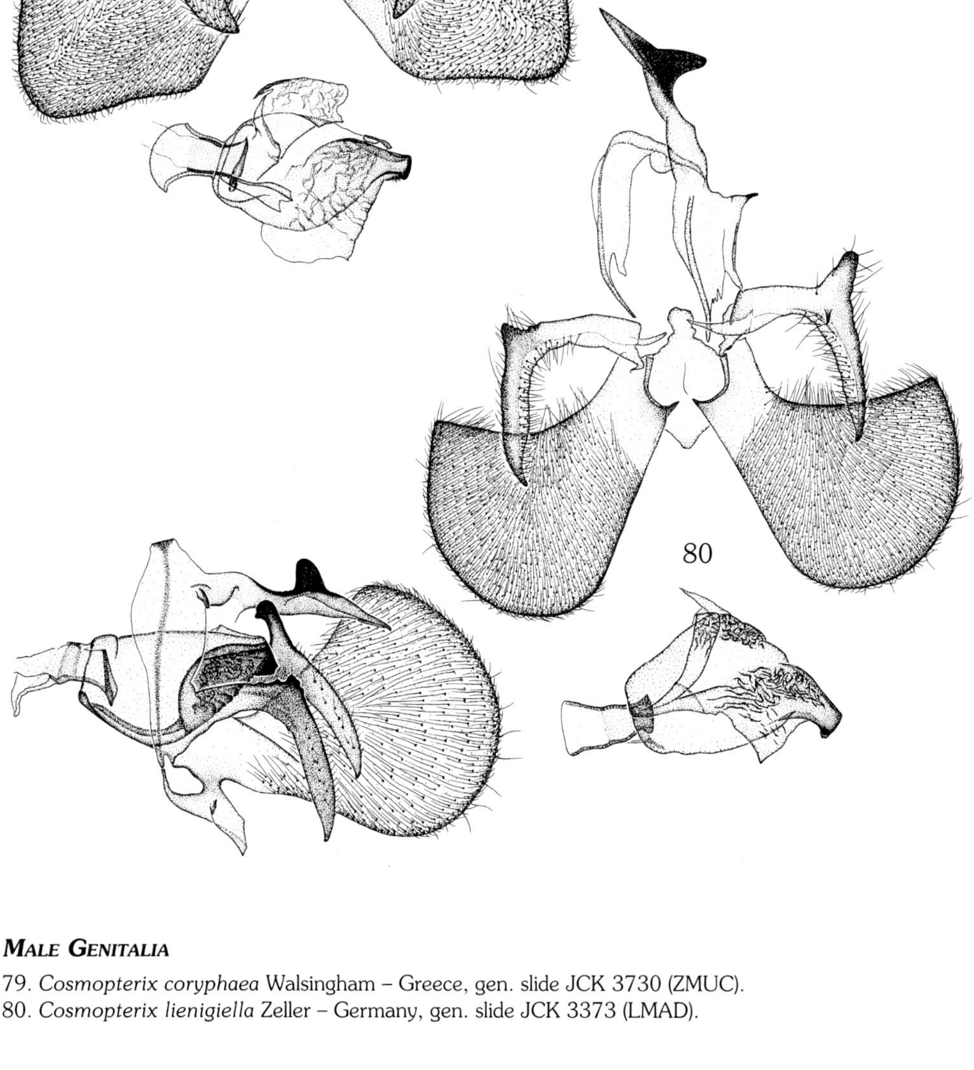

80

Male Genitalia

79. *Cosmopterix coryphaea* Walsingham – Greece, gen. slide JCK 3730 (ZMUC).
80. *Cosmopterix lienigiella* Zeller – Germany, gen. slide JCK 3373 (LMAD).

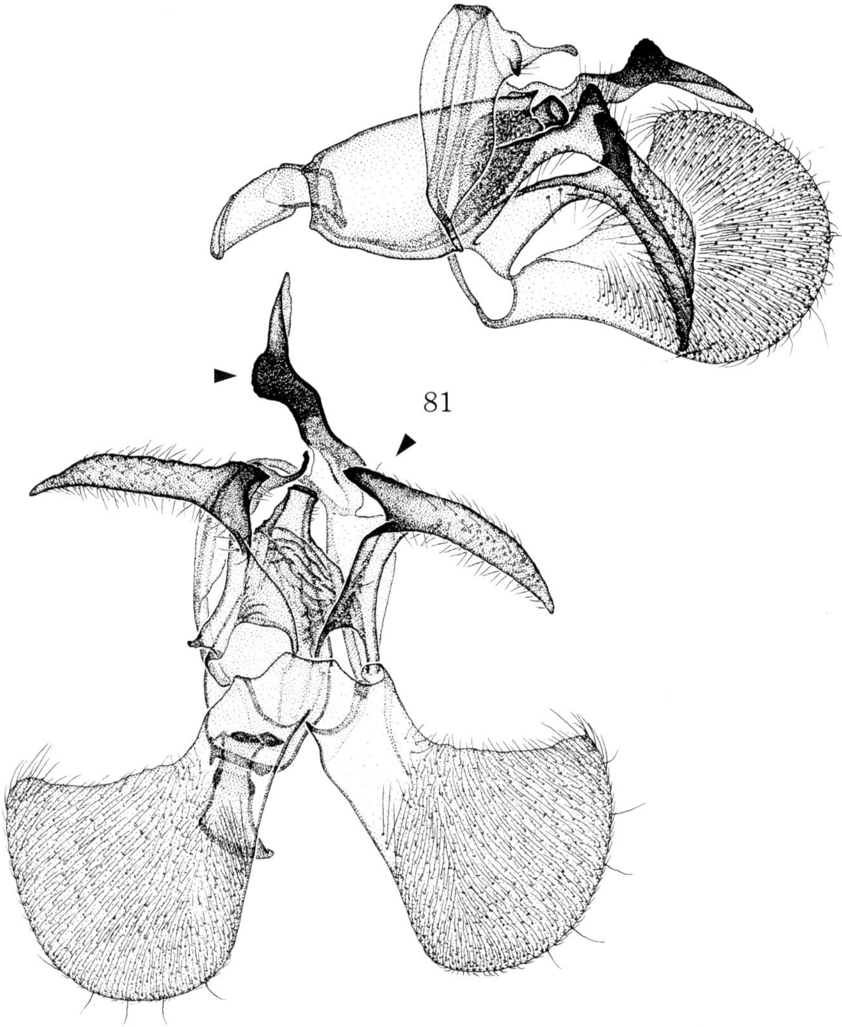

81

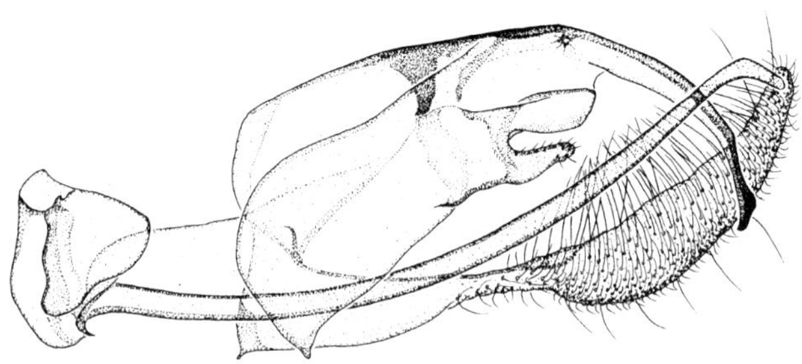

82

Male Genitalia

81. *Cosmopterix salahinella* Chrétien – Tunisia, gen. in vial (ZMUC).
82. *Labdia semicoccinea* (Stainton) – Sri Lanka, gen. slide JCK 4124 (PDW).

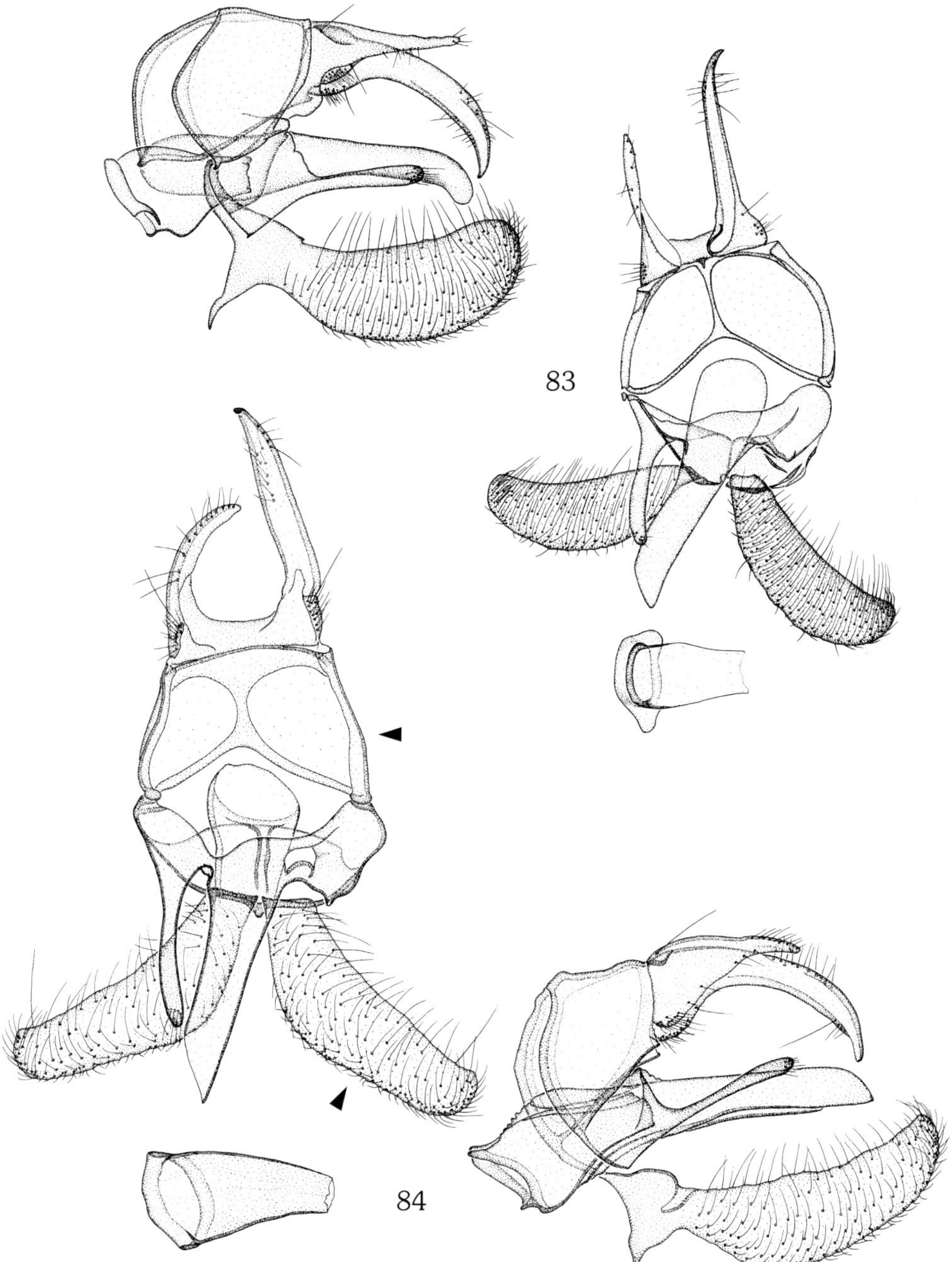

83

84

MALE GENITALIA

83. *Pyroderces argyrogrammos* (Zeller) – Spain, gen. slide JCK 3427 (VDW).
84. *Pyroderces caesaris* Gozmány – Ukraine, gen. slide JCK 2383 (KTR).

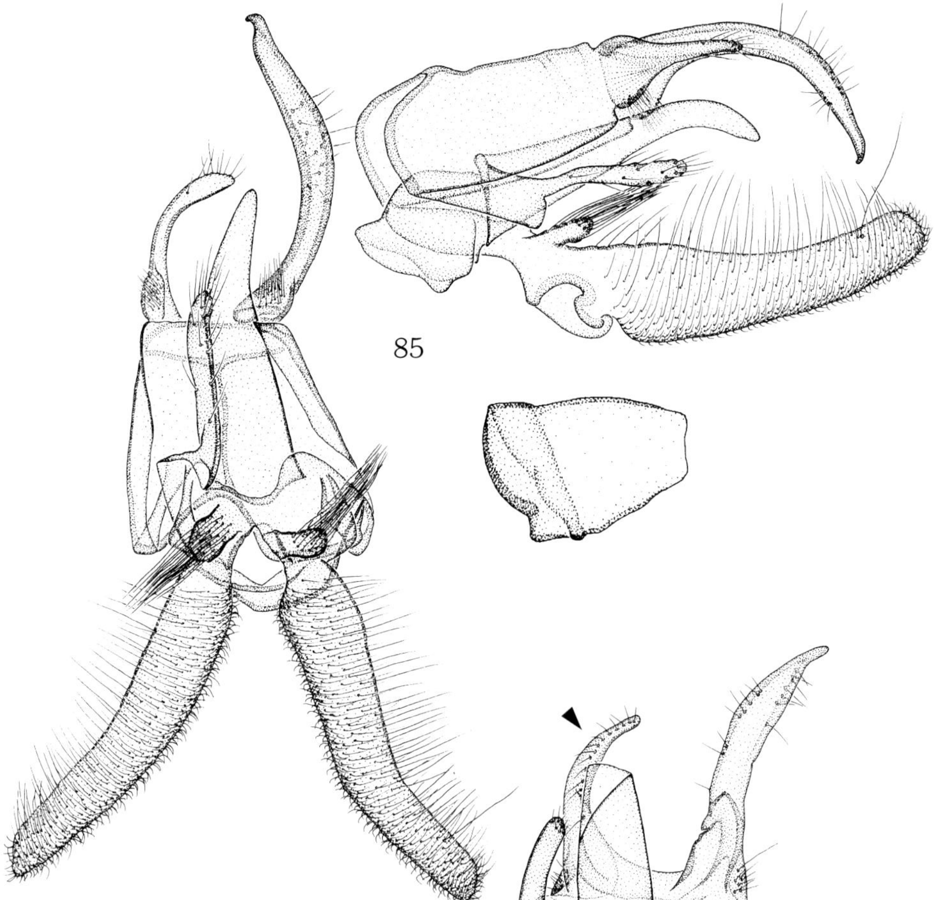

85

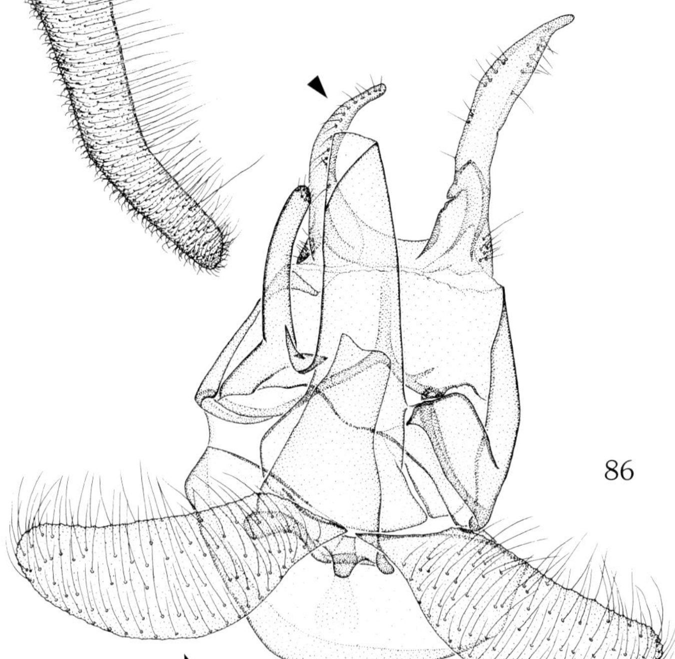

86

Male Genitalia

85. *Pyroderces klimeschi* Rebel – Spain, gen. slide JCK 4280 (ASZ).
86. *Pyroderces brosi* Riedl – holotype, Italy, gen. slide nr. 2152 (ISAK).

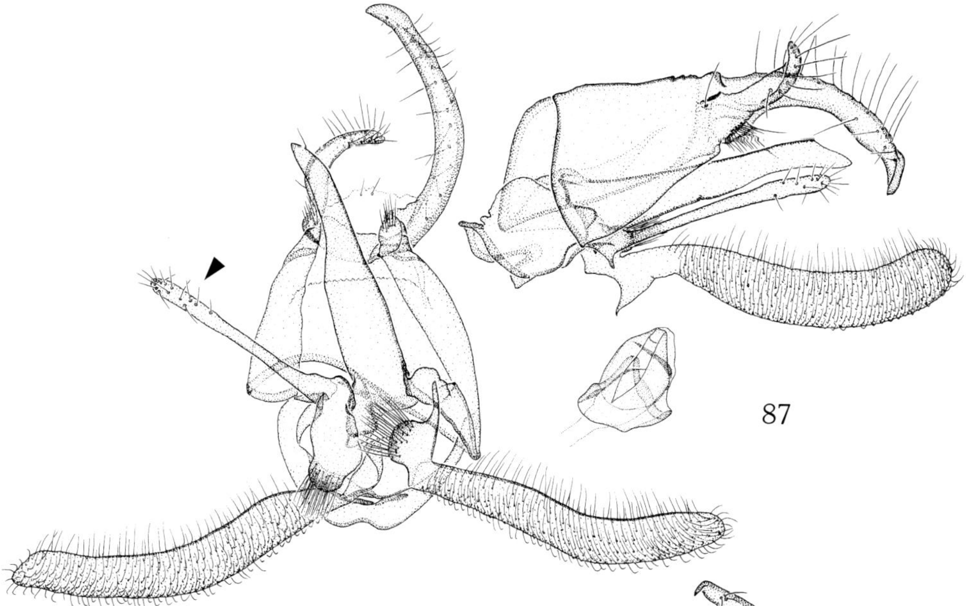

87

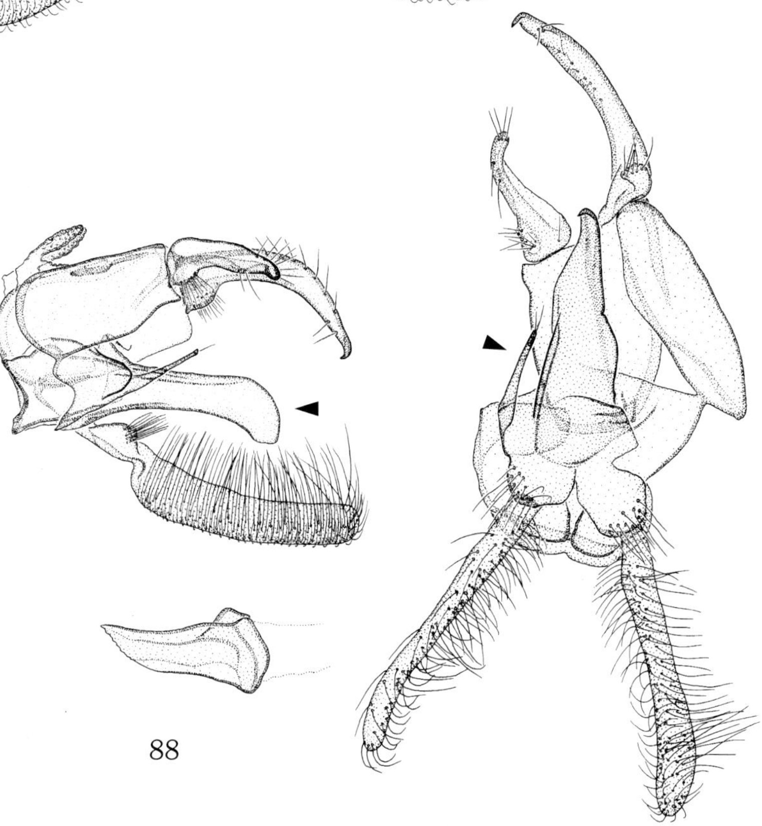

88

Male Genitalia

87. *Pyroderces tethysella* sp. n. – paratype, Spain, gen. slide JCK 4591 (WLN).
88. *Pyroderces wolschrijni* sp. n. – paratype, Spain, gen. slide JCK 4774 (WLN).

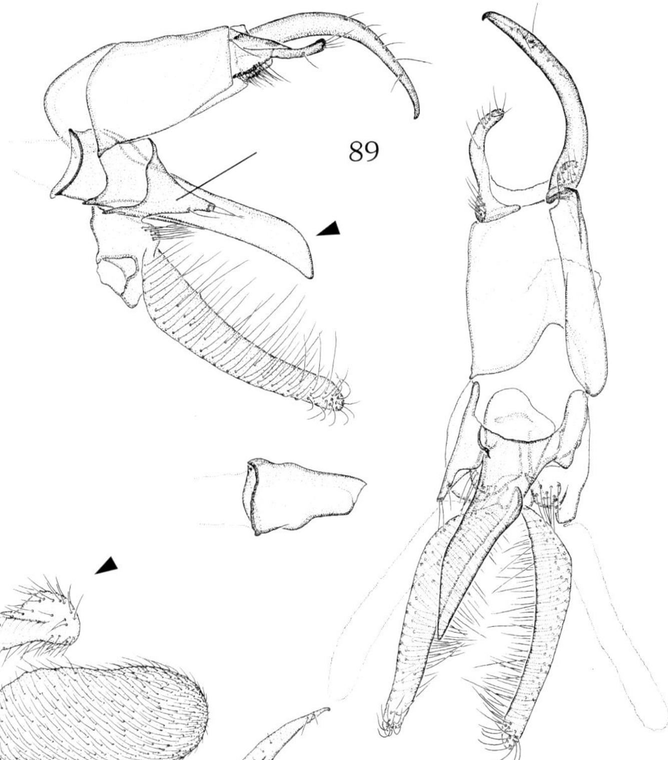

89

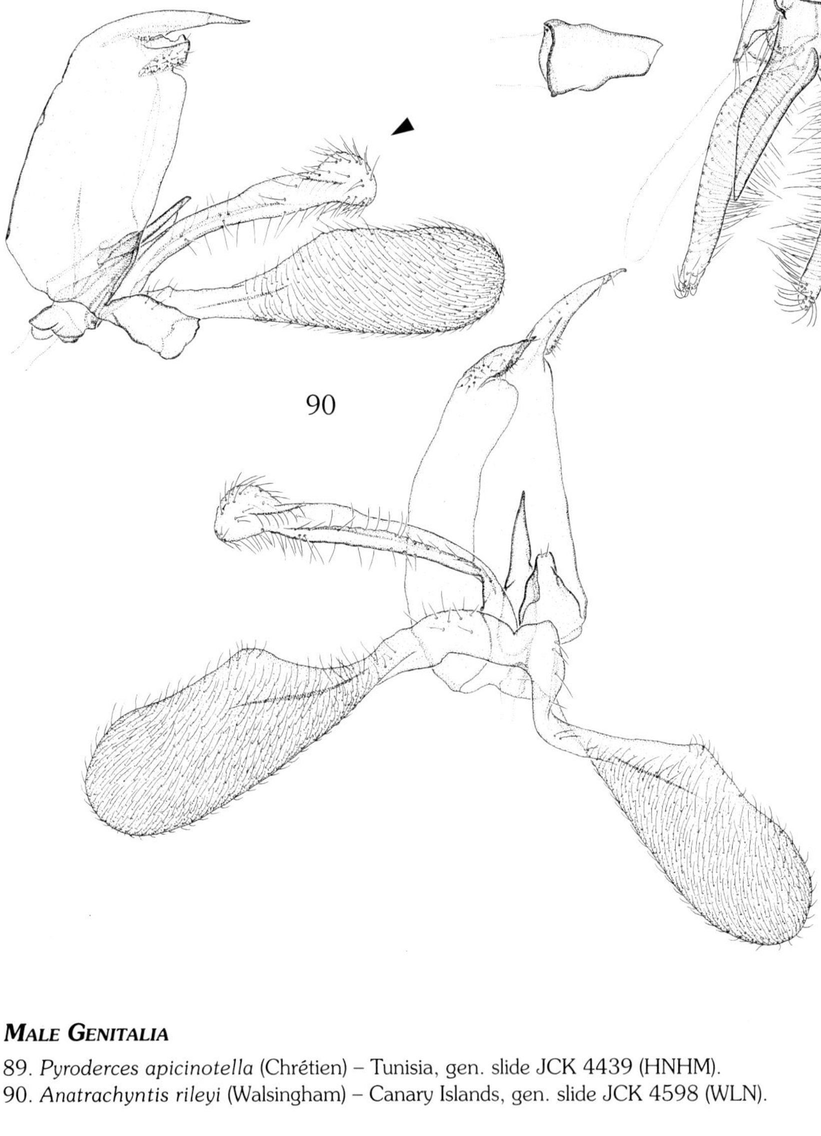

90

Male Genitalia

89. *Pyroderces apicinotella* (Chrétien) – Tunisia, gen. slide JCK 4439 (HNHM).
90. *Anatrachyntis rileyi* (Walsingham) – Canary Islands, gen. slide JCK 4598 (WLN).

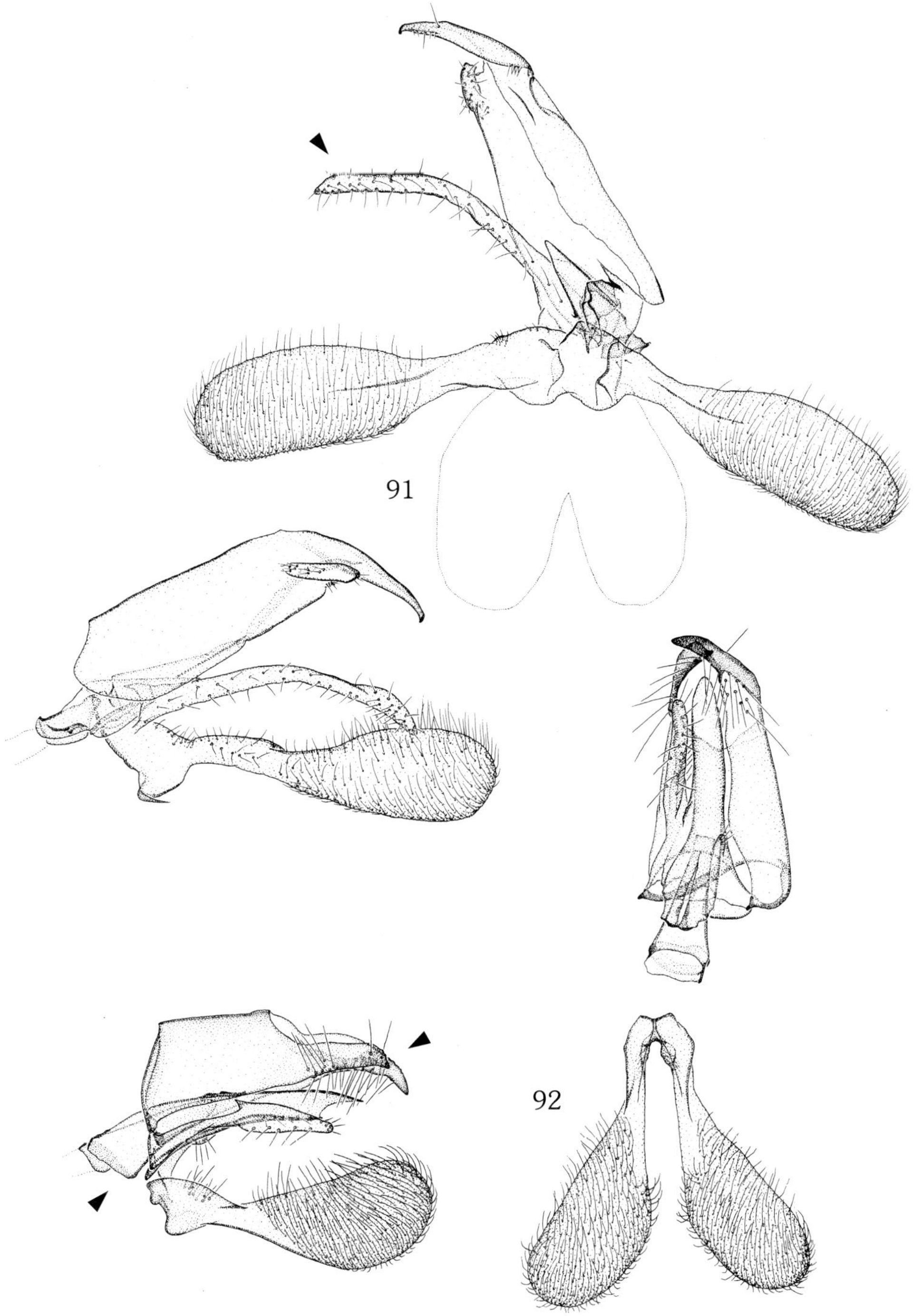

Male Genitalia

91. *Anatrachyntis badia* (Hodges) – Canary Islands, gen. slide JCK 4596 (WLN).
92. *Anatrachyntis coridophaga* (Meyrick) – Egypt, gen. slide JCK 4755 (ZFMK).

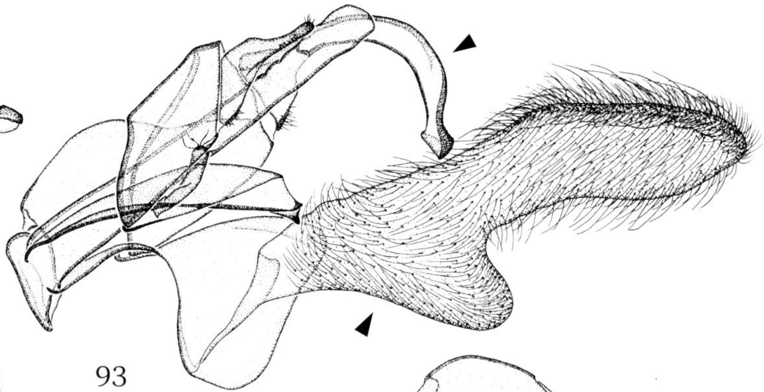

93

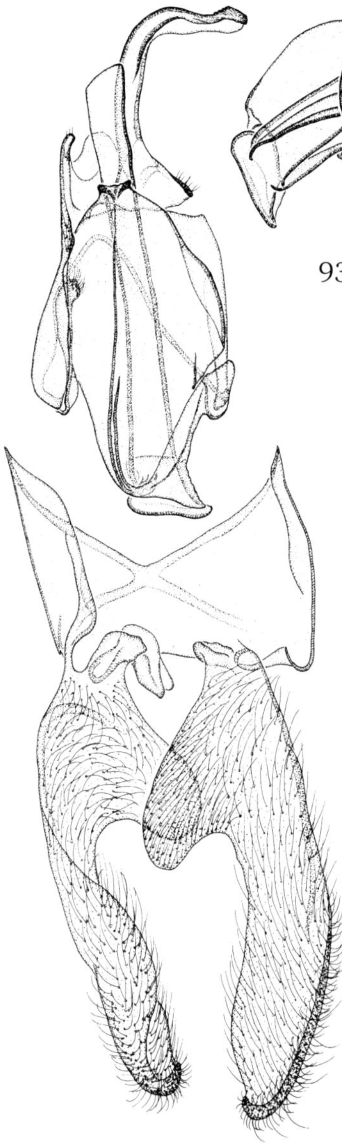

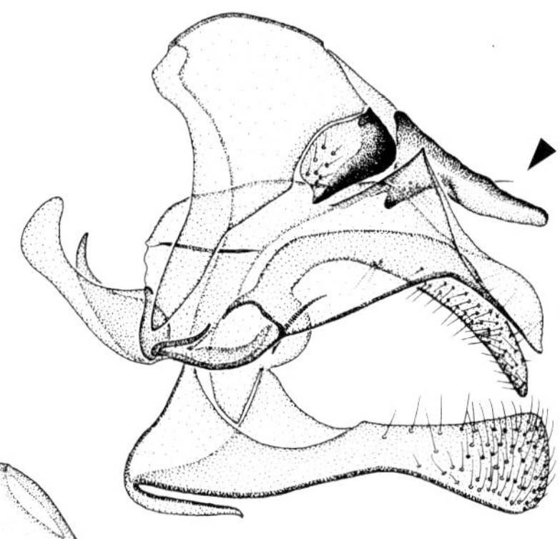

94

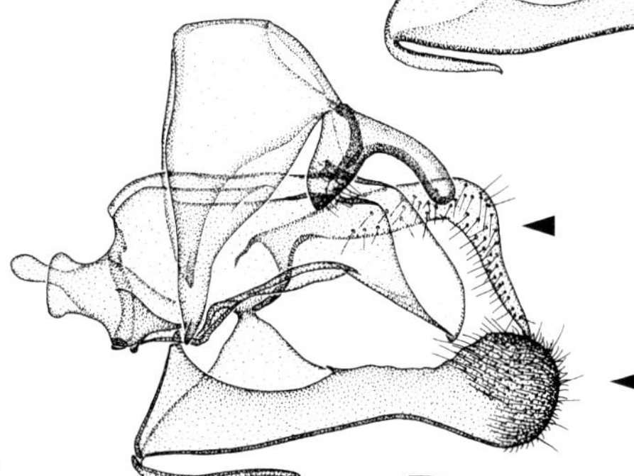

96

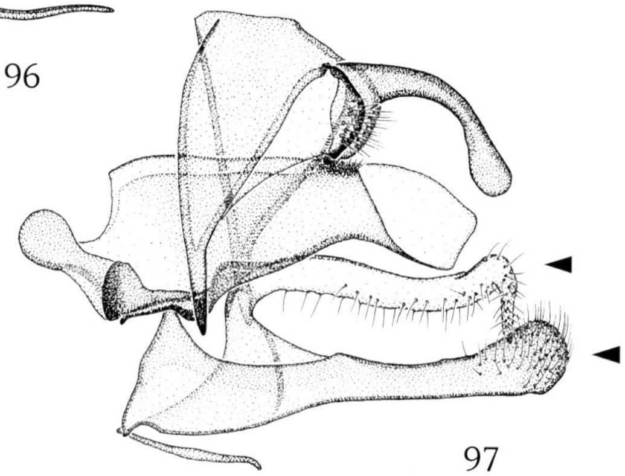

97

MALE GENITALIA

93. *Anatrachyntis simplex* (Walsingham) – Egypt, gen. in vial (ZFMK).
94. *Coccidiphila gerasimovi* Danilevsky – Georgia (Abchasia), gen. slide JCK 3973 (ZIN).
95. *Coccidiphila patriciae* J. Nel & A. Nel – unknown.
96. *Coccidiphila ledereriella* (Zeller) – France, gen. slide JCK 3858 (RMNH).
97. *Coccidiphila danilevskyi* Sinev – Spain, gen. slide JCK 2484 (KTR).

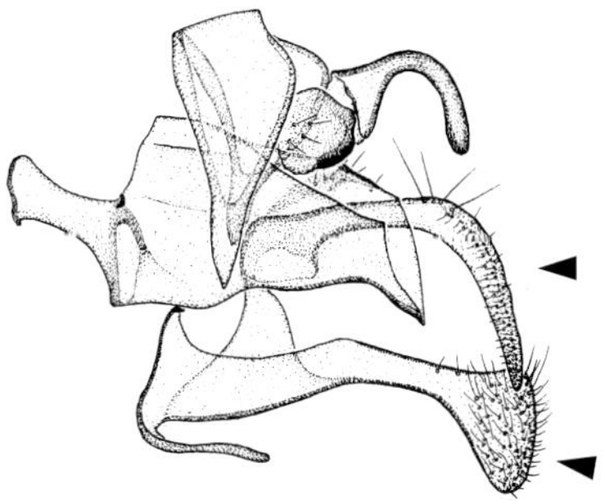

98

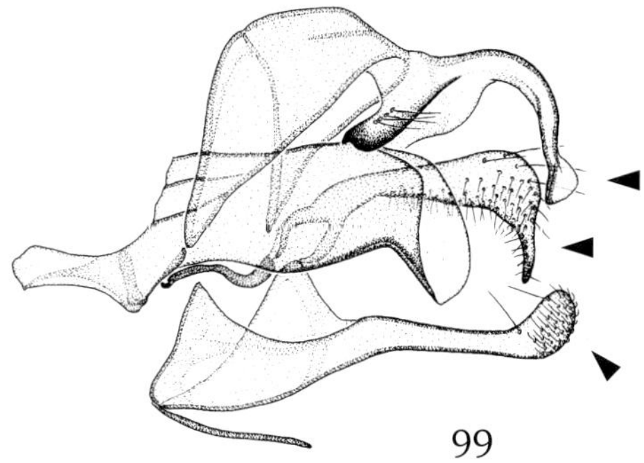

99

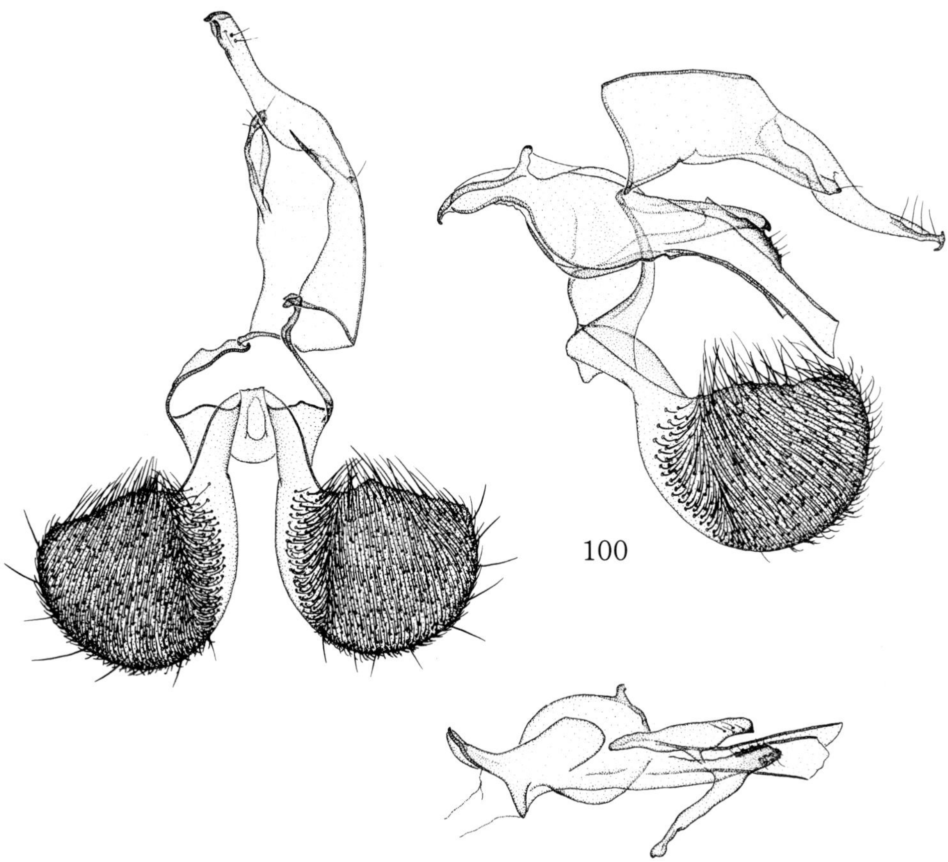

100

Male Genitalia

98. *Coccidiphila kasypinkeri* Traugott-Olsen – paratype, Canary Islands, gen. slide C 16.6.85/ETO (NHMW).
99. *Coccidiphila riedli* Traugott-Olsen – paratype, Canary Islands, gen. slide B 5.1.85 (NHMW).
100. *Stagmatophora heydeniella* (Fischer von Röslerstamm) – Germany, gen. slide JCK 3857 (RMNH).

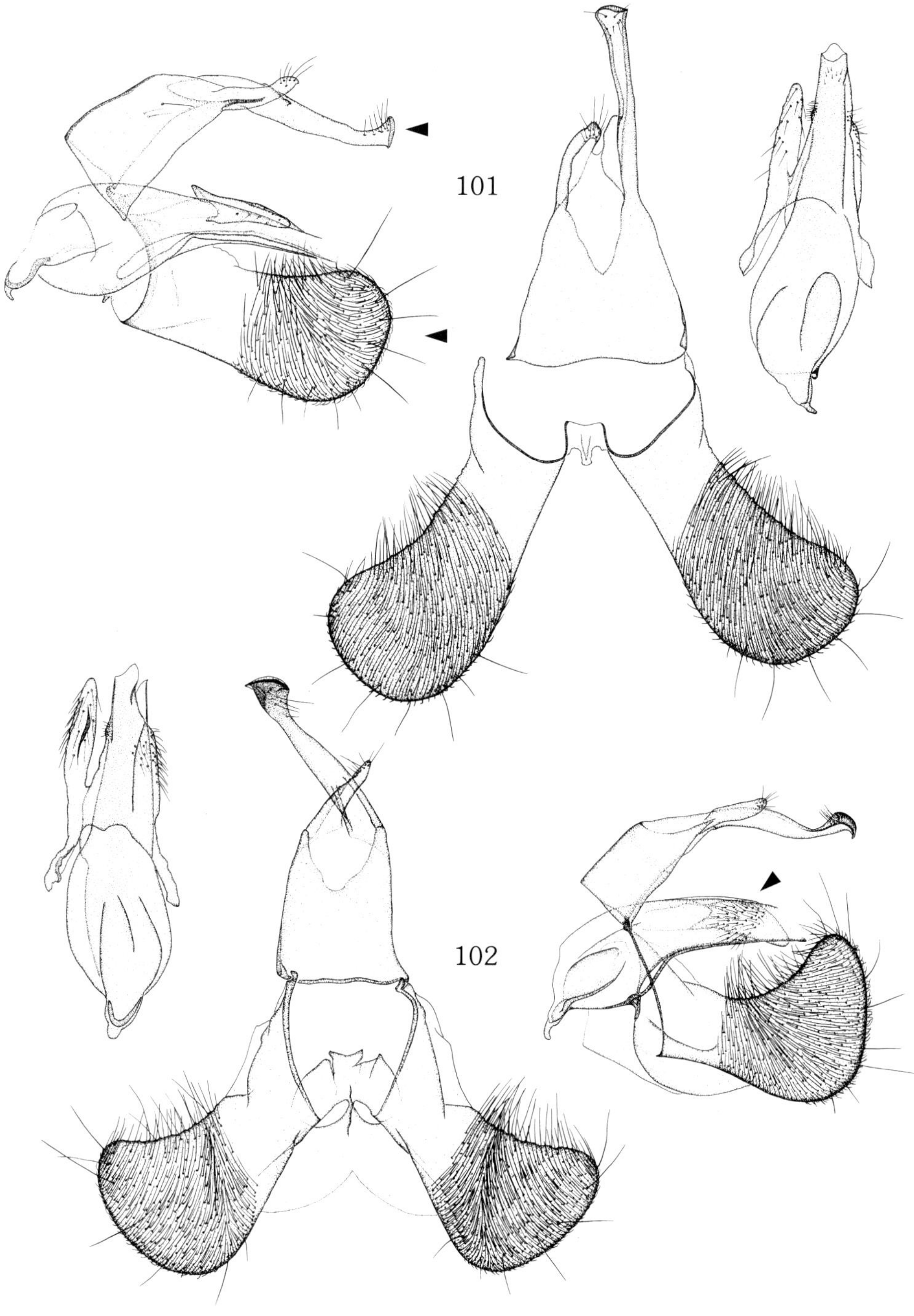

MALE GENITALIA

101. *Eteobalea anonymella* (Riedl) – Lebanon, gen. slide JCK 3586 (LMAN).
102. *Eteobalea serratella* (Treitschke) – Italy, gen. prep. JCK 2330 (KTR).

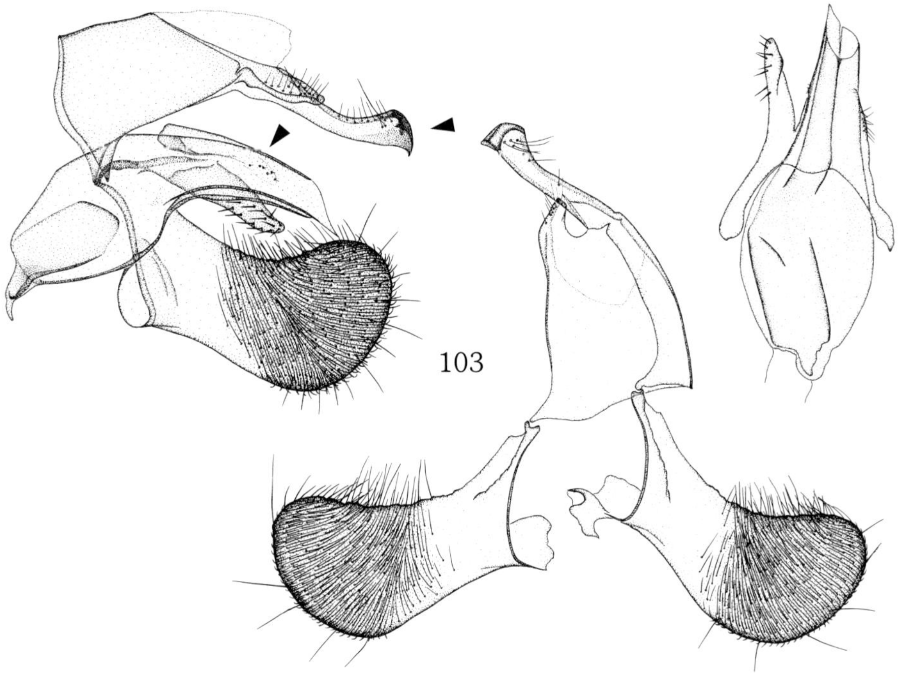

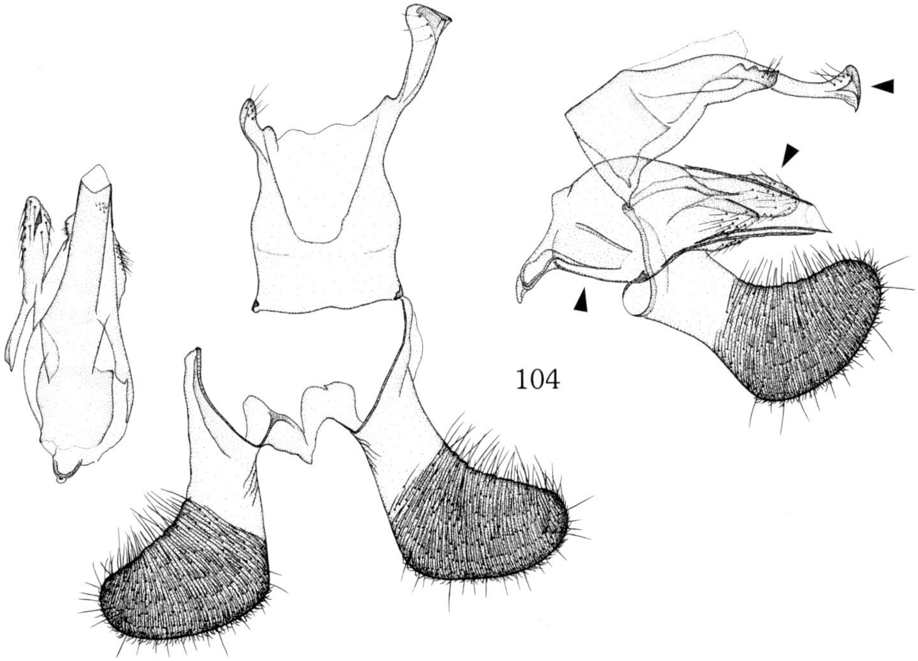

MALE GENITALIA

103. *Eteobalea intermediella* (Riedl) – Spain, gen. slide JCK 3451 (LMAN).
104. *Eteobalea beata* (Walsingham) – France, gen. slide JCK 3591 (LCS).

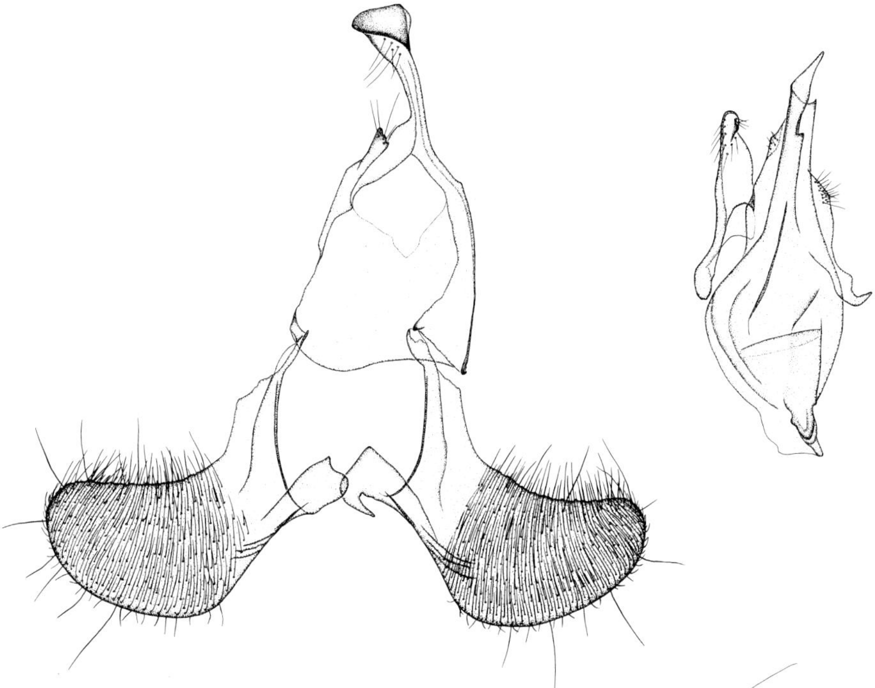

105

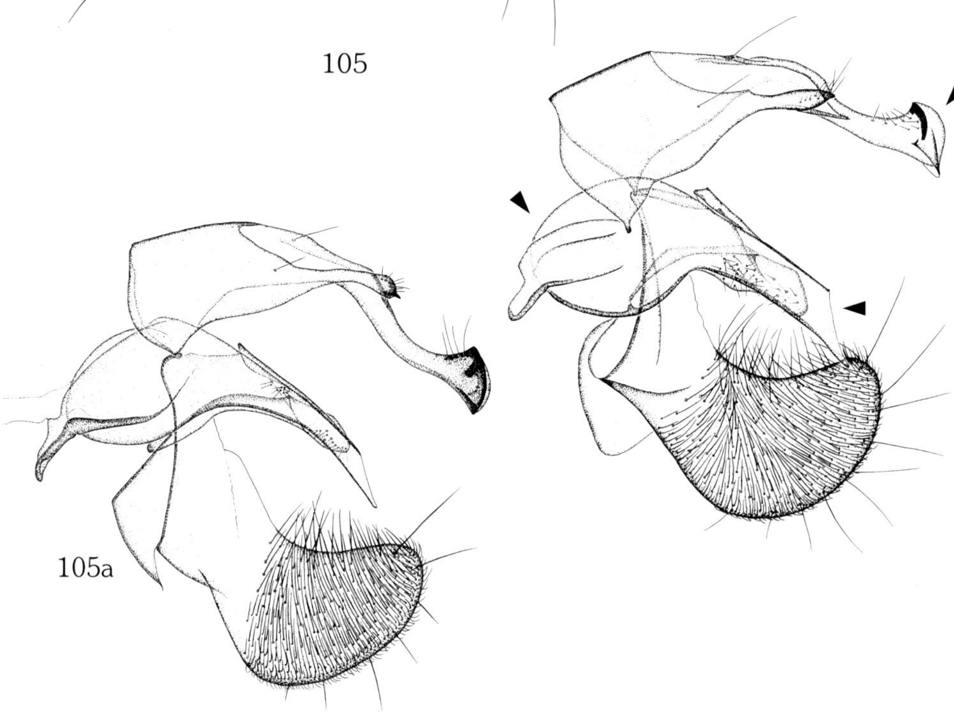

105a

MALE GENITALIA

105. *Eteobalea sumptuosella* (Lederer) – Greece, gen. slide JCK 3509 (LCS).
105a. *Eteobalea sumptuosella* (Lederer) – France, Spain, gen. slide JCK 3318 (VDW).

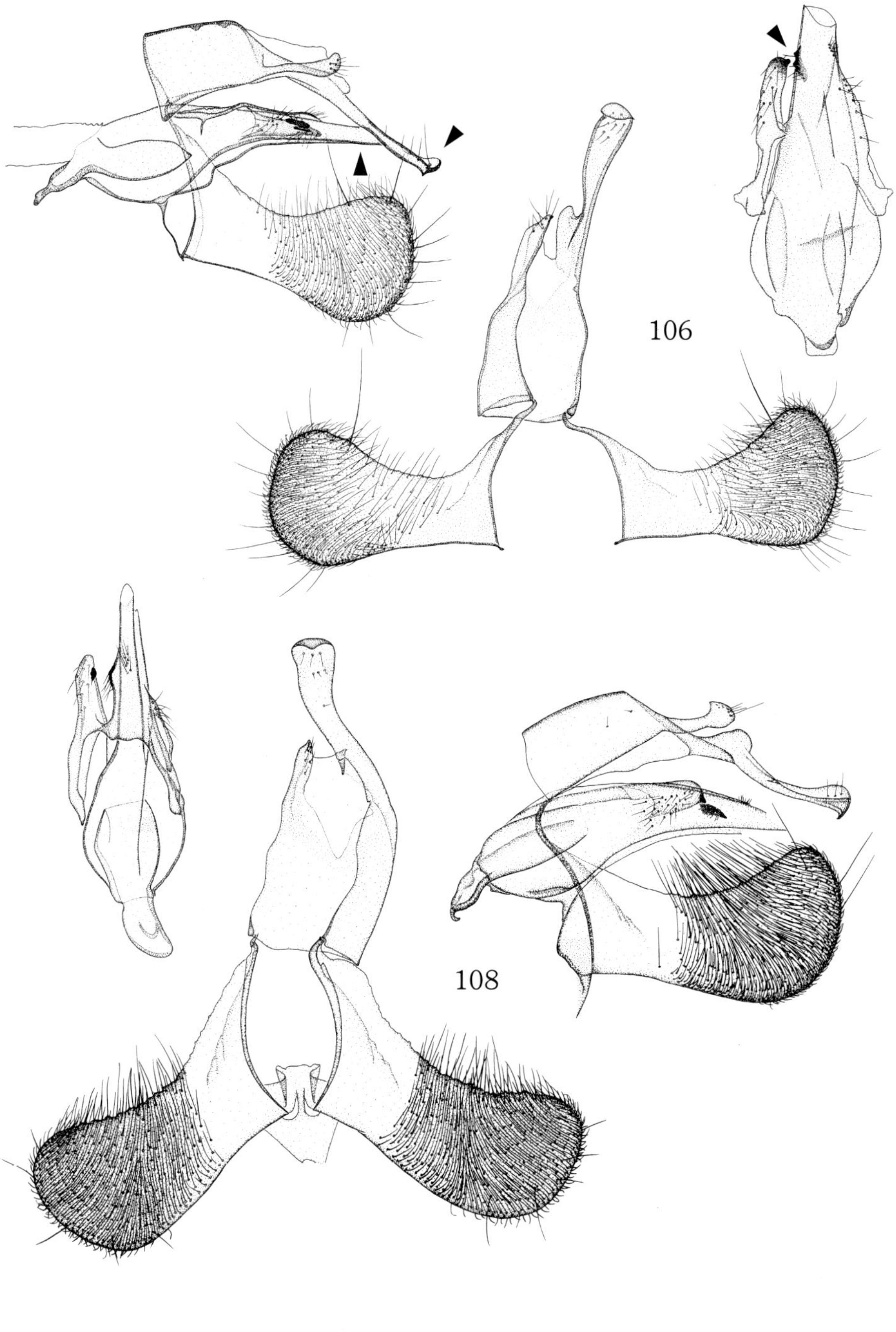

Male Genitalia

106. *Eteobalea teucrii* (Walsingham) – Spain, gen. in vial, (ZMUC).
107. *Eteobalea klisieckii* (Riedl) – unknown.
108. *Eteobalea tririvella* (Staudinger) – Slovakia, gen. slide JCK 1985 (KTR).

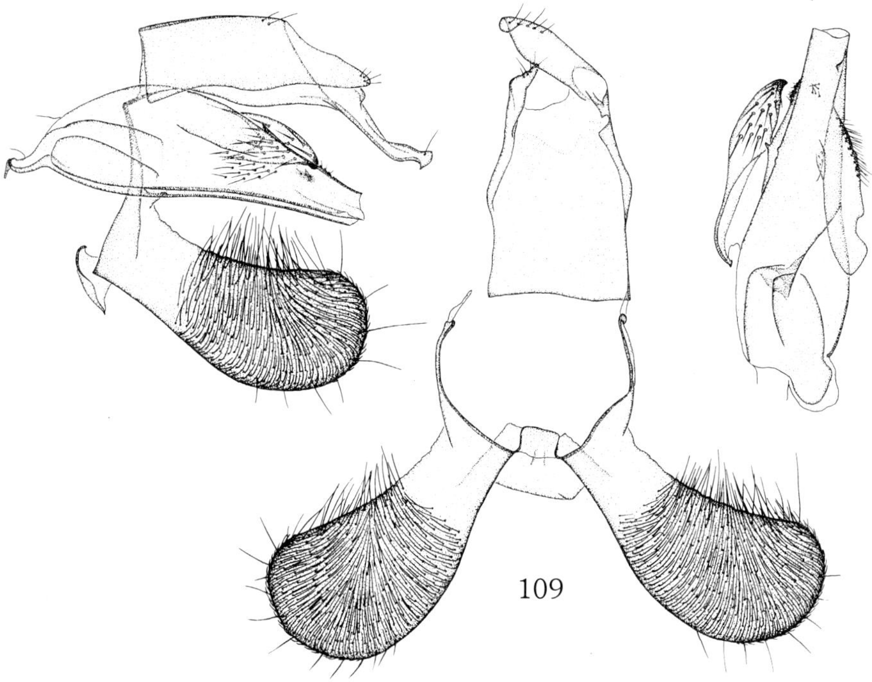

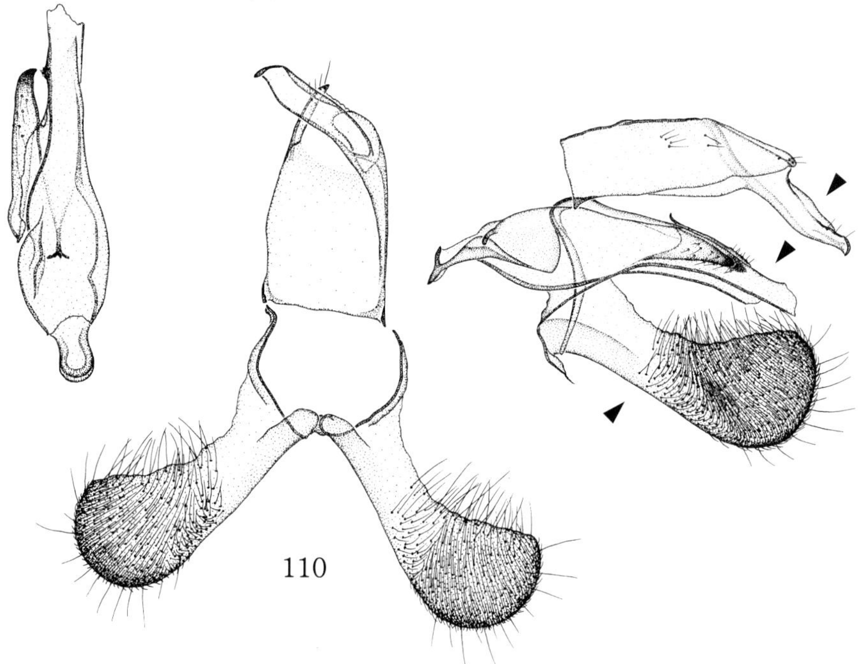

Male Genitalia

109. *Eteobalea albiapicella* (Duponchel) – Austria, gen. slide JCK 3557 (ZIN).
110. *Eteobalea alypella* (Klimesch) – Spain, gen. slide JCK 3880 (ZMUC).

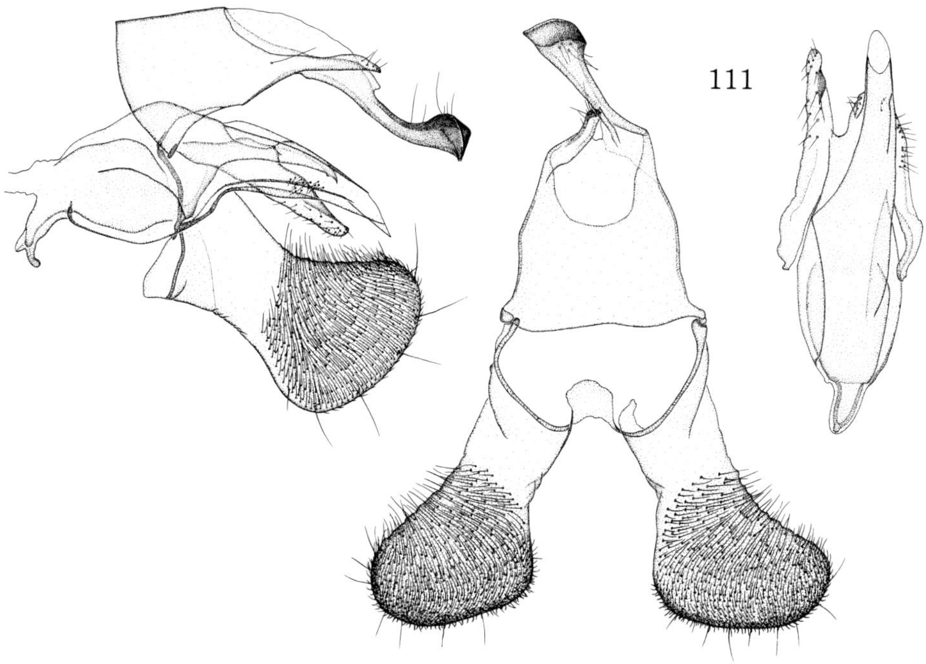

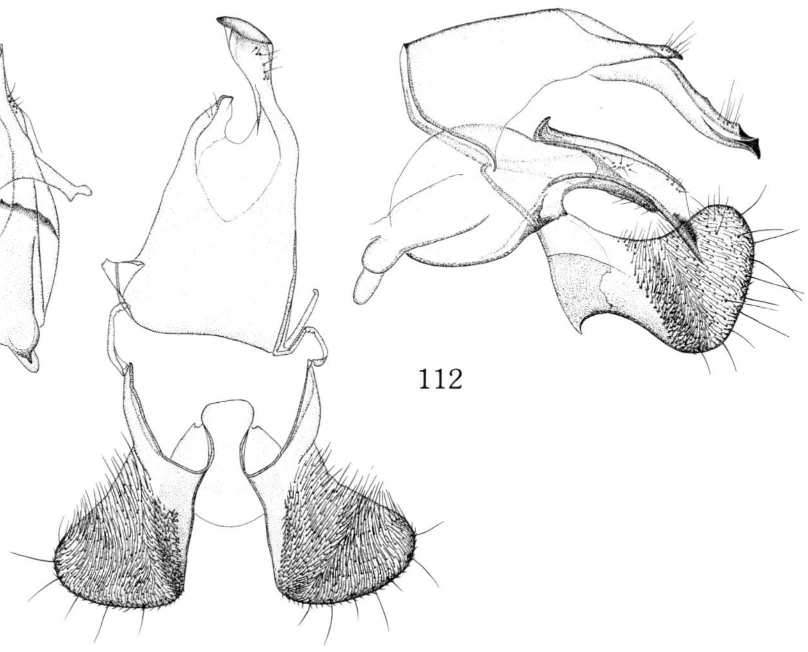

Male Genitalia

111. *Eteobalea isabellella* (Costa) – Italy, gen. slide JCK 3574 (LCS).
112. *Eteobalea dohrnii* (Zeller) – Spain, gen. slide JCK 2122 (KTR).

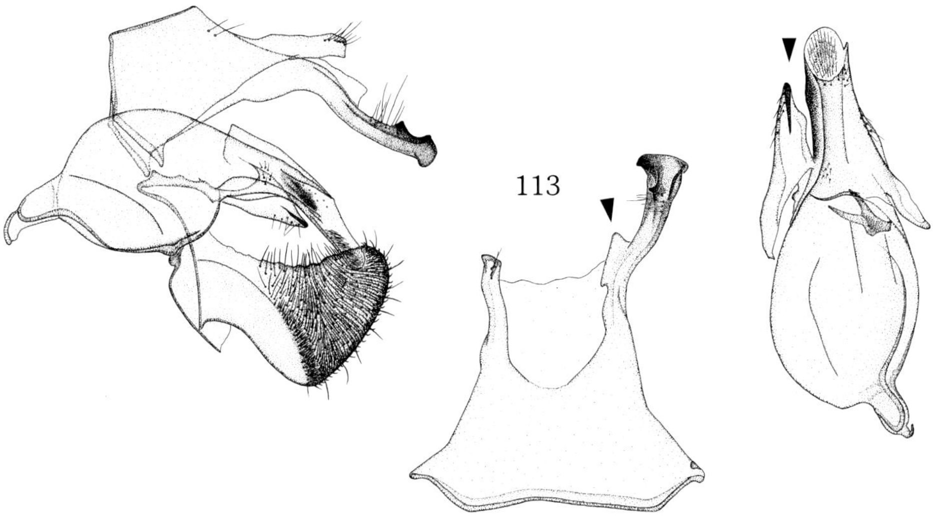

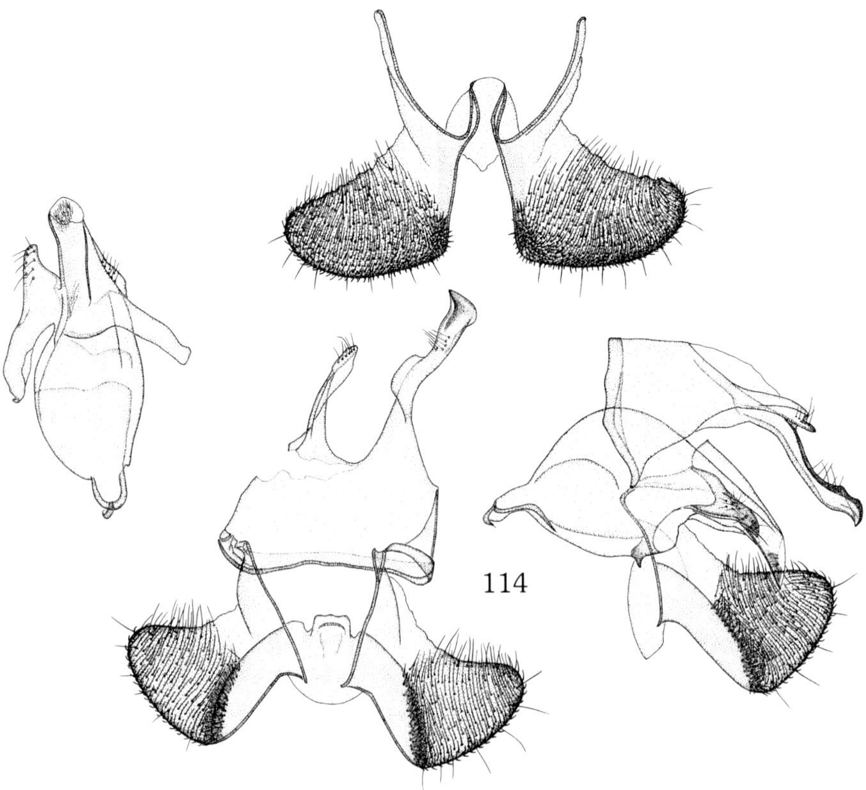

MALE GENITALIA

113. *Eteobalea siciliae* (Riedl) – Algeria, gen. slide 991/R (MNHN).
114. *Eteobalea thaumatella* (Walsingham) – Tunesia, gen. slide 982/R (MNHN).

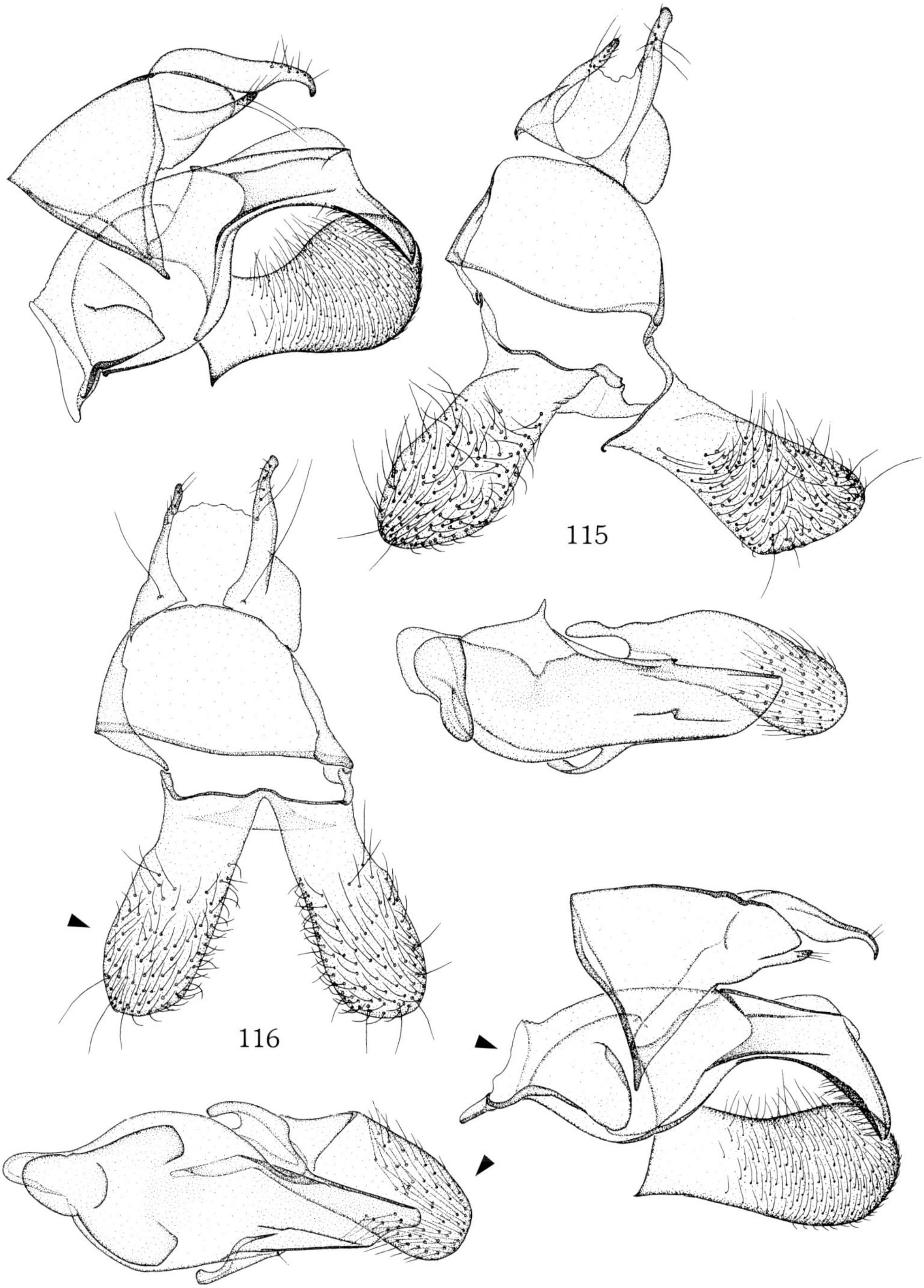

Male Genitalia

115. *Isidiella nickerlii* (Nickerl) – Austria, gen. slide JCK 3362 (LMAD).
116. *Isidiella divitella* (Constant) – Italy, gen. slide JCK 3442 (LMAD).

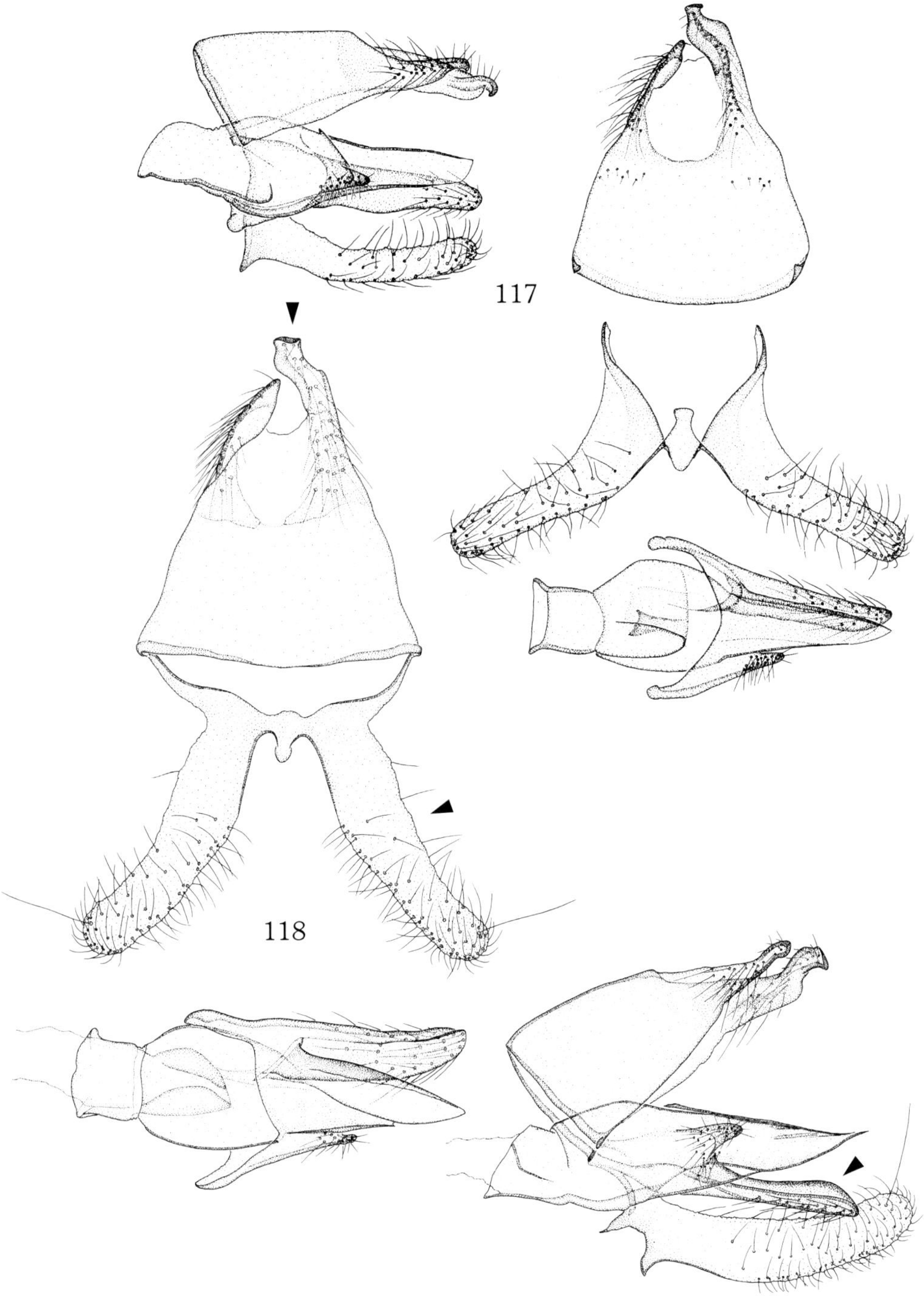

Male Genitalia

117. *Vulcaniella pomposella* (Zeller) – Greece, gen. slide JCK 3338 (LMAD).
118. *Vulcaniella grandiferella* Sinev – Ukraine (Crimea), gen. in vial (ZIN).

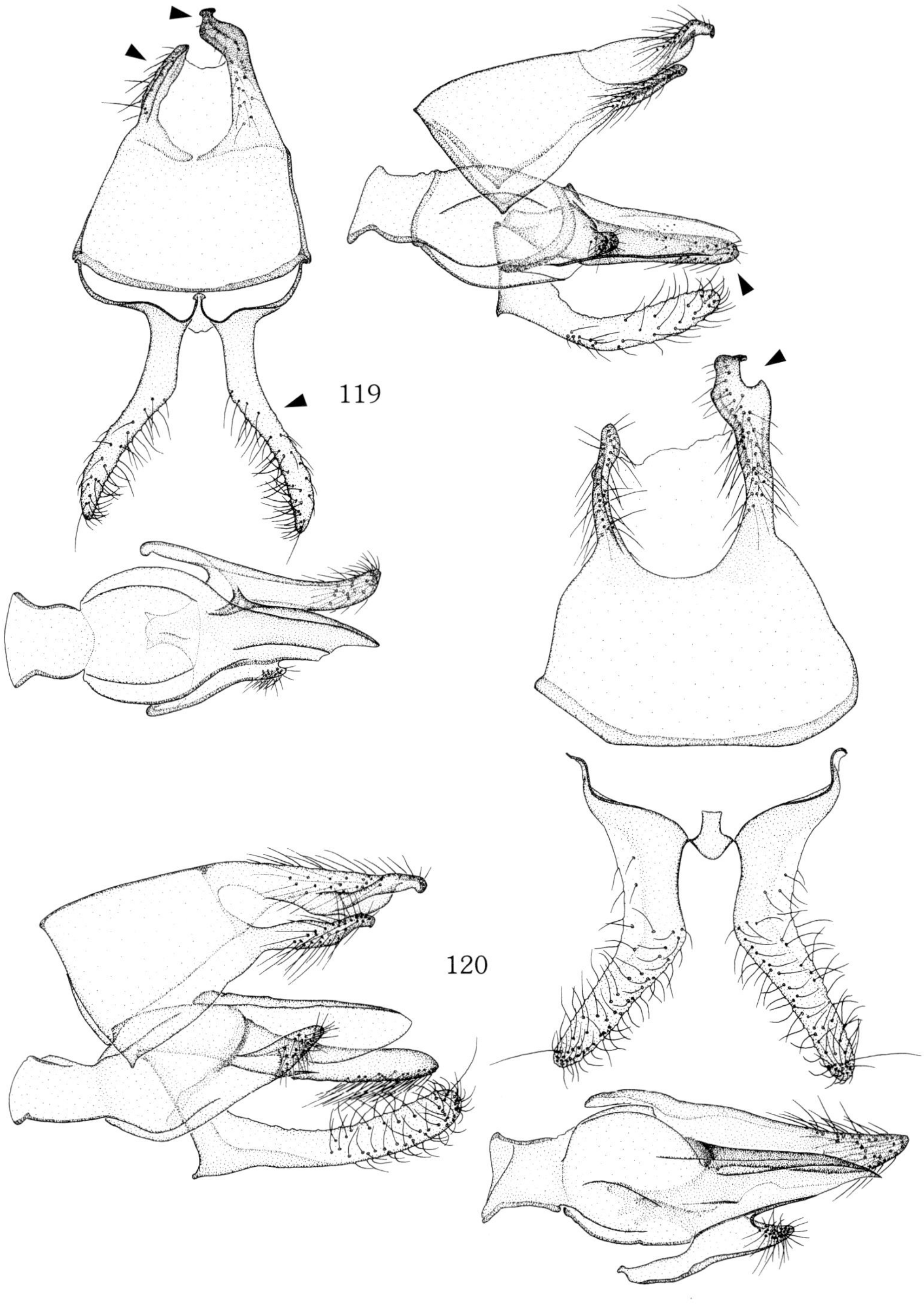

Male Genitalia

119. *Vulcaniella cognatella* Riedl – Greece, gen. slide JCK 2124 (KTR).
120. *Vulcaniella extremella* (Wocke) – Spain, gen. slide JCK 3637 (GLS).

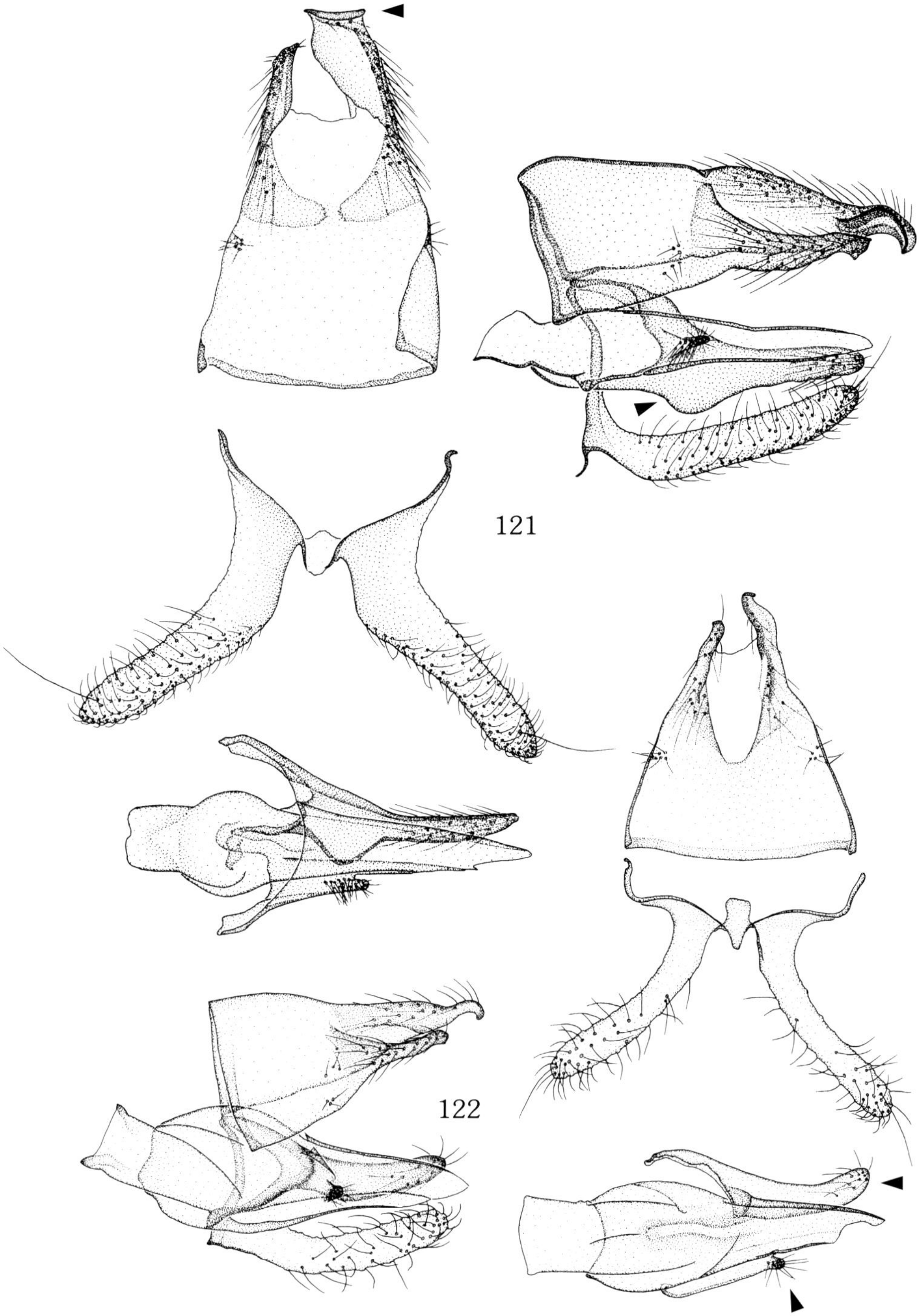

MALE GENITALIA

121. *Vulcaniella caucasica* Sinev – Turkey, gen. slide JCK 3884 (ZMUC).
122. *Vulcaniella grabowiella* (Staudinger) – Spain, gen. slide JCK 3626 (ABR).

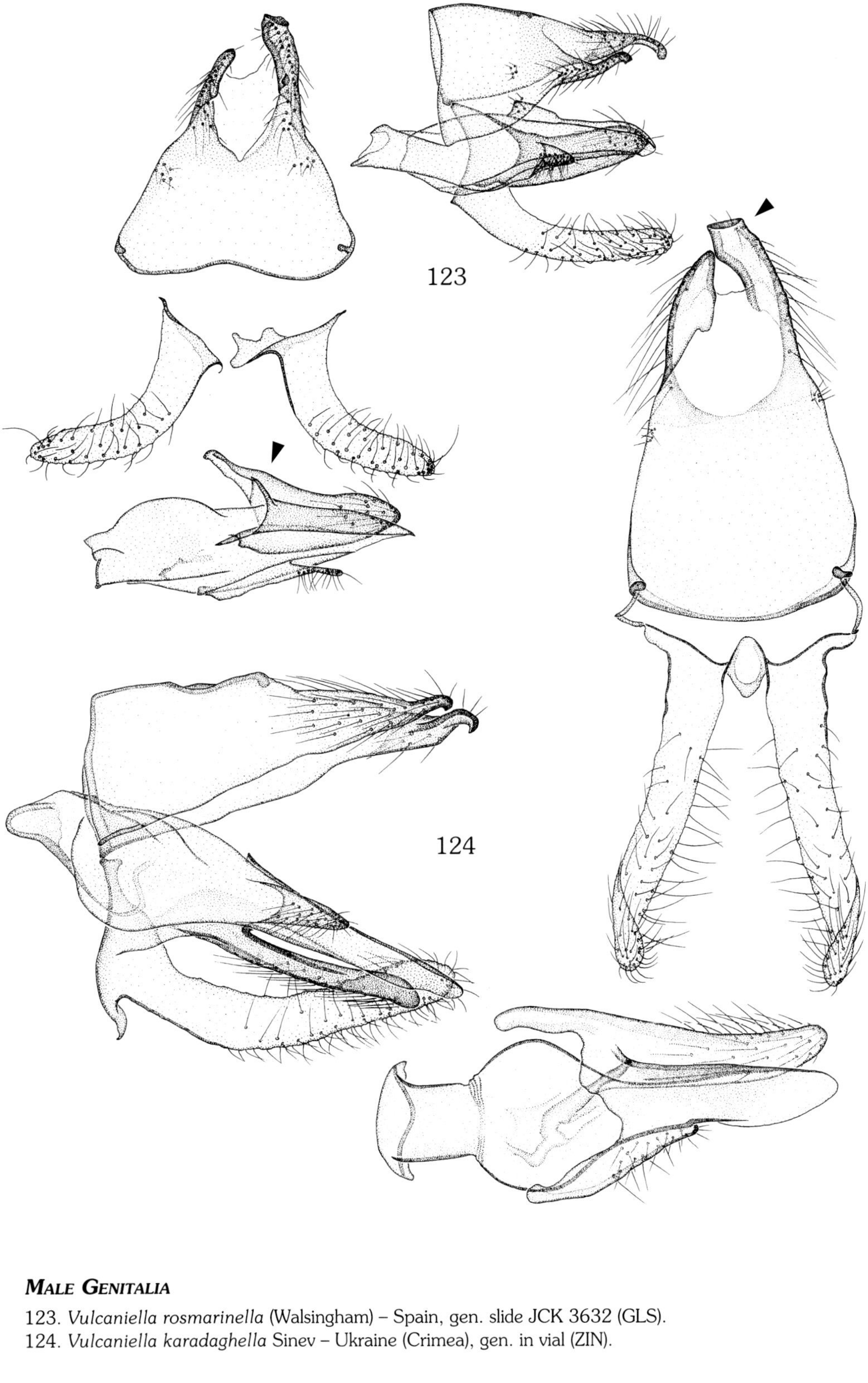

Male Genitalia

123. *Vulcaniella rosmarinella* (Walsingham) – Spain, gen. slide JCK 3632 (GLS).
124. *Vulcaniella karadaghella* Sinev – Ukraine (Crimea), gen. in vial (ZIN).

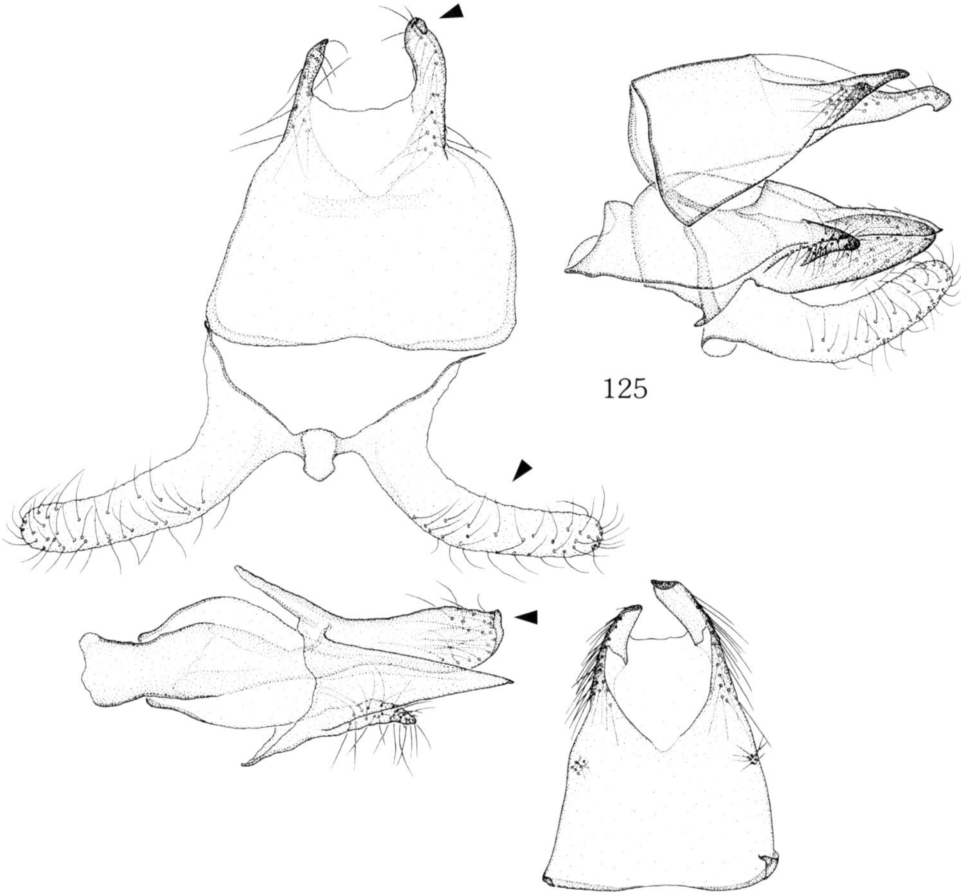

125

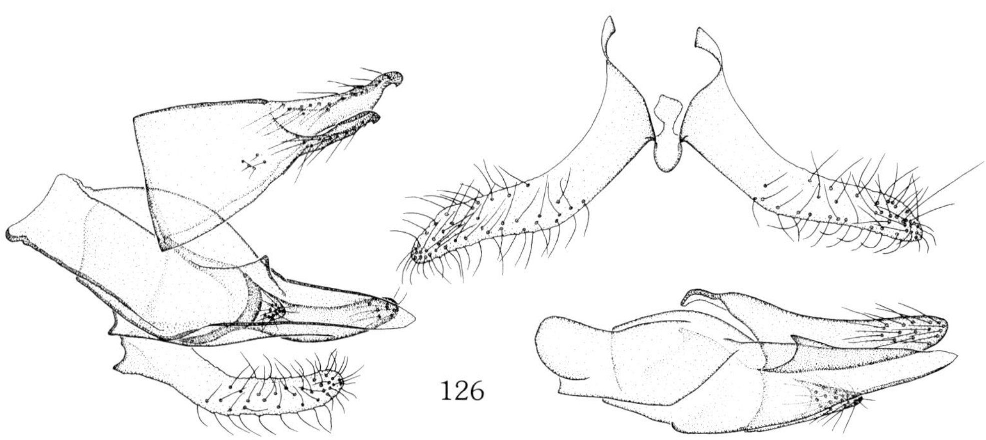

126

MALE GENITALIA

125. *Vulcaniella klimeschi* (Riedl) – Greece, gen. slide JCK 4356 (STR).
126. *Vulcaniella fiordalisa* (Petry) – Spain, gen. slide JCK 3634 (GLS).

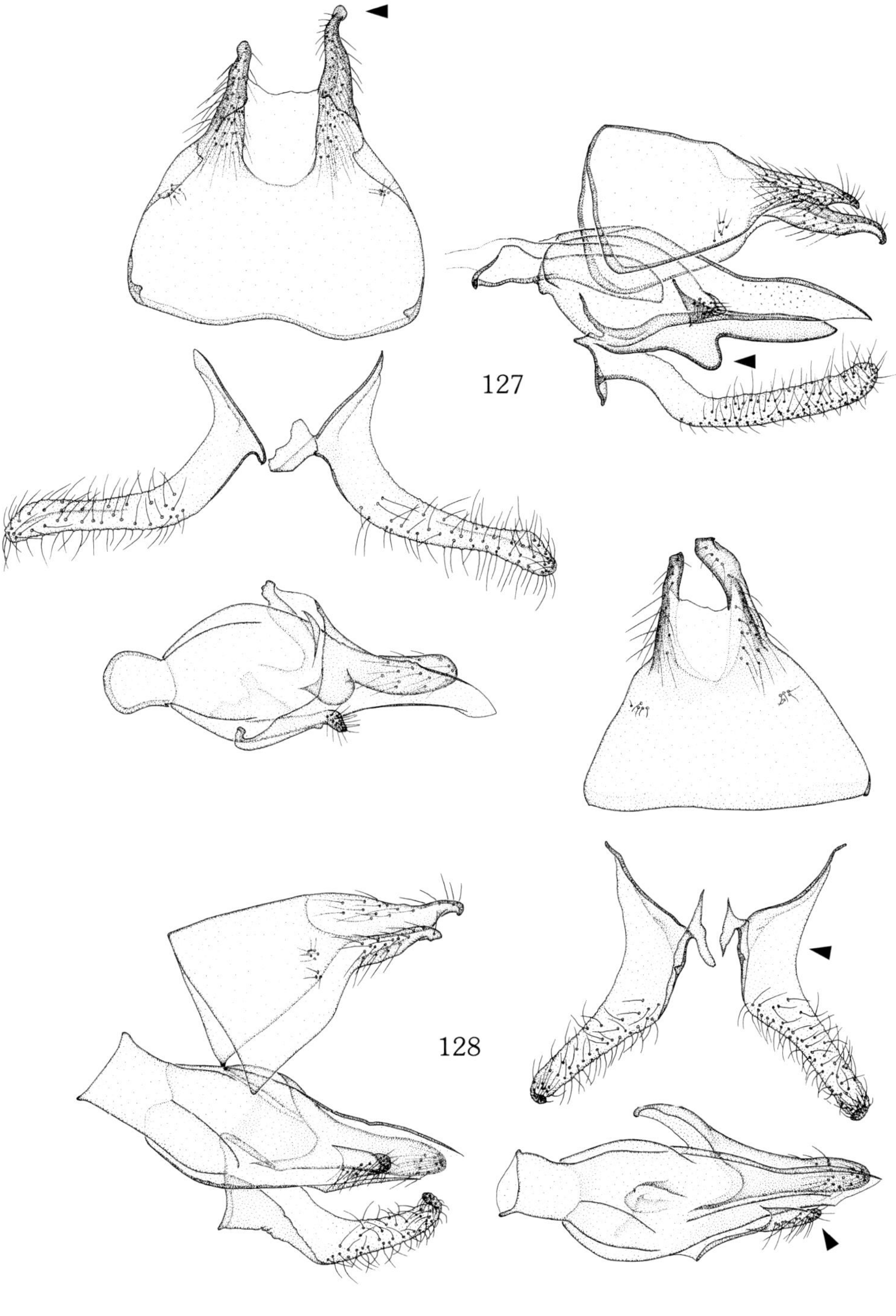

MALE GENITALIA

127. *Vulcaniella glaseri* (Riedl) – paratype, Turkey, gen. slide 738/R (NHMW).
128. *Vulcaniella gielisi* sp. n. – holotype, Spain, gen. slide JCK 3112 (RMNH).

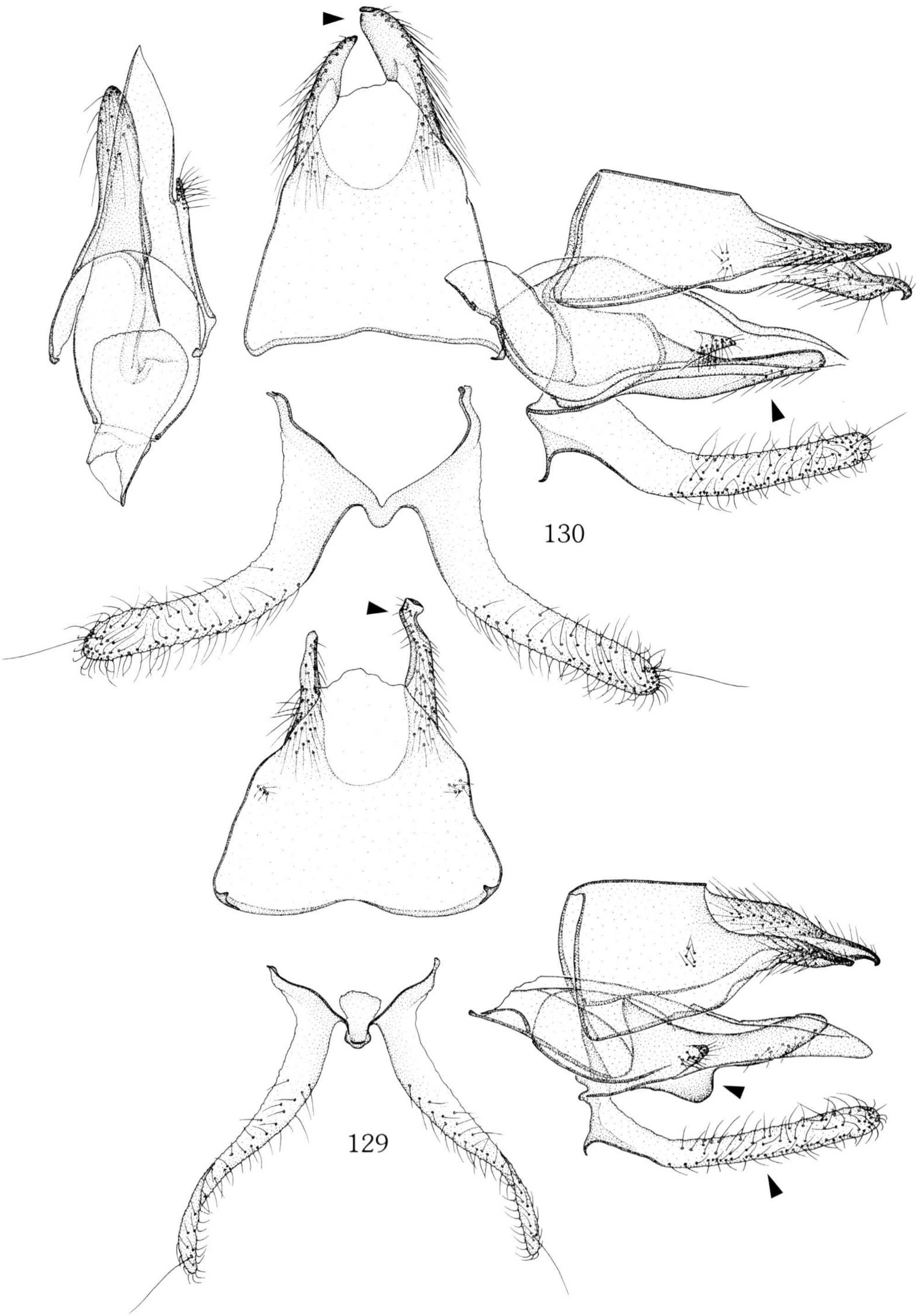

Male Genitalia

129. *Vulcaniella anatolica* sp. n. – paratype, Turkey, gen. slide JCK 3899 (ZMUC).
130. *Vulcaniella pontica* sp. n. – holotype, gen. slide JCK 3897 (ZMUC).

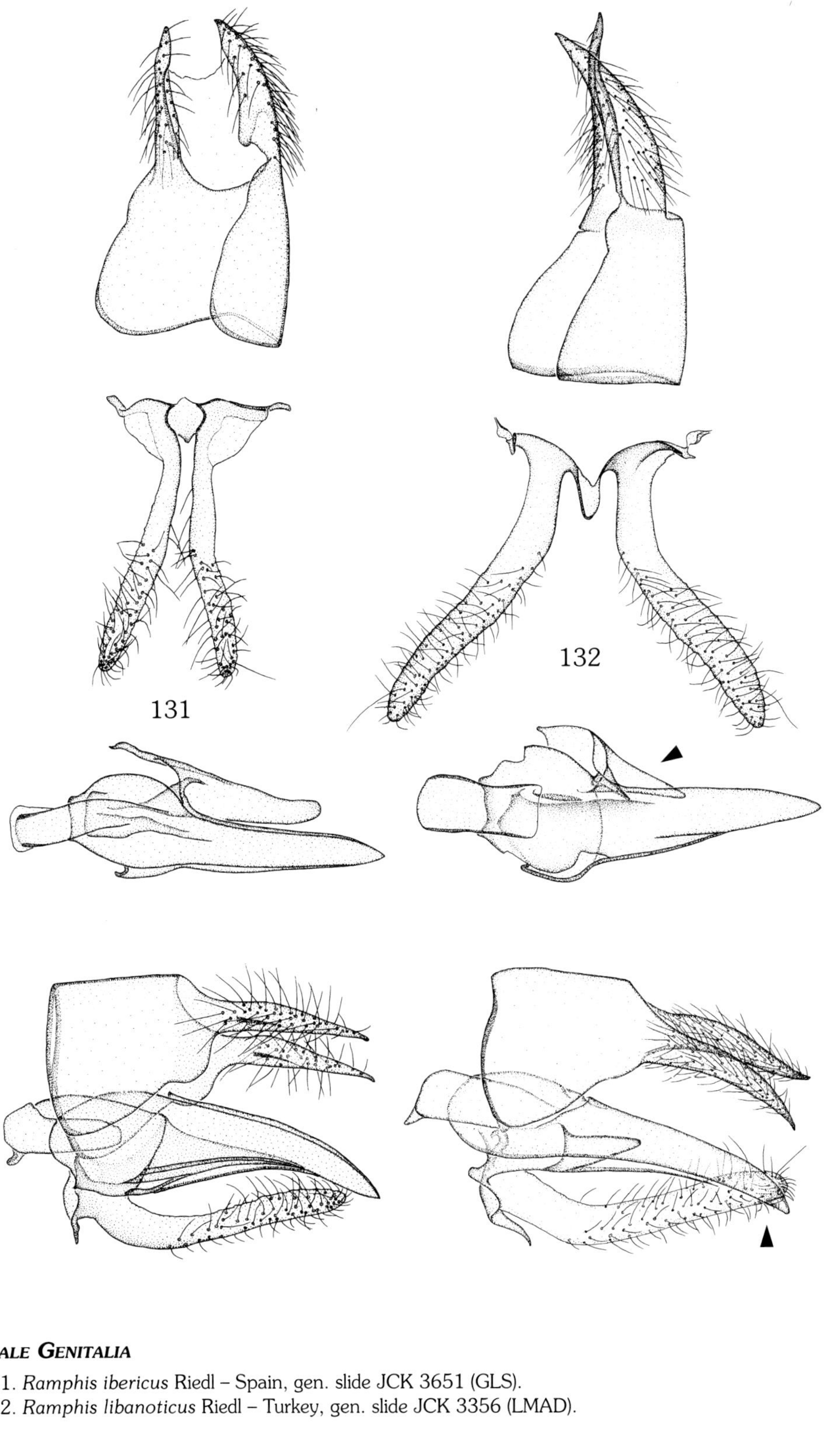

Male Genitalia

131. *Ramphis ibericus* Riedl – Spain, gen. slide JCK 3651 (GLS).
132. *Ramphis libanoticus* Riedl – Turkey, gen. slide JCK 3356 (LMAD).

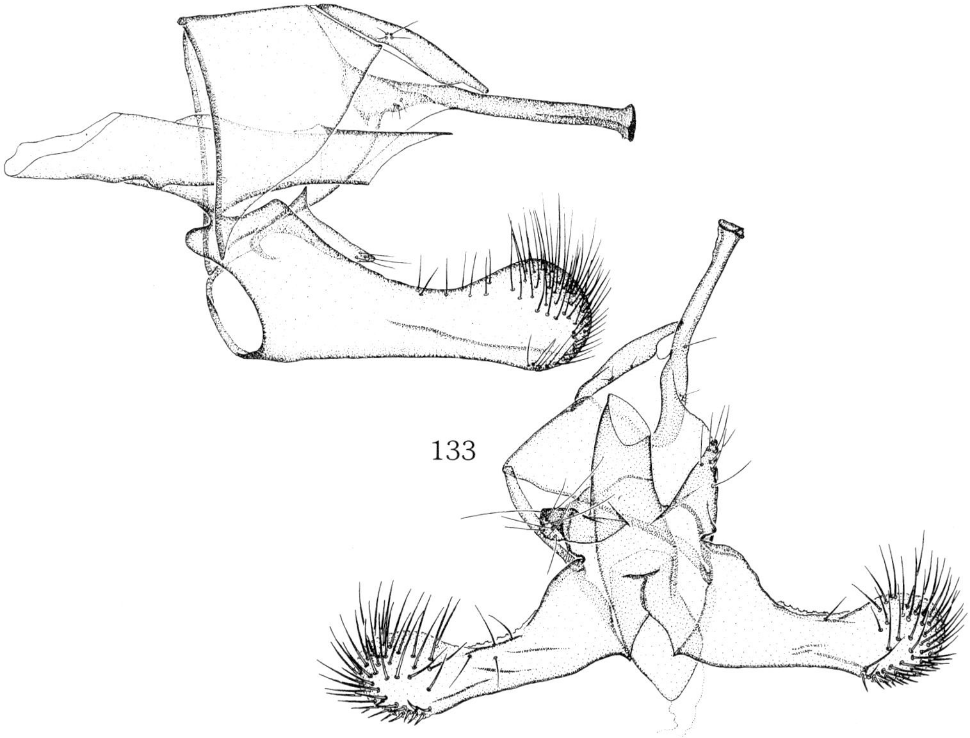

133

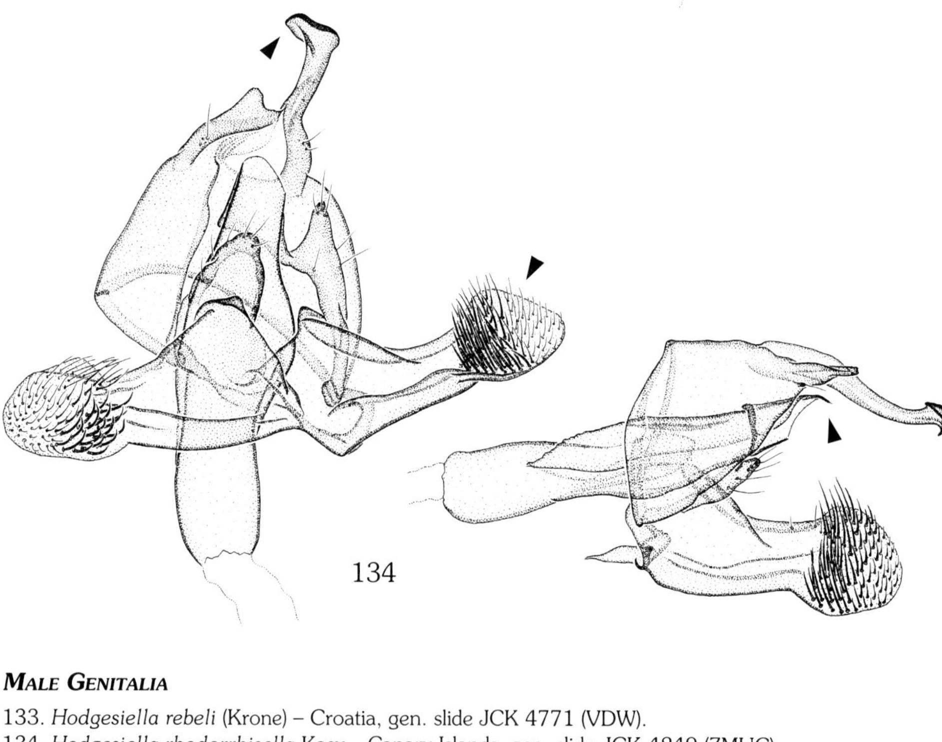

134

MALE GENITALIA

133. *Hodgesiella rebeli* (Krone) – Croatia, gen. slide JCK 4771 (VDW).
134. *Hodgesiella rhodorrhisella* Kasy – Canary Islands, gen. slide JCK 4249 (ZMUC).

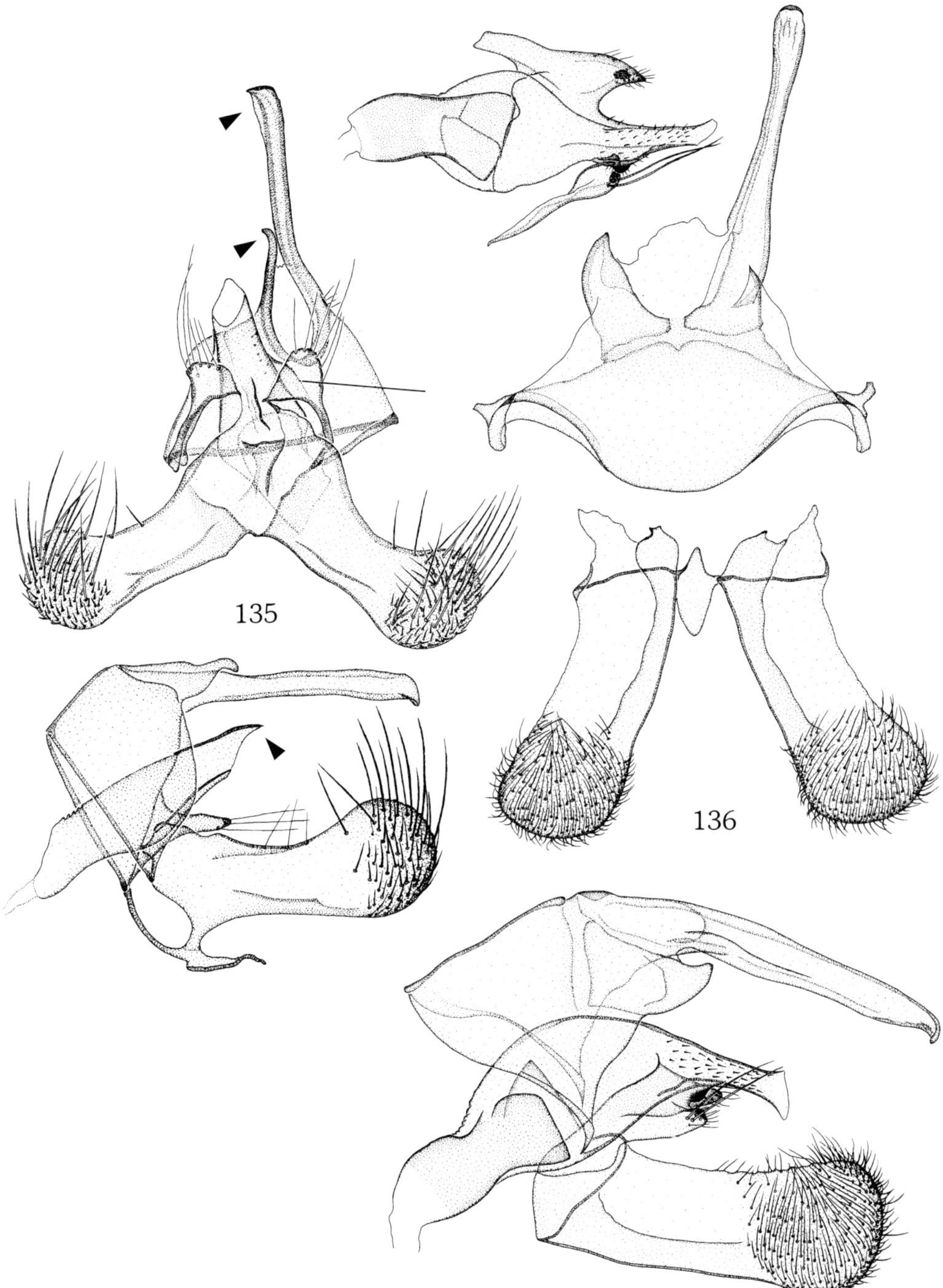

135

136

Male Genitalia

135. *Hodgesiella christophi* sp. n. – paratype, Turkey, gen. slide JCK 4668 (KTR).
136. *Tolliella fulguritella* (Ragonot) – Syria, gen. slide JCK 3659 (MNHN).

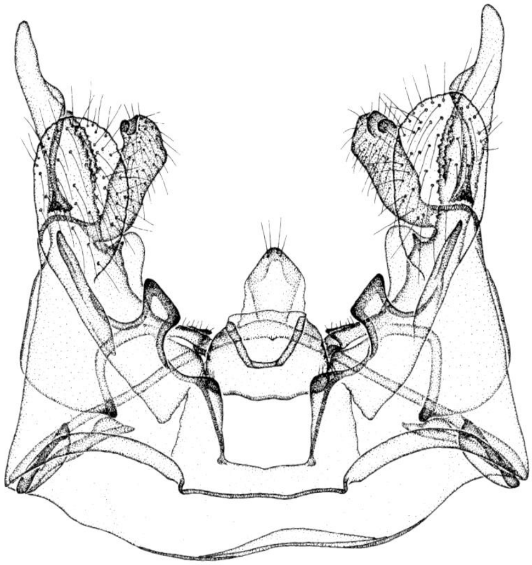

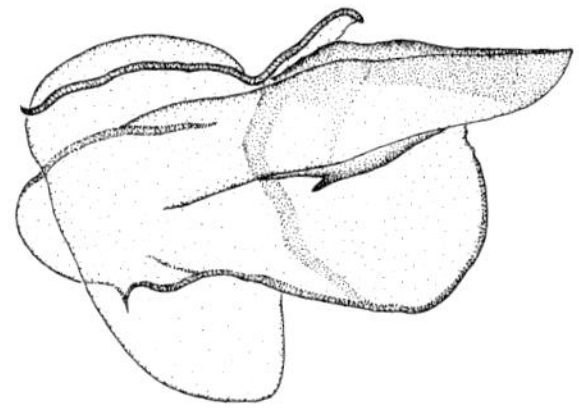

137

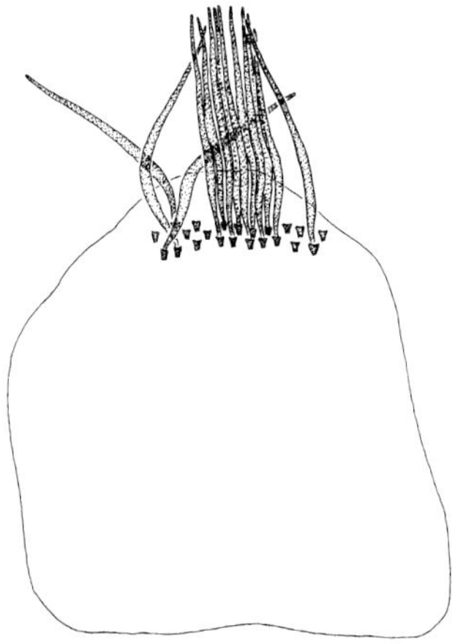

138

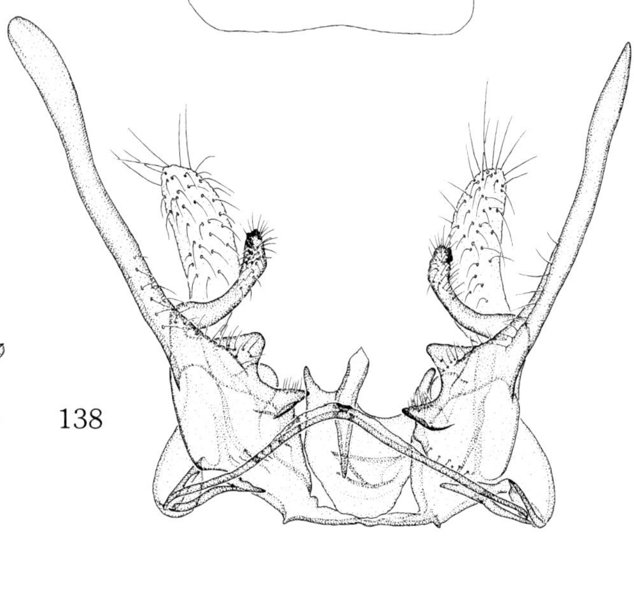

Male Genitalia

137. *Sorhagenia rhamniella* (Zeller) – Netherlands, gen. slide JCK 3813 (SRS).
138. *Sorhagenia reconditella* Riedl – Cyprus, gen. slide JCK 4623 (ZMUC).

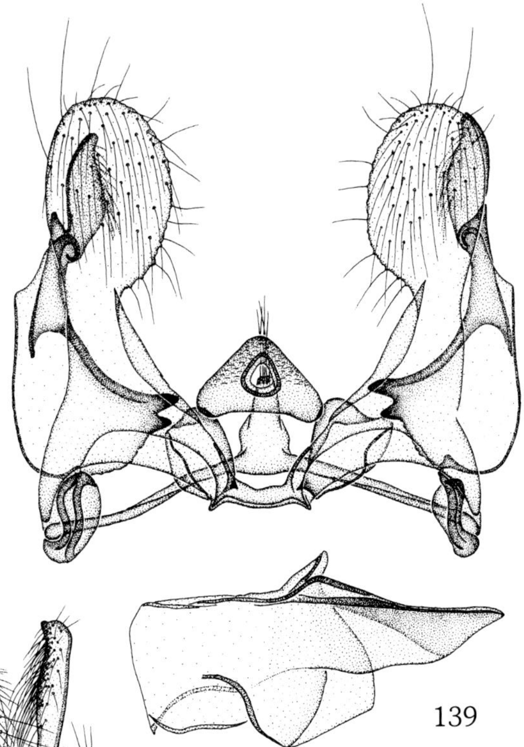

139

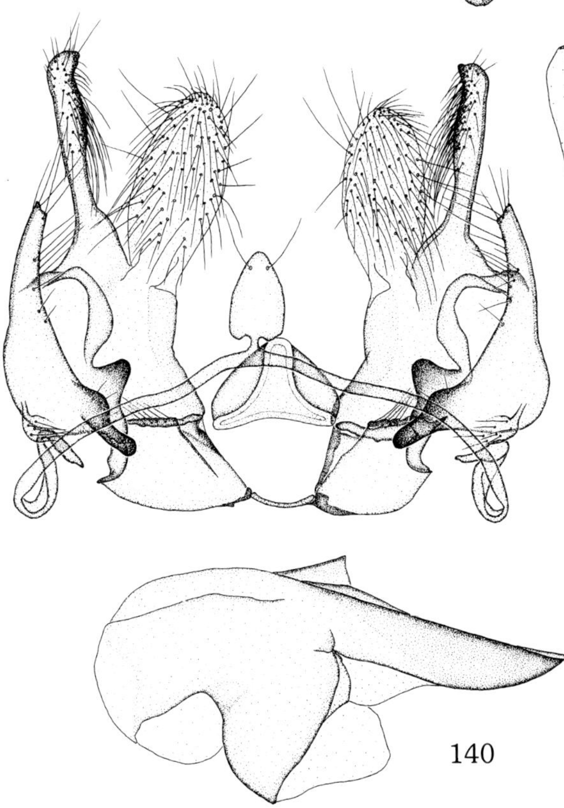

140

Male Genitalia

139. *Sorhagenia lophyrella* (Douglas) – Austria, gen. slide JCK 3867 (RMNH).
140. *Sorhagenia janiszewskae* Riedl – Germany, gen. slide JCK 3484 (LMAD).

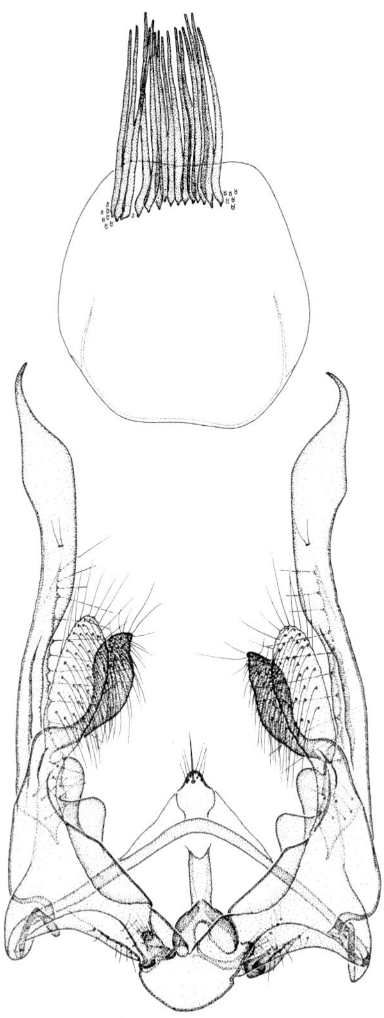

141

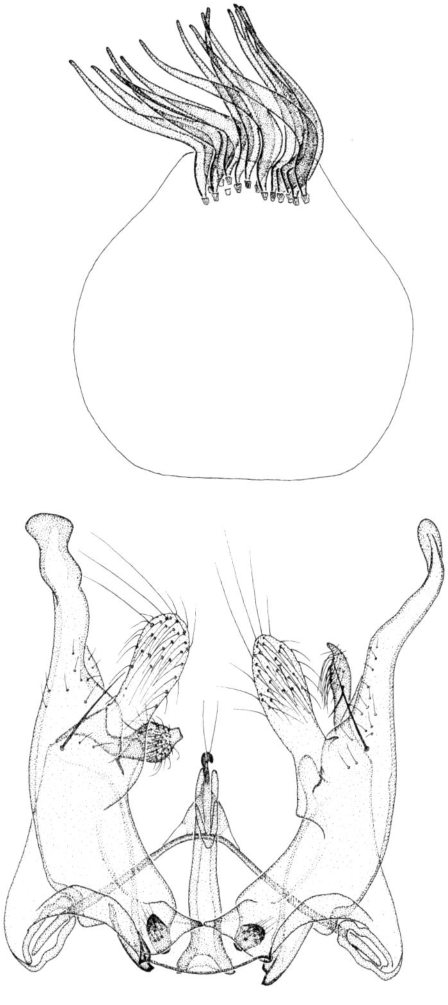

142

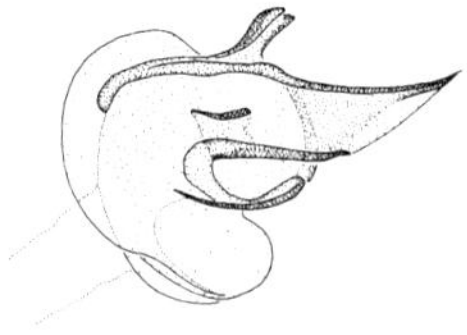

Male Genitalia

141. *Sorhagenia fibigeri* sp.n. – holotype, Turkey, gen. slide JCK 4145 (ZMUC).
142. *Sorhagenia taurensis* sp.n. – holotype, Turkey, gen. slide JCK 4129 (ZMUC).

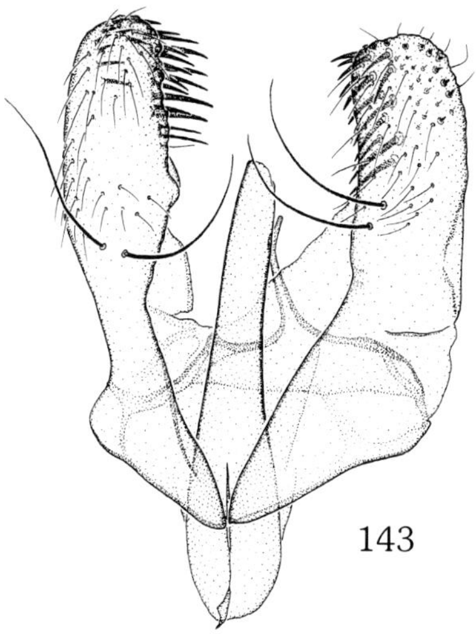

143

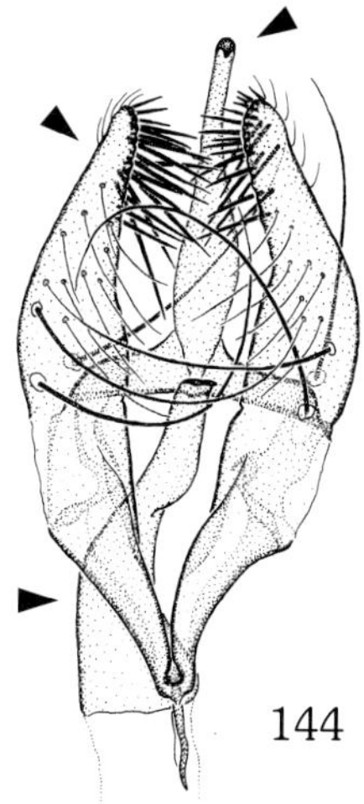

144

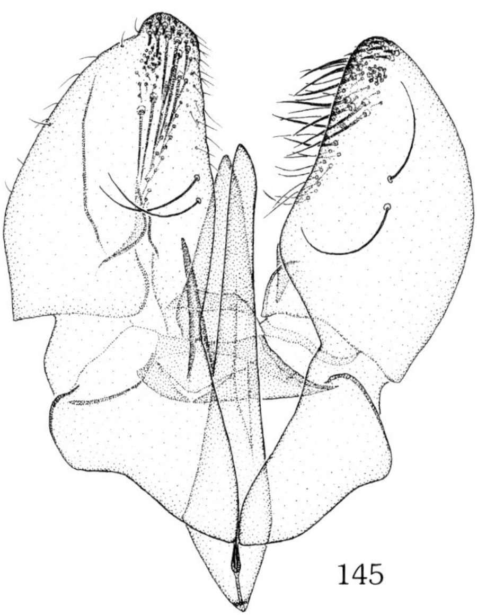

145

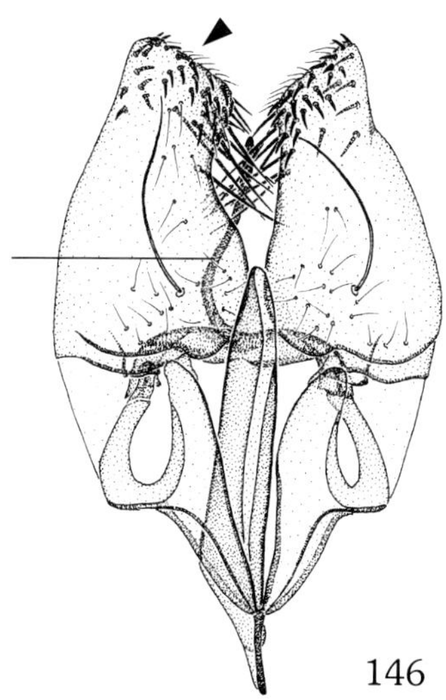

146

Male Genitalia

143. *Ascalenia vanella* (Frey) – Spain, gen. slide JCK 4161 (ZMUC).
144. *Ascalenia vanelloides* Gerasimov – Turkey, gen. slide JCK 4485 (ZMUC).
145. *Ascalenia antiqua* Meyrick – Israel, gen. in vial (LSNK).
146. *Ascalenia acaciella* Chrétien – Iran, gen. slide JCK 4271 (NHMW).
147. *Ascalenia jerichoella* Amsel – unknown.

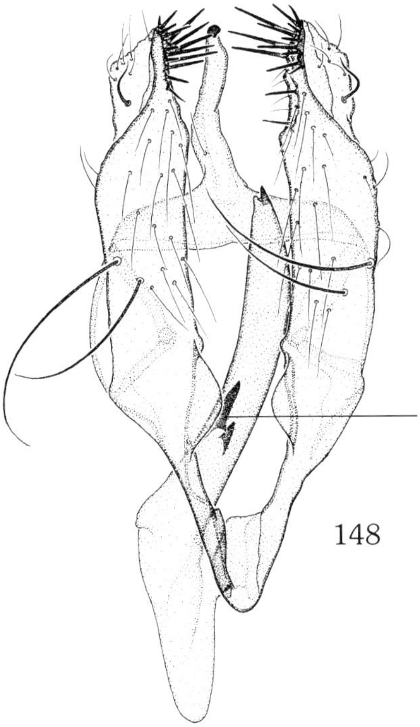

148

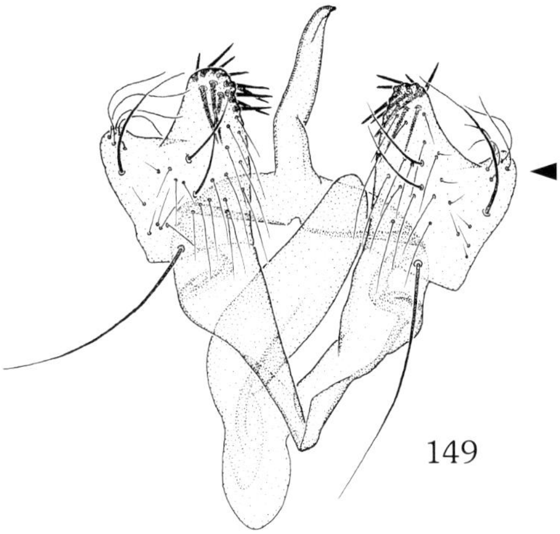

149

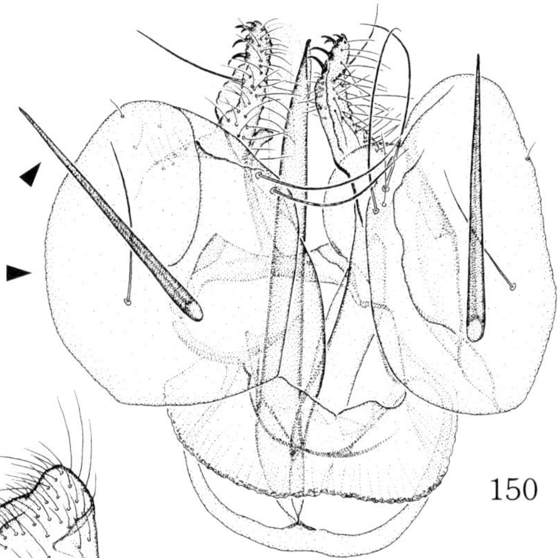

150

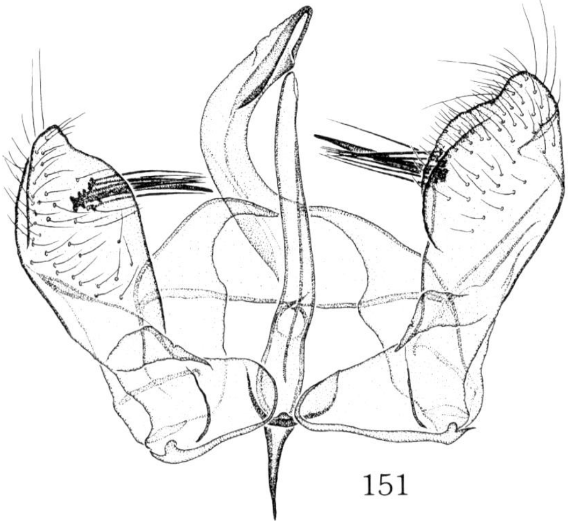

151

Male Genitalia

148. *Ascalenia viviparella* Kasy – Turkey, gen. slide JCK 4146 (ZMUC).
149. *Ascalenia echidnias* (Meyrick) – Tunisia, gen. slide BÅB 1965 (ZMUC).
150. *Ascalenia bifasciella* Chrétien – coll. De Joannis, gen. slide JCK 4480 (MNHN).
151. *Calanesia karsholti* Sinev – holotype, Tunisia, gen. in vial (ZIN).

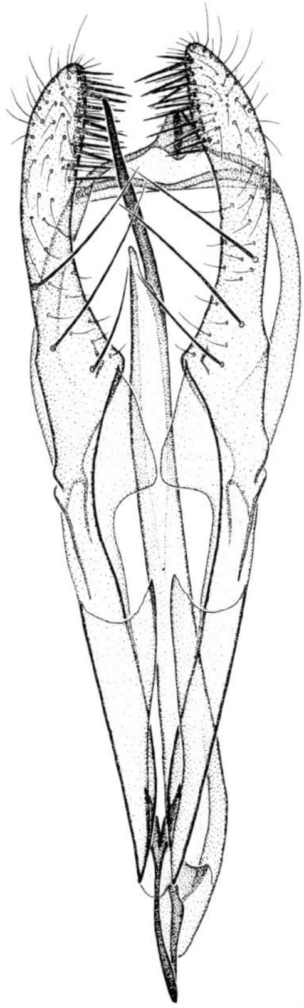

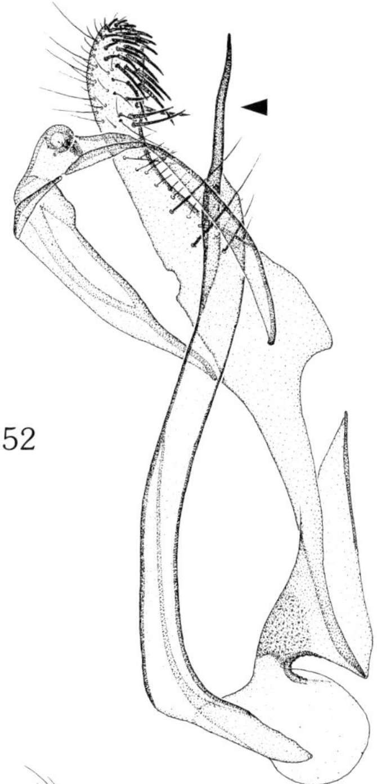

152

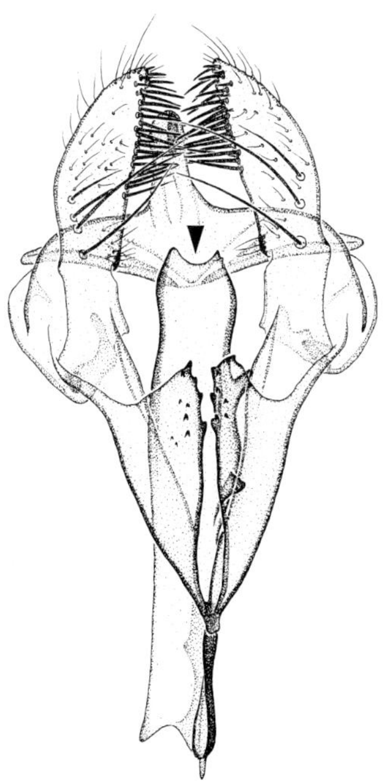

153

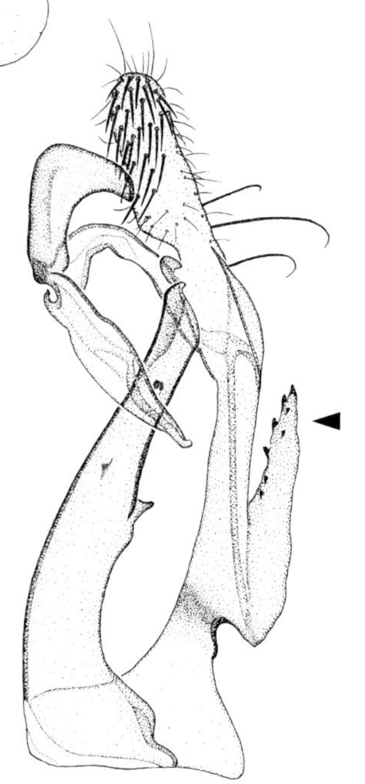

MALE GENITALIA

152. *Calycobathra acarpa* Meyrick – Morocco, gen. in vial (ZMUC).
153. *Calycobathra calligoni* Sinev – paratype, Uzbekistan, gen. in vial (ZIN).

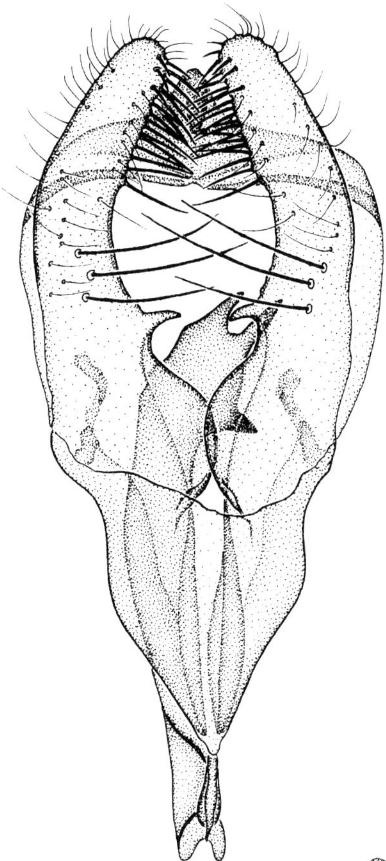

154

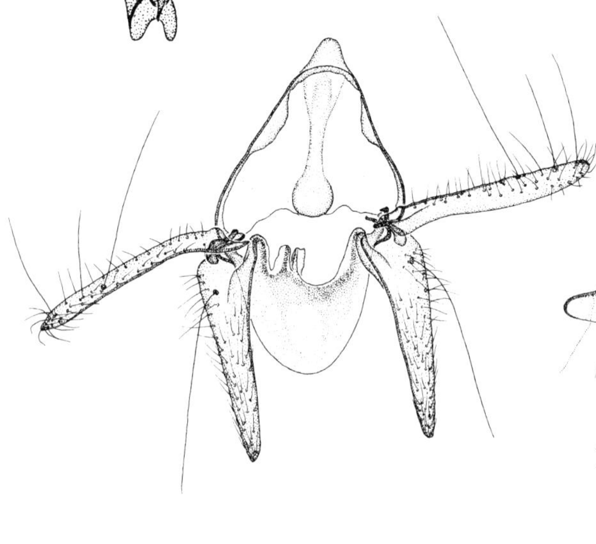

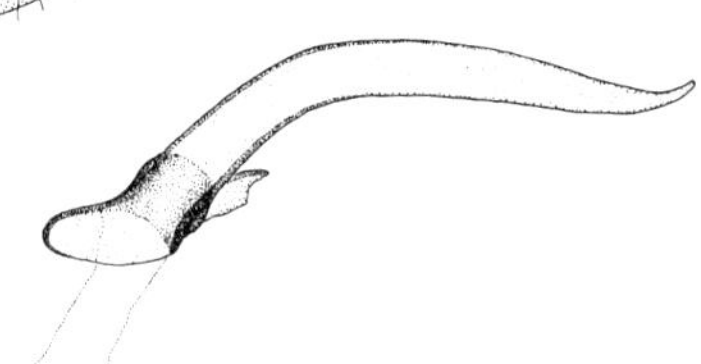

155

Male Genitalia

154. *Calycobathra variapennella* Sinev – Turkey, gen. slide JCK 4068 (ZMUC).
155. *Bifascia nigralbella* (Chrétien) – Tunisia, gen. in vial (ZMUC).

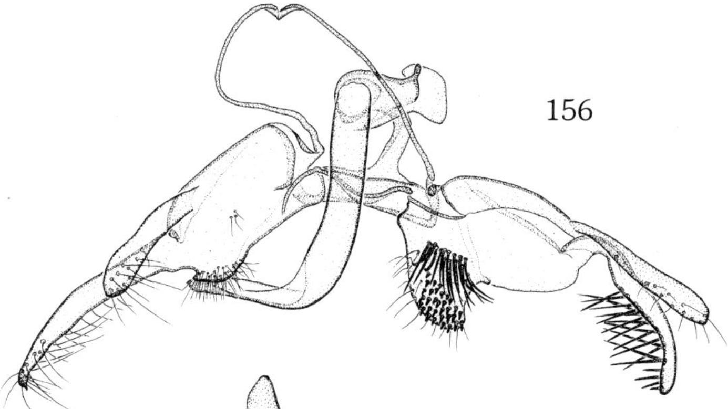

156

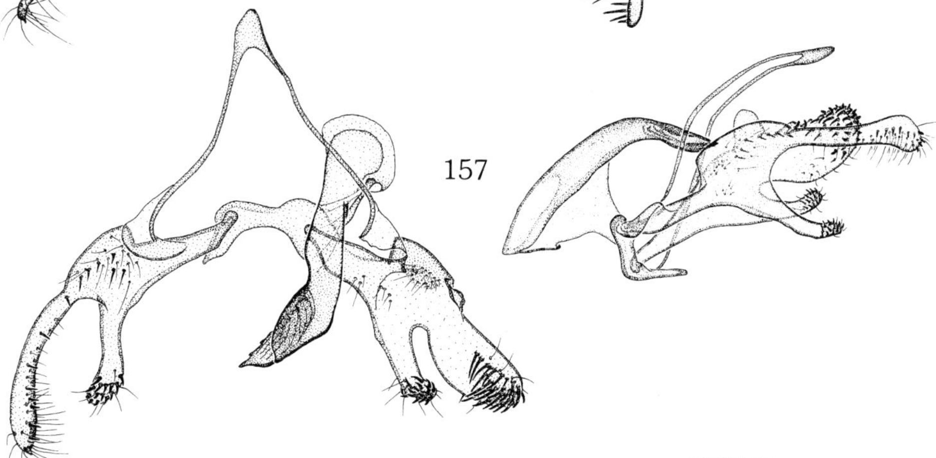

157

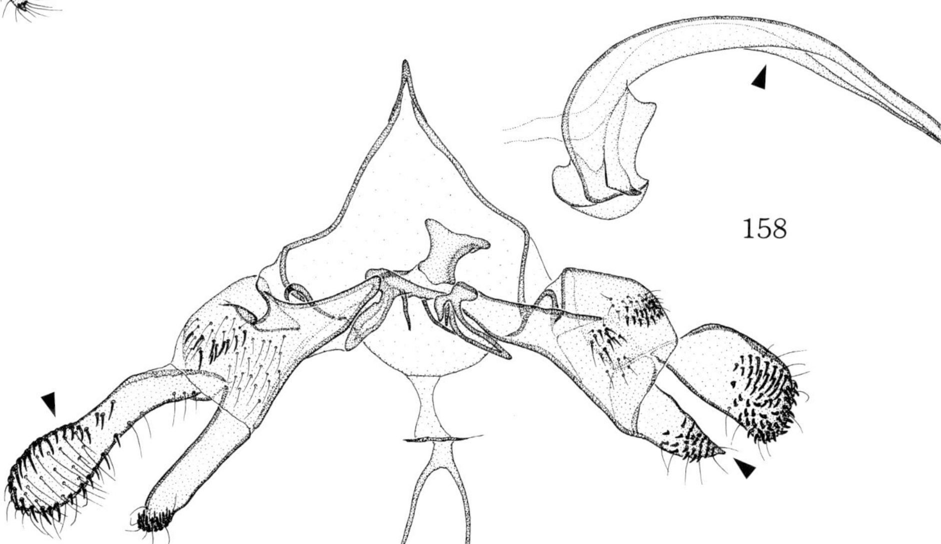

158

MALE GENITALIA

156. *Bifascioides leucomelanellus* (Rebel) – Iran, gen. in vial (NHMW).
157. *Gisilia stereodoxa* (Meyrick) – Iran, gen. in vial (NHMW).
158. *Gisilia lerautella* Gibeaux – holotype, France, gen. slide CG 2479 (GBX).

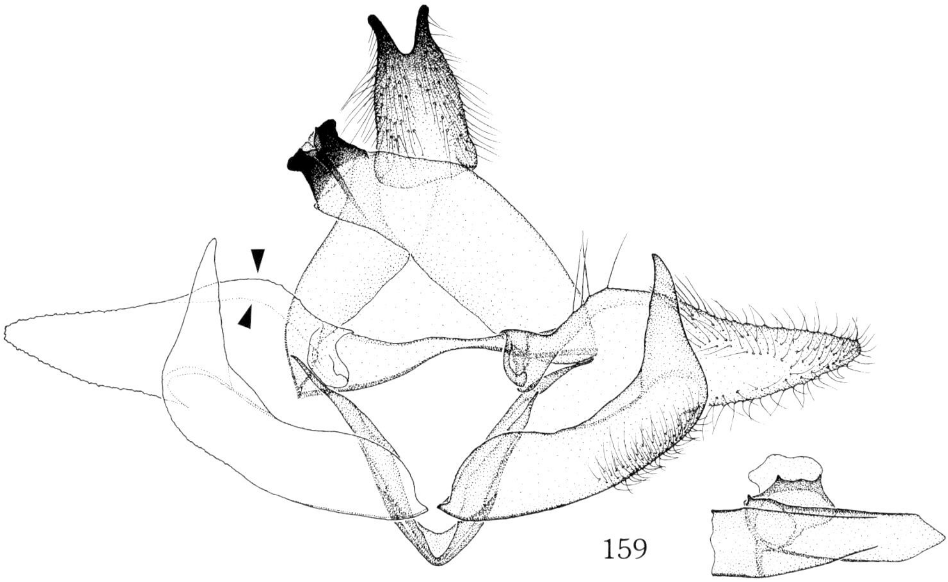

159

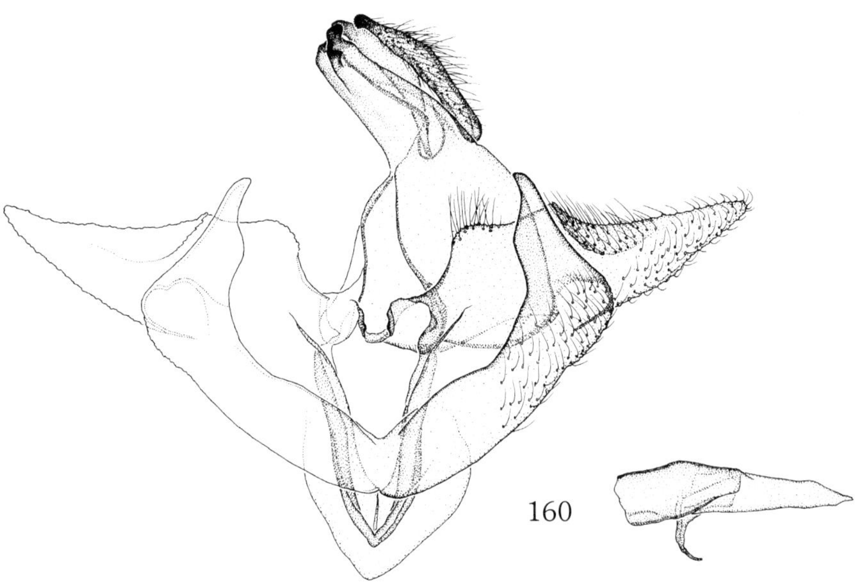

160

Male Genitalia

159. *Neomariania partinicensis* (Rebel) – Greece, gen. slide G.W. 5454 (STR).
160. *Neomariania rebeli* (Walsingham) – Madeira, gen. slide JCK 4251 (ZMUC).

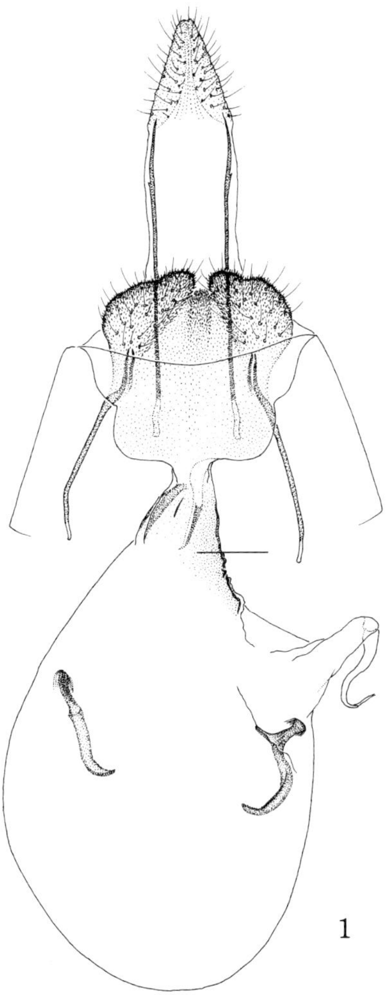
1

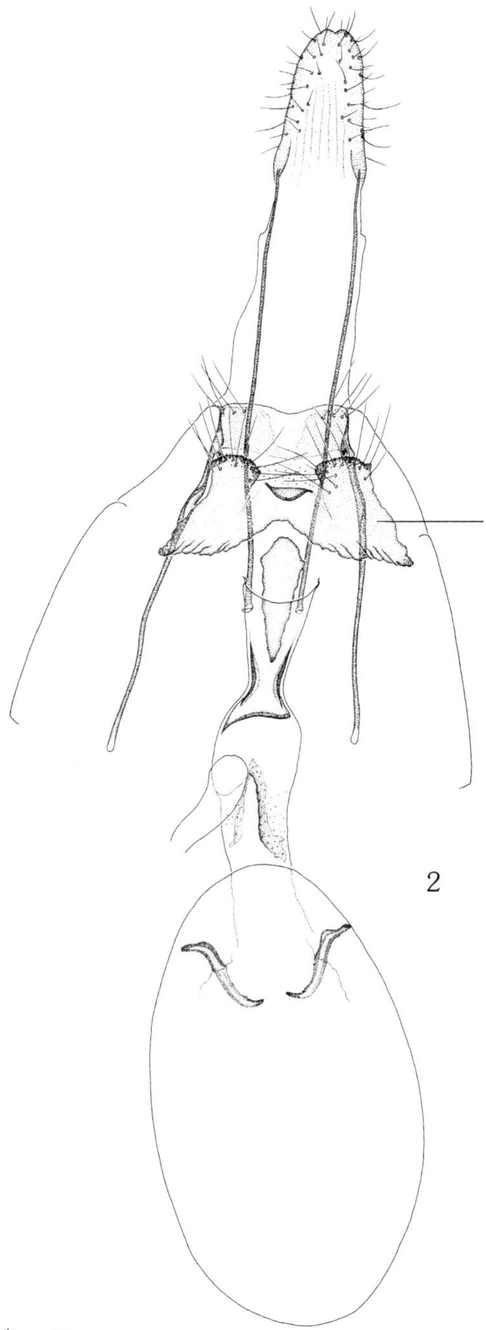
2

FEMALE GENITALIA

1. *Mompha conturbatella* (Hübner) – Netherlands, gen. slide JCK 2455 (KTR).
2. *Mompha ochraceella* (Curtis) – Netherlands, gen. slide JCK 2393 (KTR).

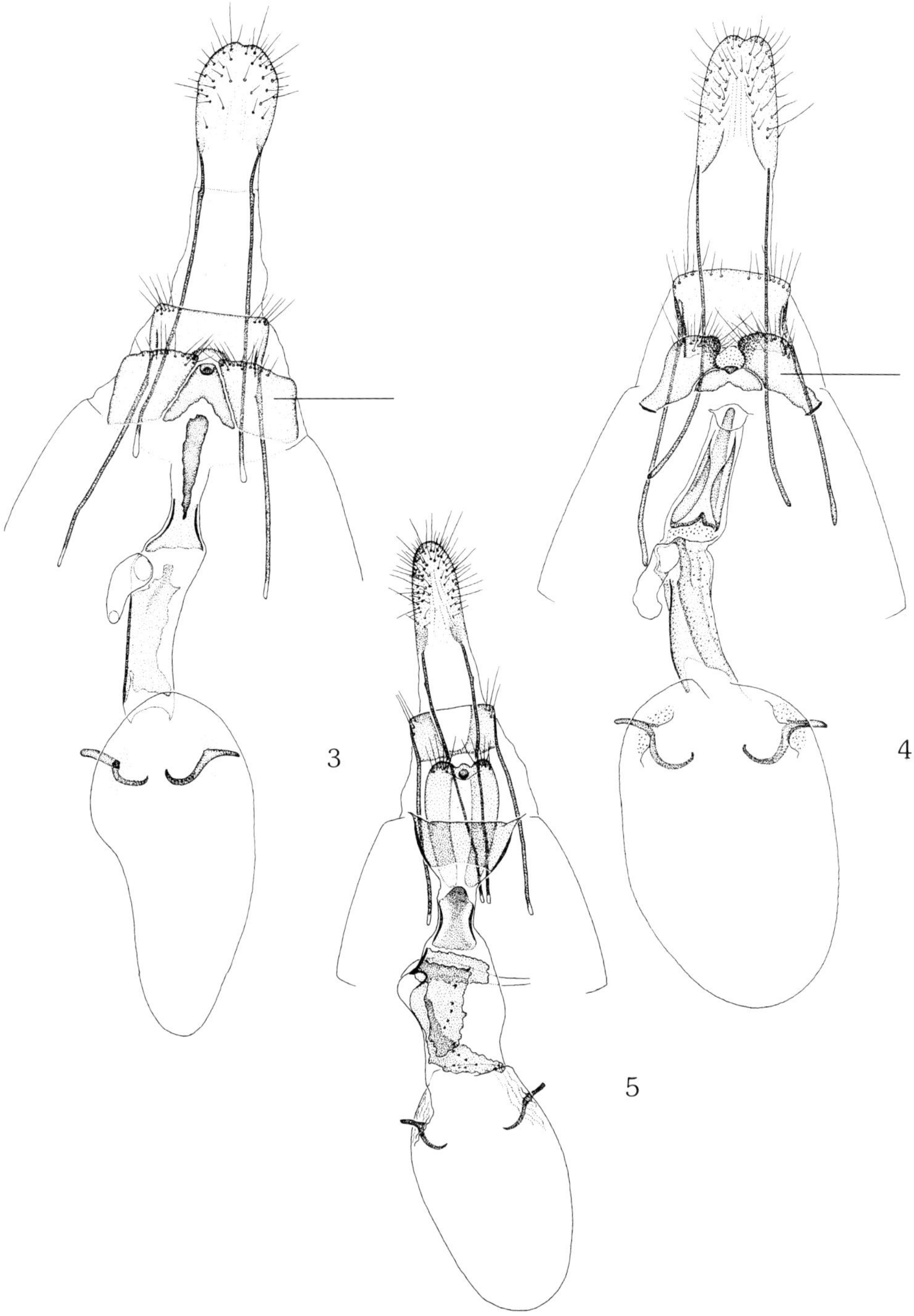

FEMALE GENITALIA

3. *Mompha lacteella* (Stephens) – Denmark, gen. slide JCK 3361 (ZMUC).
4. *Mompha propinquella* (Stainton) – Denmark, gen. slide JCK 2474 (KTR).
5. *Mompha divisella* Herrich-Schäffer – Netherlands, gen. slide JCK 3783 (RMNH).

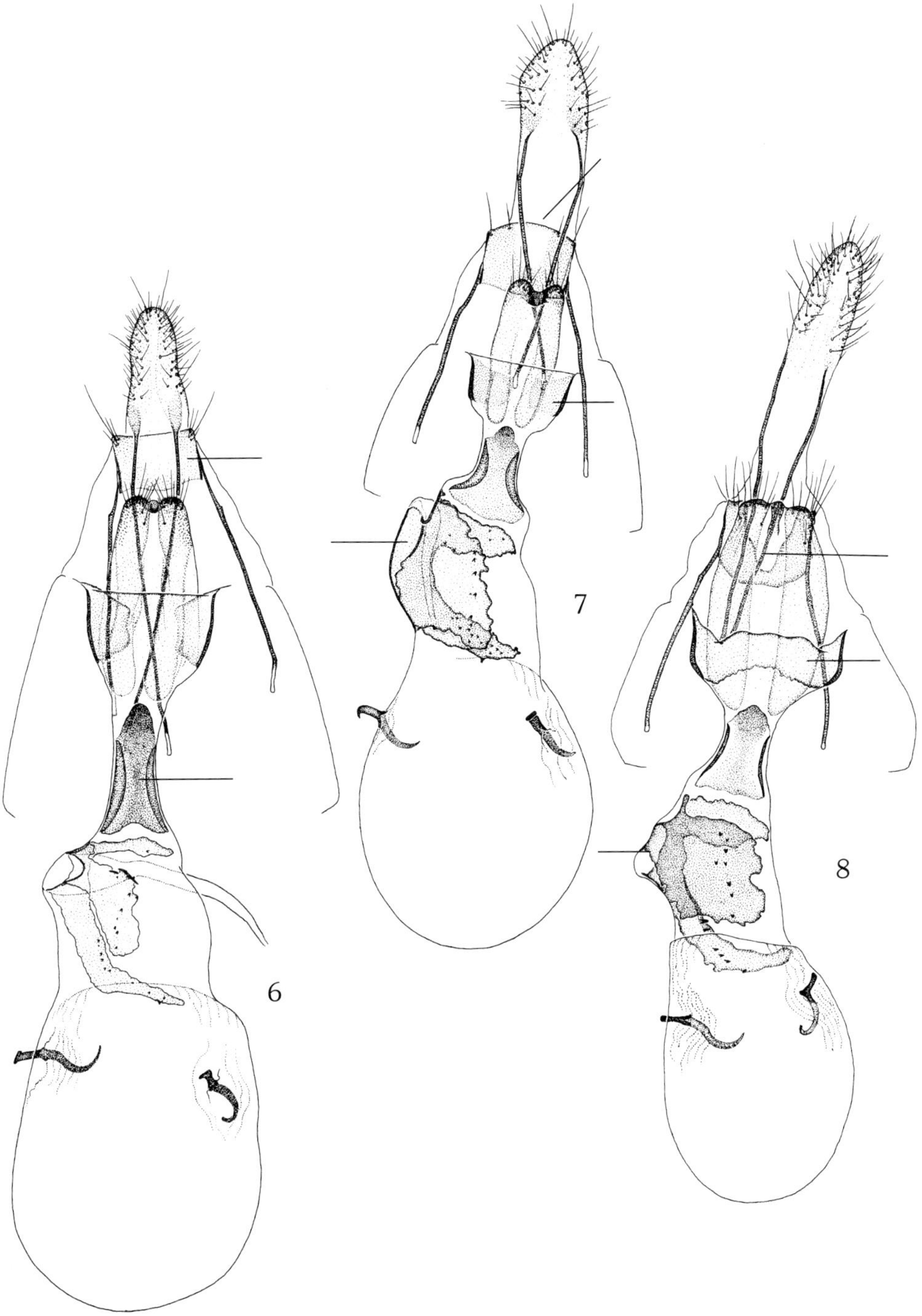

Female Genitalia

6. *Mompha jurassicella* (Frey) – U.K., gen. slide JCK 2075 (KTR).
7. *Mompha bradleyi* Riedl – Netherlands, gen. slide JCK 2115 (KTR).
8. *Mompha confusella* Koster & Sinev – paratype, Azerbaijan, gen. in vial (ZIN).

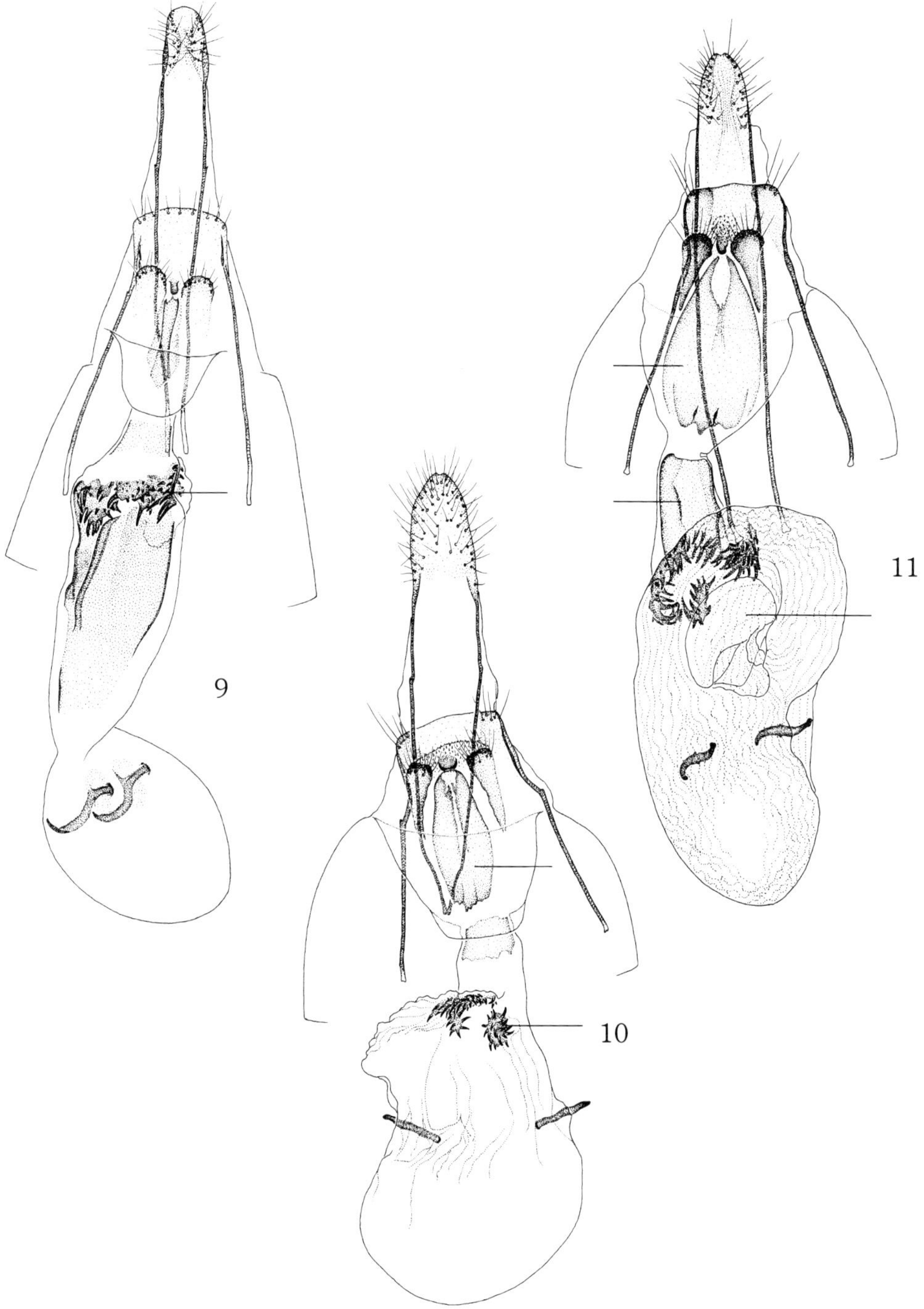

Female Genitalia

9. *Mompha subbistrigella* (Haworth) – Netherlands, gen. slide JCK 3175 (RMNH).
10. *Mompha sturnipennella* (Treitschke) – Netherlands, gen. slide JCK 2310 (KTR).
11. *Mompha meridionella* sp. n. – paratype, Greece, gen. slide JCK 3336 (ZMUC).

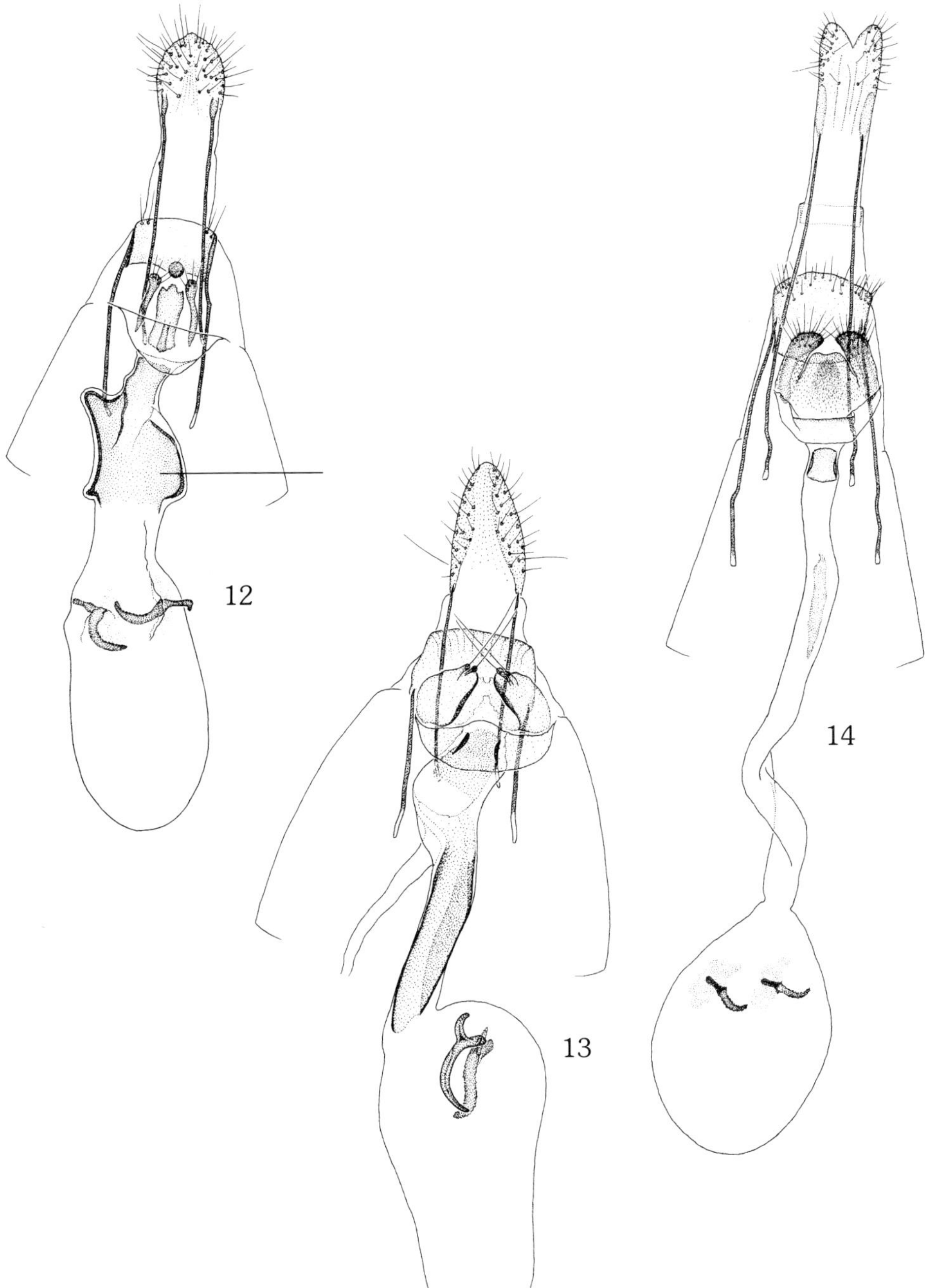

Female Genitalia

12. *Mompha epilobiella* ([Denis & Schiffermüller]) – Spain, gen. slide JCK 4029 (WLN).
13. *Mompha langiella* (Hübner) – Germany, gen. slide JCK 3208 (RMNH).
14. *Mompha idaei* (Zeller) – Germany, gen. slide JCK 3211 (RMNH).

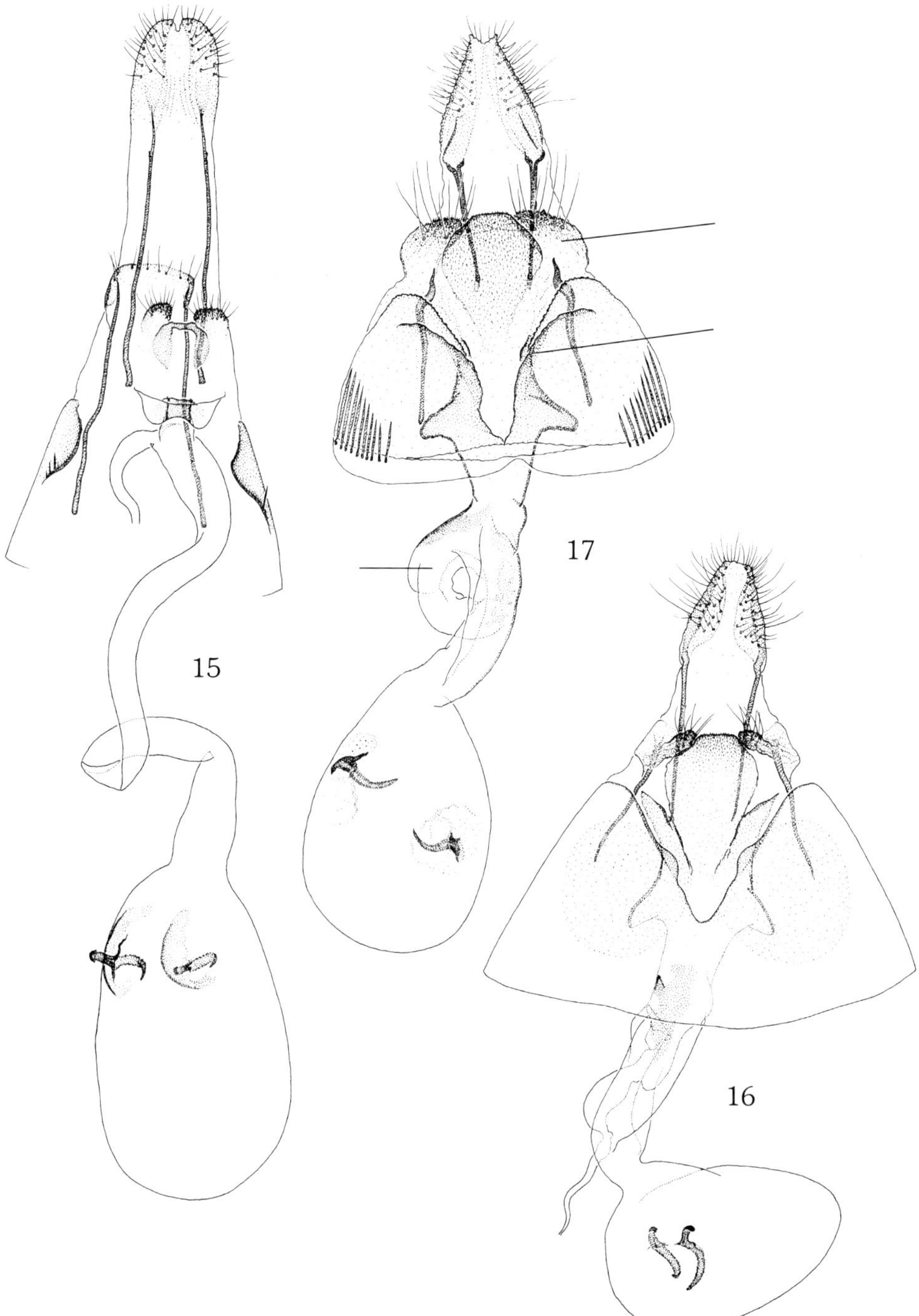

Female Genitalia

15. *Mompha miscella* ([Denis & Schiffermüller]) – Greece, gen. slide JCK 3870 (STR).
16. *Mompha locupletella* ([Denis & Schiffermüller]) – Netherlands, gen. slide JCK 2461 (KTR).
17. *Mompha maculata* Sinev – Georgia, paratype, gen. slide JCK 4370 (ZIN).

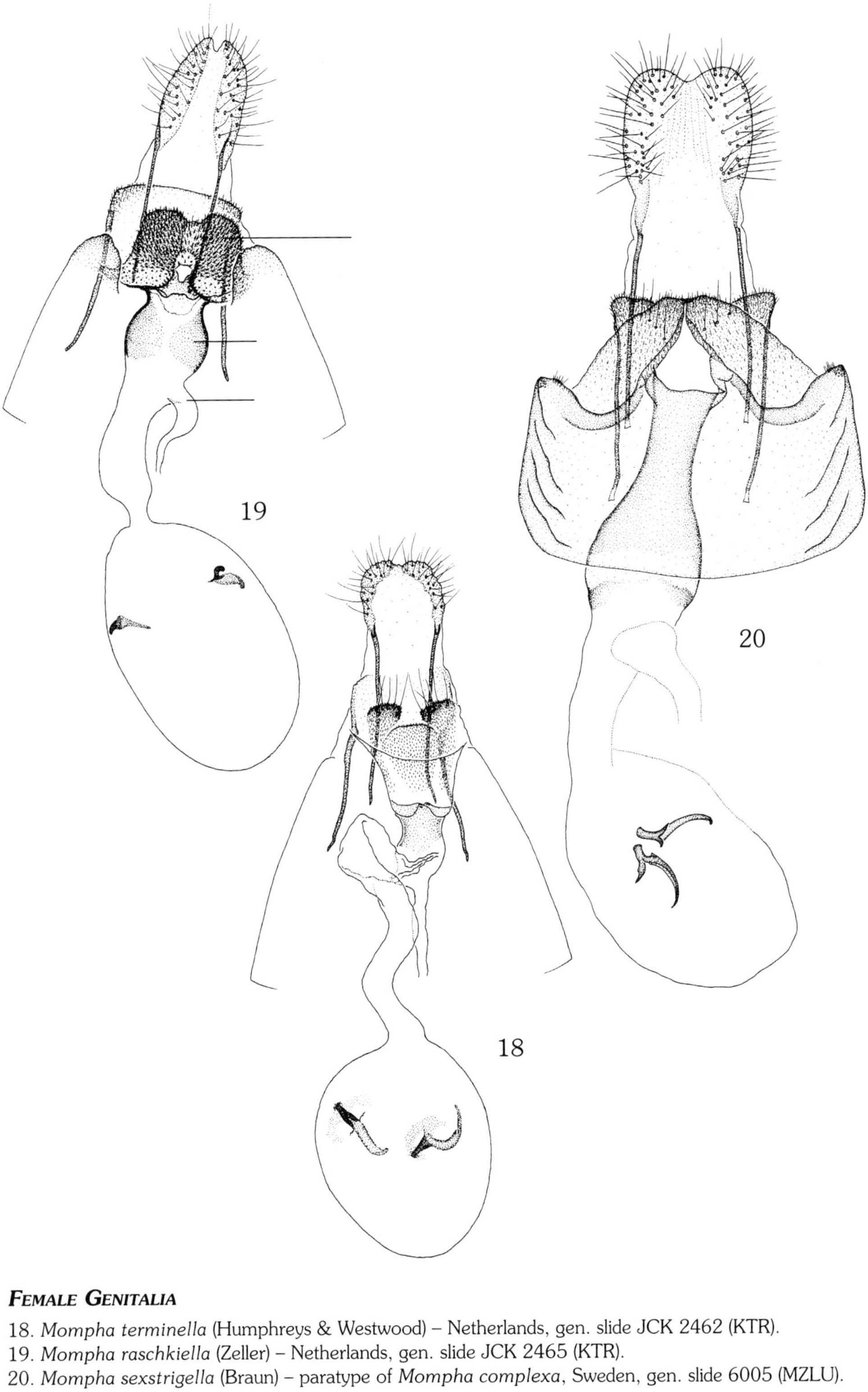

FEMALE GENITALIA

18. *Mompha terminella* (Humphreys & Westwood) – Netherlands, gen. slide JCK 2462 (KTR).
19. *Mompha raschkiella* (Zeller) – Netherlands, gen. slide JCK 2465 (KTR).
20. *Mompha sexstrigella* (Braun) – paratype of *Mompha complexa*, Sweden, gen. slide 6005 (MZLU).

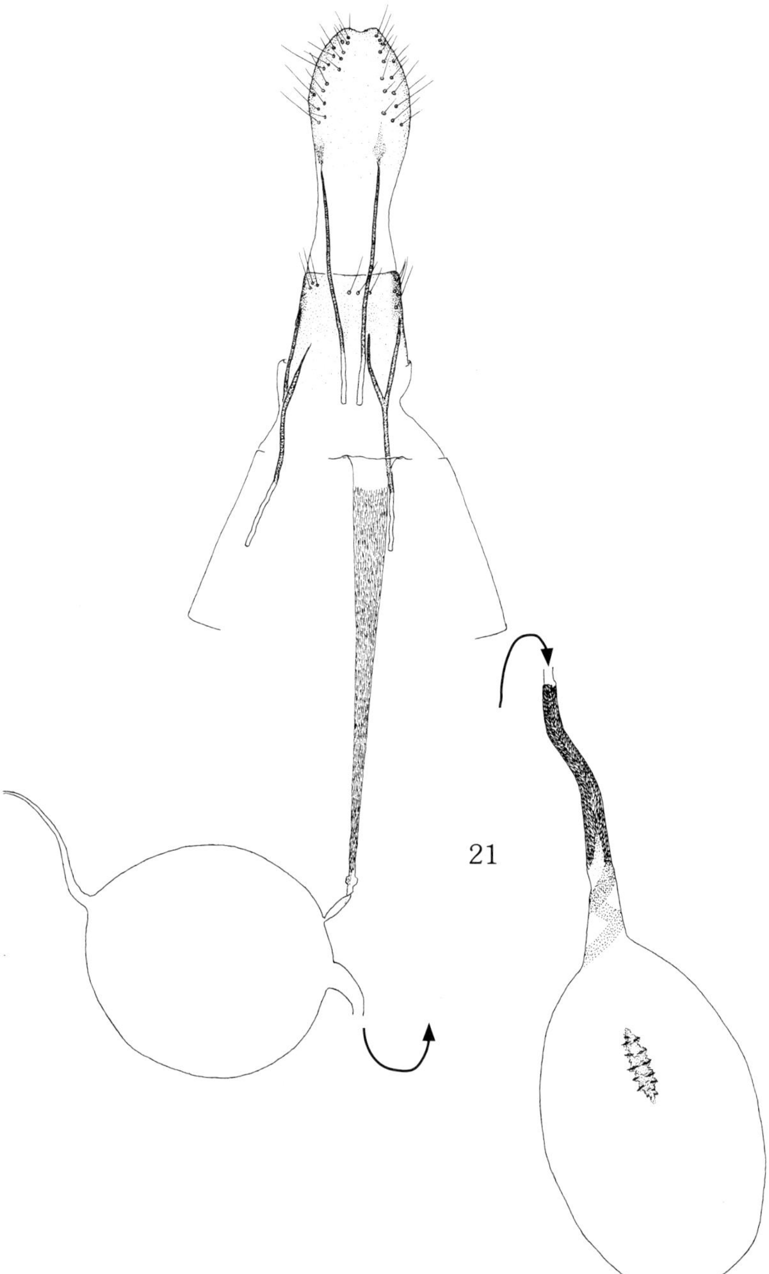

FEMALE GENITALIA

21. *Batrachedra praeangusta* (Haworth) – Netherlands, gen. slide JCK 3412 (VDW).

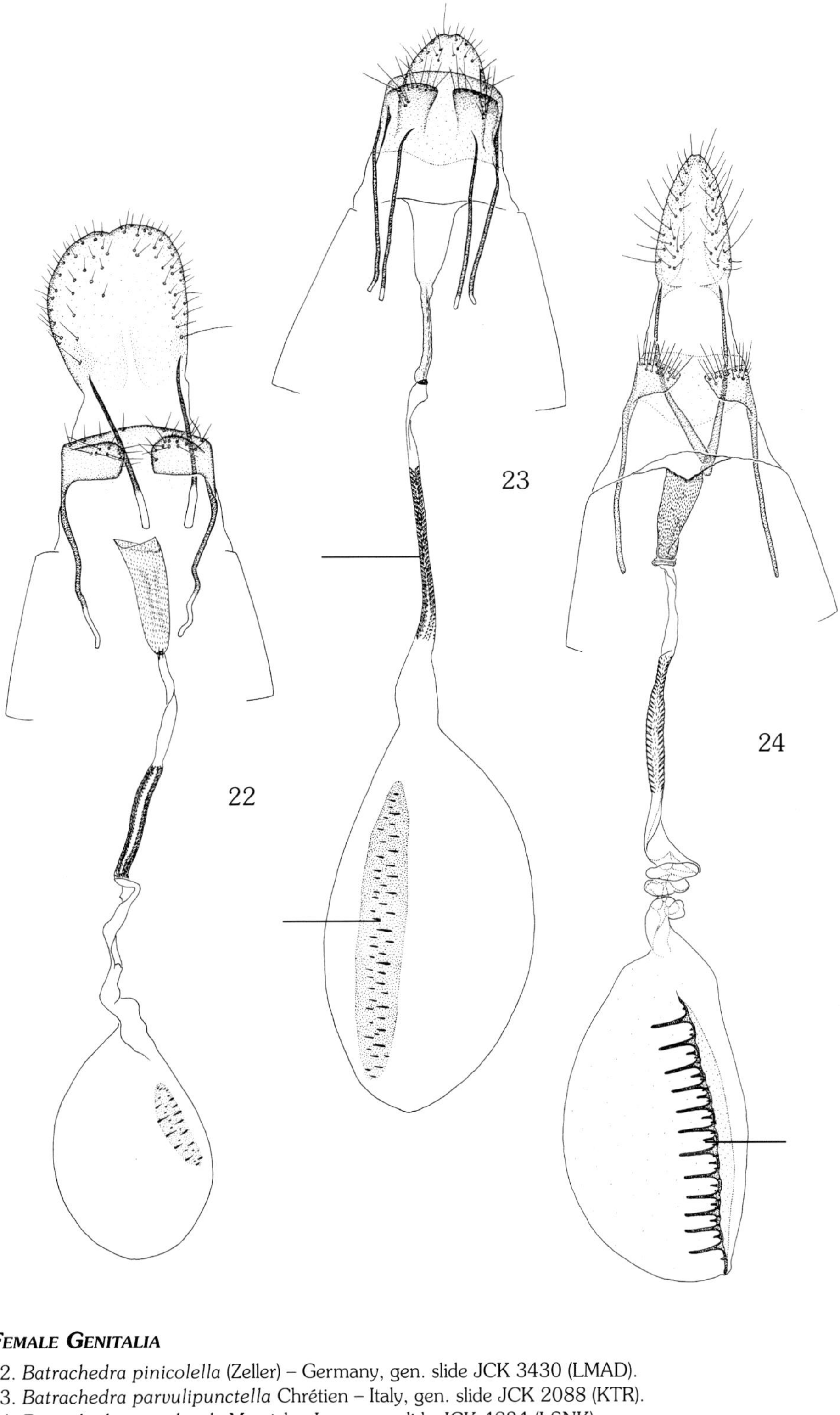

Female Genitalia

22. *Batrachedra pinicolella* (Zeller) – Germany, gen. slide JCK 3430 (LMAD).
23. *Batrachedra parvulipunctella* Chrétien – Italy, gen. slide JCK 2088 (KTR).
24. *Batrachedra amydraula* Meyrick – Iraq, gen. slide JCK 4834 (LSNK).

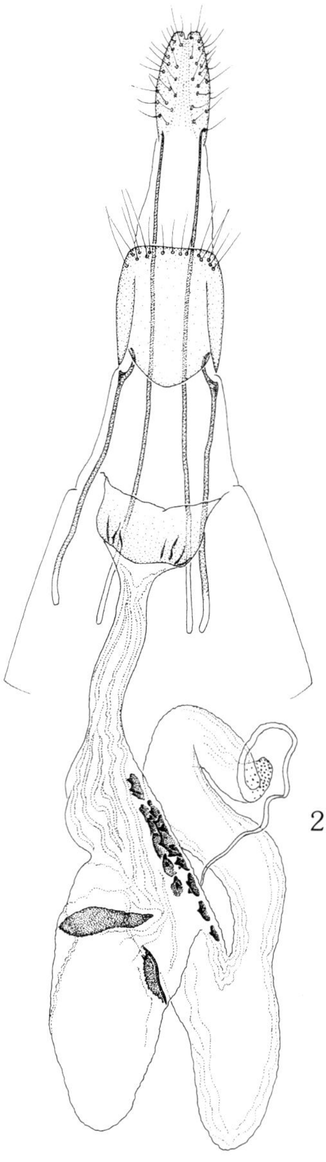

25

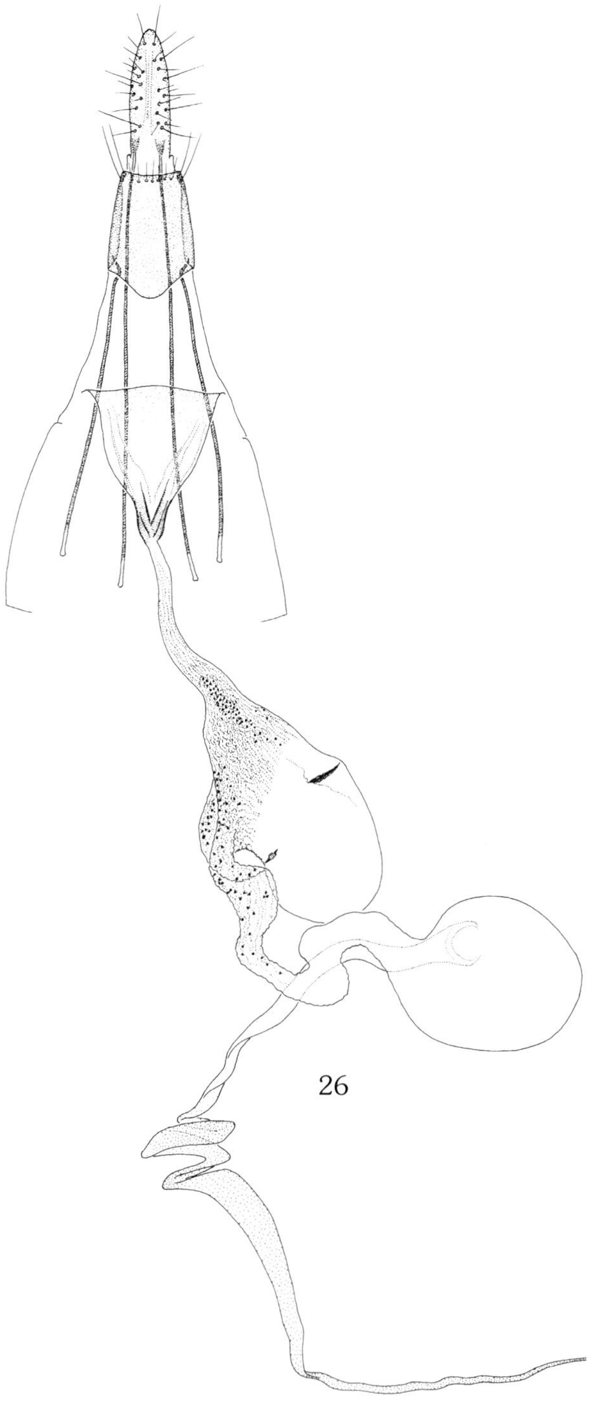

26

Female Genitalia

25. *Stathmopoda pedella* (Linnaeus) – Netherlands, gen. slide JCK 2454 (KTR).
26. *Stathmopoda auriferella* (Walker) – Israel, gen. in vial (ZFMK).

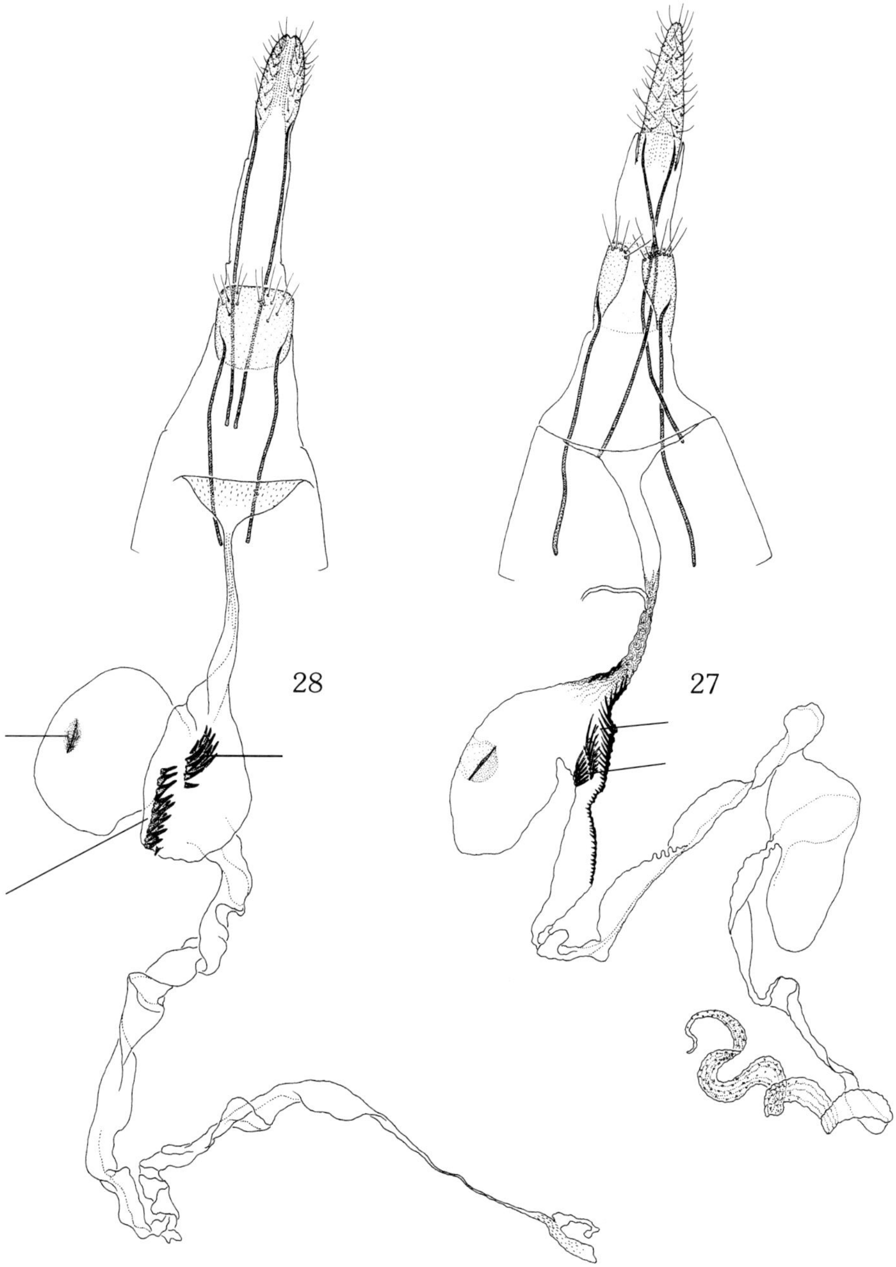

Female Genitalia

27. *Tortilia flavella* Chrétien – Morocco, gen. slide JCK 4833 (LSNK).
28. *Tortilia trigonella* (Zerny) – Morocco, gen. slide JCK 5257 (ZMAN).

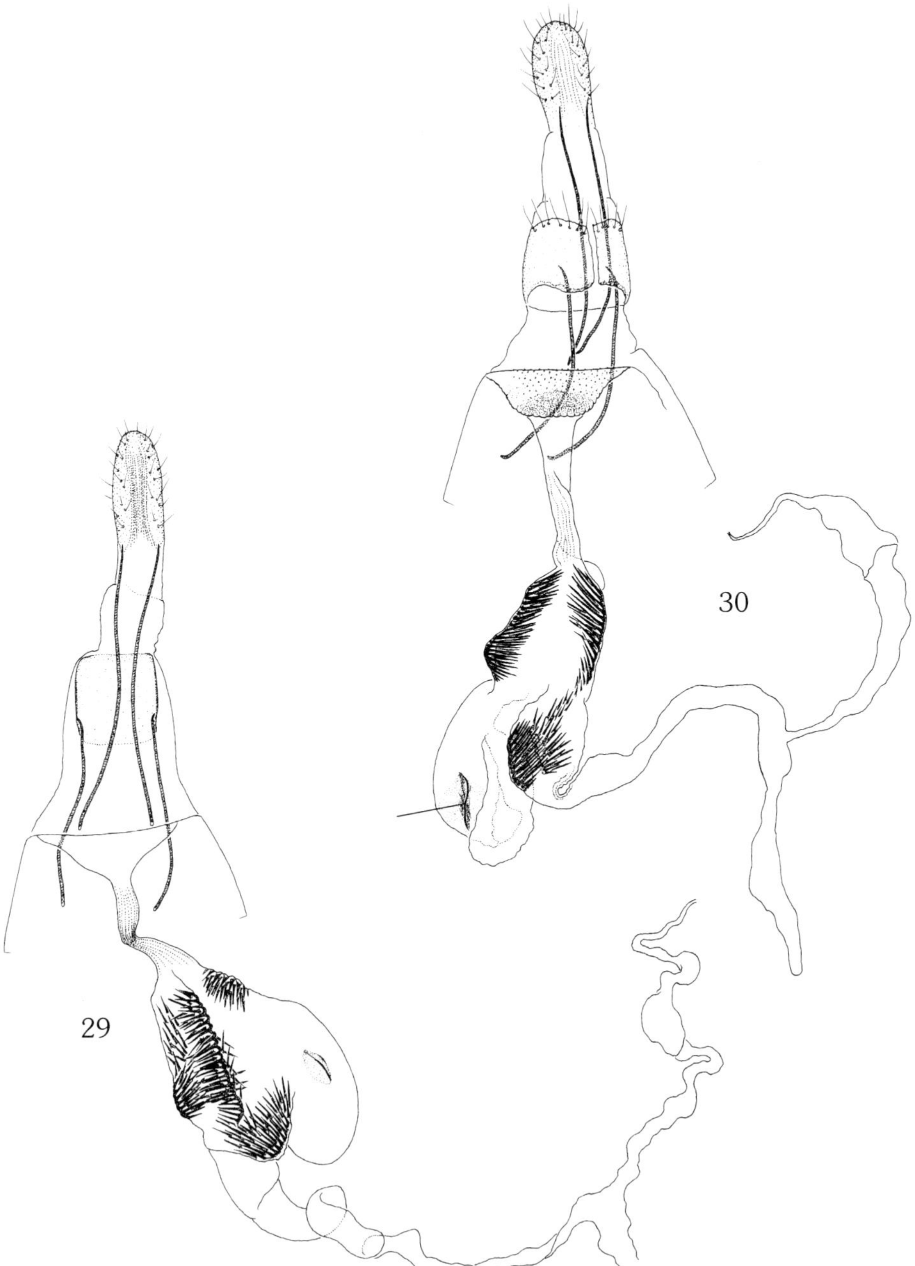

FEMALE GENITALIA

29. *Tortilia charadritis* (Meyrick) – Egypt, gen. slide BMNH 15097 (BMNH).
30. *Tortilia pallidella* Kasy – paratype, S-Iran, gen. slide Mus. Vind. 3936 (NHMW).

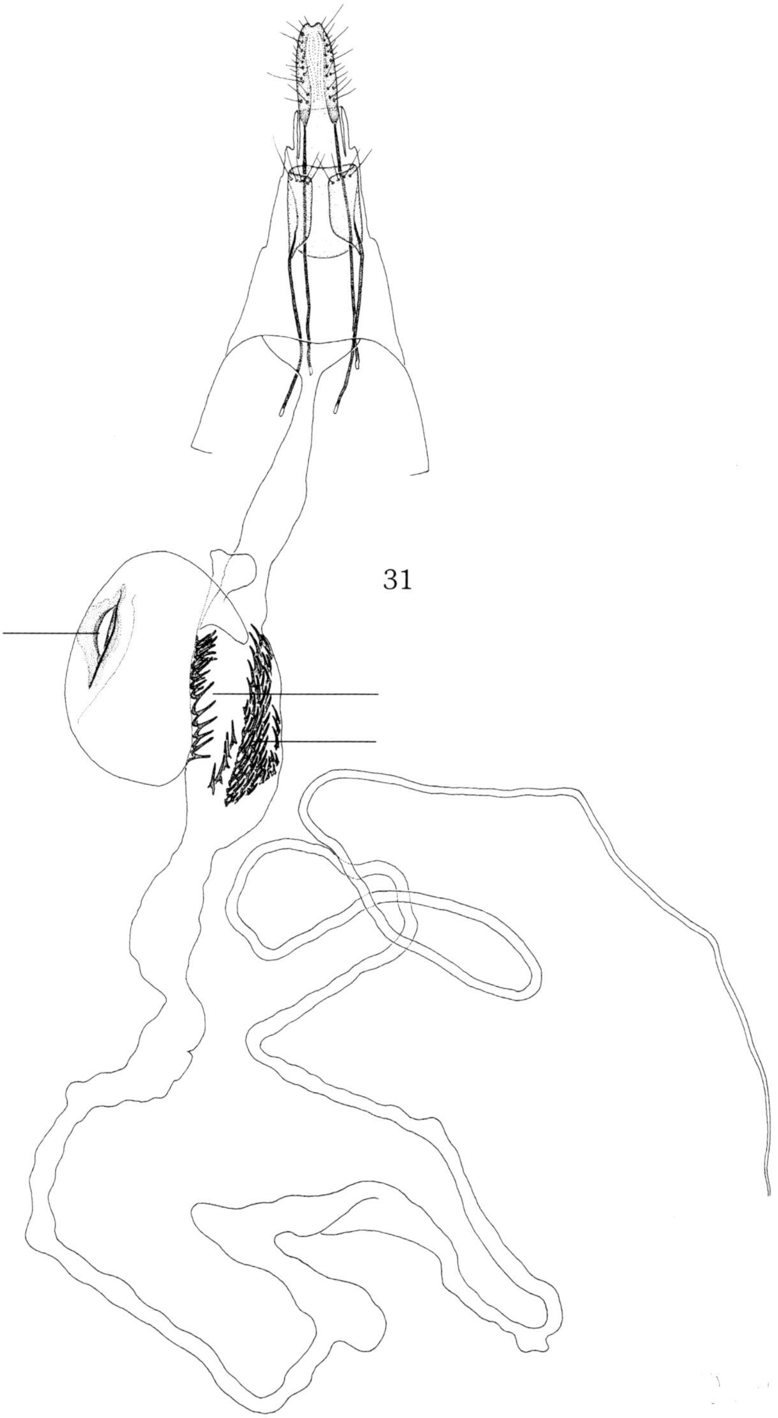

Female Genitalia

31. *Tortilia graeca* Kasy – Greece, gen. slide JCK 3661 (KTR).

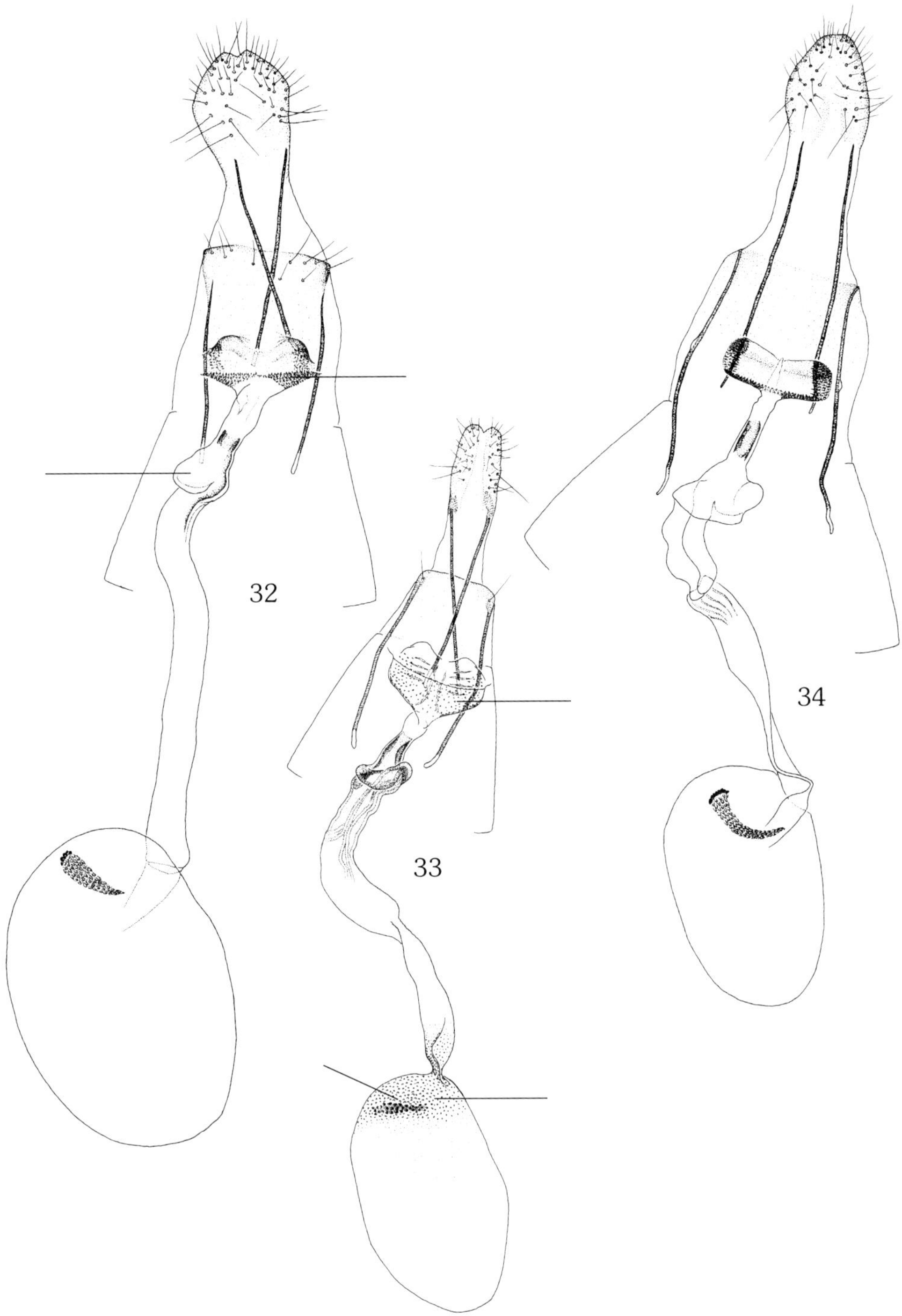

Female Genitalia

32. *Blastodacna hellerella* (Duponchel) – Germany, gen. slide JCK 3432 (LMAD).
33. *Blastodacna rossica* Sinev – Ukraine, gen. in vial (ZIN).
34. *Blastodacna atra* (Haworth) – Germany, gen. slide JCK 3434 (LMAD).

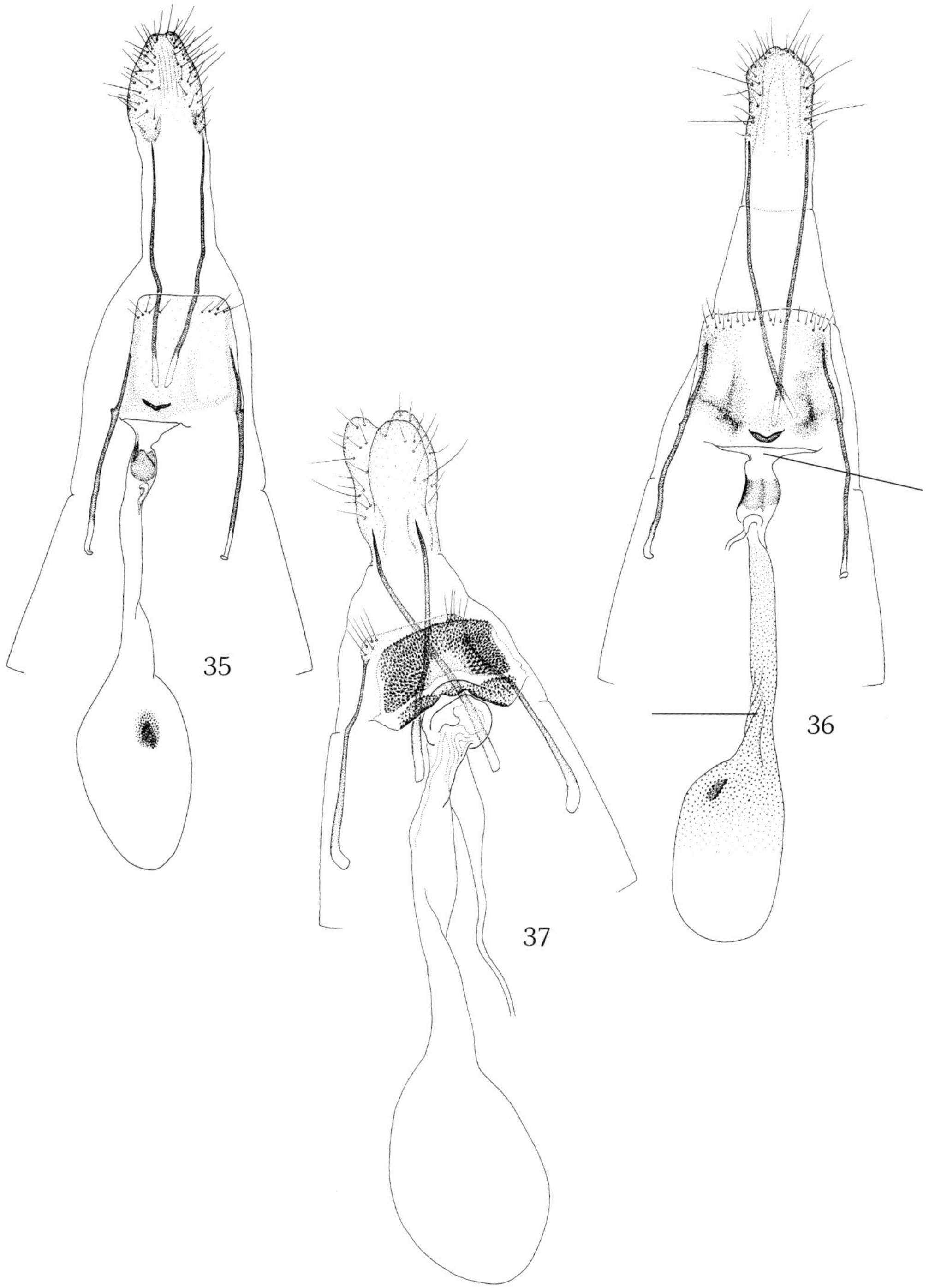

FEMALE GENITALIA

35. *Blastodacna vinolentella* (Herrich-Schäffer) – Netherlands, gen. slide JCK 3861 (RMNH).
36. *Blastodacna georgiella* Sinev – paratype, Georgia, gen. slide JCK 4274 (ZIN).
37. *Blastodacna libanotica* Diakonoff – Turkey, gen. slide JCK 4277 (TLMF).

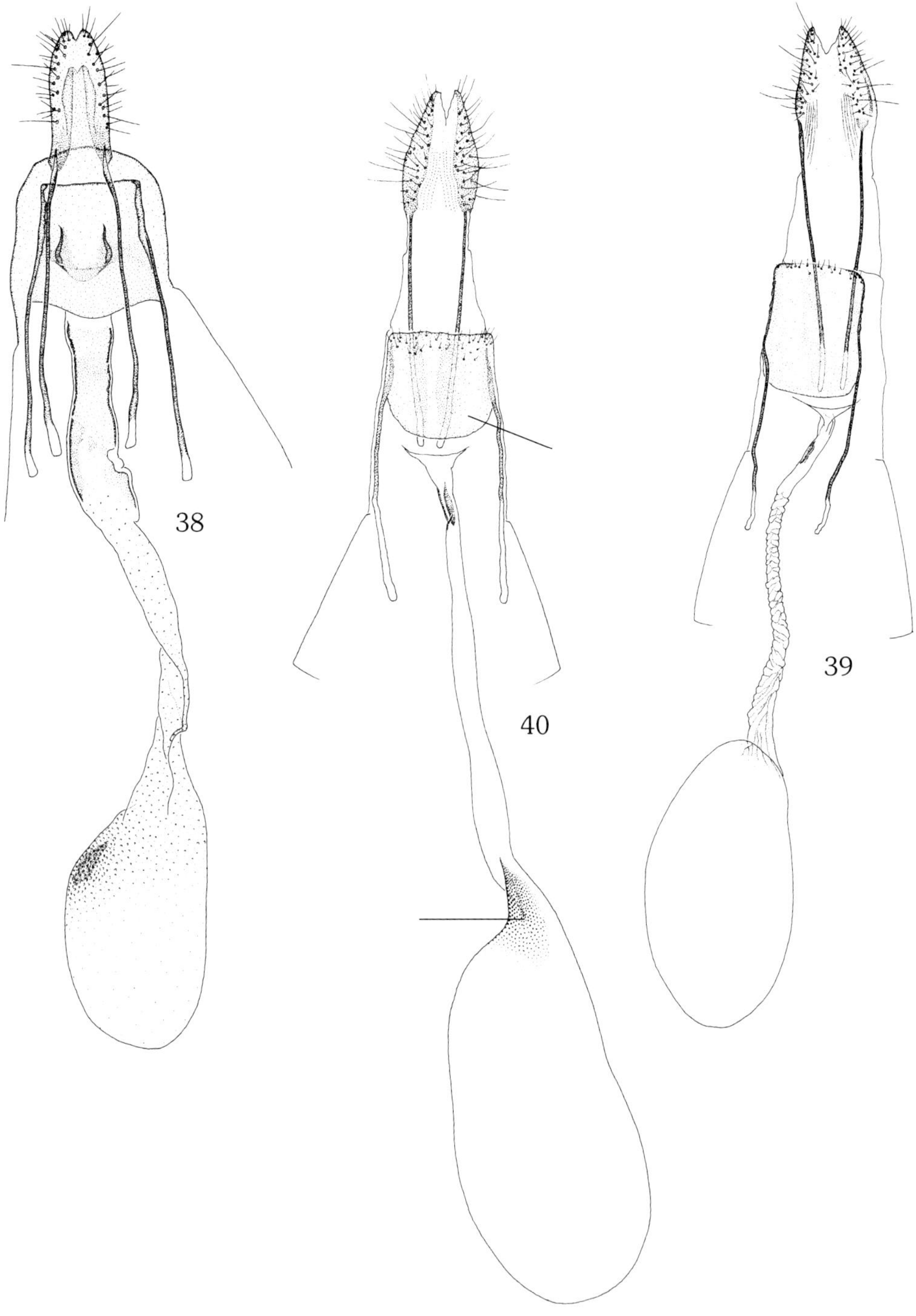

Female Genitalia

38. *Spuleria flavicaput* (Haworth) – Netherlands, gen. slide JCK 3654 (RMNH).
39. *Heinemannia laspeyrella* (Hübner) – Country unknown, gen. slide JCK 3757 (RMNH).
40. *Heinemannia festivella* ([Denis & Schiffermüller]) – Turkey, gen. slide JCK 4131 (ZMUC).

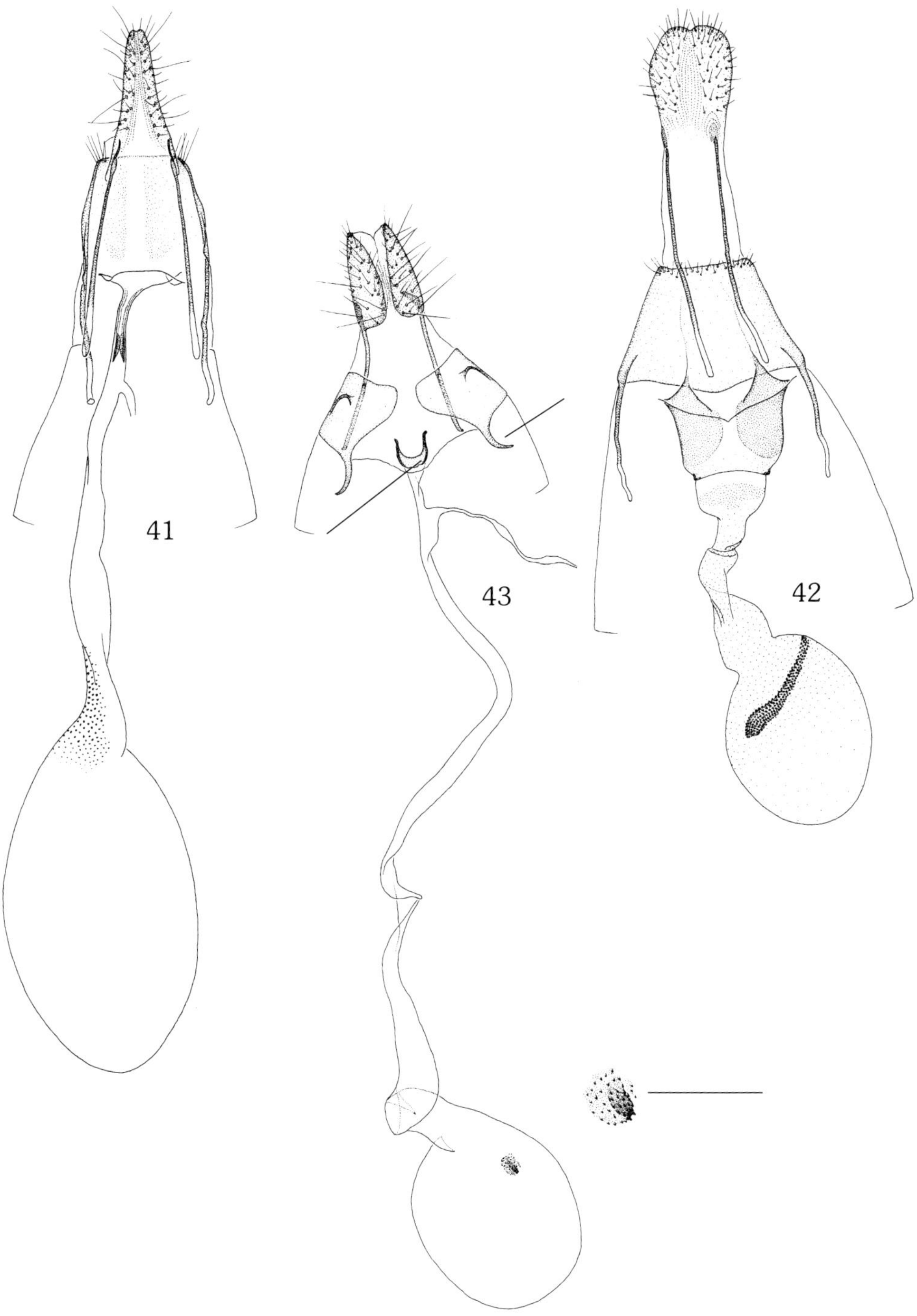

Female Genitalia

41. *Heinemannia albidorsella* (Staudinger) – France, gen. slide JCK 4127 (HSN).
42. *Dystebenna stephensi* (Stainton) – Poland, gen. slide JCK 3864 (RMNH).
43. *Haplochrois gelechiella* (Rebel) – Greece, gen. in vial (ZMUC).

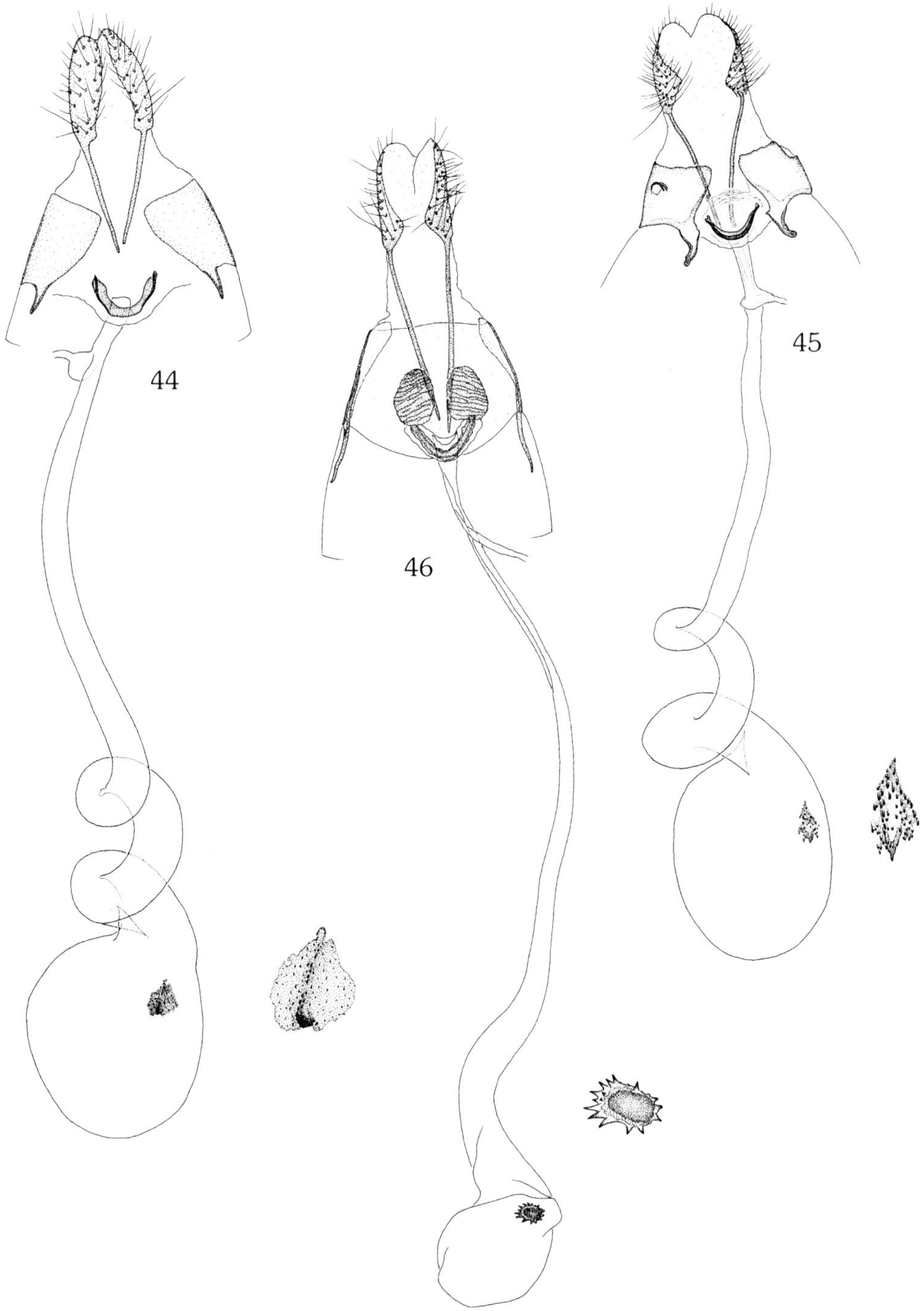

FEMALE GENITALIA

44. *Haplochrois ochraceella* (Rebel) – Spain, gen. slide JCK 4128 (GLS).
45. *Haplochrois albanica* (Rebel & Zerny) – Greece, gen. in vial (ZMUC).
46. *Haplochrois buvati* (Baldizzone) – Spain, gen. in vial (ZIN).

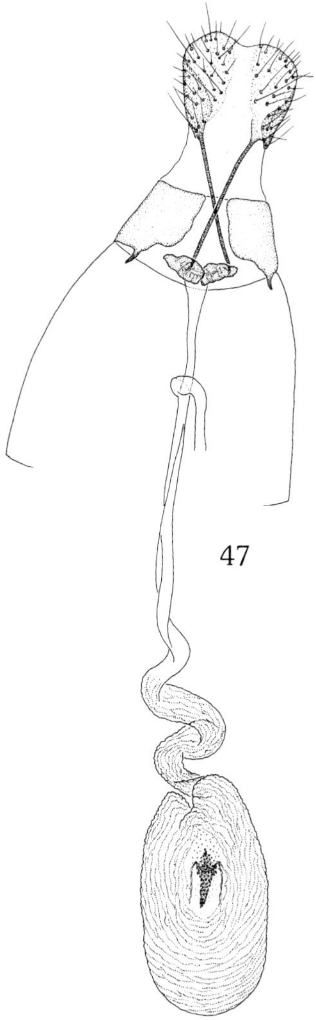

47

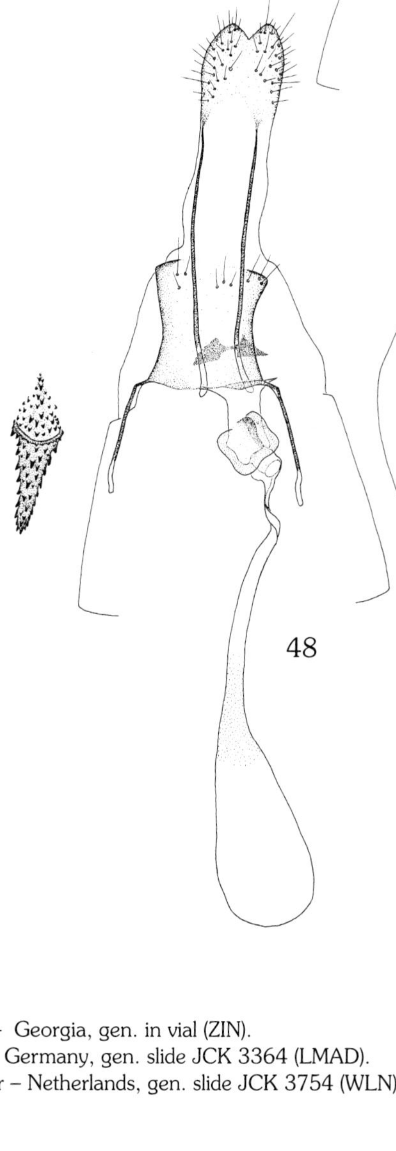

48

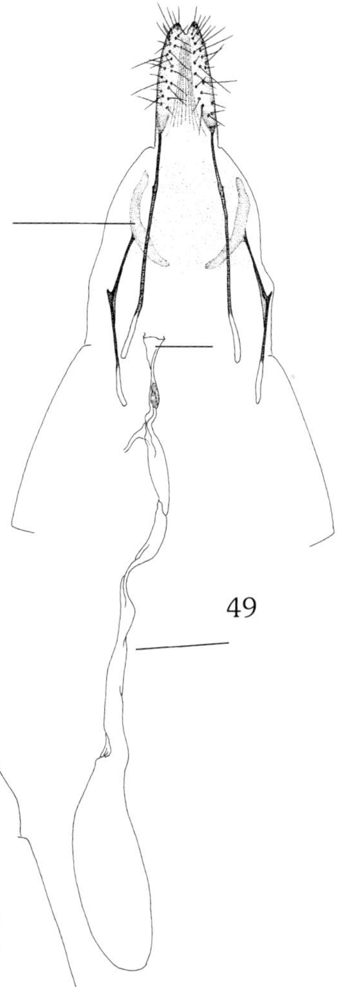

49

Female Genitalia

47. *Haplochrois theae* (Kuznetsov) – Georgia, gen. in vial (ZIN).
48. *Chrysoclista linneella* (Clerck) – Germany, gen. slide JCK 3364 (LMAD).
49. *Chrysoclista lathamella* Fletcher – Netherlands, gen. slide JCK 3754 (WLN).

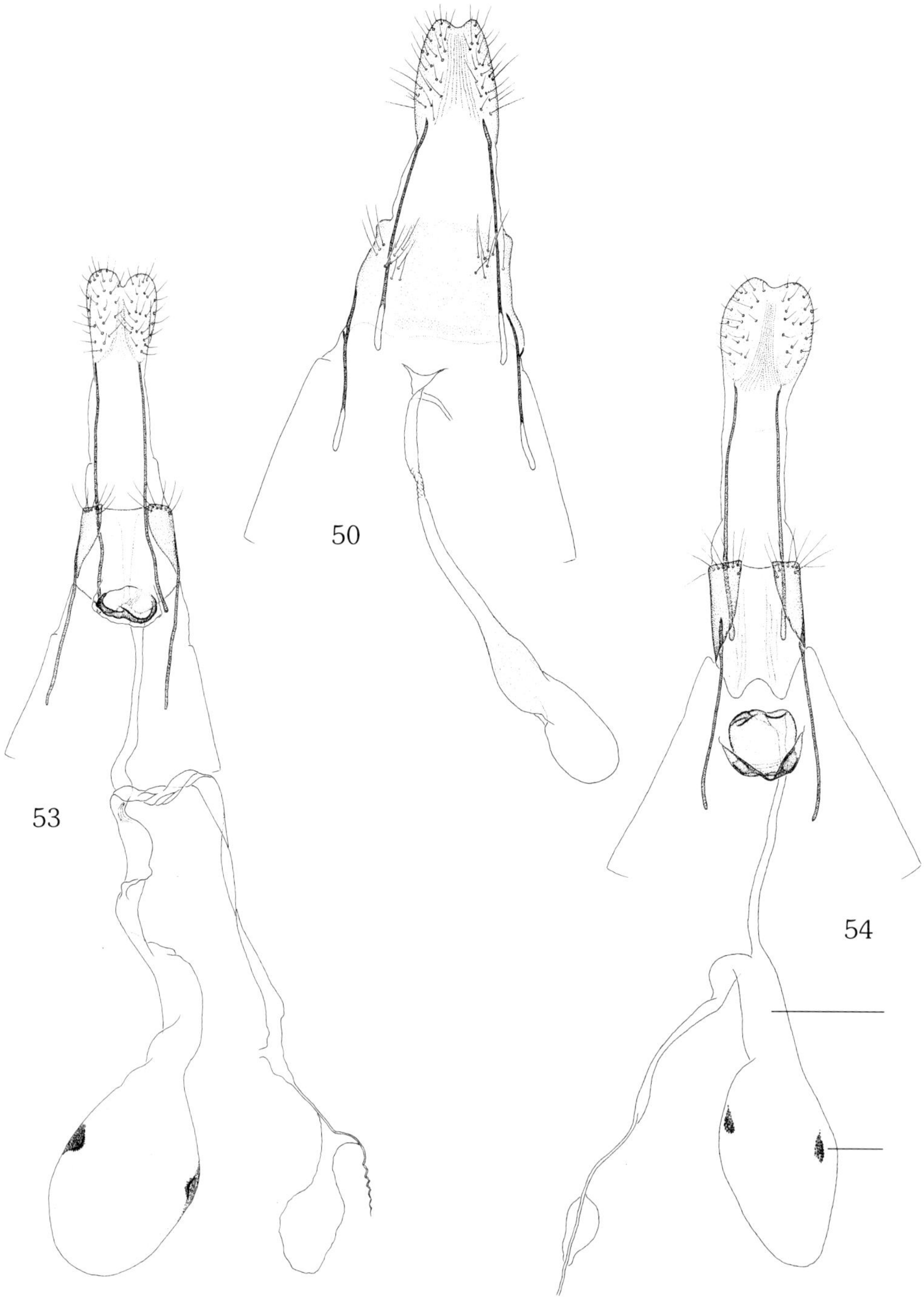

Female Genitalia

50. *Chrysoclista splendida* Karsholt – Germany, gen. in vial (ZIN).
51. *Chrysoclista zagulajevi* Sinev – unknown.
52. *Chrysoclista abchasica* (Sinev) – unknown.
53. *Pancalia leuwenhoekella* (Linnaeus) – Belgium, gen. slide JCK 3075 (VDW).
54. *Pancalia schwarzella* (Fabricius) – France, gen. slide JCK 3119 (RMNH).

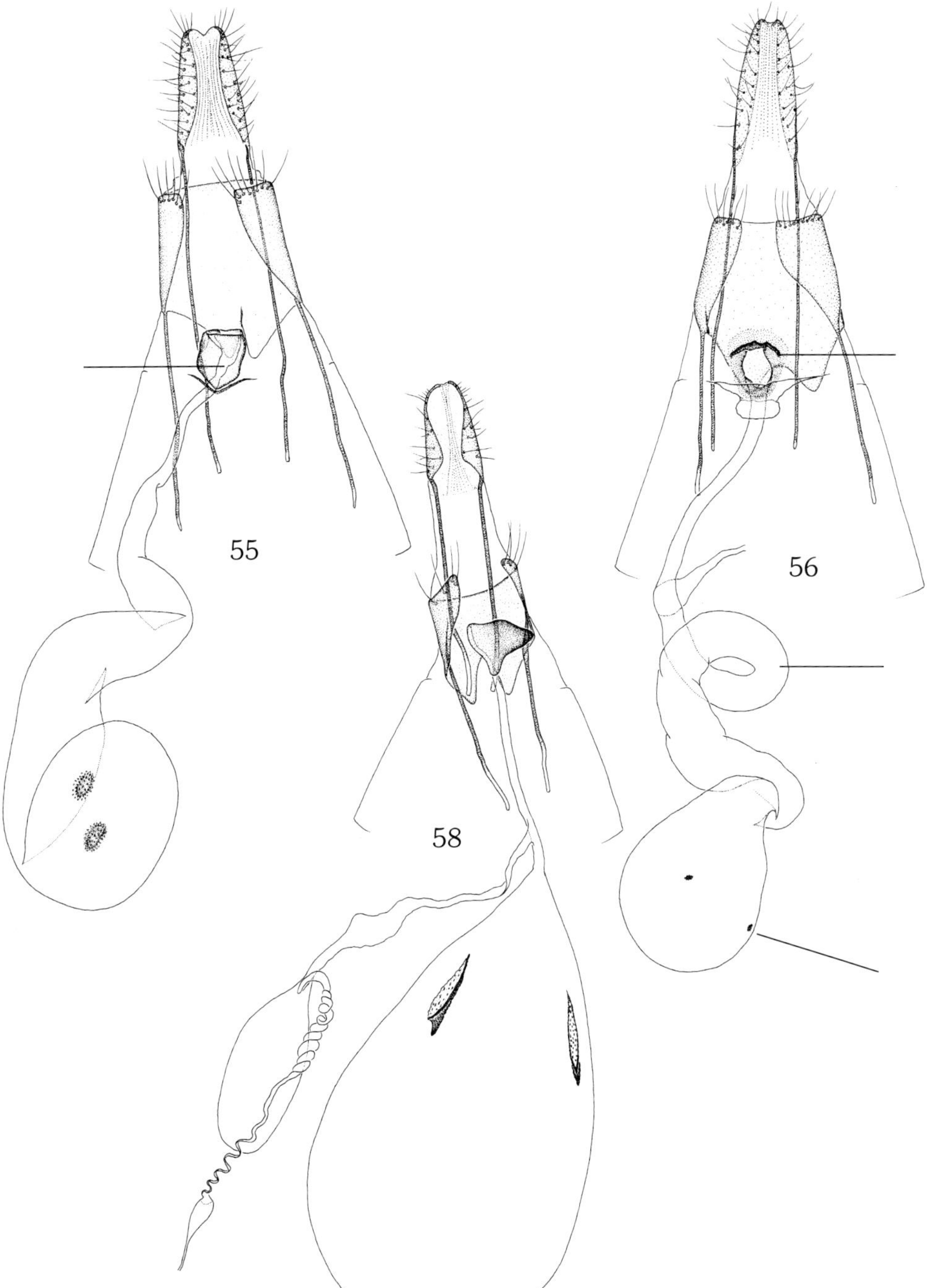

Female Genitalia

55. *Pancalia baldizzonella* Riedl – paratype, Italy, gen. slide 1414/R (BDZ).
56. *Pancalia nodosella* (Bruand) – Netherlands, gen. slide JCK 3100 (ZMAN).
57. *Euclemensia woodiella* (Curtis) – unknown.
58. *Phosphaticola gemmatella* Viette – Tunisia, gen. in vial (MNHN).

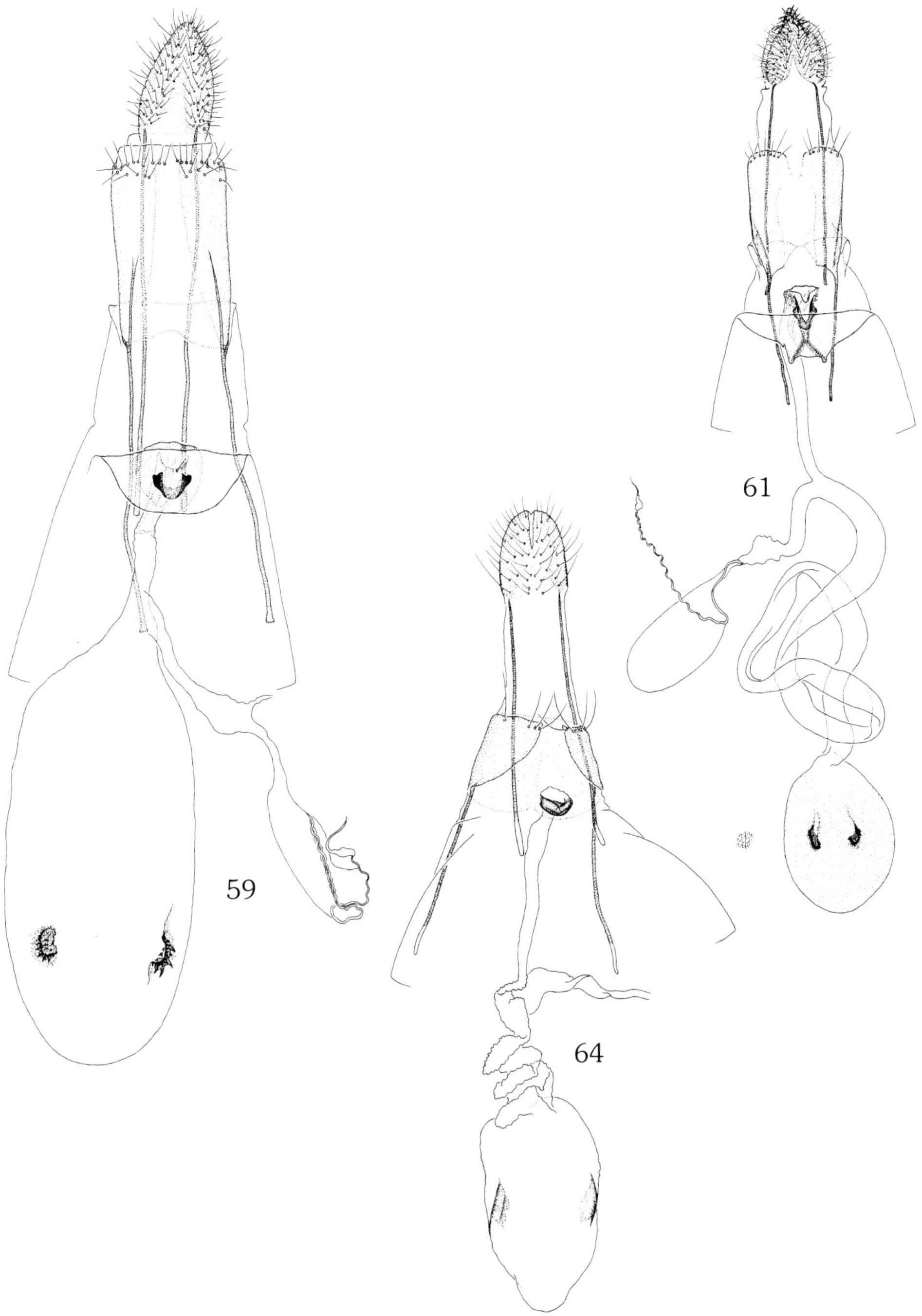

FEMALE GENITALIA

59. *Alloclita recisella* Staudinger – Spain, gen. slide JCK 4738 (ZMUC).
60. *Alloclita haifensis* Rebel – unknown.
61. *Alloclita delozona* Meyrick – Oman, gen. in vial (ZMUC).
62. *Alloclita brachygrapta* Meyrick – unknown.
63. *Alloclita deprinsi* sp. n. – unknown.
64. *Alloclita francoeuriae* Walsingham – Algeria, gen. slide BMNH 29280 (BMNH).

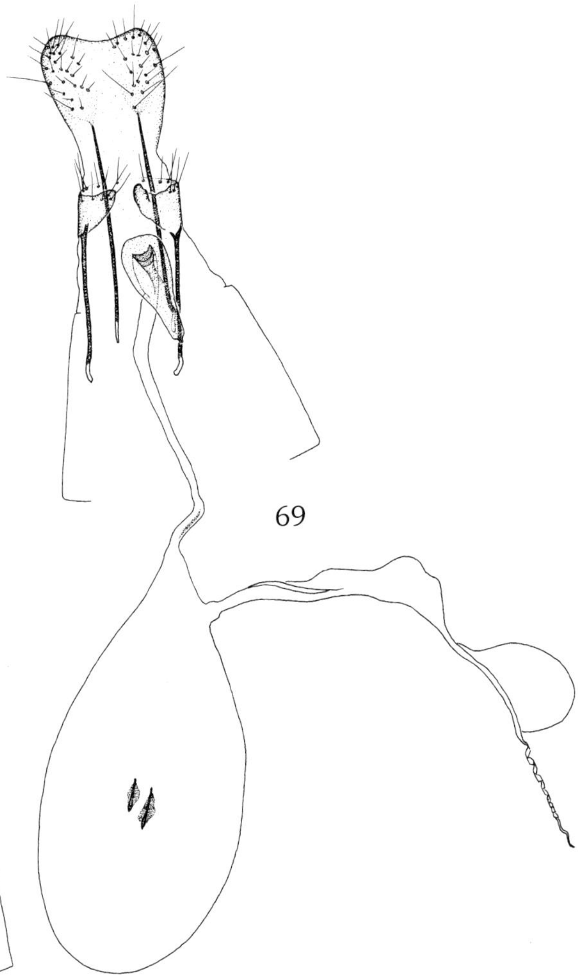

69

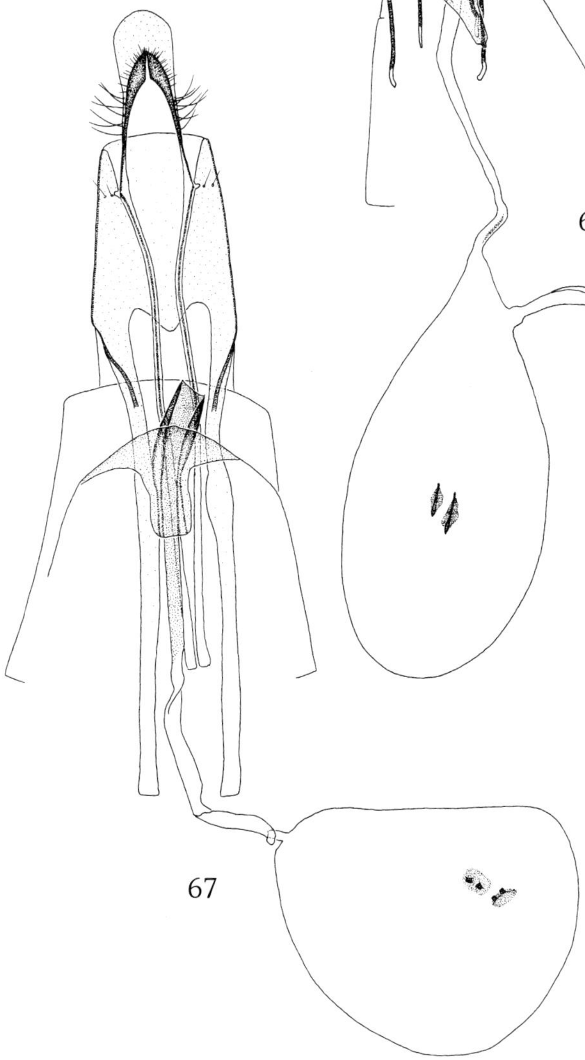

67

FEMALE GENITALIA

65. *Gibeauxiella bellaqueifontis* (Gibeaux) – unknown.
66. *Gibeauxiella reliqua* (Gibeaux) – unknown.
67. *Limnaecia phragmitella* Stainton – Netherlands, gen. slide JCK 3419 (VDW).
68. *Allotalanta autophaea* Meyrick – unknown.
69. *Cosmopterix zieglerella* (Hübner) – Germany, gen. slide JCK 3357 (LMAD).

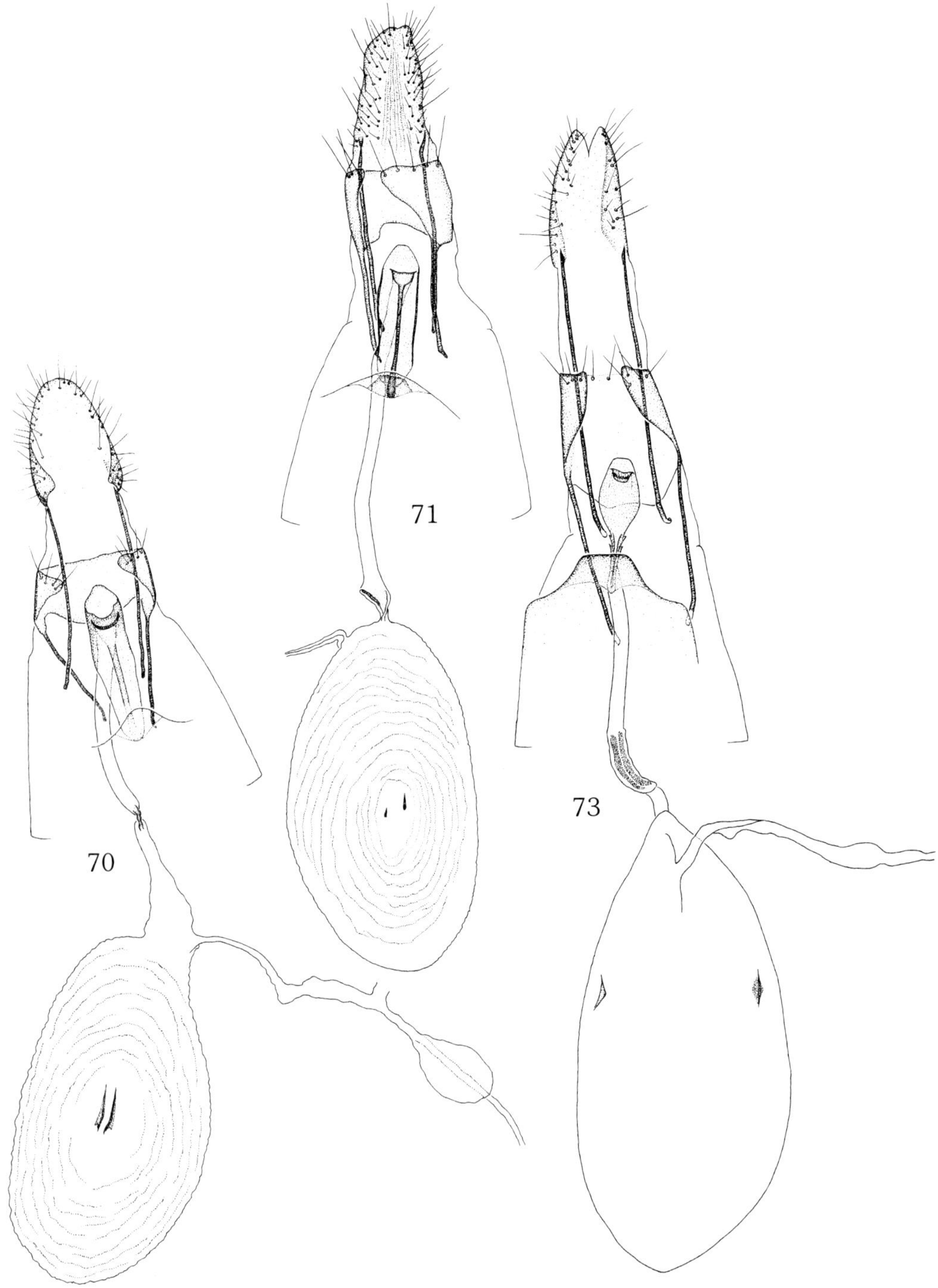

Female Genitalia

70. *Cosmopterix schmidiella* Frey – Germany, gen. slide JCK 3982 (ZIN).
71. *Cosmopterix orichalcea* Stainton – Germany, gen. slide JCK 3748 (RMNH).
72. *Cosmopterix sibirica* Sinev – unknown.
73. *Cosmopterix scribaiella* Zeller – Germany, gen. slide JCK 3372 (KTR).

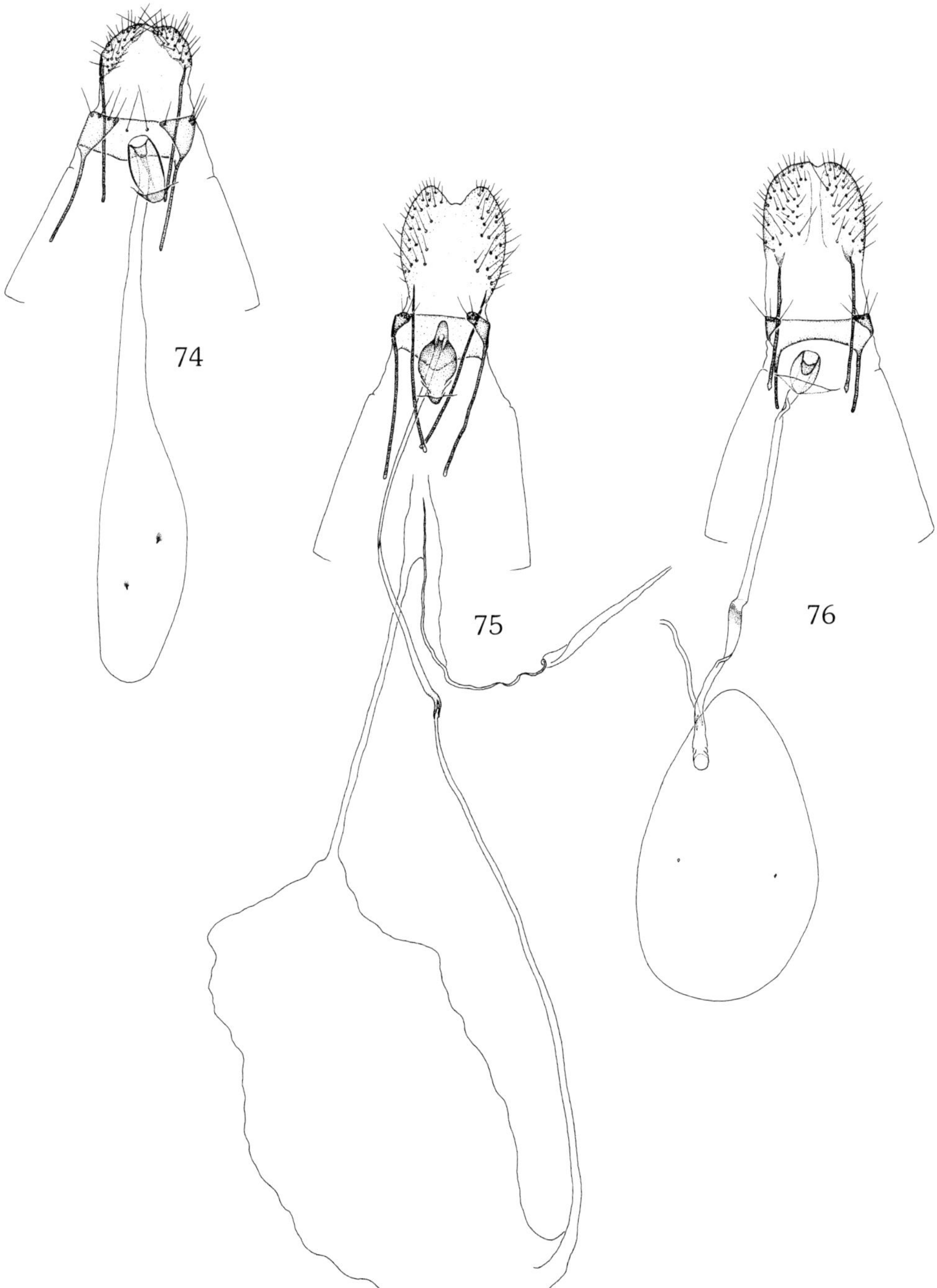

Female Genitalia

74. *Cosmopterix pulchrimella* Chambers – Croatia, gen. slide JCK 3749 (WLN).
75. *Cosmopterix attenuatella* (Walker) – Madeira, gen. slide JCK 3999 (ZMUC).
76. *Cosmopterix crassicervicella* Chrétien – Greece, gen. slide JCK 3742 (ZMUC).

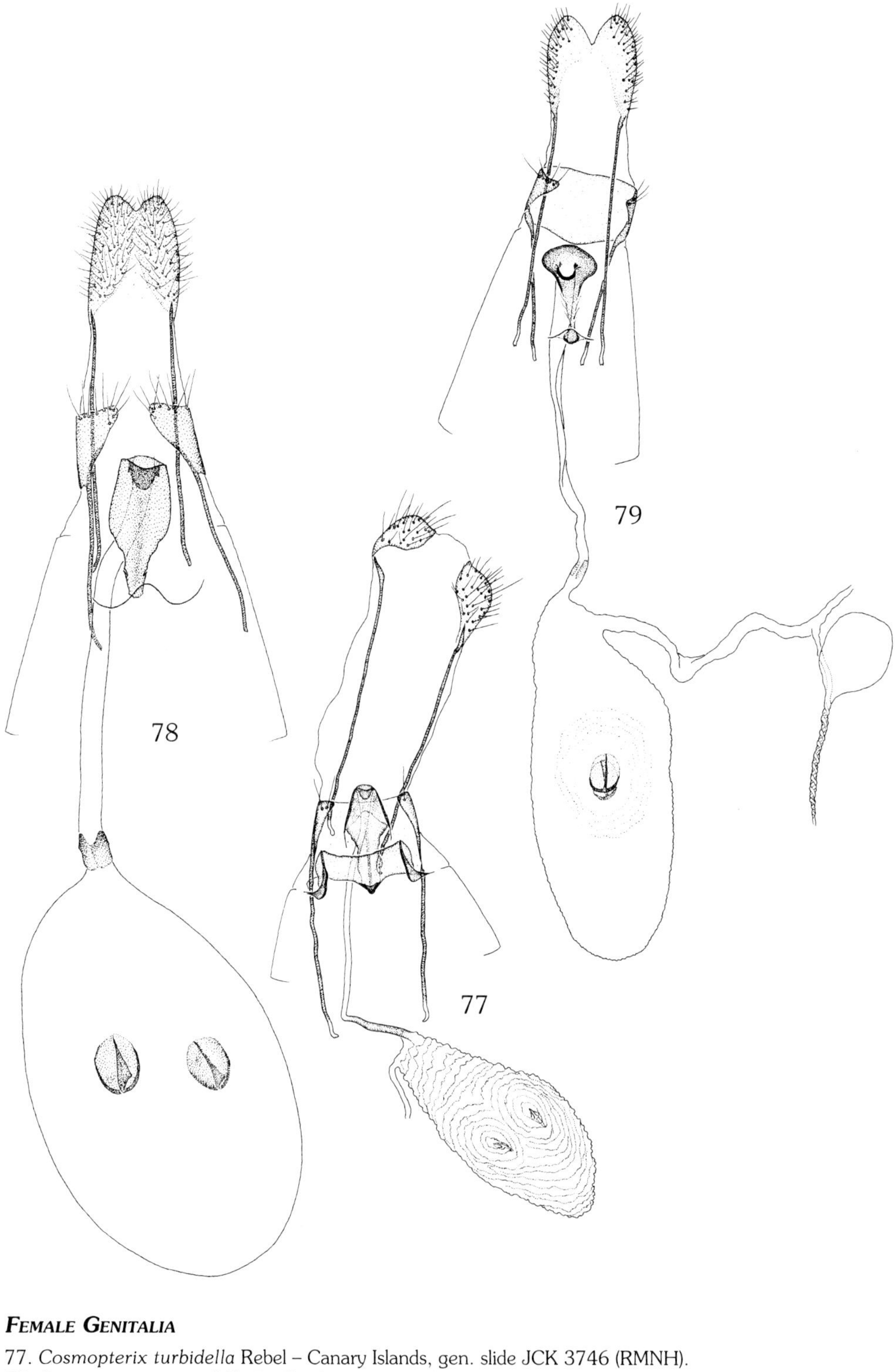

Female Genitalia

77. *Cosmopterix turbidella* Rebel – Canary Islands, gen. slide JCK 3746 (RMNH).
78. *Cosmopterix pararufella* Riedl – Greece, gen. slide GU 5339 (STR).
79. *Cosmopterix coryphaea* Walsingham – Spain, gen. slide JCK 3989 (ZMUC).

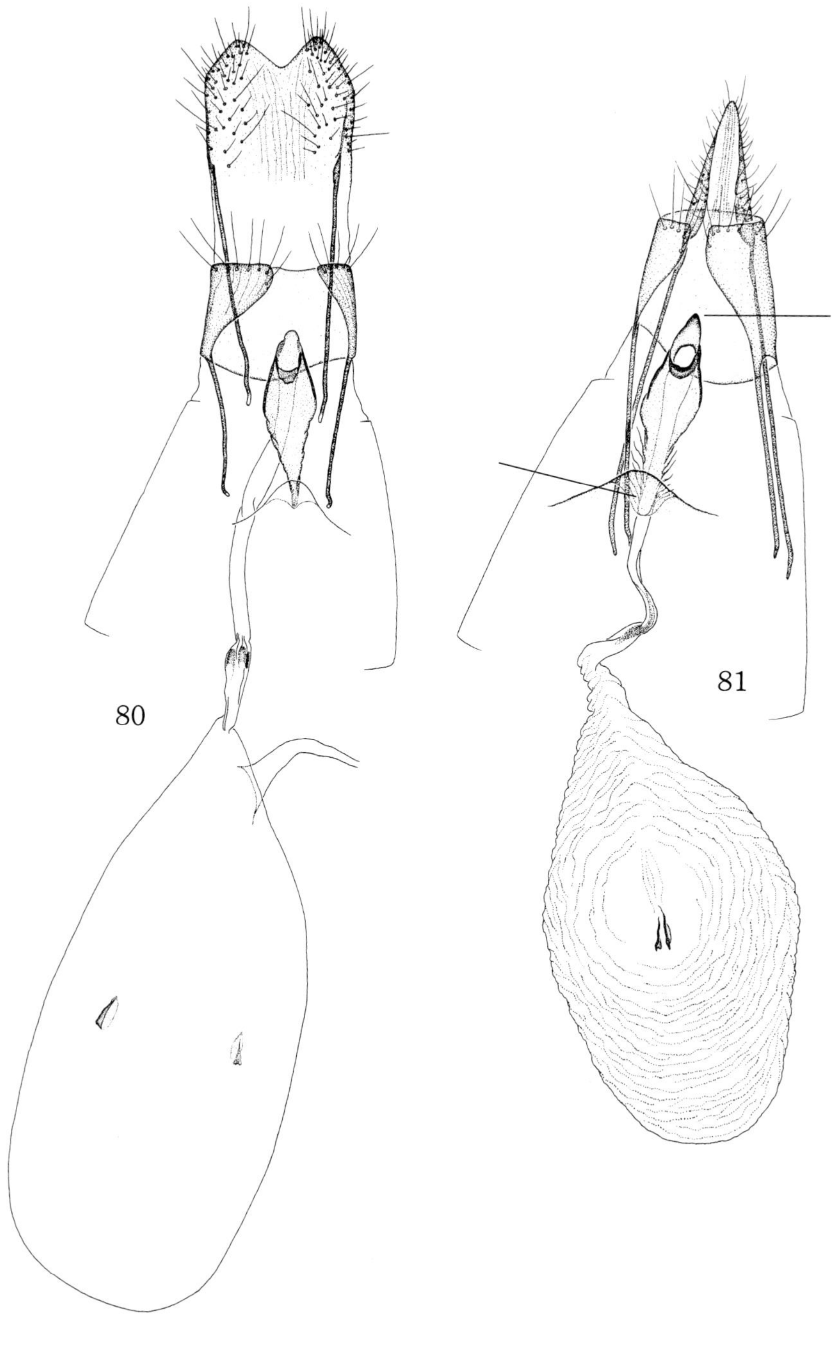

Female Genitalia

80. *Cosmopterix lienigiella* Zeller – Netherlands, gen. slide 3753 (RMNH).
81. *Cosmopterix salahinella* Chrétien – Israel, gen. in vial (LSNK).

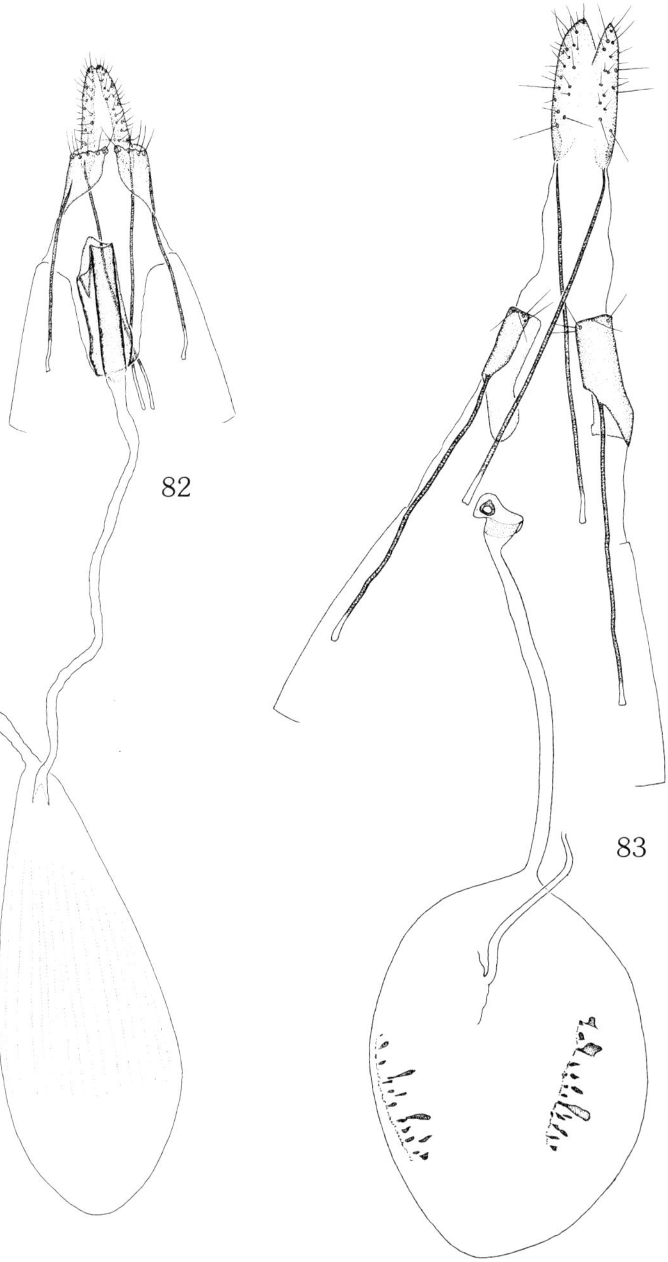

Female Genitalia

82. *Labdia semicoccinea* (Stainton) – Taiwan, gen. slide JCK 4052 (PDW).
83. *Pyroderces argyrogrammos* (Zeller) – Lebanon, gen. slide JCK 3374 (LMAD).

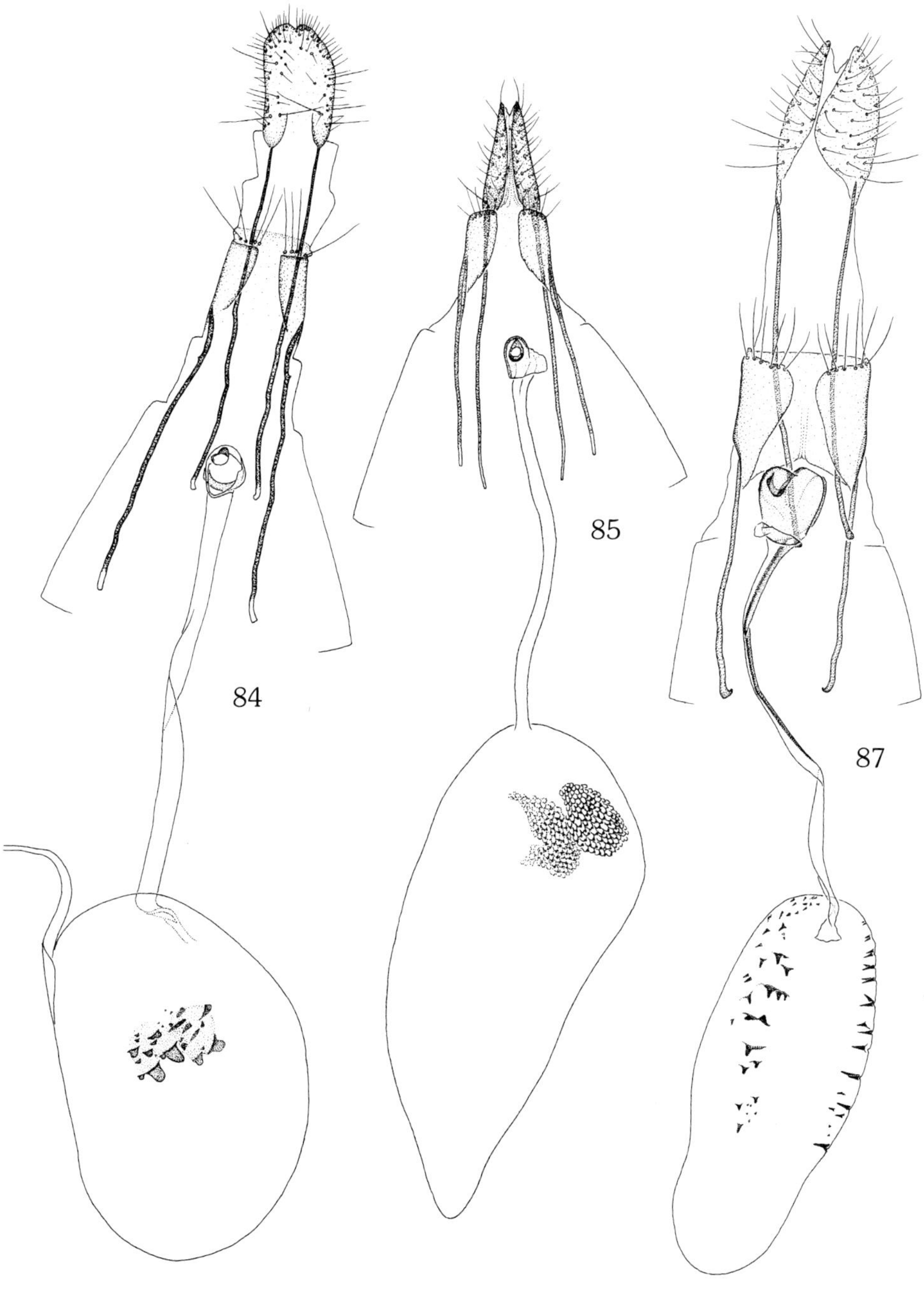

Female Genitalia

84. *Pyroderces caesaris* Gozmány – Turkey, gen. slide JCK 2131 (KTR).
85. *Pyroderces klimeschi* Rebel – Hungary, gen. in vial (LSNK).
86. *Pyroderces brosi* Riedl – unknown.
87. *Pyroderces tethysella* sp. n. – paratype, Spain, gen. slide JCK 4595 (WLN).

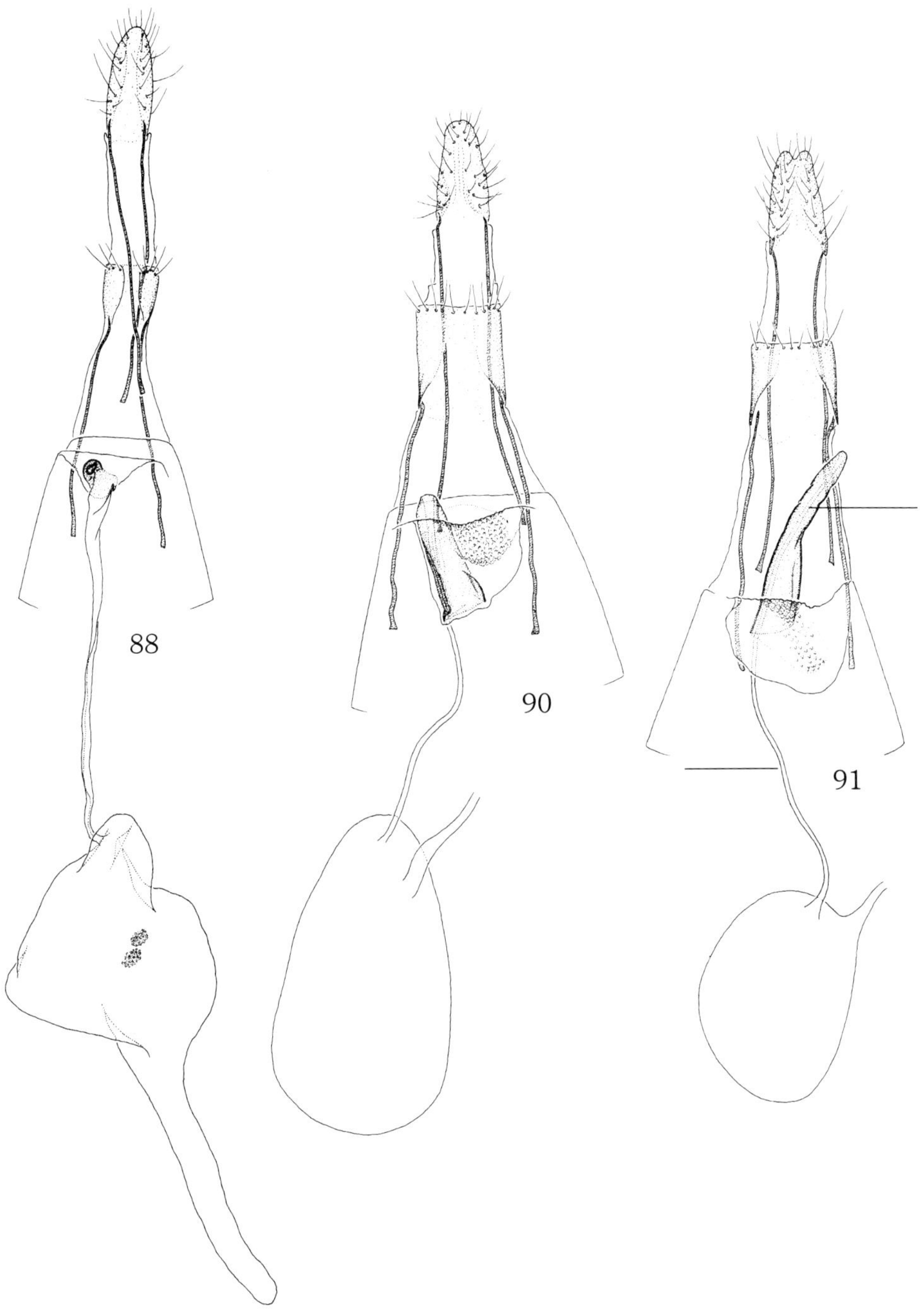

FEMALE GENITALIA

88. *Pyroderces wolschrijni* sp. n. – paratype, Spain, gen. slide JCK 4838 (WLN).
89. *Pyroderces apicinotella* (Chrétien) – unknown.
90. *Anatrachyntis rileyi* (Walsingham) – Cyprus, gen. slide JCK 4654 (ABR).
91. *Anatrachyntis badia* (Hodges) – Canary Islands, gen. slide JCK 4624 (WLN).

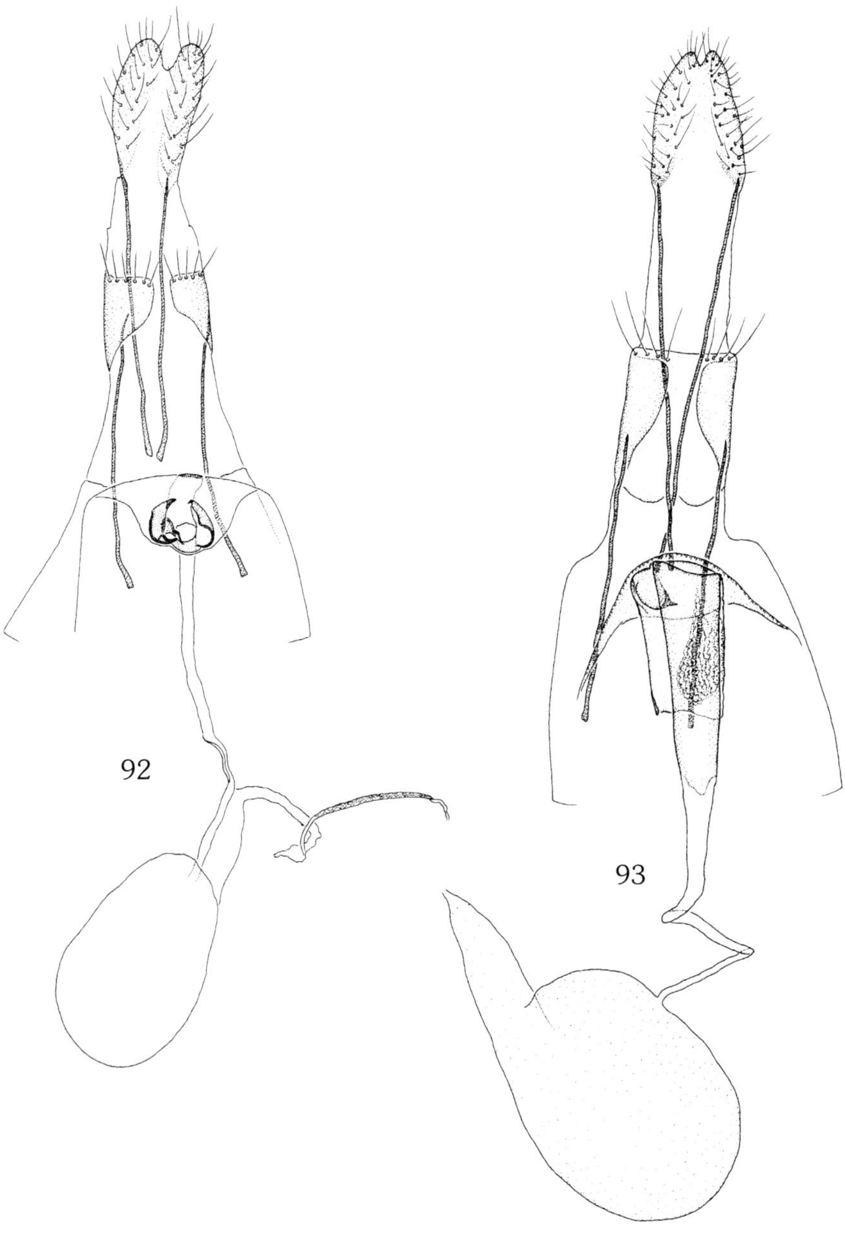

FEMALE GENITALIA

92. *Anatrachyntis coridophaga* (Meyrick) – Egypt, gen. slide JCK 4834 (ZFMK).
93. *Anatrachyntis simplex* (Walsingham) – Egypt, gen. in vial (ZFMK).

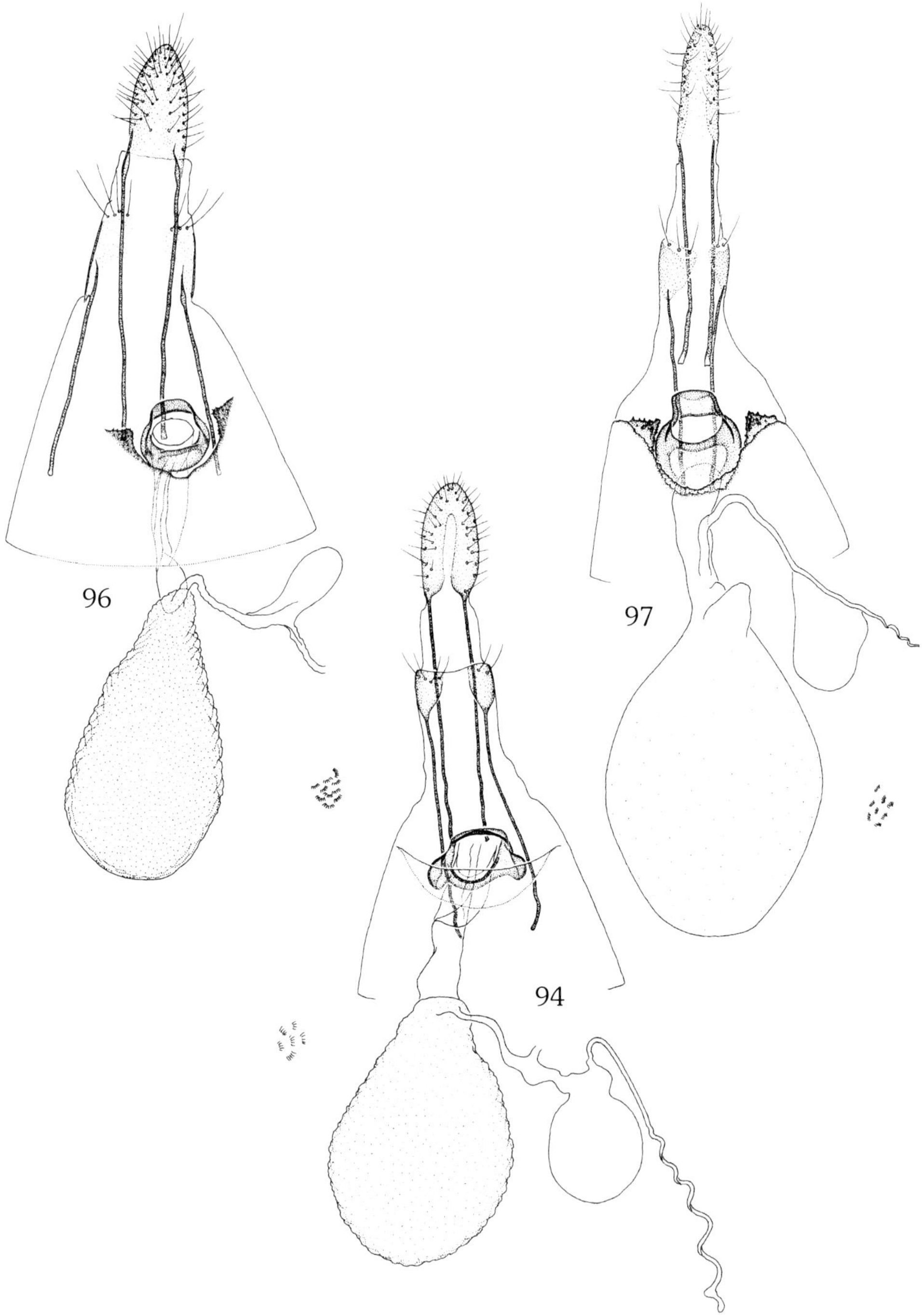

Female Genitalia

94. *Coccidiphila gerasimovi* Danilevsky – France, gen. slide JCK 3975 (VDW).
95. *Coccidiphila patriciae* J. Nel & A. Nel – not figured, see pag....
96. *Coccidiphila ledereriella* (Zeller) – France, gen. slide JCK 3855 (RMNH).
97. *Coccidiphila danilevskyi* Sinev – Spain, gen. slide JCK 2485 (KTR).

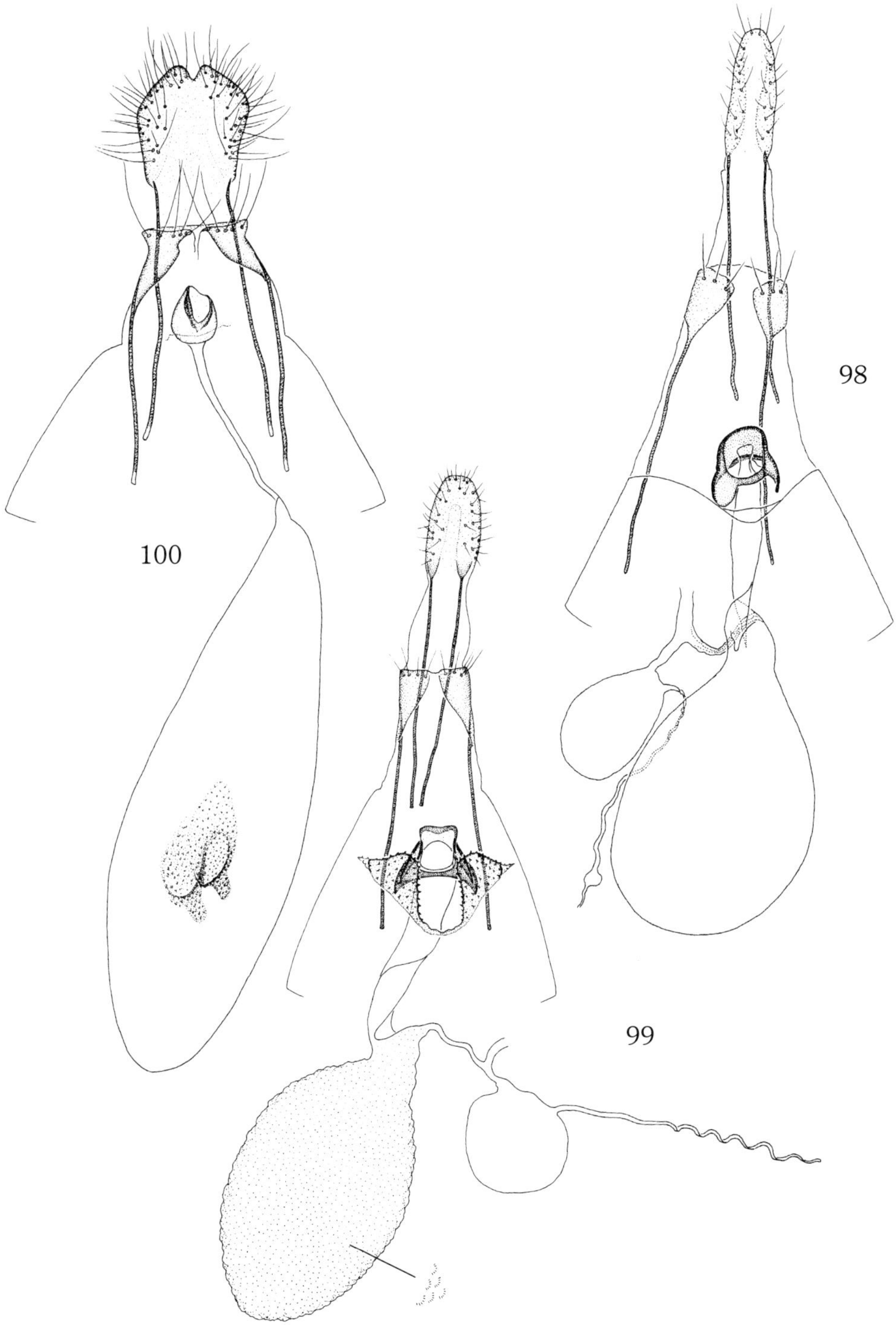

Female Genitalia

98. *Coccidiphila kasypinkeri* Traugott-Olsen – paratype, Canary Islands, gen. slide A 16.6.85 (NHMW).
99. *Coccidiphila riedli* Traugott-Olsen – paratype, Canary Islands, gen. slide A 9.1.85 (NHMW).
100. *Stagmatophora heydeniella* (Fischer von Röslerstamm) – Germany, gen. slide JCK 3365 (LMAD).

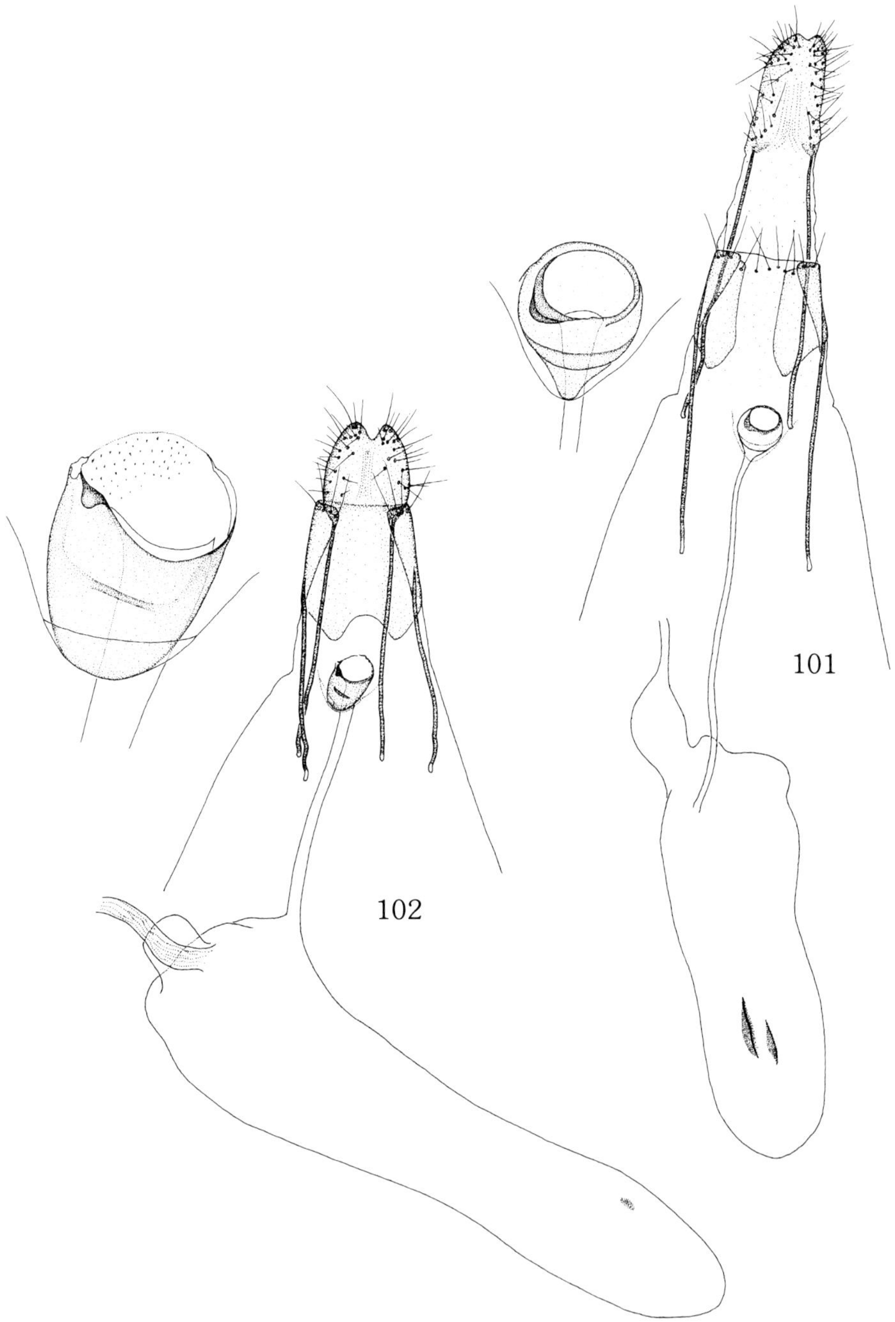

Female Genitalia

101. *Eteobalea anonymella* (Riedl) – Russia, gen. slide JCK 3563 (ZIN).
102. *Eteobalea serratella* (Treitschke) – Turkey, gen. slide JCK 3564 (ZIN).

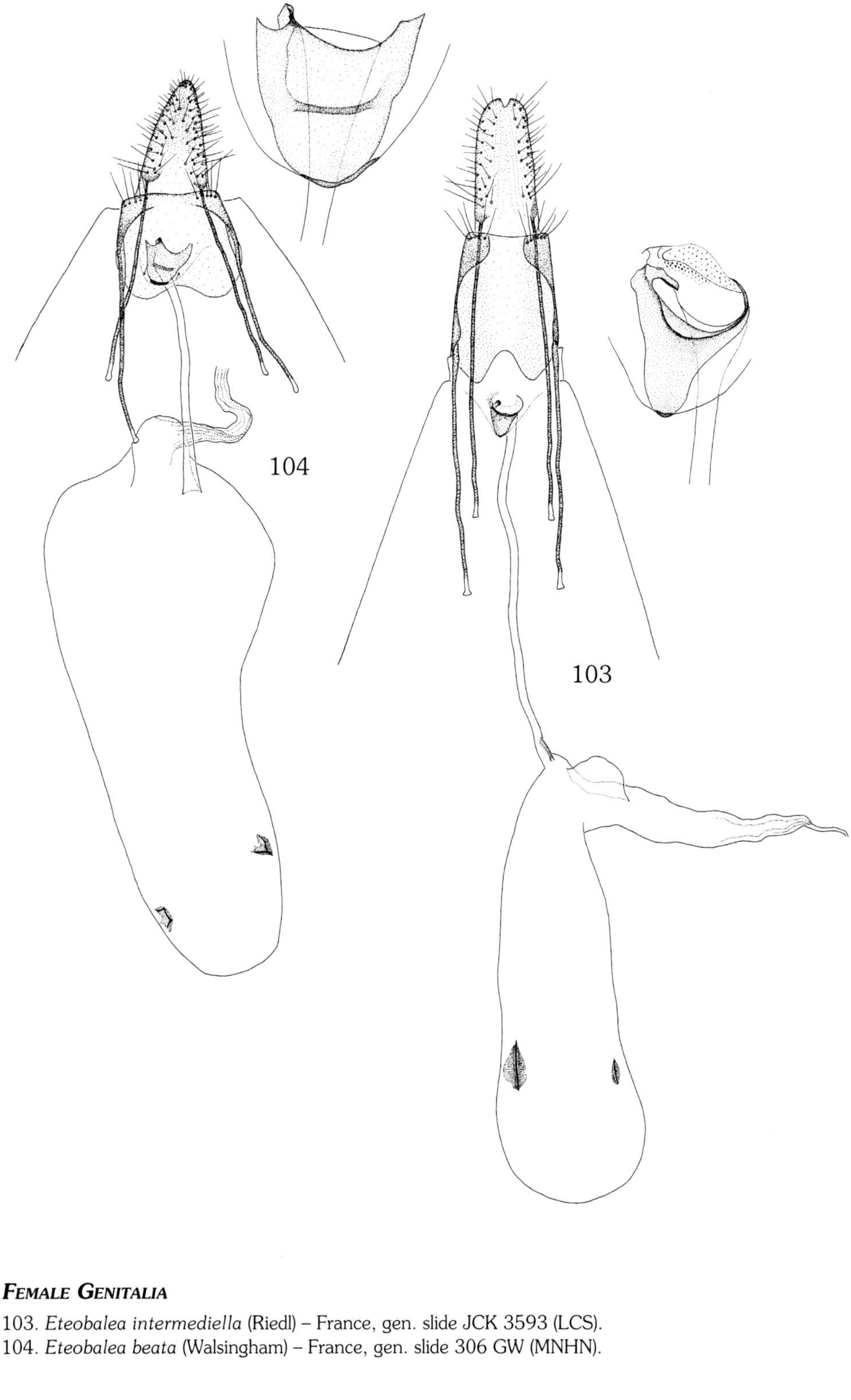

Female Genitalia

103. *Eteobalea intermediella* (Riedl) – France, gen. slide JCK 3593 (LCS).
104. *Eteobalea beata* (Walsingham) – France, gen. slide 306 GW (MNHN).

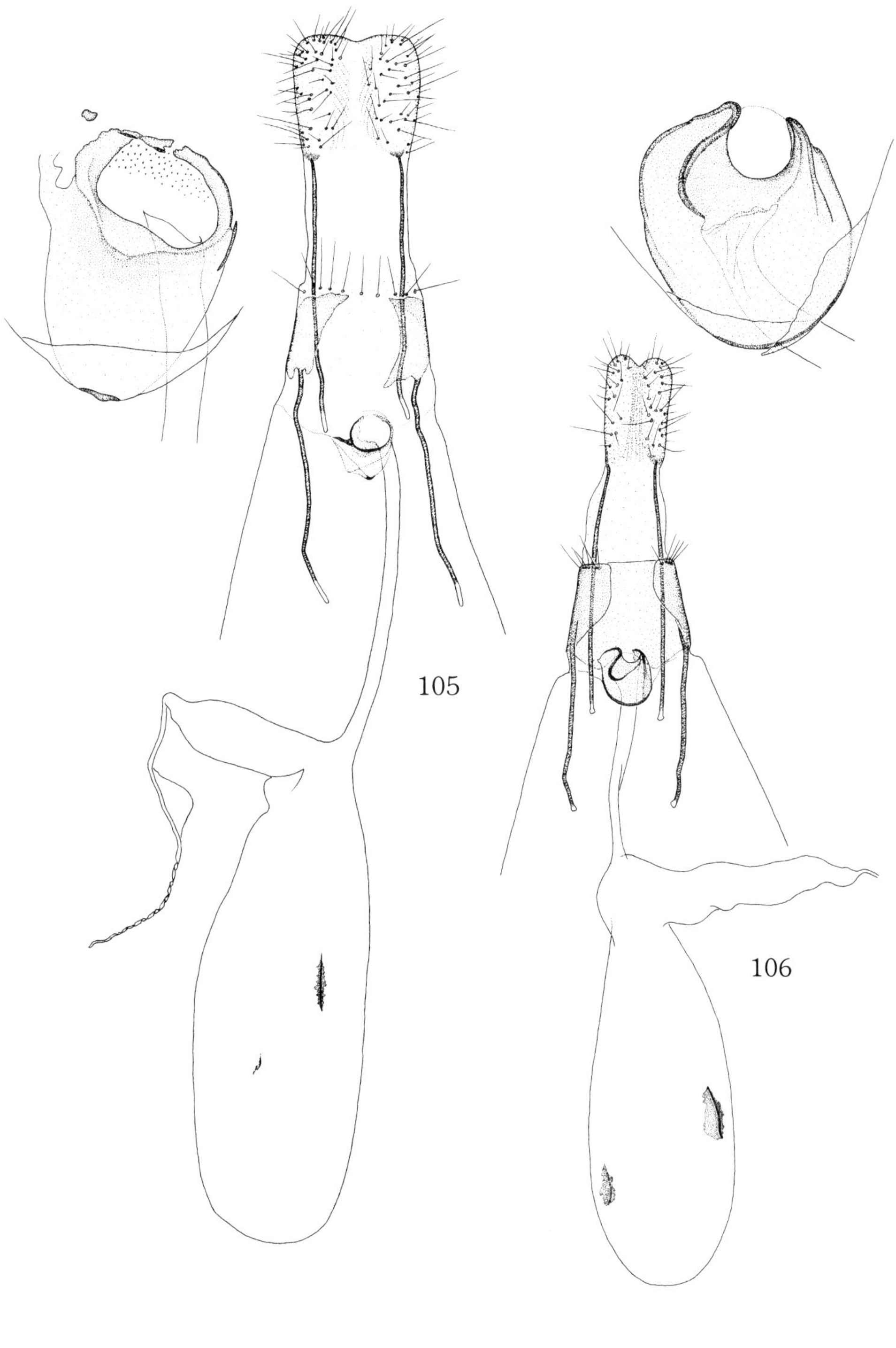

Female Genitalia

105. *Eteobalea sumptuosella* (Lederer) – Greece, gen. slide JCK 3508 (LCS).
106. *Eteobalea teucrii* (Walsingham) – Morocco, gen. slide 926 R (MNHN).

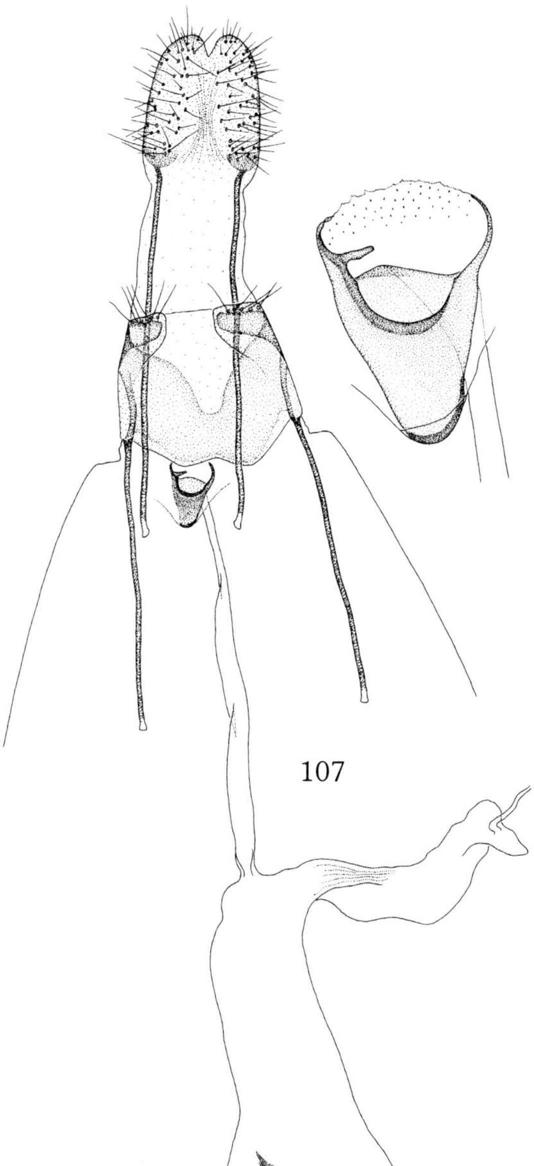

107

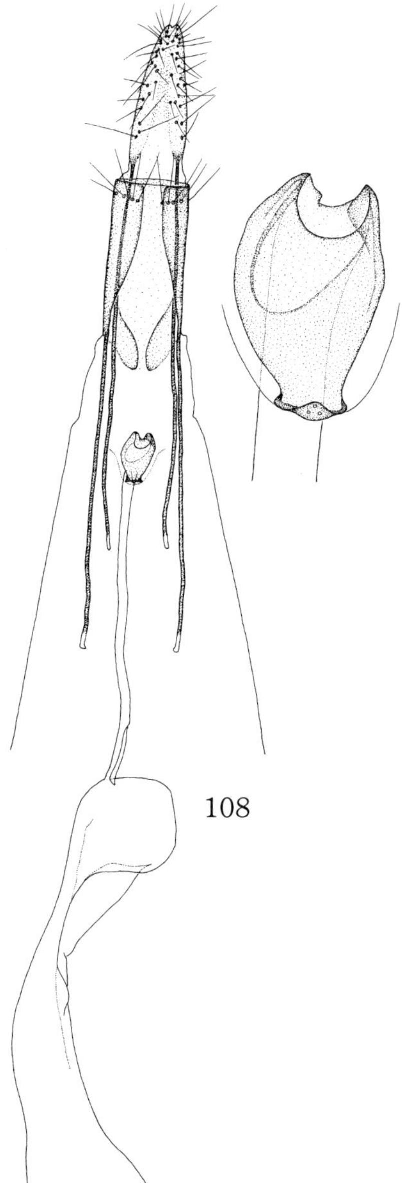

108

FEMALE GENITALIA

107. *Eteobalea klisieckii* (Riedl) – Morocco, gen. slide 924/R (MNHN).
108. *Eteobalea tririvella* (Staudinger) – Kazakhstan, gen. slide 3560 JCK (ZIN).

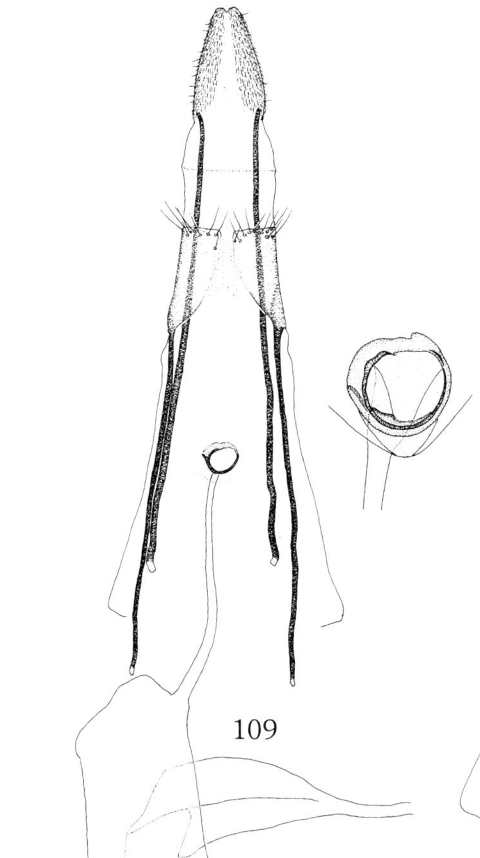
109

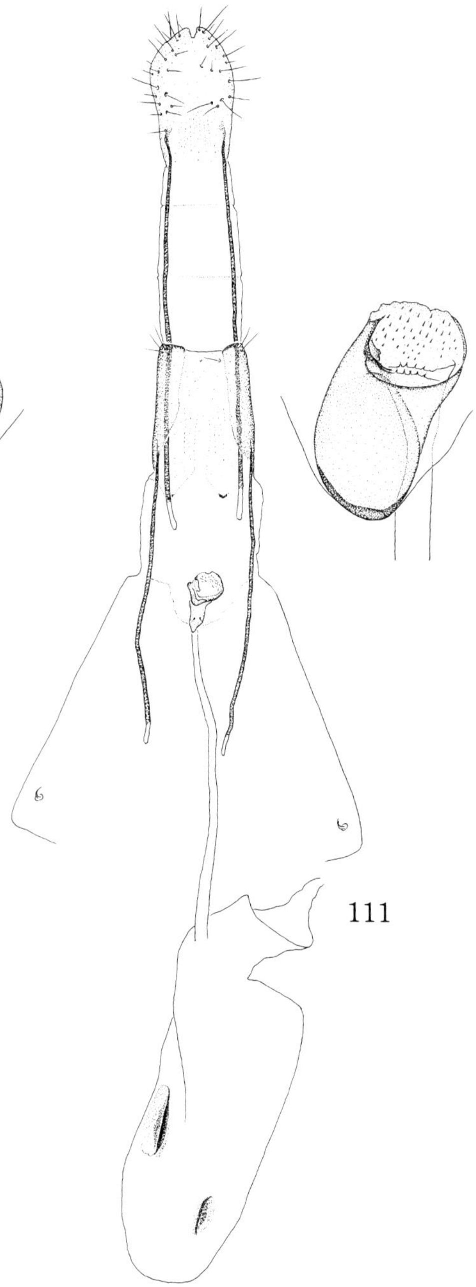
111

Female Genitalia

109. *Eteobalea albiapicella* (Duponchel) – Switzerland, gen. slide JCK 3332 (VDW).
110. *Eteobalea alypella* (Klimesch) – unknown.
111. *Eteobalea isabellella* (Costa) – Italy, gen. slide JCK 3575 (LCS).

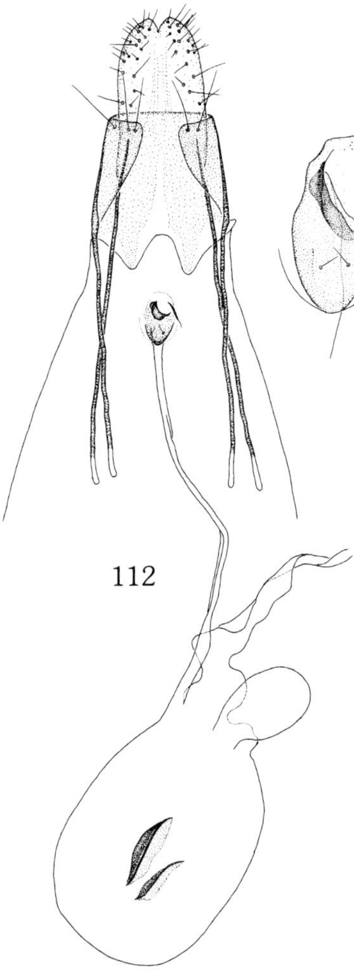

112

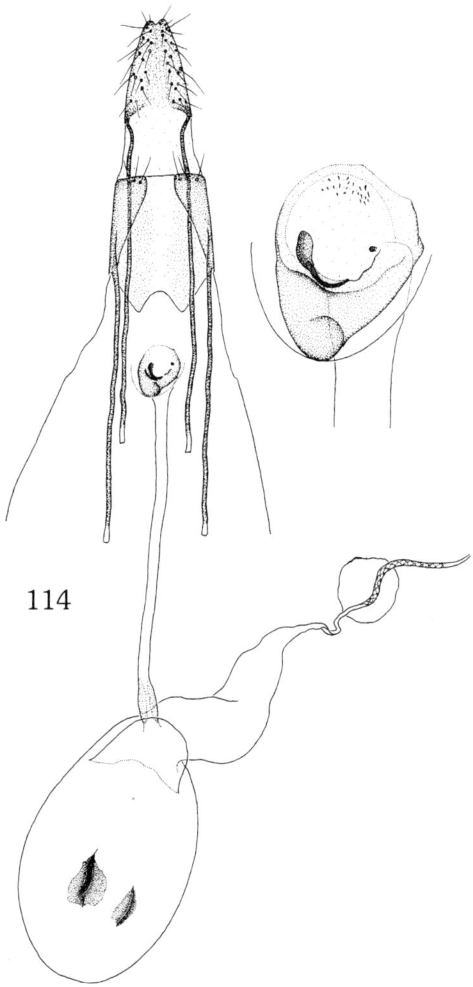

114

FEMALE GENITALIA

112. *Eteobalea dohrnii* (Zeller) – France, gen. slide JCK 3470 (ZIN).
113. *Eteobalea siciliae* (Riedl) – unknown.
114. *Eteobalea thaumatella* (Walsingham) – Tunisia, gen. slide 997/R (MNHN).

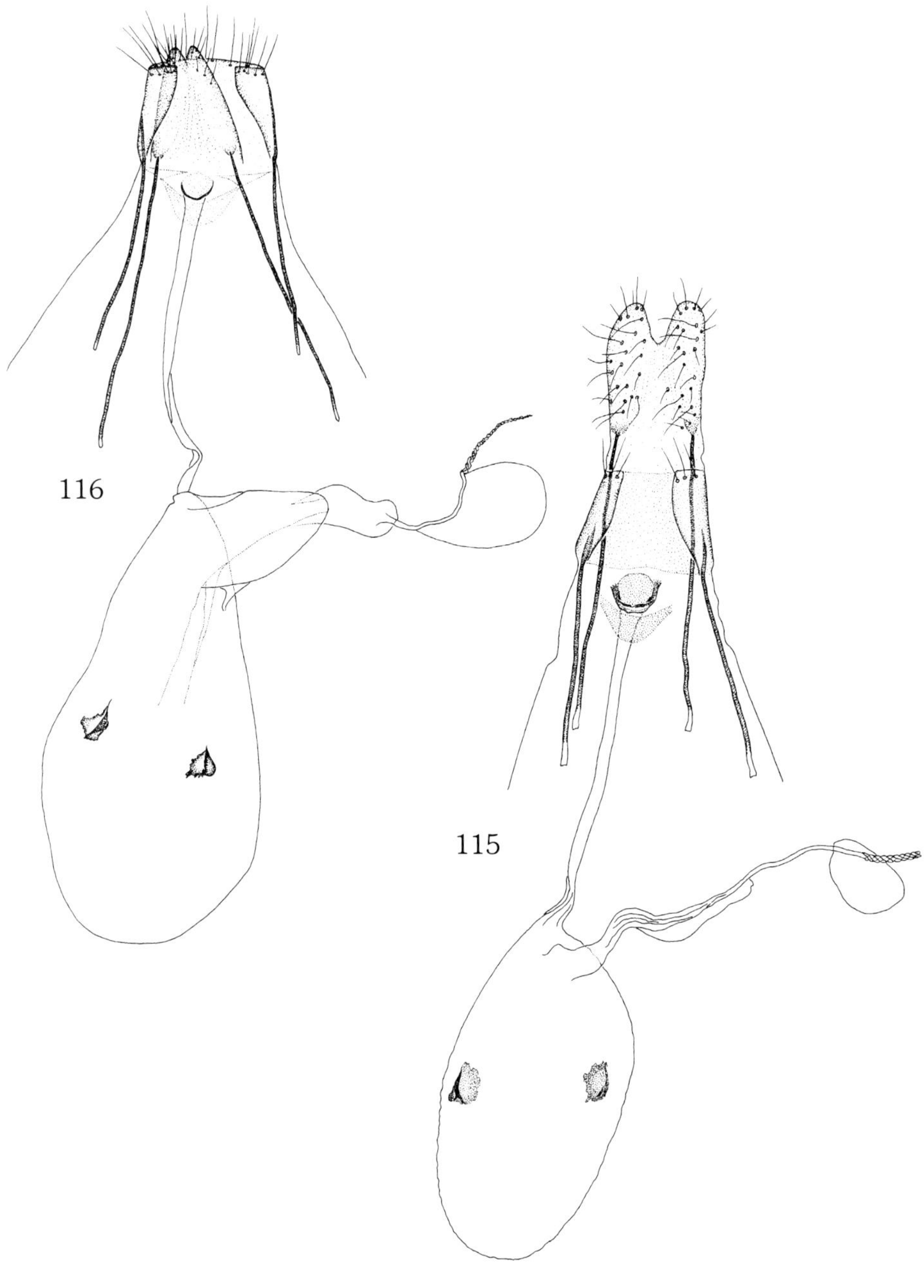

Female Genitalia

115. *Isidiella nickerlii* (Nickerl) – Czech Republic, gen. slide JCK 3568 (ZIN).
116. *Isidiella divitella* (Constant) – Italy, gen. slide JCK 3441 (LMAD).

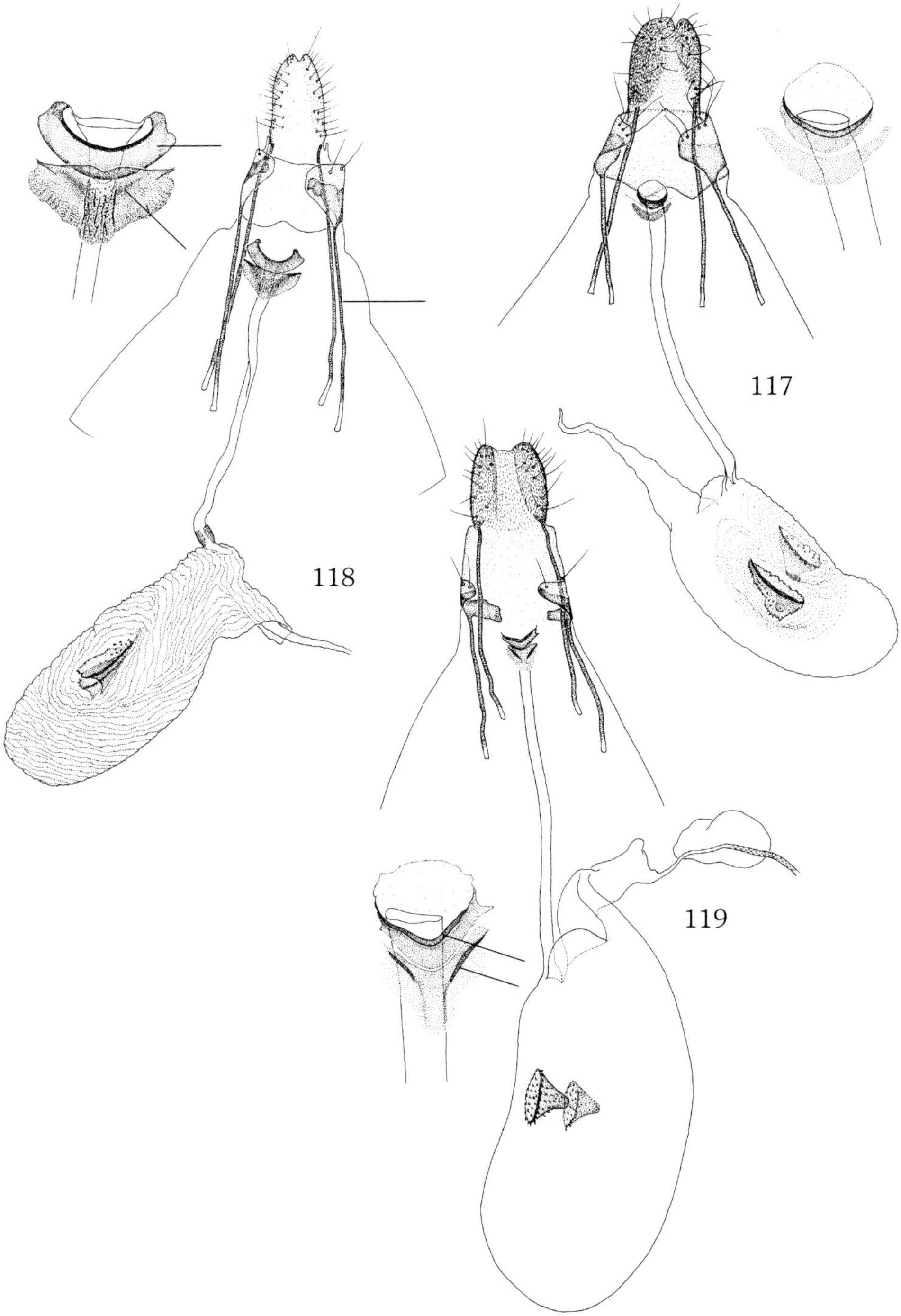

Female Genitalia

117. *Vulcaniella pomposella* (Zeller) – Poland, gen. slide JCK 3569 (ZIN).
118. *Vulcaniella grandiferella* Sinev – Ukraine (Crimea), gen. in vial (ZIN).
119. *Vulcaniella cognatella* Riedl – Greece, gen. slide JCK 2127 (KTR).

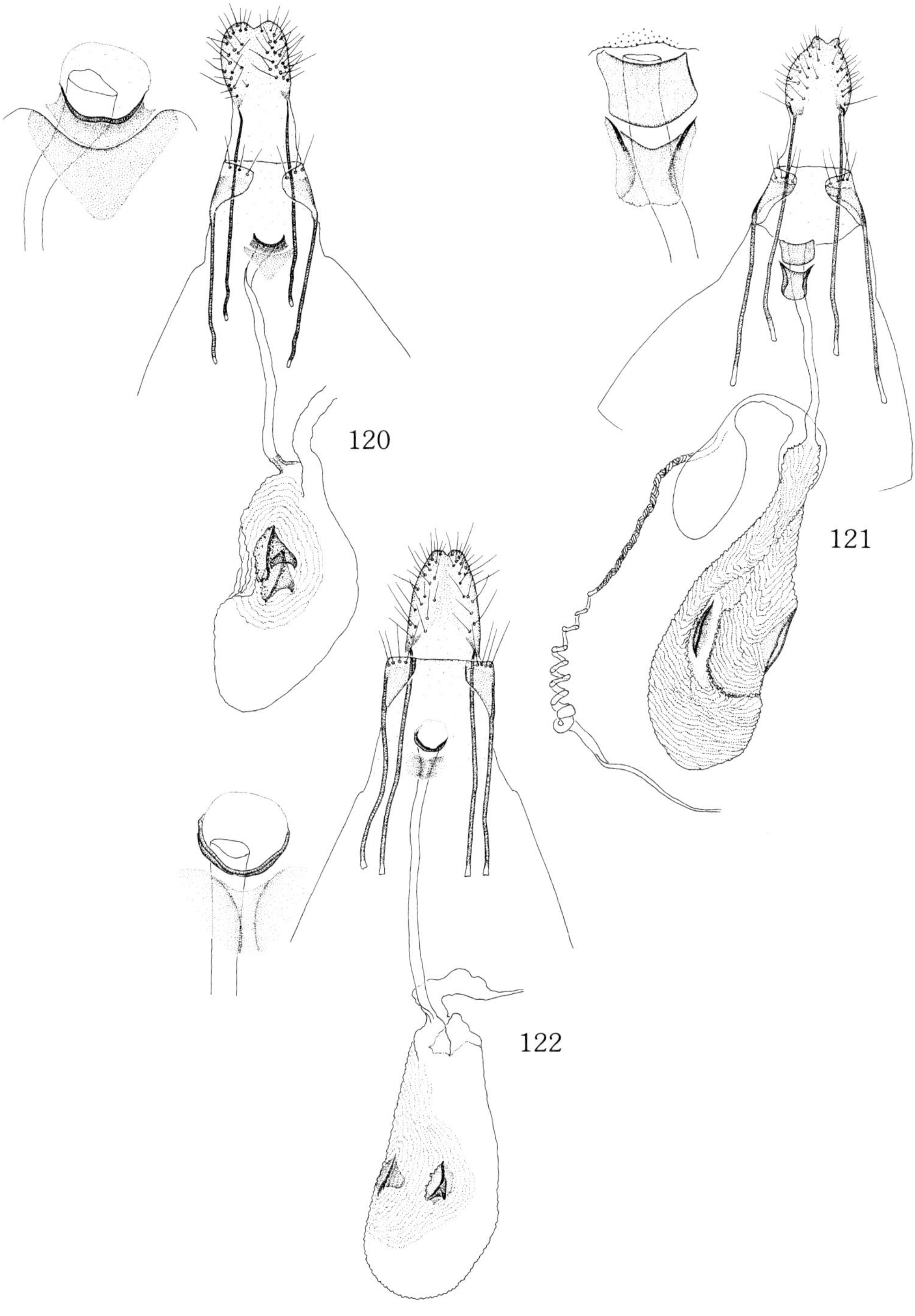

FEMALE GENITALIA

120. *Vulcaniella extremella* (Wocke) – France, gen. slide 1077 /R (MNHN).
121. *Vulcaniella caucasica* Sinev – Azerbaijan, gen. in vial (ZIN).
122. *Vulcaniella grabowiella* (Staudinger) – Spain, gen. slide JCK 3572 (ZIN).

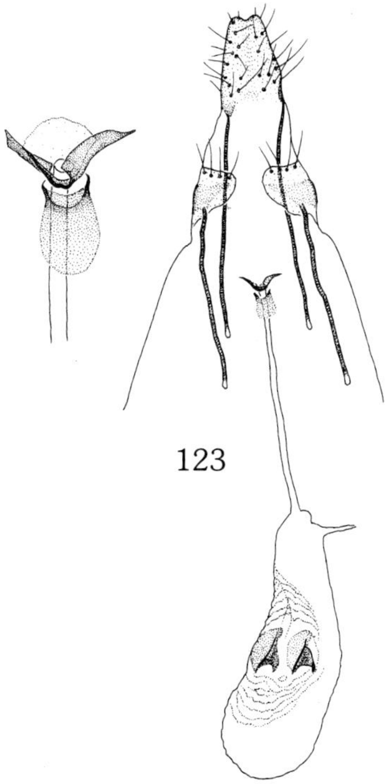

123

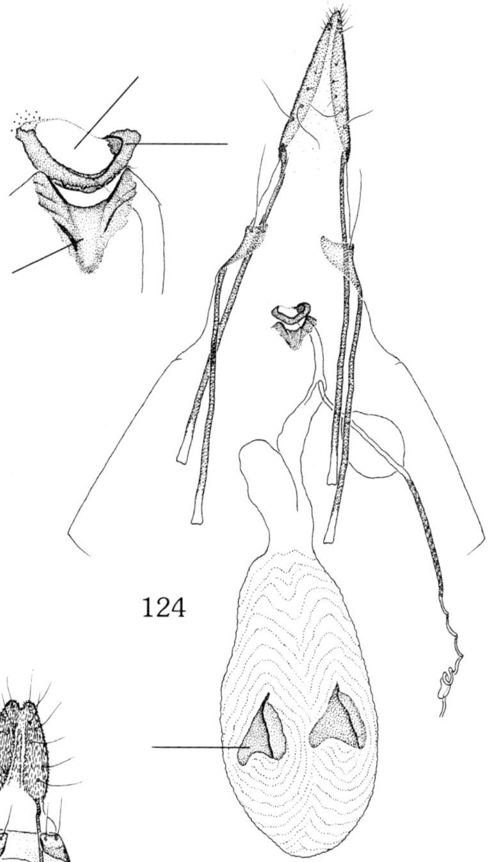

124

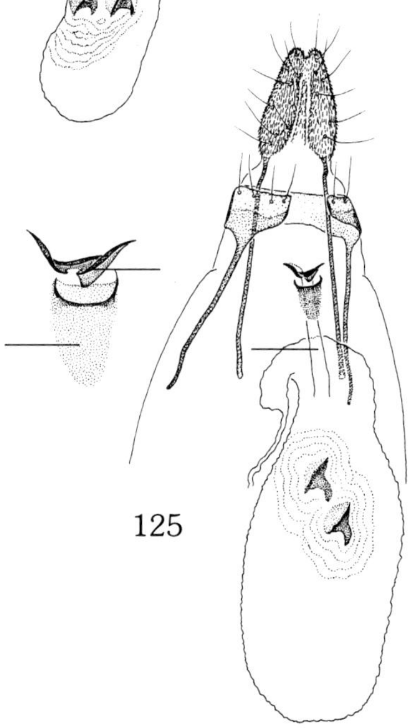

125

FEMALE GENITALIA

123. *Vulcaniella rosmarinella* (Walsingham) – France, gen. slide 1074/R (MNHN).
124. *Vulcaniella karadaghella* Sinev – Ukraine, gen. slide JCK 4275 (ZIN).
125. *Vulcaniella klimeschi* (Riedl) – paratype, Macedonia, gen. slide 524/R (ZSMC).

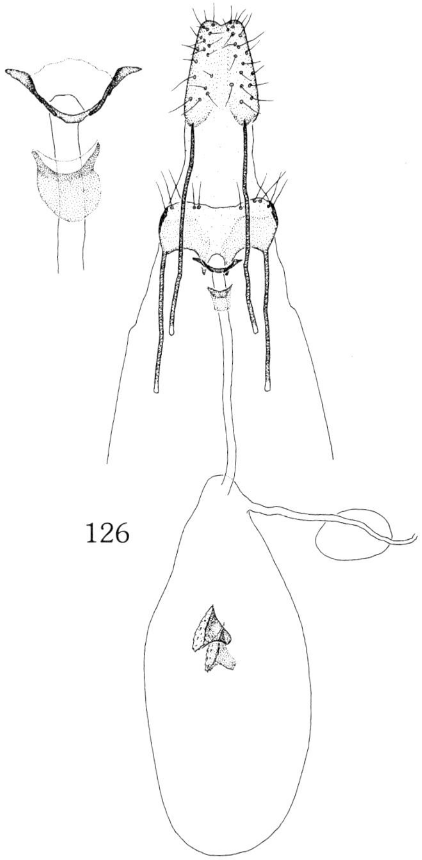

126

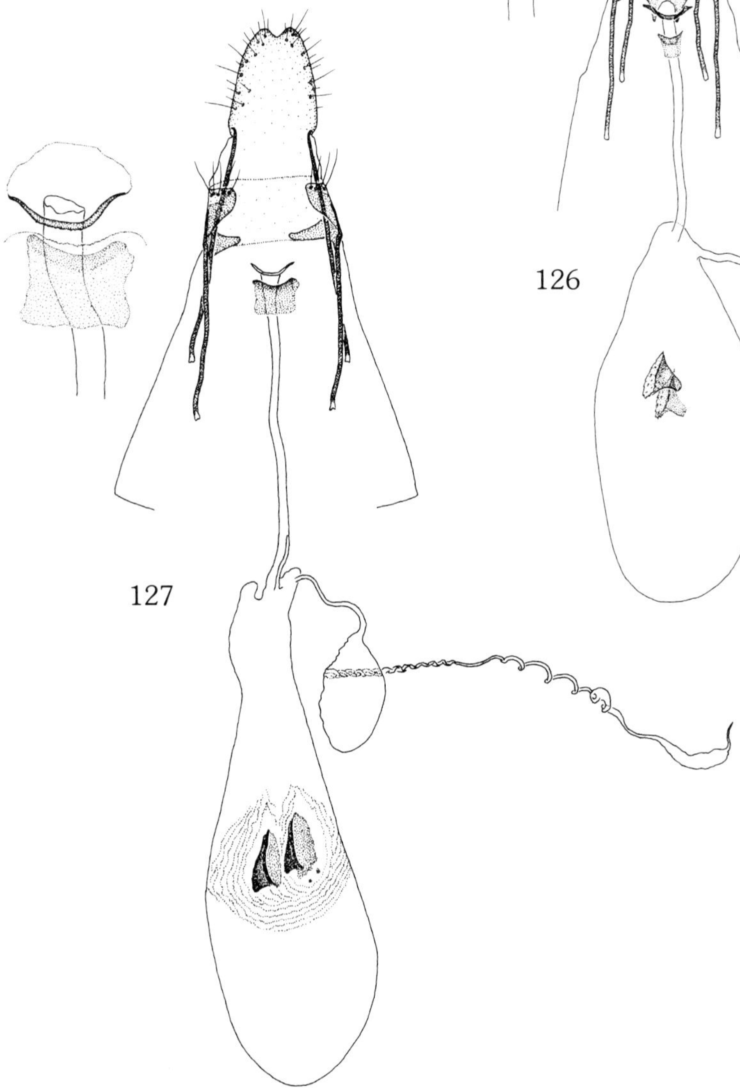

127

FEMALE GENITALIA

126. *Vulcaniella fiordalisa* (Petry) – Spain, gen. slide JCK 3633 (GLS).
127. *Vulcaniella glaseri* (Riedl) – Turkey, gen. slide JCK 3930 (ZMUC).
128. *Vulcaniella gielisi* sp. n. – unknown.
129. *Vulcaniella anatolica* sp. n. – unknown.
130. *Vulcaniella pontica* sp. n. – unknown.

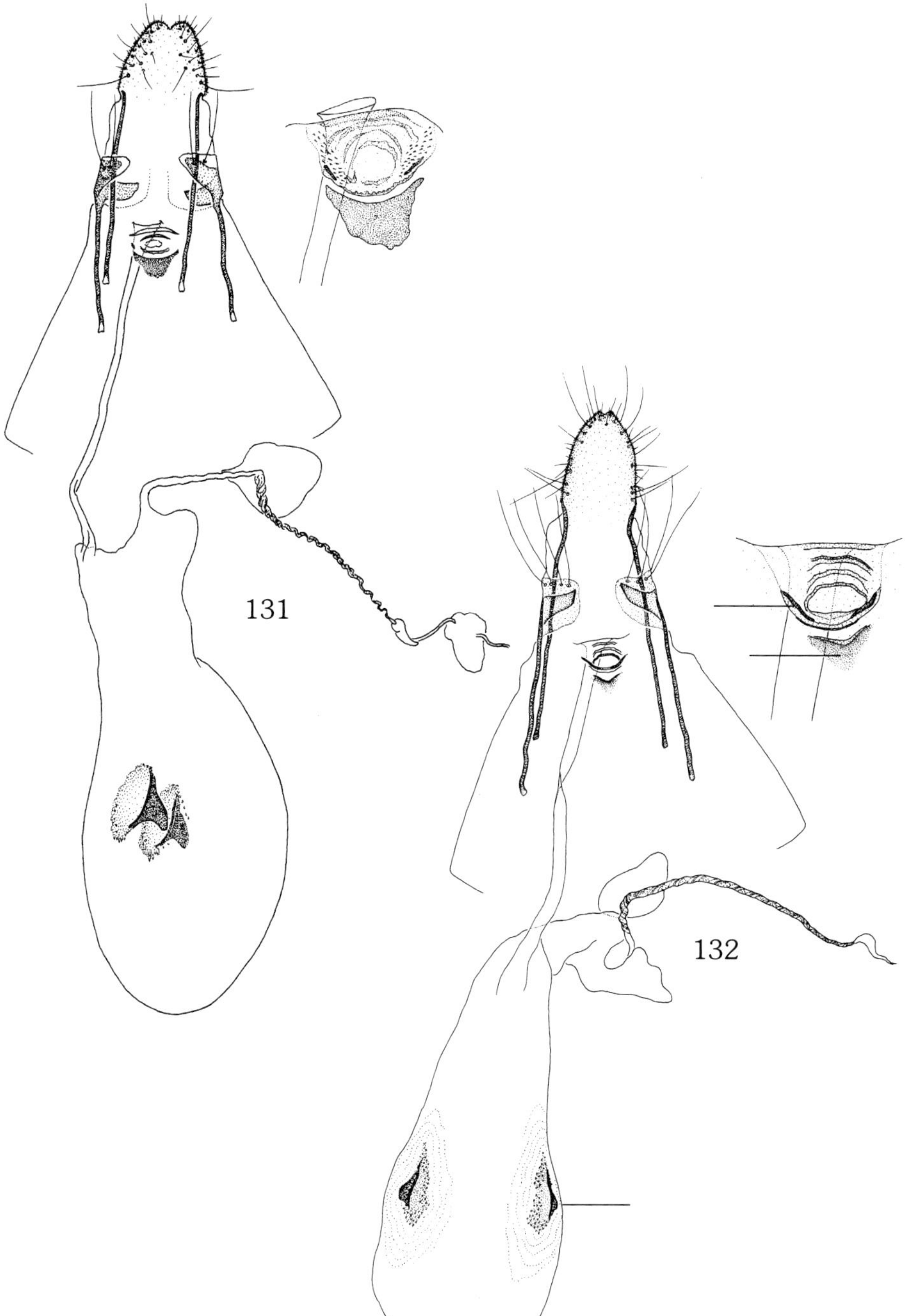

Female Genitalia

131. *Ramphis ibericus* Riedl – Spain, gen. slide JCK 3933 (ZMUC).
132. *Ramphis libanoticus* Riedl – Turkey, gen. slide JCK 2095 (KTR).

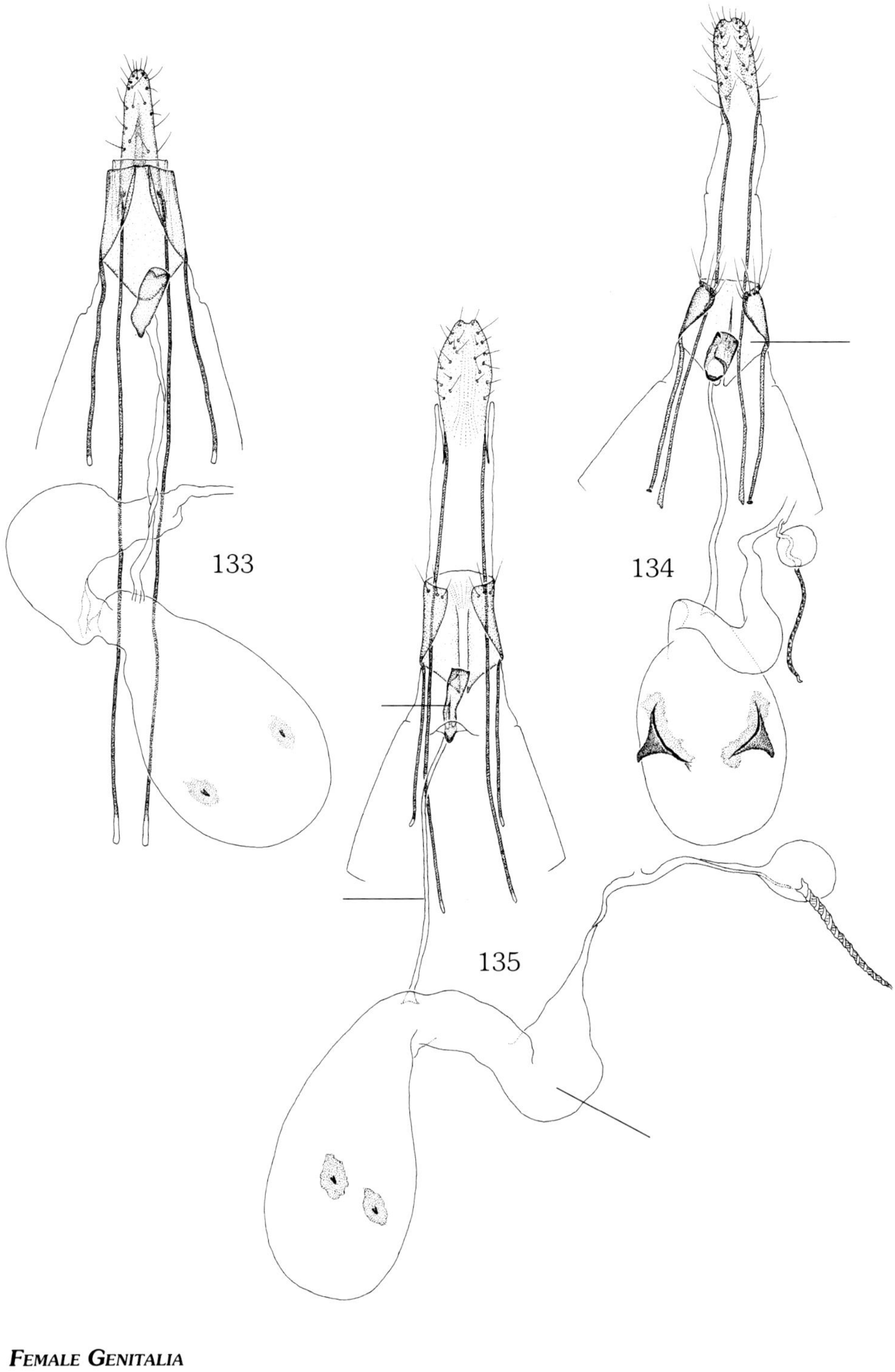

Female Genitalia

133. *Hodgesiella rebeli* (Krone) – Croatia, gen. slide 1105/R (MNHN).
134. *Hodgesiella rhodorrhisella* Kasy – Canary Islands, gen. slide JCK 4250 (ZMUC).
135. *Hodgesiella christophi* sp. n. – paratype, Turkey, gen. slide JCK 3986 (ZMUC).

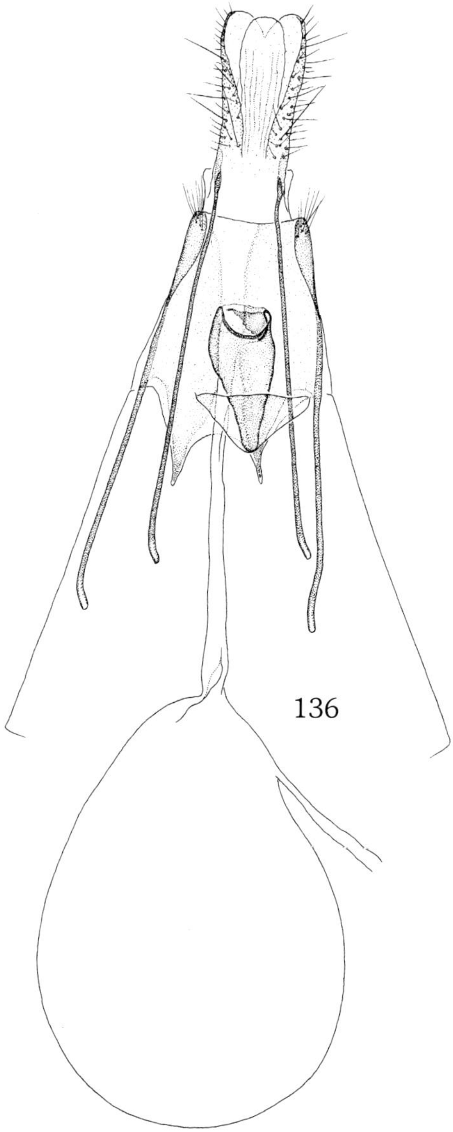

136

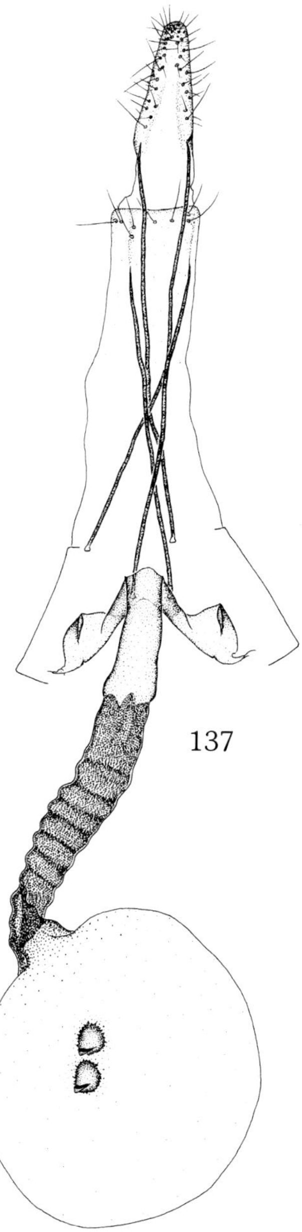

137

Female Genitalia

136. *Tolliella fulguritella* (Ragonot) – Turkey, gen. in vial (ZMUC).
137. *Sorhagenia rhamniella* (Zeller) – Austria, gen. slide JCK 3483 (LMAD).

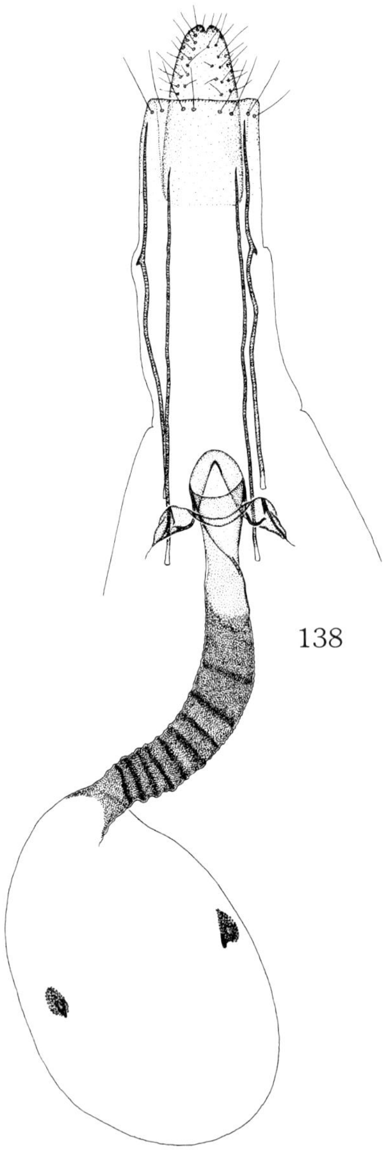

138

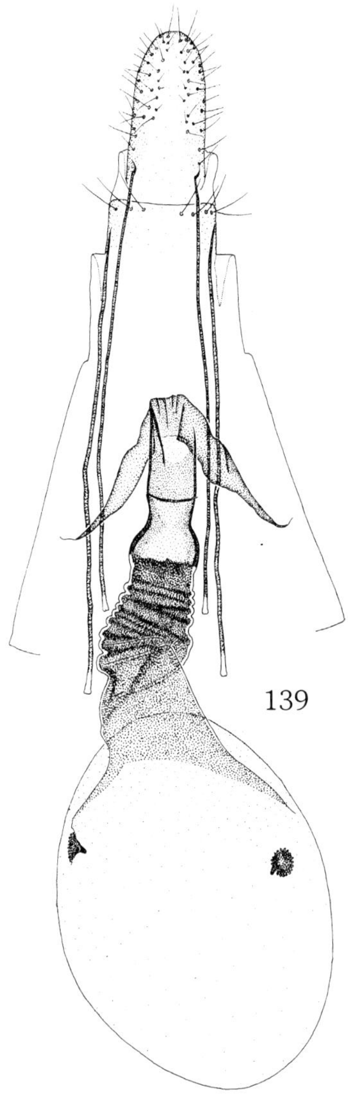

139

FEMALE GENITALIA

138. *Sorhagenia reconditella* Riedl – Greece, gen. slide JCK 2639 (KTR).
139. *Sorhagenia lophyrella* (Douglas) – Czech Republic, gen. slide 1983 (KTR).

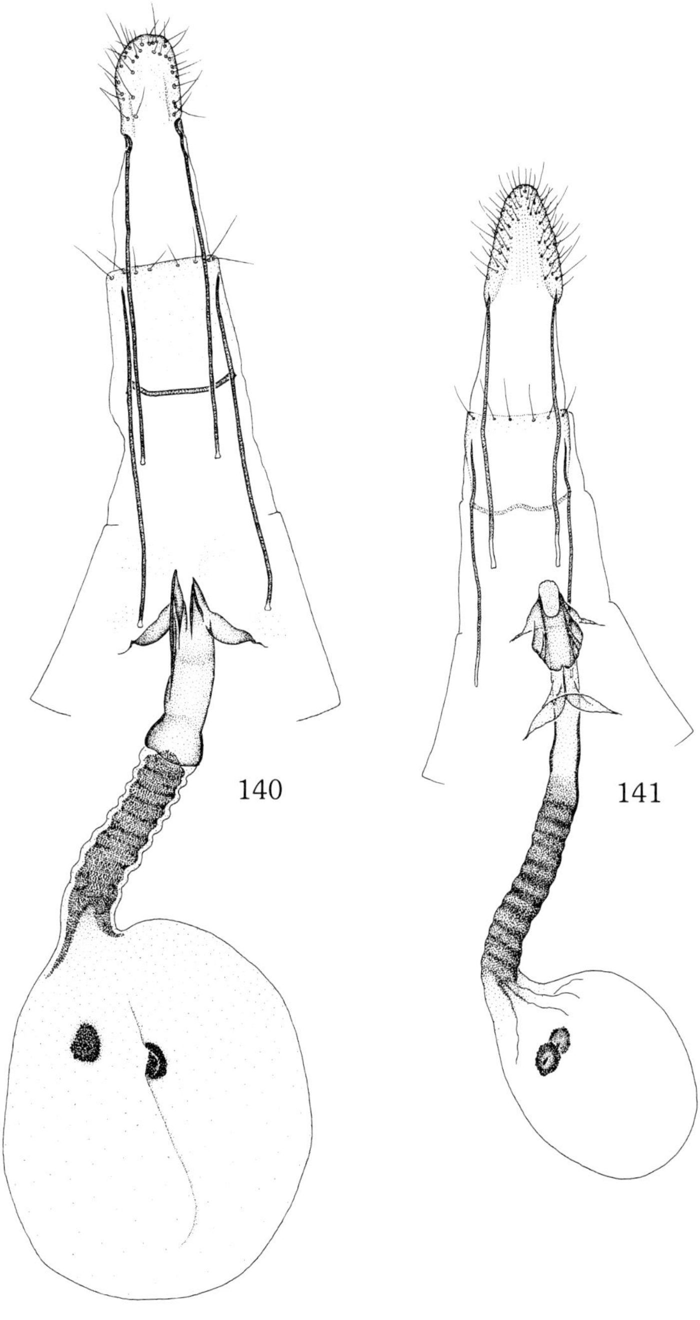

Female Genitalia

140. *Sorhagenia janiszewskae* Riedl – Germany, gen. slide JCK 3482 (LMAD).
141. *Sorhagenia fibigeri* sp.n. – paratype, Turkey, gen. slide JCK 4144 (ZMUC).

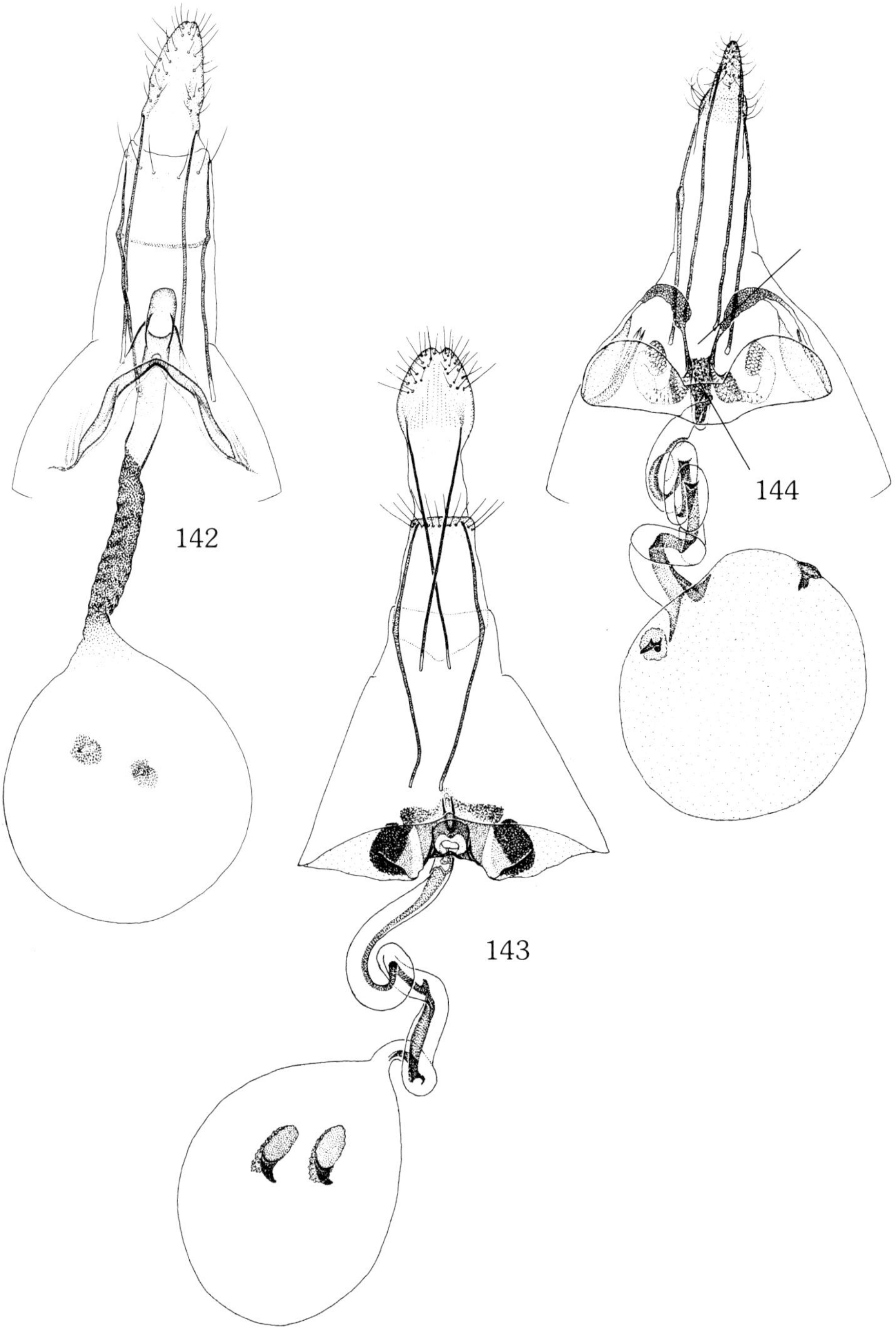

Female Genitalia

142. *Sorhagenia taurensis* sp.n. – paratype, Turkey, gen. slide JCK 4065 (RMNH).
143. *Ascalenia vanella* (Frey) – France, gen. slide JCK 3860 (RMNH).
144. *Ascalenia vanelloides* Gerasimov – Turkmenia, gen in vial (ZIN).

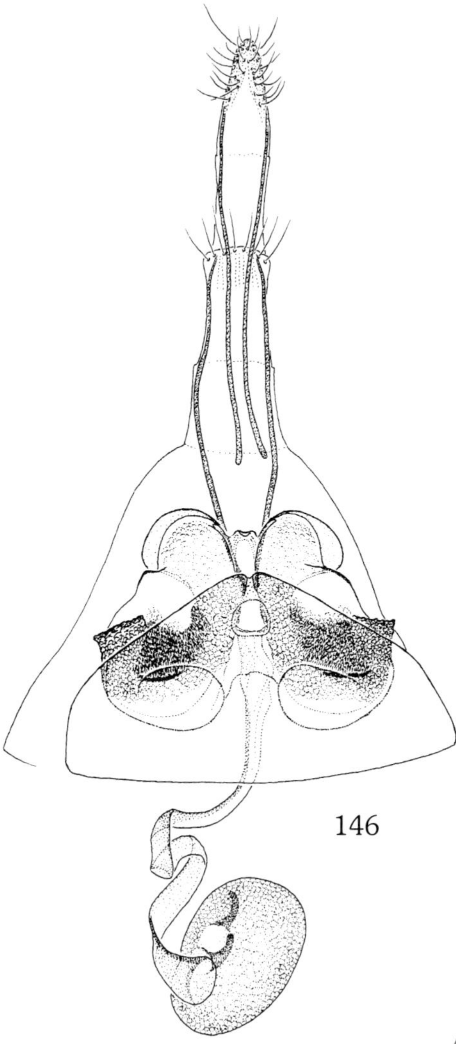

146

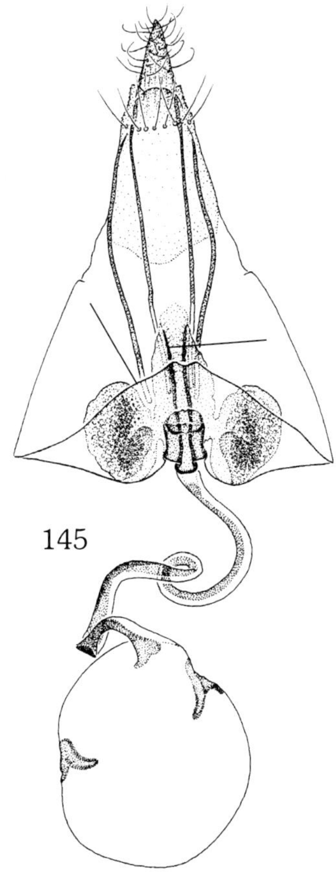

145

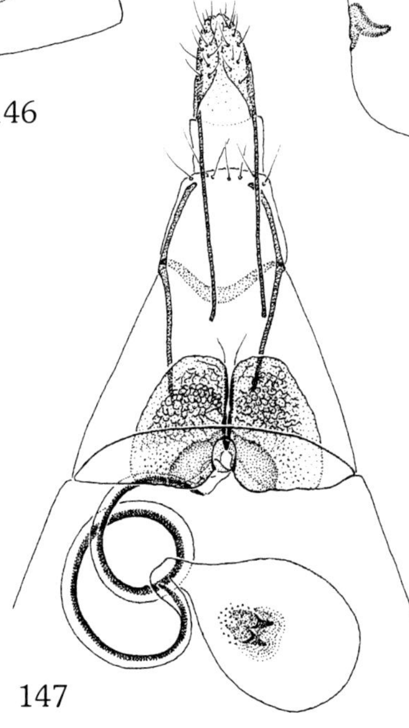

147

Female Genitalia

145. *Ascalenia antiqua* Meyrick – Israel, gen. slide BMNH 16157 (BMNH).
146. *Ascalenia acaciella* Chrétien – Algeria, gen. slide JCK 4430 (MNHN).
147. *Ascalenia jerichoella* Amsel – holotype, Israel, gen. in vial (LSNK).

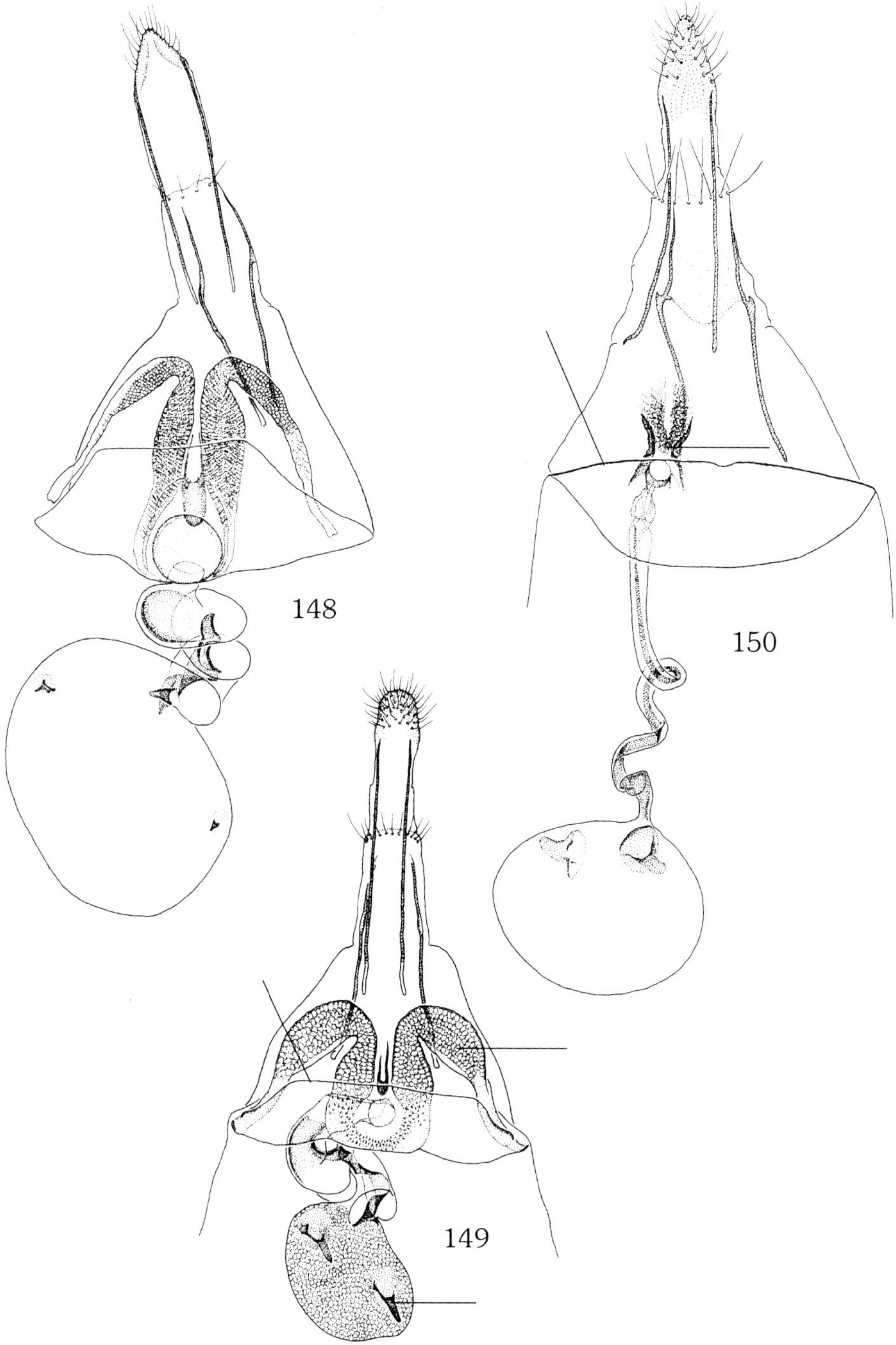

Female Genitalia

148. *Ascalenia viviparella* Kasy – Turkey, gen. slide JCK 4063 (ZMUC).
149. *Ascalenia echidnias* (Meyrick) – Turkey, gen. slide JCK 4137 (ZMUC).
150. *Ascalenia bifasciella* Chrétien – country unknown, gen. slide JCK 4481 (MNHN).

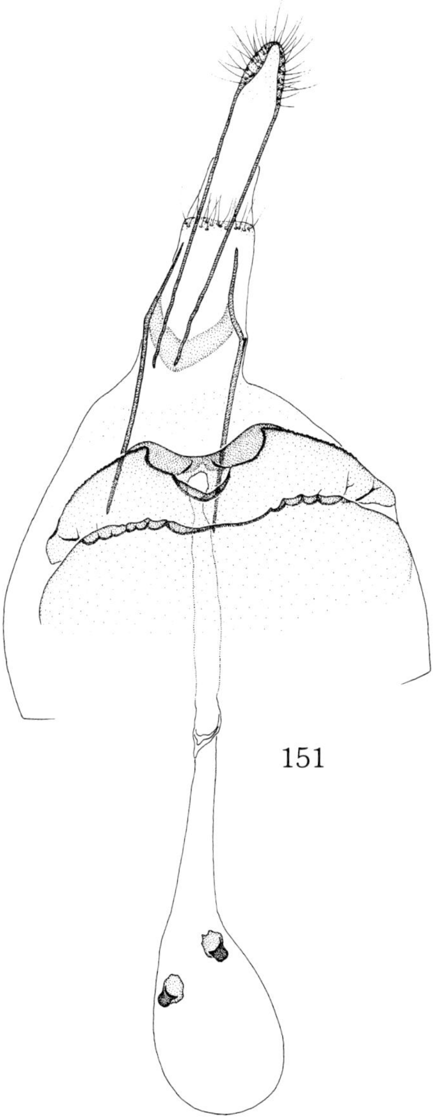

151

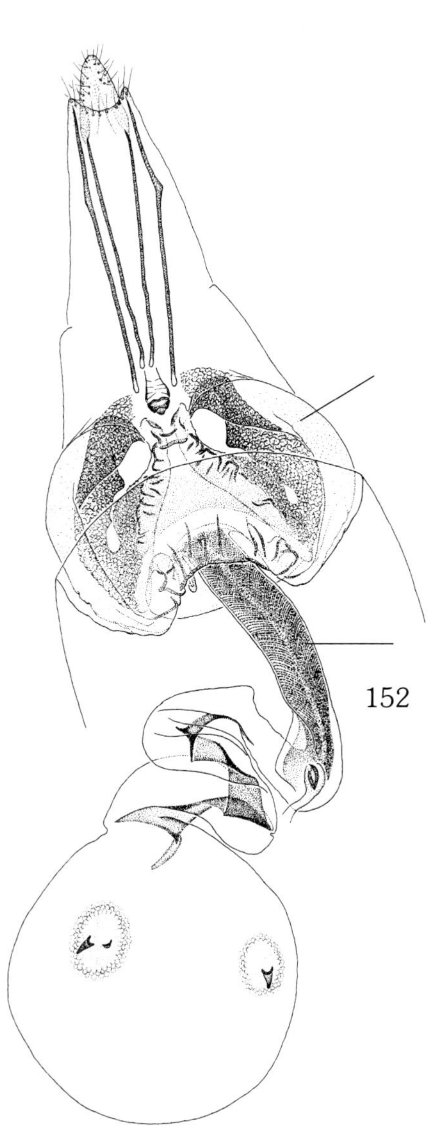

152

Female Genitalia

151. *Calanesia karsholti* Sinev – Tunisia, gen. in vial (ZIN).
152. *Calycobathra acarpa* Meyrick – Morocco, gen. in vial (ZMUC).

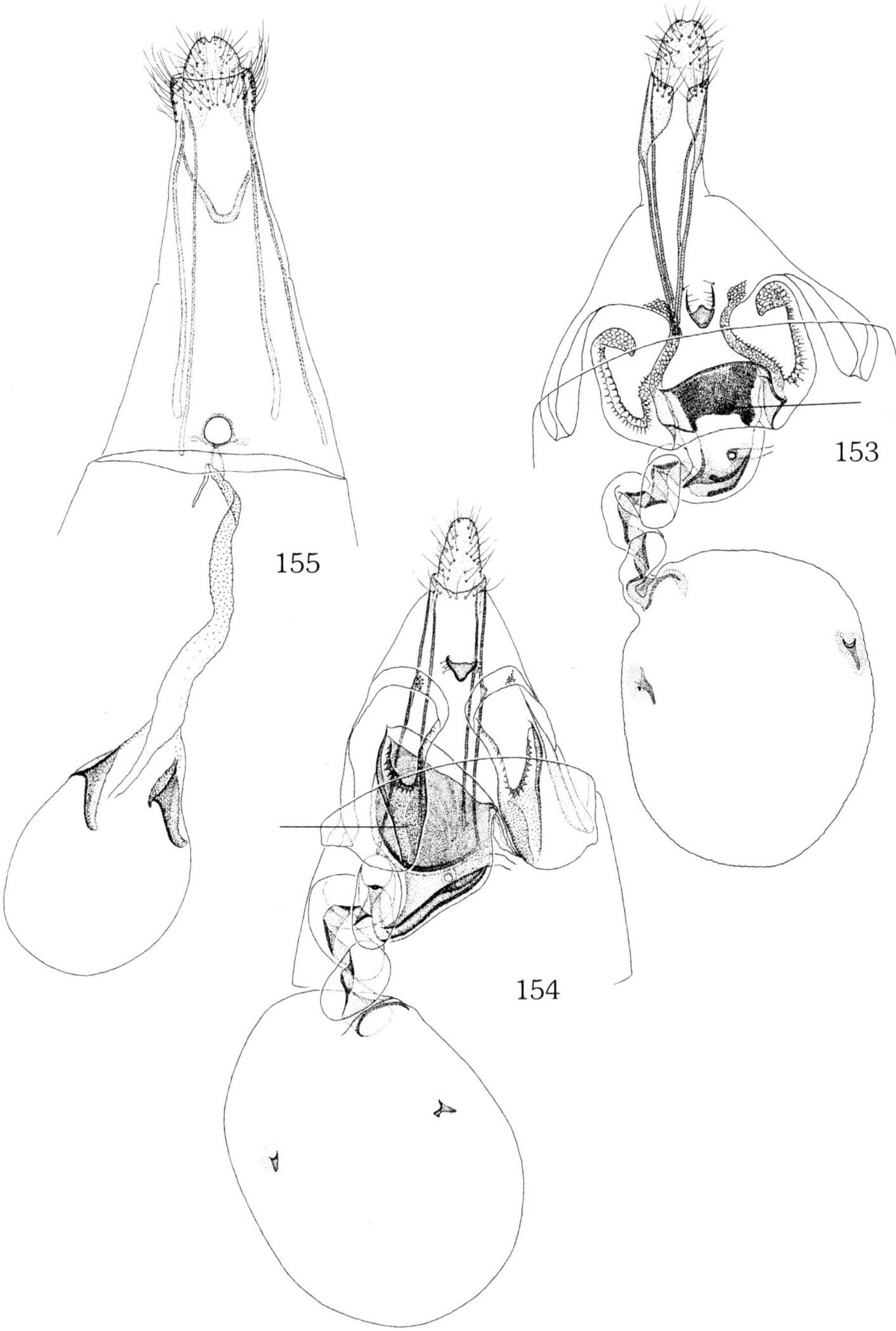

Female Genitalia

153. *Calycobathra calligoni* Sinev – Turkmenia, gen. in vial (ZIN).
154. *Calycobathra variapennella* Sinev – Kazakhstan, gen. in vial (ZIN).
155. *Bifascia nigralbella* (Chrétien) – S-Iran, gen. in vial (ZIN).

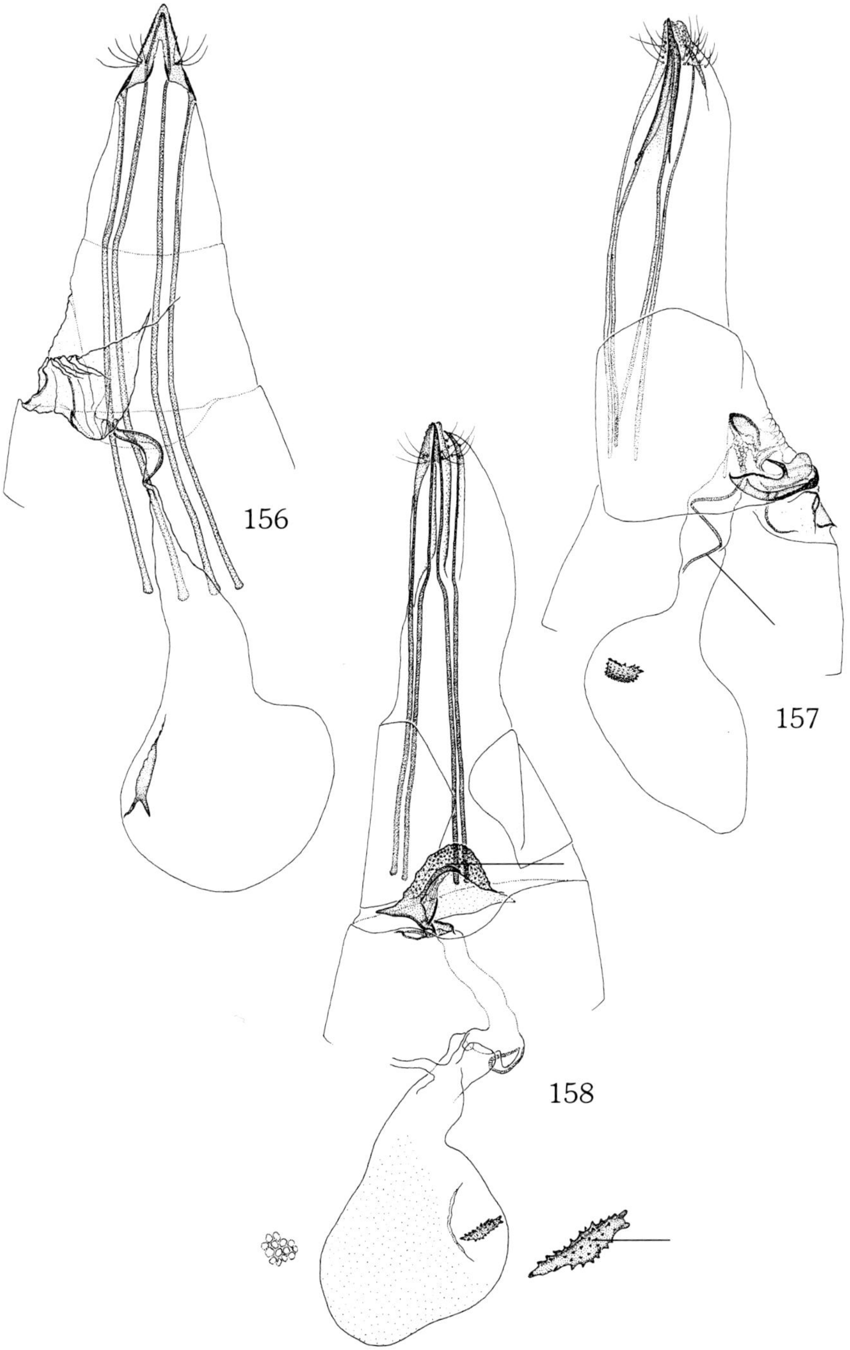

Female Genitalia

156. *Bifascioides leucomelanellus* (Rebel) – S-Iran, gen. in vial (ZIN).
157. *Gisilia stereodoxa* (Meyrick) – Iran, gen. in vial (NHMW).
158. *Gisilia lerautella* Gibeaux – paratype, France, gen. slide CG 2449 (GBX).

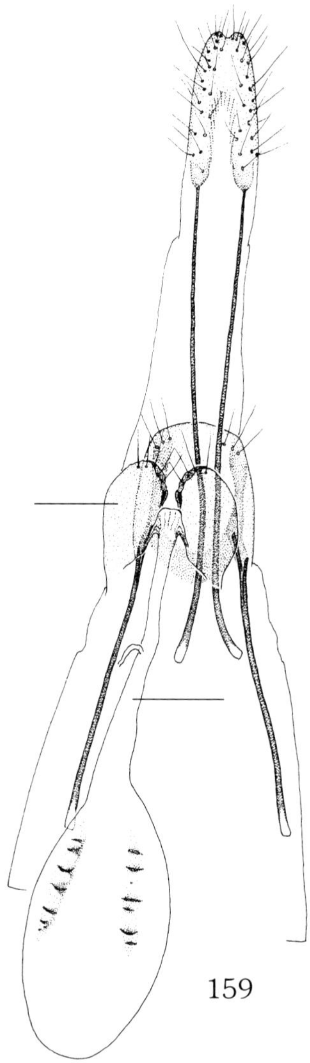

159

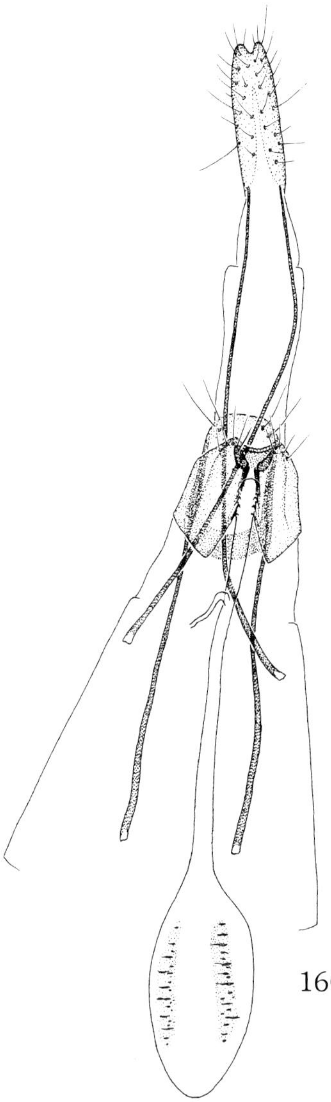

160

FEMALE GENITALIA

159. *Neomariania partinicensis* (Rebel) – Greece, gen. slide Wf. 5265 (VDW).
160. *Neomariania rebeli* (Walsingham) – Madeira, gen. slide JCK 4281 (ZMUC).

References

Agassiz, L., 1847. *Nomenclatoris zoologici index universalis*. x+1135 pp. Soloduri.

Agenjo, R., 1952. *Faunula Lepidopterologica Almeriense*. 370 pp. Madrid.

Agenjo, R., 1981. *Cosmopterix dalii*, nueva especie españolo de Cosmopterigidae, descubierta en el "Bosque del Marques", de Alella, provincia de Barcelona. – *Bol. Estac. centr. Ecol.* **10** (20): 49-54.

Amsel, H. G., 1930. Die Microlepidopterenfauna der Mark Brandenburg nach dem heutigen Stande unserer Kenntnisse. – *Dt. ent. Z. Iris* **44**: 83-132.

Amsel, H. G., 1933. Die Lepidopteren Palästinas. Eine zoogeographisch-oekologisch-faunistische Studie. – *Zoogeographica, Jena* **2** (1): 1-146.

Amsel, H. G., 1935a. Neue palästinensische Lepidopteren. – *Mitt. zool. Mus. Berlin* **20** (2): 271-319.

Amsel, H. G., 1935b. Weitere Mitteilungen über palästinensische Lepidopteren. – *Veröff. Dt. Kol. Übersee-Mus. Bremen* **1** (2): 223-278.

Amsel, H. G., 1936. Zur Kenntnis der Kleinschmetterlingsfauna Sardiniens. – *Veröff. Dt. Kol. Übersee-Mus. Bremen* **1** (3): 334-365.

Amsel, H. G., 1939. See Hartig, F. & Amsel, H. G., 1939.

Amsel, H. G., 1955. Über mediterrane Microlepidopteren und einige transkaspische Arten. – *Bull. Inst. R. Sci. nat. Belg.* **31** (83): 1-64.

Amsel, H. G., 1958. Kleinschmetterlinge aus Nordost-Arabien der Ausbeute A.S. Talhouk. – *Beitr. naturk. Forsch. SüdwDtl.* **17** (1): 61-82.

Amsel, H. G., 1959. Irakische Kleinschmetterlinge, 2. – *Bull. Soc. ent. Egypte* **43**: 41-83.

Amsel, H. G., 1961. Microlepidopteren aus SW.-Arabien der Ausbeuten H. Scott und E. B. Britton 1937/38. – *Beitr. naturk. Forsch. SüdwDtl.* **20** (1): 49-61.

Arnett, R. H., Samuelson, G. A. & Nishida, G. M., 1993. *The insect and spider collections of the world*. Edn 2. 305 pp. Gainesville.

Baer, W., 1910. Über die Verpuppungsweise von *Batrachedra pinicolella* Dup. – *Naturwiss. Z. Landw., Stuttgart* **8**: 304-305.

Balachowsky, A. S. (ed.), 1966. *Entomologie appliquée á l'agriculture*. No.2. 1057 pp. Paris.

Baldizzone, G., 1985. *Aetia buvati* n. sp. (Lepidoptera, Blastodacnidae). – *Alexanor* **14** (3): 135-138.

Barrett, C. G., 1865. Notes on the galls of *Laverna decorella*. – *Entomologist's mon. Mag.* **1**: 197-198.

Barrett, C. G., 1898. Economy of *Laverna vinolentella* H.-S. – *Entomologist's mon. Mag.* **34**: 204.

Bengtsson, B. Å., 1997. Scythrididae. – *In* Huemer, P., Karsholt, O. & Lyneborg, L. (eds): *Microlepidoptera of Europe* **2**: 1-301.

Biesenbaum, W., 2000. Bemerkenswerte Nachweise von zwei Kleinschmetterlingsarten in Baden-Württemberg (Lep., Cosmopterigidae et Gelechiidae). – *Melanargia* **12** (3/4): 81-83.

Billberg, G. J., 1820. *Enumeratio Insectorum*. 138 pp. Holmiae.

Blumberg, D., 1975. Preliminary notes on the phenology and biology of *Batrachedra amydraula* Meyrick (Lepidoptera: Cosmopterygidae), a new pest of date palms in Israel. – *Phytoparasitica* **3** (1): 55-57.

Bolov, A. P. & Sinev, S. Yu., 1989. *Batrachedra pinicolella* Dup. (Lepidoptera, Batrachedridae) – a pest of conifers in Kabardino-Balkaria. [In Russian.] – *Ent. Obozr.* **68** (3): 522-526.

Bozan, I. & Aslitürk, H., 1975. Investigations on the determination of fauna of tea plantations in the East Black sea region of Turkey. – *Ann. Plant Protect. Res.* **9**: 134-135.

Bradley, J. D., 1951. A comparative study of four European species including one new species from Britain, belonging to the genus *Mompha* Huebner (Lep.: Lavernidae). – *Entomologist's Gaz.* **2**: 173-182.

Bradley, J. D., Fletcher, D. S. & Whalley, P. E. S., 1972. Lepidoptera. – *In* Kloet, G. S. & Hincks, W. D. (eds): *A Check List of British Insects*. Edn 2. viii+153 pp. London.

Braun, A. F., 1915. New genera and species of Tineina. – *Can. Ent.* **47** (6): 188-197.

Braun, A. F., 1921. Two weeks collecting in Glacier National Park. – *Proc. Acad. nat. Sci. Philad.* **73**: 1-23.

Brown, P. A., 1997. A review of techniques used in the preparation, curation and conservation of microscope slides at the Natural History Museum, London. – *The Biology Curator* **10**: 1-33.

Bruand, Th. d'Uzelle, 1847-[1851]. Catalogue systématique et synonymique des Microlépidoptères du department du Doubs. Tinéides. – *Mem. Soc. emul. Doubs, Besançon* **3** (3): 1-102. [Published 11th March 1851 according to Viette (1977: 287)]

Bruand, Th. d'Uzelle, [1859]. Classification des Tinéides. – *Annls Soc. ent. Fr.* (3) **6** (1858): 601-702.

Busck, A., 1906. New American Tineina. – *Can. Ent.* **38**: 121-124.

Busck, A., 1909. Notes on Microlepidoptera, with descriptions of new North American species. – *Proc. ent. Soc. Wash.* **11** (2): 87-103.

Busck, A., 1934. *Tortilia viatrix*, new species. An African moth on senna imported into the United States. – *Proc. ent. Soc. Wash.* **36**: 69-70.

Butler, A. G., 1880. On a collection of Lepidoptera Heterocera from Marlborough Province, New Zealand. – *Cist. Ent.* **2**: 541-562.

Capuse, I., 1971. *Recherches morphologiques et systématiques sur la famille des Coleophoridae (Lepidoptera).* 66 pp., 46 pls. Bucarest.

Caradja, A., 1920. Beitrag zur Kenntnis der geographischen Verbreitung der Mikrolepidopteren des palaearktischen Faunengebietes nebst Beschreibung neuer Formen. 3.Theil. – *Dt. ent. Z. Iris* **31**: 75-179.

Chambers, V. T., 1875a. Tineina of the United States. – *Cincinn. Q. Jl. Sci.* **2** (3): 226-259.

Chambers, V. T., 1875b. Tineina of Colorado. – *Cincinn. Q. Jl. Sci.* **2** (4): 289-305.

Chambers, V. T., 1875c. Tineina from Canada. – *Can. Ent.* **7** (11): 209-213.

Chambers, V. T., 1877. The Tineina of Colorado. – *Bull. U.S. geol. geogr. Surv.* **3** (1): 121-142.

Chambers, V. T., 1878. Index to the described Tineina of the United States and Canada. – *Bull. U.S. geol. geogr. Surv.* **4** (1): 125-167.

Chambers, V. T., 1880. Descriptions of some Tineina, with notes on a few old species. – *J. Cincinn. Soc. nat. Hist.* **2**: 179-194.

Chrétien, P., 1896. Description de Microlépidoptères nouveaux de France et d'Algerie. – *Naturaliste* (2) **10**: 104-105.

Chrétien, P., 1902. La *Stagmatophora divitella* Cst. n'est pas cecidogéne. – *Bull. Soc. ent. Fr.* **1902**: 261-264.

Chrétien, P., 1907. Description de nouvelles espèces de Lépidoptères d'Algerie. – *Bull. Soc. ent. Fr.* **1907**: 305-308.

Chrétien, P., 1908. Description de nouvelles espèces de Microlépidoptères d'Algerie. – *Bull. Soc. ent. Fr.* **1908**: 201-203.

Chrétien, P., 1915. Contribution á la connaissance des Lépidoptères du Nord de l'Afrique. – *Annls Soc. ent. Fr.* **84** (3): 289-374.

Chrétien, P., 1917. Contribution á la connaissance de Lépidoptères du Nord de l'Afrique. Notes biologiques et critiques. – *Annls Soc. ent. Fr.* **85** (1916): 369-502.

Chrétien, P., 1922. Microlépidoptères nouveaux pour la faune française. – *Amateur Papillons* **1** (8): 119-121.

Clarke, J. F. G., 1962. Neotropical Microlepidoptera. 1. The genus *Homoeoprepes* Walsingham (Lepidoptera: Blastodacnidae). – *Proc. U. S. natn. Mus.* **113** (3457): 373-382.

Clarke, J. F. G., 1972. The Lepidoptera of Rapa Island. – *Smiths. Contr. Zool.* **56**: 1-282.

Clarke, J. F. G., 1986. Pyralidae and Microlepidoptera of the Marquesas Archipelago. – *Smiths. Contr. Zool.* **416**: 1-485.

Clemens, B., 1864. North American Micro-Lepidoptera. – *Proc. ent. Soc. Philad.* **2**: 415-430.

Clerck, C., 1759. *Icones Insectorum rariorum, cum nominibus eorum trivialibus, locisque e C.Linnaei.* [xii]+[iii] pp., 55 pls. Holmiae.

Constant, M. A., 1883. Observations sur quelques chenilles nouvelles ou imparfaitement connues. – *Annls Soc. ent. Fr.* (6) **3**: 5-20.

Constant, M. A., 1885. Notes sur quelques Lépidoptères nouveaux (3 et derniere partie). – *Annls Soc. ent. Fr.* (6) **5**: 5-16.

Costa, O. G., [1836]. *Fauna del regno di Napoli, 13. Lepidotteri.* xii+[314] pp., 21 pls. Napoli.

Curtis, J., 1827. *British Entomology,* **4**. Nos 147-194, pls 147-194. London.

Curtis, J., 1830. *British Entomology,* **7**. Nos 290-337, pls 290-337. London.

Curtis, J., 1839. *British Entomology,* **16**. Nos 722-769, pls 722-769. London.

Danilevsky, A. S., 1950. A new genus and species of a predaceous moth feeding on mealybuds, *Coccidiphila gerasimovi* Danilevsky, gen. et sp. n. (Lepidoptera, Momphidae). [In Russian.] – *Ent. Obozr.* **31** (1): 47-53.

Demokidov, K. E., 1916. On the life history of the tea moth, *Parametriotes theae* Kusn. (Lepidoptera, Tineidae). [In Russian.] – *Revue Russe d'Ent.* **15** (1915) (4): 618-626.

[Denis, I. & Schiffermüller, I.], 1775. *Ankündung eines systematischen Werkes von den Schmetterlingen der Wienergegend.* 323 pp., 3 pls. Wien.

Diakonoff, A., 1939a. Notes on Microlepidoptera II. Remarks on some species of the genus *Blastodacna* Wocke (Cosmopterygidae). – *Tijdschr. Ent.* **82**: 64-77.

Diakonoff, A., 1939b. Notes on Microlepidoptera. III. A new species of *Blastodacna* Wocke from Syria (Fam. Cosmopterygidae). – *Tijdschr. Ent.* **82**: 78-80.

Diakonoff, A. & Doets, C., 1938. Aanteekeningen over de Nederlandsche vertegenwoordigers van het geslacht *Blastodacna* Wocke (Cosmopterygidae). – *Ent. Ber., Amst.* **10**: 91-95.

Doubleday, H., 1859. *The Zoologist's synonymic list of British butterflies and moths.* 40 pp. London.

Douglas, J. W., 1846. Descriptions of ten new British moths. – *Zoologist* **4**: 1266-1270.

Douglas, W. A., Henderson, C. A. & Langston, J. M., 1962. Biology of the pink scavenger caterpillar and its control in corn. – *J. Econ. Entomol.* **55** (5): 651-655.

Dumont, C., 1932. Les Lépidoptères des gommiers du nord de l'Afrique. – *Soc. ent. Fr., Livre centen.* **1932**: 689-719.

Duponchel, P.-A.-J., 1838-[1840]. Nocturnes, 8. – *In* Godart, J.-B.: *Histoire naturelle des Lépidoptères ou Papillons de France,* **11**. 720 pp., pls 287-314. Paris.

Duponchel, P.-A.-J., 1842-[1844]. Nocturnes, supplement aux tomes quatrième et suivante. – *In* Godart, J.-B.: *Histoire naturelle des Lépidoptères ou Papillons de France*, Suppl. **4**. 554 pp., pls 51-90. Paris.

Dyar, H. G., [1903]. A list of North American Lepidoptera and key to the literature of this order of Insects. – *Bull. U. S. natn. Mus.* **52** (1902): i-xix+1-723 pp.

Dzhashi, V. S., 1958. Concerning a harmfulness of the tea moth. [In Russian.] – *Byull. vses. Inst. chaya i subtrop. Kultur* **1958** (4): 9-28.

Dzhashi, V. S., 1969. Results of investigations of the tea moth. [In Russian.] – *Subtrop. Kultury* **1969** (4): 16-34.

Ellison, R. E. & Wiltshire, E. P., 1939. The Lepidoptera of the Libanon: with notes on their season and distribution. – *Trans. ent. Soc. London* **88**: 1-56.

Emmet, A. M., 1968. *Cosmopteryx druryella* Zell. at Wicken Fen (Lep., Cosmopterygidae). – *Entomologist's Rec. J. Var.* **80**: 227-228.

Emmet, A. M., 1969. Discovery of the larva of *Sorhagenia rhamniella* Zeller. – *Entomologist's Rec. J. Var.* **81**: 20-21.

Emmet, A. M. (ed.), 1979. *A field guide to the smaller British Lepidoptera*. 271 pp. London.

Emmet, A. M., 1982. Notes on *Mompha nodicolella* Fuchs (Lep., Momphidae). – *Entomologist's Rec. J. Var.* **94**: 21-24.

Emmet, A. M. (ed.), 1988. *A field guide to the smaller British Lepidoptera* (Edn 2). 288 pp. London.

Emmet, A. M. & Johnson, P. J., 1979. See Emmet, A. M., 1979.

Emmet, A. M. & Langmaid, J. R., 1997. Microlepidoptera in Ireland, September, 1995. – *Entomologist's Gaz.* **48**: 147-155.

Eversmann, E., 1844. *Fauna Lepidopterologica Volgo-Uralensis*. xiv+633 pp. Casani.

Fabricius, J. C., 1777. *Genera Insectorum*. [16]+310 pp. Chilonii.

Fabricius, J. C., 1798. *Supplementum Entomologiae Systematicae*. [4]+572 pp. Hafniae.

Falkovitsh, M. I., 1969. Trophic relations of the desert moths (Lepidoptera) in Central Asia. – Pp. 53-88. *In: Meetings in memory of N. A. Cholodkovsky. Lectures at the 21st annual meeting, 12 April 1968*. Leningrad.

Fischer, J. Edlen von Röslerstamm, 1834-[1843]. *Abbildungen zur Berichtigung und Ergänzung der Schmetterlingskunde besonders der Microlepidopterologie*. 304 pp., 100 pls. Leipzig.

Fletcher, T. B., 1920. Life-histories of Indian insects. Microlepidoptera. IV. Cosmopterygidae, Oecophoridae, Physoptilidae, Xyloryctidae, Stenomidae and Orneodidae. – *Mem. Dep. Agric. India* (Ent.) **6** (1-9): 1-217.

Fletcher, T. B., 1929. A list of the generic names used for Microlepidoptera. – *Mem. Dep. Agric. India* (Ent.) **11**: i-ix+1-246.

Fletcher, T. B., 1933. Life-histories of Indian Microlepidoptera (2nd ser.). Cosmopterygidae to Neopseustidae. – *Imper. Council Agric. Res., Sci. Monogr.* **4**: 1-85, 77 pls.

Fletcher, T. B., 1935. *Pancalia latreillella*, Curtis. – *Entomologist's Rec. J. Var.* **47**: 82-83.

Fletcher, T. B., 1936. A new name for *Chrysoclista "bimaculella, Haworth, 1828"* (Lepidoptera: Cosmopterygidae). – *Festschr. 60. Geburtst. E. Strand* **1**: 504-506.

Fletcher, W. A., 1893. A contribution towards a life-history of *Pancalia leuwenhoekella* L. – *Entomologist's mon. Mag.* **29**: 81-82.

Forbes, W. T. M., 1931. Supplementary report on the Heterocera or moths of Porto Rico. – *In* Scientific survey of Porto Rico and the Virgin Islands, 12 (Suppl.). Insects of Porto Rico and the Virgin Islands. – *J. Dep. Agric. Porto Rico* **15** (4): 339-393.

Frey, H., 1856. *Die Tineen und Pterophoren der Schweiz*. xii+430 pp. Zürich.

Frey, H., 1860. Das Elachistiden-Geschlecht *Laverna*. – *Linnaea Entomologica* **14**: 180-205.

Frey, H., 1881. Nachträge zur Lepidopteren-Fauna der Schweiz. – *Mitt. Schweiz. ent. Ges.* **6** (4): 143-147.

Fuchs, A., 1902. Neue Geometriden und Kleinfalter des europäischen Faunengebiets. – *Stettin. ent. Ztg.* **63** (1): 317-330.

Gaedike, R., 1967. Zur systematischen Stellung einiger Gattungen der Heliodinidae/ Schreckensteiniidae sowie Revision der paläarktischen Arten der Gattung *Pancalia* Curtis, 1830. – *Beitr. Ent.* **17**: 363-374.

Gerasimov, A., 1930. Zur Lepidopteren-Fauna Mittel-Asiens. – *Ann. Mus. Zool. Acad. Sci., Leningrad* **31**: 21-48.

Geyer, C., [1832-1836]. Plate 70-71. *In* Hübner, J., 1796-1836: *Sammlung europäischer Schmetterlinge* **8**. Tineae-Schaben. [78] pp., 71 pls. Augsburg.

Gharib, A., 1968. *Batrachedra amydraula* Meyr. (Super-Famille: Gelechioidea) Momphidae (Cosmopterygidae). – *Ent. Phytopathol. appl.* 27: 63-66.

Gibeaux, C., 1986. Description de Cosmoptérygides de la forêt de Fontainebleau nouveaux pour la science (Lepidoptera Cosmopterigidae Antequerinae). – *Alexanor* **14** (4): 167-177.

Gozmány, L., 1957. Notes on the generic group *Stomopteryx* Hein., and the descriptions of some new Microlepidoptera. – *Acta zool. Acad. Sci. Hung.* **3** (1-2): 107-135.

Gozmány, L., 1960. The results of the zoological collecting trip to Egypt in 1957, of the Natural History Museum, Budapest. 8. Egyptian Microlepidoptera, 2. – *Ann. hist.-nat. Mus. natn. Hung.* **52**: 411-421.

Grabe, A., 1955. Kleinschmetterlinge des Ruhrgebietes. – *Mitt. Ruhrlandmus. Stadt Essen* **177**: i-xv+1-159.

Grote, A. R., 1878. New species of *Acopa* and *Heliothis*, and note on *Hamadryas*. – *Can. Ent.* **10**: 67-69.

Halperin, J. & Sauter, W., 1992. An annotated list with new records of Lepidoptera associated with forest and ornamental trees and shrubs in Israel. – *Israel J. Ent.* **25-26**: 105-147.

Hampson, G. F., 1918. Some small families of the Lepidoptera which are not included in the key to the families in the catalogue of Lepidoptera Phalaenae, a list of the families and subfamilies of the Lepidoptera with their types and a key to the families. – *Novit. Zool.* **25** (2): 366-394.

Hartig, F., 1951. See Hartig, F. & Amsel, H. G., 1951.

Hartig, F. & Amsel, H. G., 1939. Contributo alla conoscenza della fauna entomologica della Sardegna. Nuove forme di Lepidotteri. – *Mem. Soc. ent. ital.* **17** (1): 63-84.

Hartig, F. & Amsel, H. G., 1951. Lepidoptera Sardinica, 2. Microlepidoptera. – *Fragm. Ent.* **1** (1): 1-152.

Hartmann, A., 1880. Die Kleinschmetterlinge des europäischen Faunengebietes. – *Mitt. münch. ent. Ver.* **4**: 1-122.

Haworth, A. H., 1828. *Lepidoptera Britannica.* xxxvi+609 pp. Londini.

Heckford, R. J., 1986. Larva of *Mompha lacteella* (Stephens) on *Epilobium montanum* in April. – *Entomologist's Rec. J. Var.* **98**: 121-122.

Heckford, R. J., 1994. *Batrachedra parvulipunctella* Chrétien (Lepidoptera: Momphidae), a surprising addition to the British list. – *Entomologist's Rec. J. Var.* **45**: 261-265.

Heinemann, H. von & Wocke, M. F., [1876]. *Die Schmetterlinge Deutschlands und der Schweiz.* **2**. Kleinschmetterlinge, 2. vi+825+102 pp. Braunschweig.

Hering, E., 1889. Beiträge zur Mitteleuropäischen Micro-Lepidopteren fauna. – *Stettin. ent. Ztg.* **50** (7-9): 290-320.

Hering, M., 1927. Die Minenfauna der Canarischen Inseln. – *Zool. Jahrb., Abt. Syst.* **53**: 405-486.

Hering, M., 1931. Ein neuer *Parietaria*-Minierer, *Cosmopteryx parietariae* sp.n. (Lep.). – *Mitt. Dt. ent. Ges.* **2** (5): 67-70.

Hering, M., 1935. Minenstudien. 15. – *Z. Pflanzenkrankh. Pflanzenschutz.* **45** (1): 1-15.

Hering, M., 1957. *Bestimmungstabellen der Blattminen von Europa*, **1-3**. 1185 pp. 's-Gravenhage.

Herms, H., 1888. Lepidopterologische Beobachtungen. – *Stettin. ent. Ztg* **49**: 81-83.

Herrich-Schäffer, G. A. W., 1853-1855. *Systematische Bearbeitung der Schmetterlinge von Europa.* **5**. Die Schaben und Federmotten. 394 pp. Regensburg.

Herrich-Schäffer, G. A. W., 1857. Anzeige von Kaltenbach, Phytophagen; Eversmann, Noctuélites; Lederer, Noctuinen; Walker, Lepidopterous Insects in the coll. of the Brit. Mus.; Brauer, Neuropt. austriaca; Koch, Schmetterl. südw. Deutschl. – *Korrespbl. zool.-min. Ver. Regensburg* **11**: 58.

Hodges, R. W., 1962. A revision of the Cosmopterigidae of America North of Mexico, with a definition of the Momphidae and Walshiidae (Lepidoptera: Gelechioidea). – *Ent. Amer.* (n.s.) **42**: 1-171.

Hodges, R. W., 1966. Review of New World species of *Batrachedra* with description of three new genera (Lepidoptera, Gelechioidea). – *Trans. Amer. ent. Soc.* **92**: 585-652.

Hodges, R. W., 1978. Gelechioidea. Cosmopterigidae. – *In* Dominick, R.B., *et al.* (eds): *The Moths of America North of Mexico* **6**.1. ix + 166 pp., 6 pls. London.

Hodges, R. W., 1999. The Gelechioidea. – Pp. 130-158. *In* Kristensen, N.P. (ed.), Lepidoptera: Moths and Butterflies. *Handbook of Zoology/Handbuch der Zoologie* **1** (1999). Berlin & New York.

Hodges, R. W. , Dominick, T., Davis, D. R., Ferguson, D. C., Franclemont, J. G., Munroe, E. G. & Powell, J. A., 1983. *Check list of the Lepidoptera of America North of Mexico.* xxiv + 284 pp. London.

Hofmann, O., 1898. Drei neue Tineen-Gattungen. – *Dt. ent. Z. Iris* **10** (1897) (2): 225-230.

Hruby, K., 1964. *Prodromus Lepidopter Slovenska.* 962 pp. Bratislava.

Hübner, J., 1796-[1836]. *Sammlung europäischer Schmetterlinge.* **8**. Tineae-Schaben. [78] pp., 71 pls. Augsburg.

Hübner, J., 1816-[1826]. *Verzeichniss bekannter Schmetterlinge.* 431+72 pp. Augsburg.

Huemer, P. & Karsholt, O., 1999. Gelechiidae I (Gelechiinae: Teleiodini, Gelechiini). – *In* Huemer, P., Karsholt, O. & Lyneborg, L. (eds): *Microlepidoptera of Europe* **3**: 1-356.

Humphreys, H. N. & Westwood, J. O., 1845. *British Moths Transformations* **2**. 268 pp. London.

ICZN, 1969. Opinion 866: *Cosmopterix* Hübner, 1825 (Insecta, Lepidoptera): designation of a type-species under the plenary powers. – *Bull. zool. Nomencl.* **25** (4-5): 150-151.

ICZN, 1986. Opinion 1418: *Glyphipterix* Hübner, [1825] (Insecta, Lepidoptera): *Tinea bergstraesserella* Fabricius, 1781 designated as type species. – *Bull. zool. Nomencl.* **43** (4): 325-327.

ICZN, 1999. *International Code of Zoological Nomenclature.* Edn 4. xxix + 306 pp. London.

Itämies, J. & Kyrki, J., 1979. *Mompha* n. sp.: N Elintavoista. – *Baptria* **4**: 28-29.

Ivinskis, P., 1982. 138 Lepidoptera species new to the Lithuanian SSR, found in 1968-1982. – Pp. 28-47. *In* New and rare for the Lithuanian SSR insect species. Reports and descriptions of 1982. Vilnius.

Janse, A. J. T., 1917. *Check-list of the South African Lepidoptera Heterocera.* 219+xii pp. Pretoria.

Joannis, J. de, 1921. Note sur *Stagmatophora serratella* Tr. (Lep. Tineidae). – *Bull. Soc. ent. Fr.* **1921**: 265-267.

Karsholt, O., 1997. The genus *Chrysoclista* Stainton, 1854 in Europe (Lepidoptera, Agonoxenidae). – *Ent. Meddr.* **65**: 29-33.

Karsholt, O. & Razowski, J. (eds), 1996. *The Lepidoptera of Europe. A Distributional Checklist.* 380 pp. Stenstrup.

Kasy, F., 1954. Kleinschmetterlingsraupen als Bewohner von der Weidenholzgallmücke (*Helicomyia saliciperda* Duf.) befallenen Weidenäste und -stämmchen. – *Ent. Nachrichtenbl. österr. schweiz. Entomol.* **6** (Sonderheft): 1-4.

Kasy, F., 1965. Zur Kenntnis der Schmetterlingsfauna des östlichen Neusiedlersee-Gebietes. – *Wiss. Arb. Bgld.* **34**: 75-211.

Kasy, F., 1968a. Die Walshiidae-Gattung *Calycobathra* Meyrick (Lepidoptera, Gelechioidea). – *Ann. naturhistor. Mus. Wien* **72**: 177-195.

Kasy, F., 1968b. Ergebnisse der zoologischen Nubien-Expedition 1962. – *Ann. naturhistor. Mus. Wien* **72**: 497-525.

Kasy, F., 1969. Vorläufige Revision der Gattung *Ascalenia* Wocke (Lepidoptera, Walshiidae). – *Ann. naturhistor. Mus. Wien* **73**: 339-375.

Kasy, F., 1970. Eine neue *Stagmatophora* s.l. von den Kanarischen Inseln (Lepidoptera, Cosmopterigidae). – *Ann. naturhistor. Mus. Wien* **74**: 195-200.

Kasy, F., 1973a. Beitrag zur Kenntnis der Familie Stathmopodidae Meyrick, 1913 (Lepidoptera, Gelechioidea). – *Tijdschr. Ent.* **116** (13): 227-299.

Kasy, F., 1973b. Eine interessante *Stagmatophora* s.l. aus den World Wildlife Fund-Natuurreservat "Marchauen, Marchegg", Niederösterreich (Lepidoptera: Cosmopterigidae). – *Ann. naturhistor. Mus. Wien* **77**: 275-282.

Kasy, F., 1974. Ein weiterer Beitrag zur Kenntnis der Gattung *Ascalenia* Wocke (Lepidoptera, Walshiidae). – *Z. ArbGem. österr. Ent.* **25** (1973) (1-2): 12-15.

Kasy, F., 1981. Data on the Lepidopterous fauna of Greece (based on the collections of G.Christensen and L.Gozmány). 3. Eine neue *Tortilia* (Lepid., Oecophoridae). – *Z. ArbGem. österr. Ent.* **33** (1-2): 30-32.

Kerppola, S., Albrecht, A. & Huldén, L., 1995. Distribution maps of Microlepidoptera in Finland (Lepidoptera). [In Finnish.] – *Baptria* **20** (2a): 1-79.

Klimesch, J., 1942. Über Microlepidopteren-Ausbeuten aus der Gegend von Zaton bei Gravosa (Süddalmatien). – *Mitt. münch. ent. Ges.* **32** (2): 347-399.

Klimesch, J., 1943. *Stagmatophora extremella* Wck. (Lep., Momphidae). – *Z. wien. ent. Ges.* **28**: 65-72.

Klimesch, J., 1946. *Stagmatophora alypella* spec. nova (Lep., Momphidae). – *Z. wien. ent. Ges.* **31**: 72-74.

Klimesch, J., 1961. Ordnung Lepidoptera, 1: Pyralidina, Tortricina, Tineina, Eriocraniina und Micropterygina. – Pp. 481-789. *In*: *Die Nordost-Alpen im Spiegel ihrer Landtierwelt*, 2. Innsbruck.

Klimesch, J., 1968. *Die Lepidopterenfauna Mazedoniens.* **4**. Microlepidoptera. 201 pp. Skopje.

Klimesch, J., 1983. Beiträge zur Kenntnis der Microlepidopteren-Fauna des Kanarischen Archipels. 5. Beitrag: Choreutidae, Glyphipterigidae, Ethmiidae, Acrolepiidae, Epermeniidae, Tinaegeriidae, Momphidae, Cosmopterigidae, Walshiidae. – *Vieraea* **12** (1982) (1-2): 95-112.

Klots, A. B., 1942. A newly imported European Lavernid (Microlepidoptera). – *Bull. Brooklyn ent. Soc.* **37**: 173.

Klots, A. B., 1970. See Tuxen, S. L., 1970.

Koçak, A. O. & Seven, S., 1993. Taxonomical notes on the Codelist of Turkey, 1. [Turkish]. – *Misc. Papers* **1993** (18): 1-4.

Kollar, V., 1832. Systematisches Verzeichnis der Schmetterlinge im Erzherzogthume Oesterreich. – *Beitr. Landeskde Oesterr. Enns, Wien* **8** (2): 1-101.

Koster, J. C., 2002a. Momphidae. – Pp. 224-254. *In* Emmet, A. M. & Langmaid, J. R. (eds): *The Moths and Butterflies of Great Britain and Ireland* **4** (1). 326 pp. Harley Books, Colchester.

Koster, J. C., 2002b. Cosmopterigidae. – Pp. 255-278. *In* Emmet, A. M. & Langmaid, J. R. (eds): *The Moths and Butterflies of Great Britain and Ireland* **4** (1). 326 pp. Harley Books, Colchester.

Koster, J. C., 2002c. Review of the Nearctic species of the genus *Chrysoclista* Stainton (Lepidoptera, Agonoxenidae). – *Tijdschr. Ent.* **145**: 103-114.

Koster, J. C. & Harrison, T., 1997. Identity of *Mompha sexstrigella* (Braun), with two new Nearctic species (Lepidoptera: Momphidae). – *Holarctic Lepidoptera* **4** (1): 19-23.

Koster, J. C. & Sinev, S. Yu., 1996. A revision of the *divisella* group of the genus *Mompha*, with the description of *Mompha confusella* spec. nov. (Lepidoptera: Momphidae). – *Ent. Ber., Amst.* **56** (9): 137-148.

Krone, W., 1905. Neues über Microlepidopteren. – *Jahrb. wien. ent. Ver.* **16**: 83-85.

Krulikowsky, L. K., 1903. Petites notices lepidopterologiques. 6. [In Russian.] – *Revue Russe d'Ent.* **3** (1): 30-32.

Kuroko, H., 1960. On a new subspecies and an unrecorded species of the genus *Cosmopteryx* from Japan (Lep.; Cosmopterygidae). – *Esakia* **1**: 9-12.

Kusnezov, N. J., 1916. Description of *Parametriotes theae*, gen.n., sp.n. (Lepidoptera, Tineidae), a new enemy of the tea bush in Transcaucasia. [In Russian]. – *Revue Russe d'Ent.* **15** (1915) (4): 627-652.

Kuznetzov, V. I. & Stekolnikov, A. A., 1984. The systematic and phylogenetic relationships of the families and superfamilies of the gelechioid lepidopterans of the infraorder Papilionomorpha (Lepidoptera: Copromorhoidea, Elachistoidea, Coleophoroidea, Gelechioidea) taking into account the functional morphology of the male genitalia. [In Russian] – *Trudy Zool. Inst., Leningrad* **122**: 3-68.

Langmaid, J. R. & Young, M. R., 2002. Microlepidoptera review of 2001. – *Entomologist's Rec. J. Var.* **114**: 269-281.

Lederer, J., 1855. Beitrag zur Schmetterlings-Fauna von Cypern, Beirut und einem Theile Klein-Asiens. – *Verh. zool.-bot. Ges. Wien* **5**: 177-234.

Legrand, P., 1966. Lépidoptères des îles Seychelles et d'Aldabra. – *Mem. Mus. natn. Hist. nat. (Ser. A)* **37**: 1-210.

Leraut, P., 1980. *Liste systematique et synonymique des Lépidoptères de France, Belgique et Corse*. 334 pp. Paris.

Leraut, P., 1997. *Liste systematique et synonymique des Lépidoptères de France, Belgique et Corse*. (Edn 2). 526 pp. Wetteren.

Lhomme, L., [1948-1949]. *Catalogue des Lépidoptères de France et de Belgique.* **2** (5): 649-808. Douelle (Lot).

Lienig, F. & Zeller, P. C., 1846. Lepidopterologische Fauna von Lievland und Curland. – *Isis Oken, Leipzig* **1846** (3-4): 175-302.

Linnaeus, C., 1761. *Fauna Svecica*, **2**. [xlviii] + 578 pp. Stockholmiae.

Linnaeus, C., 1767. *Systema Naturae*, **1** (2) (Edn 2). 1327 pp. Holmiae.

Liška, J., Patocka, J., Skyva, J. & Turécni, M., 1992. Faunistic records from Czechoslovakia. – *Acta ent. Bohemoslov.* **89**: 73.

Liu Youqiao & Shen Guangpu, 1981. Cosmopterygidae. – Pp. 17-18. *In: Iconographia heterocerorum sinicorum,* Beijing.

Lucas, D., 1933. Lépidoptères nouveaux de la France occidentale et de l'Afrique du Nord. – *Bull. Soc. ent. Fr.* **38** (1933): 195-200.

Lucas, D., 1937. Contribution á la fauna des Lépidoptères de la France occidentale et de l'Afrique du Nord. – *Bull. Soc. ent. Fr.* **42** (1937): 123-128.

Lucas, D., 1939. Contribution á l'etude des Lépidoptères du Nord de l'Afrique. – *Lambillionea* **1939** (12): 206-210.

Lvovsky, A. L., 1996. A review of the genus *Lamprystica* Meyrick, 1914, with the description of a new subfamily (Lepidoptera, Agonoxenidae). – *Atalanta* **27**: 427-432.

Malicky, H., 1966. Vorläufige Mitteilung über Biologie und Verarbeitung (sic) von *Sorhagenia* Spuler (Lep., Momphidae). – *Ent. Nachrbl.* **13**: 37-42.

Malicky, H. & Sobhian, R., 1971. Untersuchungen über Bionomie, Präimaginalstadien und Verbreitung der europäischen *Sorhagenia*-Arten. – *Beitr. Ent.* **21**: 179-190.

Mann, J., 1854. Aufzählung der Schmetterlinge, gesammelt auf einer Reise im Auftrage des k.k. zoologischen Museums nach Oberkrain und dem Küstenlande in den Monaten Mai und Juni 1854, als Beitrag zur Fauna des österreichischen Kaiserstaates. – *Verh. zool.-bot. Ges. Wien* **4**: 545-596.

Mariani, M., 1932. Nota preventiva sulle *Cosmopteryx* d'Europa. Descrizione di una nuova specie e di tre nuove forme della Sicilia. – *Boll. Soc. sci. Nat. Econ., Palermo (N.S.)* **14**: 31-43.

Mariani, M., 1936. Una nuova specie di *Eustaintonia* Spul. e studio sulla morfologia comparata delle larve del genere (Lepidoptera, Momphidae). – *Boll. Soc. ent. ital.* **68**: 97-103.

Mariani, M., 1937. Nuove osservazioni su *Eustaintonia phragmitidella* Mar. (Lepid. Momphidae). – *Boll. Soc. ent. ital.* **69**: 46-48.

Mariani, M., 1943. Fauna Lepidopterorum Italiae, 1. Catalogo ragionato dei lepidotteri d'Italia, 2-3. – *Giorn. sci. Nat. Econ., Palermo* **42** (1941-1942) (3): 1-236.

Matsumura, M., 1898. Two japanese insects injurious to fruit. – *Bull. Dep. agric. Ent.* **10**: 36-40.

Matsumura, S., 1931. *6000 Illustrated Insects of Japan-Empire*. 191+1497 pp. Tokyo.

Meyrick, E., 1887. Descriptions of some exotic Micro-Lepidoptera. – *Trans. ent. Soc. London* **1887**: 269-280.

Meyrick, E., 1891. A fortnight in Algeria, with descriptions of new Lepidoptera. – *Entomologist's mon. Mag.* (2) **2** (27): 9-13, 55-62.

Meyrick, E., 1895. *A Handbook of British Lepidoptera*. vi + 843 pp. London.

Meyrick, E., 1897. Descriptions of Australian Micro-Lepidoptera. XVII. Elachistidae. – *Proc. Linn. Soc. N.S.Wales* **22**: 297-435.

Meyrick, E., 1906. Descriptions of Indian Micro-Lepidoptera, 2. – *J. Bombay nat. Hist. Soc.* **17** (2): 403-417.

Meyrick, E., 1911. Descriptions of South African Micro-Lepidoptera. – *Ann. Transv. Mus.* **3** (1): 63-83.

Meyrick, E., 1913a. Descriptions of South African Micro-Lepidoptera. – *Ann. Transv. Mus.* **3** (4): 267-336.

Meyrick, E., 1913b. *Exotic Microlepidoptera*, **1** (3-4): pp. 65-128. Marlborough.

Meyrick, E., 1914a. Descriptions of South African Micro-Lepidoptera. V. – *Ann. Transv. Mus.* **4** (4): 187-205.

Meyrick, E., 1914b. Lepidoptera Heterocera. Fam. Heliodinidae. – *In* P. Wytsman (ed.). *Genera Insectorum*, **165**. 29 pp., 2 pls. Bruxelles.

Meyrick, E., 1914c. *Exotic Microlepidoptera*, **1** (7-9): pp. 193-288. Marlborough.

Meyrick, E., 1915. *Exotic Microlepidoptera*, **1** (11): pp. 321-352. Marlborough.

Meyrick, E., 1916a. *Exotic Microlepidoptera*, **1** (18): pp. 545-576. Marlborough.

Meyrick, E., 1916b. *Exotic Microlepidoptera*, **2** (1): pp. 1-32. Marlborough.

Meyrick, E., 1917. *Exotic Microlepidoptera*, **2** (2): pp. 33-64. Marlborough.

Meyrick, E., 1919. *Exotic Microlepidoptera*, **2** (8): pp. 225-256. Marlborough.

Meyrick, E., 1922a. Lepidoptera Heterocera. Fam. Oecophoridae. – *In* P. Wytsman (ed.): *Genera Insectorum*, **180**. 224 pp., 6 pls. Bruxelles.

Meyrick, E., 1922b. *Exotic Microlepidoptera*, **2** (18-19): pp. 545-608. Marlborough.

Meyrick, E., 1924. *Exotic Microlepidoptera*, **3** (3): pp. 65-96. Marlborough.

Meyrick, E., 1925. Descriptions of twenty-four new Egyptian Microlepidoptera. – *Bull. Soc. R. Ent. Egypte, Cairo* **9** (1-3): 206-219.

Meyrick, E., 1926. *Exotic Microlepidoptera*, **3** (8): pp. 225-256. Marlborough.

Meyrick, E., 1936. *Exotic Microlepidoptera*, **5** (1-3): pp. 1-96. Marlborough.

Meyrick, E., 1939. New Microlepidoptera, with notes on others. – *Trans. R. Ent. Soc. Lond.* **89**: 47-62.

Miles, M., 1930. On the life-history of *Blastodacna atra* Haw., The Pith Moth of the Apple. – *Ann. appl. Biol., Cambridge* **17**: 775-795.

Millière, P., 1876. *Catalogue raisonné des Lépidoptères des Alpes-Maritimes* (troisième et derniere partie). pp. 251-455. Cannes.

Mitic-Mužina, N. & Bogavac, M., 1971. Biology of *Stagmatophora pomposella* Z. (Lepidoptera, Momphidae-Cosmopterigidae) and its role in reducing Mediterranean Sage (*Salvia aethiopis* L.). – *Zaštita Bilja* **22** (115-116): 349-360.

Mosher, E., 1916. A classification of the Lepidoptera based on characters of the pupa. – *Bull. Illinois State Lab. nat. Hist.* **12**: 1-120.

Müller-Rutz, J., 1934. Ueber Microlepidopteren. – *Mitt. schweiz. ent. Ges.* **16** (2): 118-128.

Nagano, K., 1916. On a new micropterous moth from Japan. – *Konchu Sekai [Insect World]* **20** (4): 136-141.

Nel, J., 2000. A propos des *Coccidiphila* Danilevsky, 1950, de France; *C. rungsella* Nel & Brusseaux, 1997, bona species, stat. restauré (Lep. Cosmopterygidae). – *Bull. Soc. ent. Fr.* **105** (1): 22.

Nel, J. & Brusseaux, G., 1997. *Coccidiphila rungsella* n. sp. (Momphidae). Note sur la repartition de quelques Gelechioidea (Lepidoptera). – *Alexanor* **19** (1996) (8): 455-459.

Nel, J. & Nel, A., 2000. Contribution à la connaissance des Lépidoptères des îles Canaries, avec la description de *Coccidiphila patriciae* n. sp. (Lepidoptera, Cosmopterigidae). – *Bull. Soc. ent. Fr.* **105**: 381-385.

Nickerl, F. A., 1864. Neue Microlepidopteren. – *Wien. ent. Monatschr.* **8** (1): 1-8.

Nye, I. W. B. & Fletcher, D. S., 1991. *The Generic Names of Moths of the World,* **6**. Microlepidoptera. xxix + 368 pp. London.

Osthelder, L., 1940. Mikrolepidopteren. – *In* Osthelder, L. & Pfeiffer, E.: *Lepidopteren-Fauna von Marasch in türkisch Nordsyrien*. Nachtrag. – *Mitt. münch. ent. Ges.* **30** (1): 107-116.

Parsons, M.S., 2002. *Cosmopterix pulchrimella* Chambers, 1875 (Lepidoptera: Cosmopterigidae) new to the British Isles. – *Entomologist's Gaz.* **53**: 93-96.

Passos de Carvalho, J., 1995. Microlepidoptera of the Madeira and Selvagem Archipelagos (Insecta, Lepidoptera). – *Boll. Mus. Mun. Funchal*, Suppl. **4**: 559-588.

Passos de Carvalho, J. & Corley, M. F. V., 1995. Additions to the Lepidoptera of Algarve, Portugal (Insecta: Lepidoptera). – *SHILAP Revta. lepid.* **23**: 191-230.

Patocka, J., 1960. *Die Tannenschmetterlinge der Slowakei*. 214 pp. Bratislava.

Patocka, J., 1997. Über die Puppen der mitteleuropäischen Schmetterlinge (Insecta: Lepidoptera): Überfamilien Yponomeutoidea, Gelechioidea, Schreckensteinioidea, Epermenioidea und Copromorphoidea. – *Linzer biol. Beitr.* **29**: 715-751.

Petry, A., 1904. Beschreibung neuer Microlepidopteren aus Korsika. – *Stettin. ent. Ztg.* **65** (2): 242-254.

Plotnikov, V. I., 1926. *Insect pests of cultural plants in Central Asia.* [In Russian.]. 292 pp. Tashkent.

Poutiers, R., 1921. *Stagmatophora serratella* Tr. (Lep. Tineidae) et *Mecinus sicardi* Hust. (Col. Curculionidae) causes de cécidies sur *Antirrhinum majus* L. – *Bull. Soc. ent. Fr.* **1921**: 269-271.

Ragonot, E.-L., 1882. *Laverna Seeboldiella* [Les description de quatre espèces nouvelles de Microlépidoptères d'Espagne]. – *Bull. Soc. ent. Fr.* **1882**: lxv-lxvii.

Ragonot, E.-L., 1895. Microlépidoptères de la Haute-Syrie recoltes par M.Ch.Delagrande, et descriptions des espèces nouvelles. – *Bull. Soc. ent. Fr.* **1895** (4): xciv-xcix.

Réal, P., 1989. Un *Coccidiphila* français nouveau pour la science, *Coccidiphila charlierella* sp. nov. (Lepidoptera, Momphidae). – *Alexanor* **15**: 461-464.

Réal, P., 1997. Une mise au point sur les *Phyllonorycter* Huebner in coll. Ozorski (Lepidoptera, Gracillariidae). – *Bull. Soc. ent. Mulhouse* **1997** (juill.-sept.): 36-44.

Rebel, H., 1896. Dritter Beitrag zur Lepidopterenfauna der Canaren. – *Ann. naturhist. Mus. Wien* **11**: 102-148.

Rebel, H., 1902. Lepidopteren aus Morea, gesammelt von Herrn Martin Holtz im Jahre 1901. – *Berl. ent. Z.* **47** (1-2): 83-110.

Rebel, H., 1903. Neue Microheteroceren aus Osterreich-Ungarn. – *Verh. zool.-bot. Ges. Wien* **53** (2): 90-103.

Rebel, H., 1911. Beitrag zur Lepidopterenfauna von Syrien. – *Verh. zool.-bot. Ges. Wien* **61** (7-8): (142)-(156).

Rebel, H., 1917. See Rebel, H. & Zerny, H., 1917.

R[a]ebel, H., 1925. *Chrysoclista bimaculella* Hw. in Oberschlesien. – *Intern. ent. Z.* **19**: 21-22.

Rebel, H., 1926. Beitrage zur Lepidopterenfauna Aegyptens. 3. – *Bull. Soc. Ent. Egypte* **10**: 179-191.

Rebel, H., 1937. Neue europäische Tortriciden und Tineiden. – *Z. österr. ent. Ver. Wien* **22** (5): 41-48.

Rebel, H., 1938. Ein neuer Kleinschmetterling aus Westungarn. – *Z. österr. ent. Ver. Wien* **23** (1): 5-6. Taf. 2.

Rebel, H., 1940. Die Lepidopterenfauna des Azorischen Archipels. – *Comment. biol. Soc. Sci. Fenn.* **8** (1): 1-59.

Rebel, H. & Zerny, H., 1917. Wissenschaftliche Ergebnisse der mit Unterstützung der kaiserlichen Akademie der Wissenschaften in Wien aus der Erbschaft Treitl von F. Werner unternommenen zoologischen Expedition nach dem anglo-ägyptischen Sudan (Kordofan) 1914. I. Lepidoptera. – *Denkschr. Akad. Wiss. Wien* **93**: 423-446.

Rebel, H. & Zerny, H., [1932]. Die Lepidopterenfauna Albaniens. – *Denkschr. Akad. Wiss. Wien, Math.-Natur. Klasse* **103** (1931) (2): 37-161.

Reutti, C., 1898. *Übersicht der Lepidopteren-Fauna des Grossherzogtums Baden*, **2**. xii+361 pp. Berlin.

Riedl, T., 1962. Beitrag zur Kenntnis der paläarktischen Momphidae (Lepidoptera), 1. *Sorhagenia rhamniella* (Zell.) und zwei neue europaische Arten der Gattung *Sorhagenia* Spul. – *Polsk. Pismo ent.* **32** (1): 69-75.

Riedl, T., 1965. Matériaux pour la connaissance des Momphidae paléarctiques (Lepidoptera), 3. Étude sur quelques Momphides européennes. – *Polsk. Pismo ent.* **35** (2): 419-468.

Riedl, T., 1966a. Matériaux pour la connaissance des Momphidae paléarctiques (Lepidoptera), 4. Momphides nouveaux ou peu connus de la region paléarctique occidentale. – *Polsk. Pismo ent.* **36** (1): 75-84.

Riedl, T., 1966b. Matériaux pour la connaissance des Momphidae paléarctiques (Lepidoptera), 7. Deux nouvelles espèces paléarctiques du genre *Stagmatophora* H.-S. – *Mitt. ent. Ges. Basel (N.F.)* **16** (4-6): 104-109.

Riedl, T., 1969. Matériaux pour la connaissance des Momphidae paléarctiques (Lepidoptera), 9. Revue des Momphidae européennes, y compris quelques espèces d'Afrique du Nord et du Proche-Orient. – *Polsk. Pismo ent.* **39** (4): 635-919.

Riedl, T., 1976. Sur quelques Momphidae (s.l.) d'Afrique du Nord (Lep. Gelechioidea). – *Annls Soc. ent. Fr. (N.S.)* **12** (1): 187-197.

Riedl, T., 1978. Etude des types de Momphidae et de Cosmopterigidae d'E.Turati (Lepidoptera). – *Polsk. Pismo ent.* **48** (4): 505-510.

Riedl, T., 1983. Records of the Lepidoptera of Greece based on the collections of G. Christensen and L. Gozmány: VIII, V. Momphidae, Blastodacnidae, Chrysopeleiidae, Cosmopterigidae. – *Annls Musei Goulandris* **6**: 313-317.

Riedl, T., 1984. Considérations sur quelques taxa du groupe-famille des Gelechioidea (Lepidoptera). – *Polsk. Pismo ent.* **54**: 145-161.

Riedl, T., 1986. Revue des lepidoptères Gelechioidea de Grece conserves au Musee Zoologique de Copenhague, y compris la redescription de femelle de *Ramphis libanoticus* Riedl (Cosmopterigidae). – *Nota lepid.* **9**: 227-228.

Riedl, T., 1990a. Considérations sur les espèces du genre *Vulcaniella* Riedl (Lepidoptera Cosmopterigidae) – *Polsk. Pismo ent.* **60** (1-2): 65-71.

Riedl, T., 1990b. Sur quelques Gelechioidea conservés au Musée Zoologique de Copenhague (Lepidoptera) (Partie II). – *Nota lepid.* **12**: 319-327.

Riedl, T., 1993. *Amneris flexiloquella* gen. et sp. n. d'Espagne (Lepidoptera, Cosmopterigidae). – *Polsk. Pismo ent.* **62** (1-4): 113-116.

Riedl, T., 1994a. Une nouvelle espèce européenne du genre *Pancalia* Stephens (Lepidoptera: Cosmopterigidae, Antequerinae). – *Nota lepid.* **17**: 87-91.

Riedl, T., 1994b. Liste des taxa de six familles des Lépidoptères Gelechioidea paléarctiques (Stathmopodidae, Batrachedridae, Agonoxenidae, Momphidae, Cosmopterigidae, Chrysopeleiidae). – *Annls Upper Silesian Mus.(Ent.)* **5**: 9-24.

Riedl, T., 1996a. Catalogue des Chrysopeleiidae paléarctiques (Lepidoptera). – *Annls Upper Silesian Mus. (Ent.)* **6-7**: 127-134.

Riedl, T., 1996b. *Euamneris* – new name for *Amneris* Riedl (Lepidoptera: Cosmopterigidae). – *Annls Upper Silesian Mus. (Ent.)* **6-7**: 299.

Riedl, T., 1997a. Stathmopodinae. – P. 83. *In* Karsholt, O. & Razowski, J. (eds.): *The Lepidoptera of Europe*. Stenstrup.

Riedl, T., 1997b. Momphidae. – P. 95. *In* Karsholt, O. & Razowski, J. (eds.): *The Lepidoptera of Europe*. Stenstrup.

Rizza, A. & Pecora, P., 1979. *Linaria dalmatica* project. – Pp. 42-53. *In: USDA Biological Control of Weeds Laboratory, Rome*. Annual Report 1979.

Robinson, G. S., 1976. The preparation of slides of Lepidoptera genitalia with special reference to the Microlepidoptera. – *Entomologist's Gaz.* **27**: 127-132.

Römer, J. J., 1794. Beschreibung und Abbildung einiger kleiner Nachtvögelchen und ihrer Raupen (*Geometra fasciaria, Noctua cucubali, Alucita epilobiella, Tinea stigmatella*). – *Schrift. Ges. naturforsch. Freunde, Berlin* **9**: 156-165, pl. 1.

Rungs, C. E. E., 1979. Catalogue raisonné des Lépidoptères du Maroc. Inventaire faunistique et observations écologiques. I. – *Trav. Inst. Sci.,Sér. Zool.* **39**: 1-222.

Saner, M, 1991. *Interactions of Linaria vulgaris Mill. and L. dalmatica (L.) Mill. (Scrophulariaceae) with Insect Herbivores*. 138 pp. Basel.

Schefer-Immel, V., 1958. Ein Massenauftreten von Kätzchenschädlingen an Aspe *(Populus tremula)*. – *Anz. Schädlingsk.* **31** (1): 7-10.

Schütze, K. T., 1931. *Die Biologie der Kleinschmetterlinge unter besonderer Berücksichtigung ihrer Nährpflanzen und Erscheinungszeiten*. 235 pp. Frankfurt a. M.

Scopoli, I. A., 1772. *Annus V. Historico-Naturalis. 5. Observationes Zoologicae*. 128 pp. Lipsiae.

Sinev, S. Yu., 1979a. A new species of the narrow-winged moth (Lepidoptera, Momphidae) from the black-sea coast of Caucasus. [In Russian.] – *Trudy Vses. ent. Obsch.* **61**: 110-111.

Sinev, S. Yu., 1979b. A new species of the genus *Calycobathra* Meyrick, 1891 (Lepidoptera, Momphidae, Walshiinae) from the Middle Asia deserts. [In Russian.] – *Trudy Zool. Inst., Leningrad* **88**: 85-88.

Sinev, S. Yu., 1979c. On the composition and systematic position of the genus *Tetanocentria* Rbl. (Lepidoptera, Momphidae). [In Russian.]– *Ent. Obozr.* **58**: 590-598.

Sinev, S. Yu., 1980. On the fauna of narrow-winged moths (Lepidoptera, Momphidae) of Mongolia. [In Russian.] – *Insects of Mongolia* **7**: 378-385.

Sinev, S. Yu., 1982. *Batrachedra praeangusta* (Haworth, 1828): Lepidoptera, Momphidae. – *In: Insect ranges in the european part of the USSR*. No.4. Leningrad: 31 (map 153).

Sinev, S. Yu., 1984a. *Pancalia latreillella* Curtis, 1830: Lepidoptera, Momphidae. – *In: Insect ranges in the european part of the USSR*. No.5. Leningrad: 35 (map 195).

Sinev, S. Yu., 1984b. *Pancalia leuwenhoekella* (Linnaeus, 1761): Lepidoptera, Momphidae. – *In: Insect ranges in the european part of the USSR*. No.5. Leningrad: 36 (map 196).

Sinev, S. Yu., 1984c. *Pyroderces argyrogrammos* (Zeller, 1849): Lepidoptera, Momphidae. – *In: Insect ranges in the european part of the USSR*. No.5. Leningrad: 37 (map 197).

Sinev, S. Yu., 1984d. New species of narrow-winged moths of the subfam. Chrysopeleiinae (Lepidoptera, Momphidae) from Middle Asia. [In Russian.] – *Ent. Obozr.* **63** (2): 326-335.

Sinev, S. Yu., 1985a. A review of the genus *Pancalia* Stephens (Lepidoptera, Cosmopterigidae) in the fauna of the USSR. [In Russian.] – *Ent. Obozr.* **64** (4): 804-822.

Sinev, S. Yu., 1985b. New species of the genus *Cosmopterix* Hb. (Lepidoptera, Cosmopterigidae) from the Far East of the USSR. [In Russian.] – *Trudy Zool. Inst., Leningrad* **134**: 73-94.

Sinev, S. Yu., 1986. A list of the narrow-winged moths (Lepidoptera, Momphidae s.l.) in the fauna of the USSR. [In Russian.] – *Trudy Vses. ent. Obshch.* **67**: 19-74.

Sinev, S. Yu., 1988a. New taxa of the Blastodacninae moth subfamily (Lepidoptera, Momphidae s.l.) of the USSR fauna. [In Russian.] – *Vestn. Zool.* **1988** (5): 15-21.

Sinev, S. Yu., 1988b. A review of bright-leged moths (Lepidoptera, Stathmopodidae) in the fauna of the USSR. [In Russian.] – *Trudy Zool. Inst., Leningrad* **178:** 104-133.

Sinev, S. Yu., 1989a. New taxa of the narrow-winged moths (Lepidoptera: Blastodacnidae, Cosmopterigidae, Chrysopeleiidae) from Middle-Asia. [In Russian.] – *Trudy Zool. Inst., Leningrad* **200**: 3-26.

Sinev, S. Yu., 1989b. Adaptive radiation and modes of speciation among the onagraceous narrow-winged moths (Lepidoptera, Momphidae). [In Russian.] – *Trudy Zool. Inst., Leningrad* **202**: 106-133.

Sinev, S. Yu., 1990. A new genus and species of narrow-winged moths of the family Chrysopeleiidae (Lepidoptera, Gelechioidea) from Northern Africa. [In Russian.] – *Zool. Zhurn.* **69** (2): 140-142.

Sinev, S. Yu., 1992. On the system and phylogeny of the Gelechioidea s.l. (Lepidoptera). [In Russian.] – *Ent. Obozr.* **71** (1): 143-159.

Sinev, S. Yu., 1993a. On the biology of a few narrow-winged moths from the subfamily Cosmopteriginae (Lepidoptera, Cosmopterigidae). [In Russian.] – *Trudy Zool. Inst., St.Petersburg* **193**: 73-83.

Sinev, S. Yu., 1993b. A brief review of the genus *Alloclita* Stgr. (Lepidoptera, Cosmopterigidae), with description of a new species from Mongolia. [In Russian.] – *Trudy Zool. Inst., St.Petersburg* **248**: 132-138.

Sinev, S. Yu., 1993c. A review of the narrow-winged moths of the genus *Sorhagenia* Spul. (Lepidoptera, Chrysopeleiidae) in the Palaearctic fauna. [In Russian.] – *Trudy Zool. Inst., St. Petersburg* **255**: 42-63.

Sinev, S. Yu., 1994a. Fam. Stathmopodidae. – Pp. 276-278. *In: Insects and mites – pests of agricultural plants.* Vol. **3**: Lepidoptera, pt. 1. St.Petersburg.

Sinev, S. Yu., 1994b. Fam. Blastodacnidae. – Pp. 279-284. *In: Insects and mites – pests of agricultural plants.* Vol. **3**: Lepidoptera, pt. 1. St.Petersburg.

Sinev, S. Yu., 1994c. Fam. Batrachedridae. – Pp. 284-285. *In: Insects and mites – pests of agricultural plants.* Vol. **3**: Lepidoptera, pt. 1. St.Petersburg.

Sinev, S. Yu., 1994d. Fam. Cosmopterigidae. – Pp. 285-289. *In: Insects and mites – pests of agricultural plants.* Vol. **3**: Lepidoptera, pt. 1. St.Petersburg.

Sinev, S. Yu., 1996. Geographic distribution and nomenclature of nine species of Momphidae (Lepidoptera) discovered in Asia and North America. – *Acta Zool. Fennica* **200**: 9-17.

Sinev, S. Yu., 1997a. A brief review of the genus *Coccidiphila* Dan. (Lepidoptera, Cosmopterigidae) with a description of a new species from west Mediterranean. [In Russian.] – *Ent. Obozr.* **76** (1): 203-209.

Sinev, S. Yu., 1997b. A review of the narrow-winged moths of the genus *Cosmopterix* Hb. (Lepidoptera, Cosmopterigidae) of Palaearctic region. [In Russian.] – *Ent. Obozr.* **76** (4): 813-829.

Sinev, S. Yu., 1999. Notes on the synonymy of the narrow-winged moths (Lepidoptera: Agonoxenidae, Cosmopterigidae, Momphidae) of Palaearctic. [In Russian.] – *Ent. Obozr.* **78** (1): 138-149. [English translation in *Ent. Rev.* **79** (1): 30-39.]

Sinev, S. Yu., 2002a. World catalogue of cosmopterigid moths (Lepidoptera: Cosmopterigidae). – *Proc. Zool. Inst., St.Petersburg* **293**: 1-184.

Sinev, S. Yu., 2002b. Morphological principles for the revision of taxonomic structure of the narrow-winged gelechioid moths (Lepidoptera, Gelechioidea). – *Proc. Zool. Inst., St.Petersburg* **296**: 125-134.

Sircom, J., 1848. Descriptions of *Lophoptilus staintoni* and *Microsetia bedellella*, two new British moths of the family Tineidae. – *Zoologist* **6**: 2037.

Snellen, P. C. T., 1901. Beschrijvingen van nieuwe exotische Tortricinen, Tineinen en

Pterophorinen benevens aanteekeningen over reeds bekend gemaakte soorten. – *Tijdschr. Ent.* **44** (2): 67-98.

Sorhagen, L., 1885. Aus meinem entomologischen Tagebuche (Fortsetzung). – *Berl. ent. Z.* **29**: 81-108.

Sorhagen, L., 1886. *Die Kleinschmetterlinge der Mark Brandenburg und einiger angrenzenden Landschaften*. x+367 pp. Berlin.

Spuler, A., 1898. See Reutti, C., 1898.

Spuler, A., 1910. *Die Schmetterlinge Europas*, **2**. 523 pp. Stuttgart.

Stainton, H. T., 1849. *Catalogue of the British Tineidae & Pterophoridae*. iv + 32 pp. London.

Stainton, H. T., 1851. *A Supplementary Catalogue of the British Tineidae & Pterophoridae*. iv + 28 pp. London.

Stainton, H. T., 1854. *Insecta Britannica*. Lepidoptera: Tineina. viii + 313 pp., 10 pls. London.

Stainton, H. T., 1859. Descriptions of twenty-five species of Indian Micro-Lepidoptera. – *Trans. ent. Soc. London* (N. S.) **5** (1858-1861): 111-126.

Stainton, H. T., 1861. Lepidoptera. New British species in 1860. – *Entomologist's Annual* **1861**: 82-92.

Stainton, H. T., 1868. Observations on Tineina. – *Entomologist's Annual* **1868**: 134-155.

Stainton, H. T., 1869. *The Tineina of Southern Europe*. viii+370 pp. London.

Stainton, H. T., 1870a. *The natural history of the Tineina*, **12**. 259 pp. London.

Stainton, H. T., 1870b. *The natural history of the Tineina*, **13**. 259 pp. London.

Staudinger, O., 1859. Diagnosen nebst kurze Beschreibung neuer andalusischer Lepidopteren. – *Stettin. ent. Ztg* **20** (7-9): 211-259.

Staudinger, O., 1870-1871. Beschreibung neuer Lepidopteren des europäischen Faunengebiets. – *Berl. ent. Z.* **14**: 97-132, 193-208 (1870); 273-330 (1871).

Staudinger, O., 1877. Neue Lepidopteren des europäischen Faunengebiets aus meiner Sammlung. – *Stettin. ent. Ztg* **38** (4-6): 175-208.

Staudinger, O. & Rebel, H., 1901. *Catalog der Lepidopteren des palaearctischen Faunengebietes*. **2**. 368 pp. Berlin.

Staudinger, O. & Wocke, M. F., 1861. *Catalog der Lepidopteren Europa's und der angrenzenden Länder*. xvi+192 pp. Dresden.

Stepanov, E. M., 1966. Concerning the native land of the tea moth. – *Subtrop. Kultury* **1966** (3): 160.

Stephens, J. F., 1829. *Insecta Haustellata*. 388 pp. London.

Stephens, J. F., 1834. *Illustrations on British entomology*. Insecta Haustellata, **4**. 436 pp., 23-40 pls. London.

Sterling, P. H., 1988. The life history of *Mompha lacteella* (Steph.). – *Br. J. Ent. nat. Hist.* **1**: 126.

Sterling, P. H. & Langmaid, J. R., 1989. *Mompha subdivisella* Bradley (Lep.: Momphidae): the mystery revealed. – *Entomologist's Gaz.* **40** (3): 199-201.

Suire, I., 1940. Les premiers etats de *Batrachedra ledereriella* Z. (Lep. Cosmopteryginae) (note preliminaire). – *Bull. Soc. ent. Fr.* **45** (6): 65-69.

Svensson, I., 1982. Four new species of Microlepidoptera from northern Europe. – *Ent. scand.* **13**: 297-298.

Szöcs, J., 1981. Angaben über die minierenden Motten aus Budapest und Umgebung. – *Folia ent. Hung.* **42** (**34**): 209-220.

Talhouk, A. S., 1948. A Note on *Blastodacna libanotica* Dkff. (Lepidoptera: Cosmopterigidae). – *Bull. Soc. Fouad 1er Ent.* **32**: 41-45.

Traugott-Olsen, E., 1986. Description of two new species of *Coccidiphila* Danilevsky, 1950, the Canary Islands, and records of *C. gerasimovi* Danilevsky, 1950, from the Canary Islands to Israel (Lepidoptera: Momphidae). – *SHILAP Revta. lepid.* **14** (56): 31-46.

Treitschke, F., 1833. *Die Schmetterlinge von Europa (Forsetzung des Ochsenheimer'schen Werkes).* **9** (2). 294 pp. Leipzig.

Turati, E., 1924. Spedizione lepidotterologica in Cirenaica 1921-1922. – *Atti Soc. ital. Sci. nat.* **63**: 17-191.

Turati, E., 1927. Novita di Lepidotterologia in Cirenaica. 2. – *Atti Soc. ital. Sci. nat.* **66**: 313-344.

Turati, E., 1929. Eteroceri di Tripolitania. – *Boll. Lab. Zool. gener. appl., Portici* **23**: 98-128.

Turati, E., 1930. Novita di Lepidotterologia in Cirenaica. – *Atti Soc. ital. Sci. nat.* **69**: 46-92.

Tutin, T. G., Heywood, V. H., Burges, N. A., Moore, D. M., Valentine, D. H., Walters, S. M. & Webb, D. A. (eds), 1964-1980. *Flora Europaea.* 5 vols. Cambridge.

Tutt, J.W., 1905. *Practical hints for the field lepidopterist*, 3. 166 pp. London.

Tuxen, S. L. (ed.), 1970. *Taxonomist's glossary of genitalia in insects.* Edn 2. 359 pp. Copenhagen.

Ulenberg, S. A. & Frankenhuyzen, F. van, 1987. Entomologie. Lepidoptera: vlinders. – *Jaarb. Plantenziektek. Dienst* **1986**: 35.

Utech, L., 1962. Blattminen und Pflanzengallen aus Albanien und dem Kaukasus. – *D. ent. Z.* **9**: 229-235.

Varenne, T. & Nel, J., 2001. *Anatrachyntis rileyi* (Walsingham, 1882), un microlépidoptère ravageur nouveau pour la France et l'Europe (Lep. Cosmopterigidae). – *Bull. Soc. ent. Fr.* **106**: 176.

Vieira, V., 1997. Lepidoptera of the Azores Islands. – *Bol. Mus. Mun. Funchal* **49**: 5-76.

Viette, P., 1951. Les Microlépidoptères de C. Dumont. – *Rev. franç. Lepid.* **13** (3-4): 56-59.

Vives Moreno, A., 1992. *Catálogo systemático y sinonimico de los lepidopteros de la peninsula Iberica y Baleares (Insecta: Lepidoptera).* 378 pp. Madrid.

Wakely, S., 1944. Notes on the genus *Mompha*. – *Proc. Trans. S. Lond. ent. nat. Hist. Soc.* **1944-1945**: 81-84.

Walker, F., 1864. *List of the specimens of lepidopterous insects in the collection of the British Museum.* Pt. 30. Tineites: 837-1096. London.

Walsingham, T., 1882. Notes on Tineidae of North America. – *Trans. Amer. ent. Soc.* **10** (3) (1882-1883): 165-204.

Walsingham, T., 1891a. African Micro-Lepidoptera. – *Trans. ent. Soc. London* **1891** (1): 63-132.

Walsingham, T., 1891b. Micro-Lepidoptera collected near Cannes, 1890. – *Entomologist's mon. Mag.* (2) **2** (27): 137-152.

Walsingham, T., 1894. Catalogue of the Pterophoridae, Tortricidae and Tineidae of the Madeira Islands, with notes and descriptions of new species. – *Trans. ent. Soc. London* **1894** (4): 535-555.

Walsingham, T., 1897. Revision of the West-Indian Micro-Lepidoptera, with descriptions of new species. – *Proc. zool. Soc. London* **1897**: 54-182.

Walsingham, T., 1905. Algerian Microlepidoptera. – *Entomologist's mon. Mag.* (2) **16** (41): 124-128.

Walsingham, T., 1906. Description of a new Tineid moth infesting cotton-pods in Egypt. – *Ann. Mag. nat. Hist.* (7) **18** (103): 178-179.

Walsingham, T., 1907a. Algerian Microlepidoptera. – *Entomologist's mon. Mag.* (2) **18** (43): 147-154.

Walsingham, T., 1907b. Descriptions of new species of *Stagmatophora*, H.-S. (Lepidoptera: Tineina). – *Entomologist's mon. Mag.* (2) **18** (43): 177-181.

Walsingham, T., 1907c. Microlepidoptera. – Pp. 469-759. *In* Sharp, D. (ed.): Fauna Hawaiiensis. Vol. **1**, pt. 5. Cambridge.

Walsingham, T., 1908. Microlepidoptera of Tenerife. – *Proc. zool. Soc. London* **1907**: 911-1034.

Werneburg, A., 1864. *Beiträge zur Schmetterlingskunde*, **1**. viii+595 pp. Erfurt.

Wocke, M. F., 1871. Microlepidoptera. – *In* Staudinger, O. & Wocke, M. F.: *Catalog der Lepidopteren des europäischen Faunengebiets*, **2**. pp. 201-346. Dresden.

Wocke, M. F., [1876]. See Heinemann, H.von & Wocke, M. F., [1876].

Wollaston, T. V., 1879. Notes on the Lepidoptera of St.Helena with descriptions of new species. – *Ann. Mag. nat. Hist.* (5) **3** (18):415-441.

Wood, W., 1887. The larva of *Batrachedra pinicolella*. – *Entomologist's Rec. J. Var.* **24**: 126-127.

Yang, Chin-Kun, 1977. *Moths of North China*, **1**. 299 pp., 12 pls. Beijing.

Zagulajev, A.K., 1981. Lepidoptera from nests, holes and some anthropogenic habitats. [In Russian]. – *Ent. Obozr.* **60** (3): 577-597.

Zagulajev, A. K. & Liao, Din-si, 1959. A new genus and species of predaceous moths, *Lacciferophaga yunnanea* Zagulajev, gen. et sp.n. (Lepidoptera, Momphidae) attacking lac insect in Southern China. [In Russian]. – *Acta ent. Sinica* **9** (4): 306-315.

Zangheri, S., 1966. Recenti contributi allo studio dei Lepidotteri galligeni. – *Marcellia* **33**: 49-59.

Zeller, P. C., 1839. Versuch einer naturgemaessen Eintheilung der Schaben. – *Isis Oken, Leipzig* **1839** (3): 169-220.

Zeller, P. C., 1847a. Verzeichniss der vom Professor Dr. Loew in der Türkei und Asien gesammelten Lepidoptera. – *Isis Oken, Leipzig* **1847** (1): 3-39.

Zeller, P. C., 1847b. Bemerkungen über die auf einer Reise nach Italien und Sicilien beobachteten Schmetterlingsarten, 10. – *Isis Oken, Leipzig* **1847** (12): 881-914.

Zeller, P. C., 1850. Verzeichniss der von Herrn Jos. Mann beobachteten Toscanischen Microlepidoptera. – *Stettin ent. Ztg* **11** (6): 195-212.

Zeller, P. C., 1868. The larva of *Stathmopoda pedella* at Stettin. – *Entomologist's mon. Mag.* **4**: 233-234.

Zerny, H., 1935. Die Lepidopterenfauna des Grossen Atlas in Marokko und seiner Randgebiete. – *Mém. Soc. Sci. nat. Maroc.* **42**: 1-163.

Zetterstedt, J. W., 1838-1840. *Insecta lapponica*. 1140 pp. Lipsiae.

Zimmerman, E. C., 1978. *Insects of Hawaii*, **9**. Microlepidoptera. 1936 pp. Honolulu.

Zhang, Bin-Cheng, 1994. *Index of economically important Lepidoptera*. 599 pp. Cambridge.

Index to entomological genus names

The numbers refer to the page numbers. Numbers in bold refer to the main treatment. Synonyms are in italics.

Index to entomological species names

The numbers in bold refer to the species numbers, other numbers refer to page numbers. Synonyms are in italics.

Index to host-plants

The numbers refer to the species numbers.